The Treatment of Pharmaceutical Wastewater

The Treatment of Pharmaceutical Wastewater

Innovative Technologies and the Adaptation of Treatment Systems

Edited by

Afzal Husain Khan

Civil Engineering Department, College of Engineering, Jazan University, Jazan, Saudi Arabia

Nadeem A. Khan

Department of Civil Engineering, Mewat Engineering College, Nuh, India

Mu. Naushad

Department of Chemistry, College of Science, King Saud University, Riyadh, Saudi Arabia

Hamidi Abdul Aziz

School of Civil Engineering, Universiti Sains Malaysia, George Town, Malaysia

ELSEVIER

Elsevier
Radarweg 29, PO Box 211, 1000 AE Amsterdam, Netherlands
The Boulevard, Langford Lane, Kidlington, Oxford OX5 1GB, United Kingdom
50 Hampshire Street, 5th Floor, Cambridge, MA 02139, United States

ISBN: 978-0-323-99160-5

For information on all Elsevier publications
visit our website at https://www.elsevier.com/books-and-journals

Publisher: Matthew Deans
Acquisitions Editor: Chiara Giglio
Editorial Project Manager: John Leonard
Production Project Manager: Fizza Fathima
Cover Designer: Vicky Pearson Esser

Typeset by STRAIVE, India

Contents

Contributors xi
About the Editors xvii
Preface xix
Acknowledgments xxiii

1 Introduction: Occurrences, sources, and methods of pharmaceutical wastewater treatment 1
Pallavi Chakraborty
1 General introduction 1
2 Occurrences and sources 1
3 Sources of pharmaceuticals 3
4 Analytical techniques used to determine pharmaceutical compounds 7
5 Toxicological effect of pharmaceuticals 7
6 Classical methods of pharmaceutical wastewater treatment 9
7 Conclusions 12
References 12

2 Current situation of pharmaceutical wastewater around the globe 19
Ahmad Saud Masood, Md. Sajid Ali, Mohammad Saood Manzar, Nadeem A. Khan, and Afzal Husain Khan
1 Introduction 19
2 Characteristics of pharmaceutical wastewater 20
3 Source and occurrence of pharmaceuticals in water 25
4 Source, occurrence, and fate of pharmaceuticals in the environment 26
5 Health hazards of discharged pharmaceuticals 27
6 Pharmaceutical wastewater treatment options 28
7 Hospital wastewater management in developed countries 33
8 Hospital wastewater management in developing countries 38
9 Conclusions 44
Conflict of interest 44
References 44
Further reading 51

3 **Recent trends in graphene-based materials for pharmaceuticals wastewater treatment** **53**
Hasara Samaraweera and Afzal Husain Khan
1 Introduction **53**
2 Graphene-based materials **55**
3 Removal of pharmaceuticals using graphene-based materials **61**
4 Knowledge gaps **63**
5 Conclusions **63**
References **64**

4 **Fate of common pharmaceuticals in the environment** **69**
Omar Fawzi Suleiman Khasawneh, Puganeshwary Palaniandy, and Hamidi Abdul Aziz
1 Introduction **69**
2 Common PhACs in the environment and their physiochemical properties **71**
3 Metabolism and transformation processes **110**
4 Occurrence of common pharmaceuticals in the environment **114**
5 Fate of common pharmaceuticals in the environment **125**
6 Conclusions **134**
References **134**

5 **Analytical techniques for the detection of pharmaceuticals in the environment** **149**
Mohd Sharizal Mohd Sapingi, Omar Fawzi Suleiman Khasawneh, Puganeshwary Palaniandy, and Hamidi Abdul Aziz
1 Introduction **149**
2 Preanalysis protocols **151**
3 Analytical techniques **160**
4 Conclusions **173**
References **173**

6 **Treatment innovation using solar/UV** **179**
Waldemar Studziński and Maciej Przybyłek
1 Introduction **179**
2 Solar radiation **180**
3 Advanced oxidation processes **181**
4 Summary **198**
References **199**

7 Treatment innovation using biological methods in combination with physical treatment methods **217**
Azmatullah Noor, Shamsul Rahman Mohamed Kutty, Mohamed Hasnain Isa, Izharul Haq Farooqi, Augustine Chioma Affam, Abdullahi Haruna Birniwa, and Ahmad Hussaini Jagaba
1 Introduction **217**
2 Pharmaceutical wastewater: Existing contaminants, toxicity, and effects **220**
3 Pharmaceutical wastewater treatment **223**
4 Hybrid technologies **227**
5 Conclusions **237**
References **237**

8 Application of hybrid advanced oxidation and adsorption processes for pharmaceutical wastewater treatment **247**
Ivanna Lapchuk, Alexander Shyichuk, and Tetiana Tatarchuk
1 Introduction **247**
2 Purification of pharmaceutical wastewaters by advanced oxidation processes **249**
3 Adsorption as a method of pharmaceutical substances removing **259**
4 Hybrid oxidation and adsorption processes **260**
5 Conclusions **266**
Acknowledgments **266**
References **267**

9 Microalgae-based bioremediation of pharmaceuticals wastewater **277**
Sasmitha Aulia Zahra, Imran Ahmad, Norhayati Abdullah, Koji Iwamoto, and Ali Yuzir
1 Introduction **277**
2 Removal mechanisms of pharmaceutical contaminants (PCs) by microalgae **278**
3 Factors influencing the removal of PCs **291**
4 Different approaches for the removal of PCs by microalgae phytoremediation **293**
5 Integrated processes for removal of PCs **295**
6 Concluding remarks and future perspectives **298**
Acknowledgments **301**
References **301**

10 An overview of recent progress in membrane-based treatment for pharmaceutical wastewaters **311**
Amir Aslam and Saif Ullah Khan
1 Introduction **311**
2 Adverse effects on human health **312**
3 Membrane-based separation processes (MBSPs) **313**
4 Conclusions **323**
References **324**

11 Sustainable and eco-friendly treatment of pharmaceuticals wastewater **329**
Md. Eman Talukder, Md. Nahid Pervez, Mst. Monira Rahman Mishu, K.M. Faridul Hasan, Faizan Shafiq, Yaping Zhao, Yingjie Cai, Hongchen Song, Md. Ibrahim H. Mondal, George K Stylios, Vincenzo Naddeo, Afzal Husain Khan, and Alberto Figoli
1 Introduction **329**
2 Pharmaceutical wastewater characteristics **330**
3 Conventional pharmaceutical wastewater treatments **331**
4 Advanced pharmaceutical wastewater treatment **335**
5 Conclusions and future perspectives **342**
References **342**

12 Natural polymer-based sustainable adsorbents for pharmaceutical wastewater treatment **347**
Javed Alam, Arun Kumar Shukla, Fekri Abdulraqeb Ahmed Ali, Omar Daoud, and Mansour Alhoshan
1 Introduction **347**
2 Biopolymers at a glance: Classification and mechanisms of PhCs adsorption **351**
3 Cross-linked biopolymers: Especially cross-linked chitosan **353**
4 Conclusions **358**
Acknowledgment **359**
References **359**

13 Innovative and eco-friendly technologies for the upgradation of pharmaceutical wastewater treatment processes **367**
Abdullahi Haruna Birniwa, Umar Ali, Shamsul Rahman Mohamed Kutty, Ahmad Hussaini Jagaba, and Azmatullah Noor
1 Introduction **367**
2 Strategies for removal of pharmaceutical products from wastewater **370**
3 Bioremediation **379**
4 Membranes separation **380**

5 Removal of pharmaceuticals in advanced oxidation processes (AOPs) **382**
6 Constructed wetlands (CWs) **387**
7 Treatment of pharmaceutical residues in wastewater **387**
8 Conclusions **388**
References **389**

14 Biohazardous effect associated with various pharma-effluent discharge in a biotic system **399**
Vivek Sharma, Monalisha Ghosh Dastidar, and Sharmili Roy
1 Introduction **399**
2 Types of effluent exposures, their sources, and impact **400**
3 Conclusions **413**
Acknowledgments **413**
References **413**

Index **423**

Contributors

Norhayati Abdullah Department of Chemical and Environmental Engineering, Malaysia-Japan International Institute of Technology (MJIIT), Universiti Teknologi Malaysia, Kuala Lumpur, Malaysia

Augustine Chioma Affam Civil Engineering Department, School of Engineering and Technology, University of Technology Sarawak, Sibu, Sarawak, Malaysia

Imran Ahmad Algae and Biomass Research Laboratory, Malaysia-Japan International Institute of Technology (MJIIT), Universiti Teknologi Malaysia, Kuala Lumpur, Malaysia

Javed Alam King Abdullah Institute for Nanotechnology, King Saud University, Riyadh, Saudi Arabia

Mansour Alhoshan King Abdullah Institute for Nanotechnology; Chemical Engineering Department, College of Engineering, King Saud University; K.A. CARE Energy Research and Innovation Center, Riyadh, Saudi Arabia

Fekri Abdulraqeb Ahmed Ali K.A. CARE Energy Research and Innovation Center, Riyadh, Saudi Arabia

Md. Sajid Ali Department of Pharmaceutics, College of Pharmacy, Jazan University, Jazan, Saudi Arabia

Umar Ali Department of Chemistry, Sule Lamido University, Kafin-Hausa, Nigeria

Amir Aslam Department of Civil Engineering, Zakir Husain College of Engineering & Technology, Aligarh Muslim University, Aligarh, India; Department of Civil and Environmental Engineering, TU Munich, Munich, Germany

Hamidi Abdul Aziz School of Civil Engineering, Engineering Campus, Universiti Sains Malaysia, Nibong Tebal, Pulau Pinang, Malaysia

Abdullahi Haruna Birniwa Department of Chemistry, Sule Lamido University, Kafin-Hausa, Nigeria

Yingjie Cai Hubei Provincial Engineering Laboratory for Clean Production and High-Value Utilisation of Bio-based Textile Materials, Wuhan Textile University, Wuhan, Hubei, China

Pallavi Chakraborty Department of Biochemical Engineering and Biotechnology, Indian Institute of Technology, Delhi, Delhi, India

Omar Daoud Chemical Engineering Department, College of Engineering, King Saud University, Riyadh, Saudi Arabia

Monalisha Ghosh Dastidar Research School of Engineering, College of Engineering and Computer Science, Australian National University, Canberra, ACT, Australia

Izharul Haq Farooqi Department of Civil Engineering, Zakir Husain College of Engineering and Technology, Aligarh Muslim University, Aligarh, Uttar Pradesh, India

Alberto Figoli Institute on Membrane Technology (CNR-ITM), Rende, Italy

K.M. Faridul Hasan Hubei Provincial Engineering Laboratory for Clean Production and High-Value Utilisation of Bio-based Textile Materials, Wuhan Textile University, Wuhan, Hubei, China

Mohamed Hasnain Isa Civil Engineering Programme, Faculty of Engineering, Universiti Teknologi Brunei, Gadong, Brunei Darussalam

Koji Iwamoto Department of Chemical and Environmental Engineering, Malaysia-Japan International Institute of Technology (MJIIT), Universiti Teknologi Malaysia, Kuala Lumpur, Malaysia

Ahmad Hussaini Jagaba Department of Civil and Environmental Engineering; Centre of Urban Resource Sustainability, Institute of Self-Sustainable Building, Universiti Teknologi PETRONAS, Bandar Seri Iskandar, Perak Darul Ridzuan, Malaysia; Department of Civil Engineering, Abubakar Tafawa Balewa University, Bauchi, Nigeria

Afzal Husain Khan Civil Engineering Department, College of Engineering, Jazan University, Jazan, Saudi Arabia

Nadeem A. Khan Department of Civil Engineering, Mewat Engineering College, Nuh, India

Saif Ullah Khan Department of Civil Engineering, Zakir Husain College of Engineering & Technology, Aligarh Muslim University, Aligarh, India

Omar Fawzi Suleiman Khasawneh School of Civil Engineering, Engineering Campus, Universiti Sains Malaysia, Nibong Tebal, Pulau Pinang, Malaysia

Shamsul Rahman Mohamed Kutty Department of Civil and Environmental Engineering; Centre of Urban Resource Sustainability, Institute of Self-Sustainable Building, Universiti Teknologi PETRONAS, Bandar Seri Iskandar, Perak Darul Ridzuan, Malaysia

Ivanna Lapchuk Educational and Scientific Center of Material Science and Nanotechnology, Vasyl Stefanyk Precarpathian National University, Ivano-Frankivsk, Ukraine

Mohammad Saood Manzar Department of Environmental Engineering, College of Engineering, Imam Abdulrahman Bin Faisal University, Dammam, Saudi Arabia

Ahmad Saud Masood Politecnico di Milano, Milan, Italy

Mst. Monira Rahman Mishu Faculty of Nutrition and Food Science, Patuakhali Science and Technology University, Patuakhali, Bangladesh

Md. Ibrahim H. Mondal Polymer and Textile Research Lab, Department of Applied Chemistry and Chemical Engineering, Rajshahi University, Rajshahi, Bangladesh

Vincenzo Naddeo Sanitary Environmental Engineering Division (SEED), Department of Civil Engineering, University of Salerno, Fisciano, Italy

Azmatullah Noor Department of Civil and Environmental Engineering; Centre of Urban Resource Sustainability, Institute of Self-Sustainable Building, Universiti Teknologi PETRONAS, Bandar Seri Iskandar, Perak Darul Ridzuan, Malaysia

Puganeshwary Palaniandy School of Civil Engineering, Engineering Campus, Universiti Sains Malaysia, Nibong Tebal, Pulau Pinang, Malaysia

Md. Nahid Pervez Hubei Provincial Engineering Laboratory for Clean Production and High-Value Utilisation of Bio-based Textile Materials, Wuhan Textile University, Wuhan, Hubei, China; Sanitary Environmental Engineering Division (SEED), Department of Civil Engineering, University of Salerno, Fisciano, Italy

Maciej Przybyłek Pharmacy Faculty, Collegium Medicum of Bydgoszcz, Department of Physical Chemistry, Nicolaus Copernicus University, Bydgoszcz, Poland

Sharmili Roy Division of Oncology, School of Medicine, Stanford University, Stanford, CA, United States

Hasara Samaraweera Department of Chemistry, Mississippi State University, Mississippi State, MS, United States

Mohd Sharizal Mohd Sapingi School of Civil Engineering, Engineering Campus, Universiti Sains Malaysia, Nibong Tebal, Pulau Pinang, Malaysia

Faizan Shafiq Hubei Provincial Engineering Laboratory for Clean Production and High-Value Utilisation of Bio-based Textile Materials, Wuhan Textile University, Wuhan, Hubei, China

Vivek Sharma Department of Chemical Engineering, Indian Institute of Technology Bombay, Mumbai, Maharashtra, India

Arun Kumar Shukla King Abdullah Institute for Nanotechnology, King Saud University, Riyadh, Saudi Arabia

Alexander Shyichuk Department of Chemistry, Vasyl Stefanyk Precarpathian National University, Ivano-Frankivsk, Ukraine; Faculty of Chemical Technology and Engineering, Bydgoszcz University of Science and Technology, Bydgoszcz, Poland

Hongchen Song Water Science Center, Guangzhou Institute of Advanced Technology, Chinese Academy of Sciences, Guangzhou, China

Waldemar Studziński Faculty of Chemical Technology and Engineering, Division of Food Analytics and Environment, Department of Engineering and Chemical and Food Analytics, Bydgoszcz University of Science and Technology, Bydgoszcz, Poland

George K Stylios Research Institute for Flexible Materials, School of Textiles and Design, Heriot-Watt University, Galashiels, United Kingdom

Md. Eman Talukder Hubei Provincial Engineering Laboratory for Clean Production and High-Value Utilisation of Bio-based Textile Materials, Wuhan Textile University, Wuhan, Hubei

Tetiana Tatarchuk Faculty of Chemistry, Jagiellonian University, Kraków, Poland

Ali Yuzir Department of Chemical and Environmental Engineering, Malaysia-Japan International Institute of Technology (MJIIT), Universiti Teknologi Malaysia, Kuala Lumpur, Malaysia

Sasmitha Aulia Zahra Algae and Biomass Research Laboratory, Malaysia-Japan International Institute of Technology (MJIIT), Universiti Teknologi Malaysia, Kuala Lumpur, Malaysia

Yaping Zhao School of Ecological and Environmental Science, Shanghai Key Laboratory for Urban Ecological Process and Eco-Restoration, East China Normal University, Shanghai, China

About the Editors

Afzal Husain Khan is a lecturer in the Department of Civil Engineering at Jazan University, Saudi Arabia. He received his BE in civil engineering and MTech in environmental engineering from Aligarh Muslim University, India. He is pursuing his PhD from Universiti Sains Malaysia, Malaysia. His research interests are focused on wastewater treatment with an emphasis on removal of pharmaceuticals using physicochemical, biological, and advanced methods. His other interests are municipal solid waste management and durability and strength studies of concrete. He is a guest associate editor for *Frontiers in Sustainability* and *Frontiers in Environmental Chemistry*. His recent articles on treatment technologies for hospital wastewater treatment have been published in prestigious peer-reviewed journals including the *Journal of Cleaner Production*, *Recent Trends in Analytical Chemistry*, *Chemosphere*, *Environmental Research*, and *Journal of Water Processing*. Khan has been involved in many projects funded under the Deanship of Scientific Research (DSR) in Saudi Arabia. He has attended numerous national and international conferences and workshops. He is a life member of the Institution of Engineers, India. He is also a reviewer for prestigious peer-reviewed journals, including those published by Nature, Elsevier, Springer, and MDPI. He has more than 5 patents and 1000 citations with an H-index of 17 to his Google Scholar account.

Nadeem A. Khan is an assistant professor in the Department of Civil Engineering, Mewat Engineering College, Haryana, India, where he was also head of the department (HoD) from 2014 to 2017. He completed his PhD from Jamia Millia Islamia, New Delhi, in 2022. He received his MTech in 2011 with honors in environmental engineering from Aligarh Muslim University, India. He is an editorial board member of the Nature journal *Scientific Reports* and has worked on special issues for the Springer journal *Discover Water* and a Hindawi journal (IF = 4.2). His recent accomplishments include 15 Coursera course completion certificates. He was awarded the Best Young Scientist Award (Male) by LWT India and GECL for his scientific contribution to environmental engineering. Dr. Khan was associated with the Ministry of Environment and Forest (India) for 2 years as a junior research fellow (JRF) from 2009 to 2011 and has also worked with UNICEF as a research Assistant (RA) on the topic of GIS mapping and water quality parameter estimation in 2012. He has attended several conferences, workshops, and seminars and has been involved in several research projects. He has published 4 books with international publishers and 63 papers in SCI journals, and has been the editor of 5 books. He has also published more than 10 patents in civil engineering and water treatment. He is also active in social welfare activities.

Mu. Naushad is a full professor in the Department of Chemistry, College of Science, King Saud University, Saudi Arabia. He received his BS, MS, and PhD degrees from Aligarh Muslim University, Aligarh, India. His research is mainly focused on wastewater treatment. He is the author/coauthor of more than 450 refereed journal publications, several US patents, 25 book chapters, and 25 books published by Springer, Nova, CRC, and Elsevier. His work is widely recognized and has more than 26,500 citations with an H-index of 86 according to Google Scholar Citations. He has been included in the list of Highly Cited Researchers 2019, 2020, and 2021, which is one of the best achievements for an academician. He has been listed in the top 2% of scientists in the world issued by Stanford University, United States. He is an editor/editorial member of several reputed journals published by Nature, Elsevier, Springer, and MDPI. He has also served as an associate editor for *Environmental Chemistry Letters* (Springer). At present, he is associate editor for the *Journal of Water Process Engineering* (Elsevier) and managing editor for *Nanotechnology for Environmental Engineering* (Springer). Dr. Naushad is the recipient of several prestigious awards, including the Scientist of the Year award 2015, NESA, Delhi, India; Almarai Award 2017, Saudi Arabia, and Best Research Quality award 2019 (first position), King Saud University. He is a fellow of the Royal Society of Chemistry (FRSC).

Hamidi Abdul Aziz is a professor in environmental engineering in the School of Civil Engineering of Universiti Sains Malaysia. Professor Aziz received his PhD degree in civil engineering from the University of Strathclyde in Scotland in 1992. He is a guest editor of a special issue on landfill leachate management and control for the *International Journal of Environment and Waste Management* (*IJEWM*). Professor Aziz currently serves as the editor-in-chief of the *International Journal of Scientific Research in Environmental Sciences* (*IJSRES*). He also serves as the managing editor of the *International Journal of Environment and Waste Management* (*IJEWM*) and the *International Journal of Environmental Engineering* (*IJEE*). Professor Aziz's research focuses on alleviating problems associated with water pollution issues from industrial wastewater discharges and from solid waste management via landfilling, such as landfill leachate. Advanced oxidation processes are also one of the topics of his research focus. In addition, he has a strong interest in the biodegradation and bioremediation of oil spills.

Preface

Worldwide, communities are confronted with the dual challenges of declining water quality and limited access to water resources. Eliminating harmful organic and inorganic impurities in water bodies and recycling wastewater as a secondary water supply are both critical steps in solving the problem of water shortage. There is substantial social and economic backing, in both developed and developing countries, for efforts to increase water quality. If they are to contribute significantly to the development of industry and the betterment of humanity, methods for filtering pharmaceutical wastewater should be both cost-effective and environmentally friendly. Applying cutting-edge technologies would considerably improve the existing pharmaceutical wastewater treatment processes. Over the past two decades, academics have given increased attention to the application of new technology in environmental contexts. The application of modern technologies in pharmaceutical wastewater treatment has tremendous promise for improving current procedures.

For most of us, the first thing that springs to mind when we hear the word "hospital" is where individuals may go to get better and be assured of good health, while also receiving care and benefiting from research into novel treatments and cures for any existing conditions. These aims necessitate the use of a wide array of chemicals by hospital staff for therapeutic and diagnostic purposes, as well as for cleaning and sanitizing rooms and medical equipment. Particularly in terms of effluents, hospital wastewater will always contain some amount of harmful substances. Examples of such pharmaceuticals include antibiotics, anesthetics, analgesics, antidepressants, cytostatics, gastrointestinal agents, hormonal agents, etc. Even though the presence of these compounds in water is currently unregulated, growing international concern has been expressed over their use, ingestion, and ultimate fate in the water cycle. Considerable attention has been paid over the past 15 years to the subject of healthcare institutions' wastewater effluent. Studies have primarily focused on the chemical, physical, and microbiological characteristics of pharmaceutical wastewater, the effectiveness of conventional treatments in removing targeted pharmaceuticals, and upgrading treatment options for improving their removal rates.

It is generally acknowledged that research has only dealt with a small fraction of the thousands of compounds used in medications. The chemicals of interest have been chosen based on several factors, including the accessibility of relevant analytical methods, consumption data, and the outcomes of prior studies. It's also common knowledge that specific chemicals were prioritized while developing monitoring systems, and that certain compounds were routinely placed on analyte lists, particularly in the early stages of research. This significance of these substances was typically established through past research, which sparked further interest in succeeding eras. Over the past few years, attention has turned from classic pollutants to novel ones as advances in analytical tools have made it possible to track a broader diversity of substances. The occurrence and detection procedures applying to various pharmaceuticals, the accuracy of direct measurements, the inconsistencies in detected concentrations, and the reliability and accuracy of detection techniques options worldwide have all been widely discussed.

The main goal of this book is to act as a reference for the sharing of updated information about work being done on occurrence, sources, detection strategies, and innovations in pharmaceutical wastewater treatment. Accordingly, this book is designed inclusively of four categories: occurrence, sources, detection strategies, and innovations in pharmaceutical wastewater. All chapter overviews of each main category, including descriptions of specific interventions, are outlined. Furthermore, we offer our deepest thanks to all the authors who agreed to join in this editorial effort, who spent their time developing their research contribution, and, most all, for sharing their expertise and discoveries with other readers.

This book opens with a comprehensive introduction to occurrences, sources, and the classical methods of pharmaceutical wastewater treatment. Chapter 2 introduces treatment scenarios in various countries, including methodologies, technologies, issues encountered, and advanced treatment processes for pharmaceutical wastewater. Chapter 3 reports recent trends in graphene-based materials for pharmaceuticals wastewater treatment. The fate of common pharmaceuticals that are used widely detected in the environment is provided in Chapters 4 and 5. Chapter 6 covers the application of solar/UV for the treatment of pharmaceutical wastewater. Chapter 7 details treatment innovations using biological methods in combination with physical treatment methods for the treatment of pharmaceutical wastewater. Chapter 8 introduces the application of hybrid advanced oxidation and adsorption processes related to pharmaceutical wastewater treatment. Chapter 9 deals with the practical and cost-effective treatment process using microalgae-based bioremediation for pharmaceutical wastewater removal. Chapter 10 describes the recent progress in the membrane-based treatment of pharmaceutical wastewater. Chapter 11 reviews the sustainable and eco-friendly treatment of pharmaceutical wastewater. Chapters 12 and 13 detail various natural polymer-based sustainable adsorbents for pharmaceutical wastewater treatment and innovative technologies to upgrade existing treatment units for pharmaceuticals wastewater removal. Finally, Chapter 14 presents the hazardous effects associated with effluent discharge in a biotic system as the concluding chapter.

The editors would also like to thank friends, colleagues, collaborators, students, and family members for their assistance in bringing this book to its present state and format. Our sincere gratitude goes to all the expert reviewers who took the time to read the whole manuscript and provide useful comments and criticism without conflict of interest. Their skilled and quick assistance has enabled the book to meet global standards. The editors' special gratitude goes to Elsevier's editorial staff (John Leonard, Nirmala Arumugam, Indhumathi Mani, and others), who assisted us through to publication.

Afzal Husain Khan
Nadeem A. Khan
Mu. Naushad
Hamidi Abdul Aziz

Acknowledgments

First of all, we give thanks to God for providing us with good health and the opportunity to complete this book successfully.

We offer our heartfelt thanks to Elsevier, for accepting this book proposal and special thanks to the Elsevier team members who supported and guided us to complete this book. We are very thankful to all contributing authors for their excellent academic contributions throughout this book. We have great pleasure in acknowledging various publishers and authors for granting us permission to use their figures and tables. We would, nevertheless, like to offer our deep apologies to any copyright holder if, unknowingly, their right has been infringed.

Afzal Husain Khan would like to thank Dr. Marei bin Hussein Al-Qahtani (President), and Dr. Ahmad Abutaleb (Dean), College of Engineering, Jazan University for their support.

Nadeem A. Khan expresses his gratitude to Prof. Sirajuddin Ahmed, JMI, India, Prof. Izharul Haq Farooqi, AMU, India, and Prof. Imran Ali, JMI, India, for their constant encouragement and invaluable support.

Prof. Mu. Naushad would like to thank Prof. Abdullah Salman Al-Salman (Vice Rector) KSU, Prof. Zeid A. ALOthman (Dean, College of Science) KSU, and Prof. Abdullah A. Al-Kahtani (HOD, Chemistry Department) KSU, who have always promoted me for academic excellence.

Prof. Hamidi Abdul Aziz would like to thank the Dean of the School of Civil Engineering, Universiti Sains, Malaysia, for his kind support and encouragement.

Afzal Husain Khan

Nadeem A. Khan

Mu. Naushad

Hamidi Abdul Aziz

Introduction: Occurrences, sources, and methods of pharmaceutical wastewater treatment

1

Pallavi Chakraborty
Department of Biochemical Engineering and Biotechnology, Indian Institute of Technology, Delhi, Delhi, India

1. General introduction

In recent years, the increasing population of the world has triggered the need to provide access to good quality freshwater for a sustainable livelihood. On the other hand, rapid industrialization has impacted the waterbodies by uncontrolled effluent discharge. Pharmaceuticals are manufactured for use in the treatment and prohibition of diverse diseases in humans and animals. After completing their role in the body, pharmaceuticals are usually excreted and deposited in aquatic systems; furthermore, they are also released in effluents from industries, and disposition of unused medicines also contribute to elevating their presence in the environment. The pharmaceutical discharge from different sources has impacted the water system and led to an increase in the toxicity level of water. On the other hand, the polluted water systems affect humans, animals, and marine organisms directly. As humans and animals can't survive without water, pure, fresh water is our utmost requirement for having a good, healthy life.

Considering the negative impacts of pharmaceuticals on waterbodies, treating pharmaceutical wastewater is an urgent requisite and focusing on that may lead to a better environment. There are different conventional methods available for treating pharmaceutical wastewater. Various biological, physical, and chemical methods are available to treat pharmaceutical wastewater. However, it's impossible to achieve 100% removal by any treatment. Certain modifications are currently being incorporated, producing hybrid systems that are being developed to achieve better results. The effect of pharmaceutical wastewater is serious enough to be a matter for concern, but combination this of different methods could be a solution.

2. Occurrences and sources

The occurrence of pharmaceuticals and their byproducts in the environment is a matter of concern. Different pharmaceuticals are being found in various samples, including sewage water, surface water, soil, seawater, and industrial effluents (Nikolaou, Meric, & Fatta, 2007).

The Treatment of Pharmaceutical Wastewater. https://doi.org/10.1016/B978-0-323-99160-5.00004-7

Pharmaceuticals are mainly used as human medicine and antibiotics are used as veterinary medicine too. These two sectors comprise most of the pharmaceutical usages and, in Europe for example, some pharmaceutical products are consumed in high bulk amounts (>100 tons) (Jones, Voulvoulis, & Lester, 2002). Pharmaceuticals and their by-products enter the environment by disposal and excretion from humans and animals. In addition, due to an improper elimination from wastewater treatment plants, effluent containing leftover pharmaceutical products and various organic toxic compounds is released, which gets mixed up in ground and surface water (Fent, Weston, & Caminada, 2006; Giger et al., 2003a; Moldovan, Schmutzer, & Alder, 2006; Wong, Mok, & Fan, 2007).

As illustrated in Fig. 1, pharmaceuticals and their metabolites can enter the environment mainly via excretion and disposal in wastewater. Due to improper elimination from wastewater-treatment plants, many organic compounds' toxic residues, along with pharmaceutics, are present in surface waters (Fent et al., 2006; Giger et al., 2003b; Nikolaou et al., 2007; Wong et al., 2007). Wastewater from manufacturers, landfill leachates, and deposition of medicines in the environment contribute as sources of human pharmaceuticals, along with wastewaters from municipalities and hospitals (Johnson, Williams, & Matthiessen, 2006; Kay, Blackwell, & Boxall, 2005).

Over the last few years, applications of pharmaceuticals have multiplied in line with the advances in of medical science. Approximately 3000 compounds are used as medicine, and the annual production amount exceeds hundreds of tons (Calisto & Esteves, 2009; Kümmerer, 2009; Sarmah, Meyer, & Boxall, 2006). After their

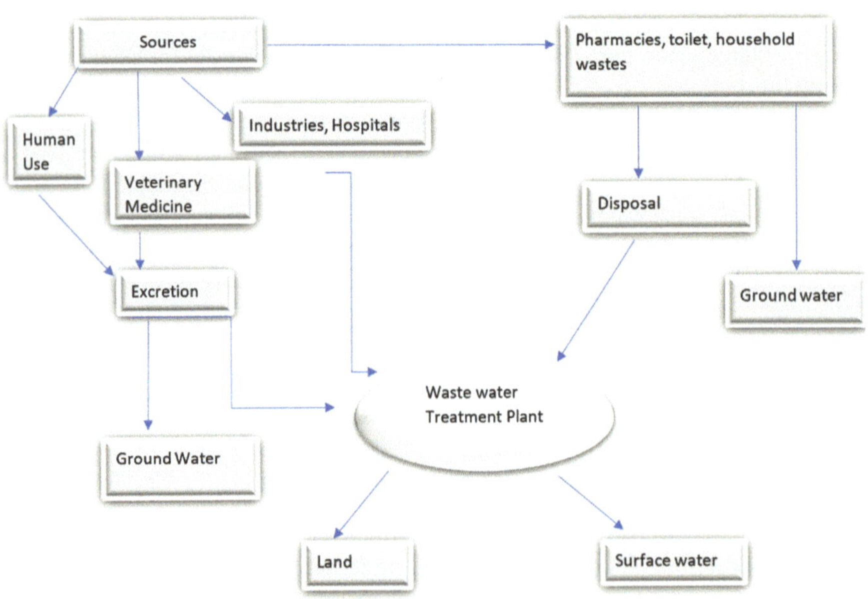

Fig. 1 Schematic representation of sources of pharmaceutical wastewater in the environment.

use, large amounts of pharmaceuticals are discharged into the water system (Kümmerer, 2009). They are present in urban and livestock agricultural wastewater and surface water (Campagnolo et al., 2002; Kolpin et al., 2002; Larsson, de Pedro, & Paxeus, 2007; Pedrouzo, Reverté, Borrull, Pocurull, & Marcé, 2007; Sim et al., 2011; Vanderford, Pearson, Rexing, & Snyder, 2003). Various pharmaceutical products are used in daily life by humans. Large quantities of veterinary pharmaceuticals are used in livestock farming to prevent and treat diseases and also to increase weight. In Korea, veterinary pharmaceuticals are used more than in any other country (Choi et al., 2008). Large amounts of pharmaceuticals are produced by pharmaceutical manufacturers and dispensed in hospitals (Li et al., 2008; Lindberg, Jarnheimer, Olsen, Johansson, & Tysklind, 2004). Pharmaceuticals mainly enter the water environment after excretion into in the form of treated or untreated wastewater. Thus, wastewaters from municipal wastewater treatment plants (MWWTPs), livestock farms, hospitals, and pharmaceutical manufactures represent known sources of pharmaceuticals discharged into the water environment (Matamoros, Hijosa, & Bayona, 2009; Zhang, Geißen, & Gal, 2008). However, there is a lack of information about pharmaceuticals being discharged from other wastewater treatment plants, such as livestock, hospital, and pharmaceutical industries (Sim et al., 2011).

Nonsteroidal antiinflammatory drugs (NSAIDS) have been found in wastewater; therefore, before potentially releasing them into the water system, municipal wastewater plants should check their level in the treated water. Their amount is usually higher in sewage plant water, which is usually due to the adsorption of micropollutants into sludge granules (Ekpeghere, Kim, Sung-Hee, Kim, & Oh, 2016). Successful removal depends on the type of treatment plant and other operating parameters like pH and temperature (Samaras, Stasinakis, Mamais, Thomaidis, & Lekkas, 2013).

In other studies, high levels of nonsteroidal antiinflammatory drugs, antimicrobials and antiepileptics have been observed in wastewater. If these substances are released continuously in the environment, it can create ecotoxicological effects. It also increases the chances of creating new resistant bacteria and establishing a pathogenic reservoir, which can harm public health (Luo et al., 2010).

Table 1 depicts the most commonly found pharmaceuticals in the world (Halling-Sørensen et al., 1998; Hirsch, Ternes, Haberer, & Kratz, 1999; Nikolaou et al., 2007), and Fig. 2 depicts some of the structures of the pharmaceuticals most used in daily life (Wishart et al., 2006).

3. Sources of pharmaceuticals

Sources of pharmaceutical pollution can be very difficult to detect, but broadly, there are two types of sources. The first is point source pollution; this kind of source can more easily be tracked down, for example, sewage treatment plants, wastewater. The second type of source is diffuse pollution; it is really difficult to find the origin of these sources (Li, 2014).

Point source consists of wastewater sources, domestic pathways, sewage treatment plants, municipal landfills, and septic tanks. In a study, it has been noticed that

Table 1 The most commonly found pharmaceuticals in the world (Halling-Sørensen et al., 1998; Hirsch et al., 1999; Nikolaou et al., 2007).

Antiinflammatory drugs	Lipid regulators	Beta blockers	Steroids and hormonal medicines	Antibiotics	Other common drugs
Aspirin	Bezafibrate	Metoprolol	17-β-estradiol	Ofloxacin	Cancer therapeutics
					Cyclophosphamide,
Diclofenac		Propranolol	Estrone	Penicillin	Ifosfamide
Ibuprofen		Nadolol	17α-ethinyl estradiol	Streptomycin	Diuretics
Acetaminophen	Gemfibrozil	Atenolol	Diethylstilbestrol	Oxytetracycline	Furosemide
Metamizole	Clofibric acid	Sotalol	Diethylstilbestrol acetate	Chlortetracycline	Antiepileptics
Codeine	Fenofibrate	Betaxolol		Lincomycin	Carbamazepine
Indomethacin				Amoxicillin	Antidepressant
Naproxen				Spiramycin	Tranquilizers
Phenazone				Sulfamethoxazole	

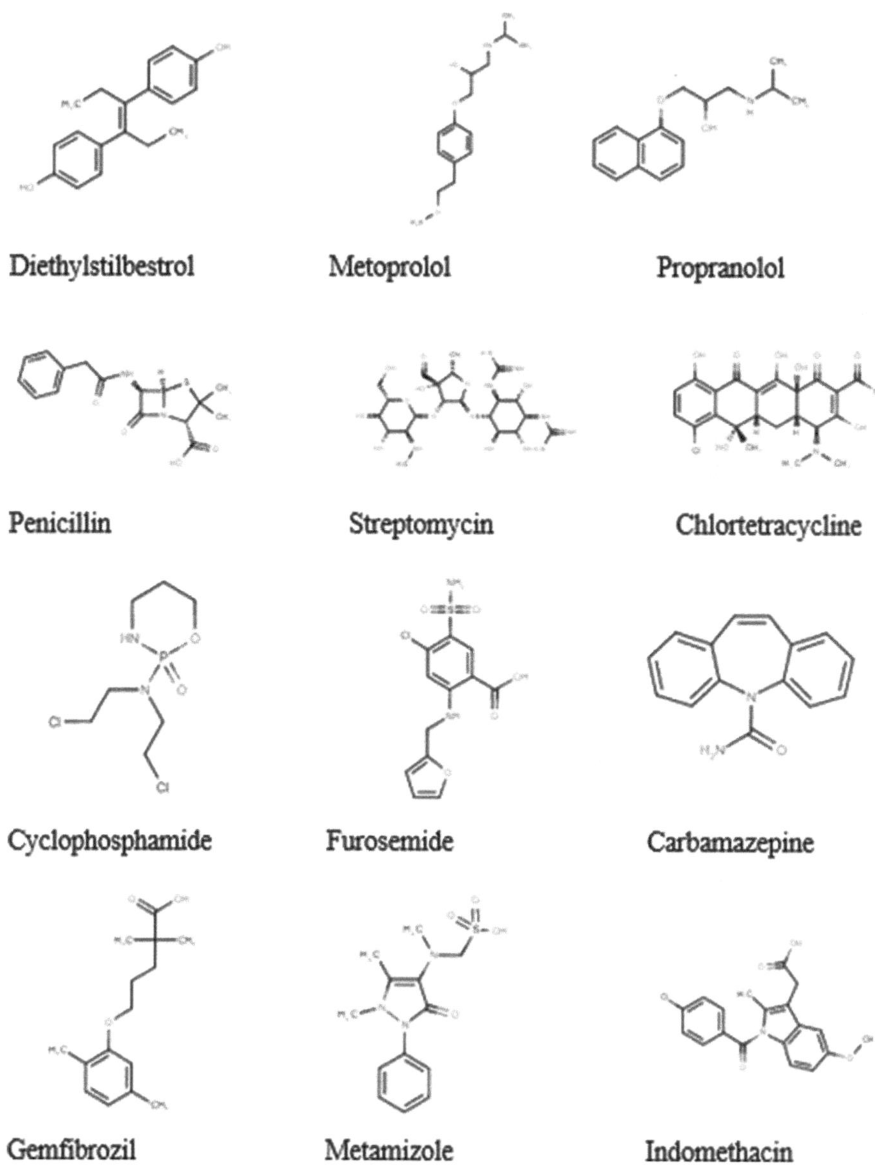

Fig. 2 Structures of a few commonly used pharmaceuticals.

municipal landfills are often filled with hazardous elements, and wastes including pharmaceutical compounds and hormones (Buszka et al., 2009). In another study by Barnes et al., they also noticed pharmaceutical compounds in municipal landfills and also in groundwater. The municipal landfills generate organic wastewater, which leaches and mixes up with the groundwater and sometime even with surface water (Barnes et al., 2004).

Wastewater sources are one of the major contributors of pharmaceutical compounds in the aquatic environment (Glassmeyer et al., 2005). There are several pharmaceutical compounds in wastewater sources, as reported by different researchers (Batt, Kostich, & Lazorchak, 2008; Petrovic, Gros, & Barcelo, 2006). The domestic pathway serves as one of the main sources of pharmaceutical compounds in wastewater, along with hospitals and industries (Ternes, 1998). Hospital wastewater is rich in pharmaceutical compounds, including antiepileptics and antiinflammatories. Most of these compounds and personal care products enter the water system as sewage water (Daughton & Ternes, 1999).

Sewage treatment plants are another major concern as various studies have reported their inefficiency in removing pharmaceutical compounds completely. In fact, the removal of such compounds can be insignificant in influent and effluent discharge from treatment plants (Snyder, 2008; Vieno, Härkki, Tuhkanen, & Kronberg, 2007).

Septic tanks are also a source of pharmaceutical wastewater. Septic tanks are small sewage tanks that mainly collect the generated wastes from households. In a study by Godfrey et at., they found at least 22 pharmaceutical compounds in septic tanks and most of them are above detection limit. Furthermore, septic tanks pose a threat to groundwater, as hazardous compounds can leak and pollute the environment.

Diffuse pollution includes sewage sludge as one of the most important pathways of pharmaceutical wastewater (Harrison, Oakes, Hysell, & Hay, 2006). Sewage sludge, also known as biosolids, is the debris of wastewater after treatment. Application of sewage sludge on the ground surface can lead to soil pollution, as it enables the leakage of pharmaceutical compounds into the soil (Lapworth, Baran, Stuart, & Ward, 2012). Another source is the groundwater-surface water interface, which allows the exchange of groundwater with surface water (Lapworth et al., 2012). In a review work, it has been noted that rivers and canals contain high amounts of pharmaceutical compounds (Pal, Gin, Lin, & Reinhard, 2010). Consequently, by natural and artificial exchange of surface water with groundwater, contaminants can pass from the former to the later (Drewes, 2009).

3.1 Occurrences of pharmaceuticals in India

India is a diverse nation, not only in terms of culture and people, but also in nature and climate. India mostly has four different seasons and climate varies with regions. Pharmaceutical compounds are found in all the different regions, although very few studies have been conducted. Some of them are discussed here.

In 2014, Singh et al. discussed the presence of pharmaceuticals and associated risk assessment in Northern India. His work primarily focused on Lucknow and Ghaziabad City, where they tested samples from waste water streams and analyzed them using high-performance liquid chromatography (HPLC). They found that a large number of hazardous pharmaceuticals are present in Ghaziabad, including the drugs listed in Table 2, ranged below the detection limit. Trimethoprim had the highest concentration in Ghaziabad and carbamazepine in Lucknow (Singh et al., 2014).

Table 2 Drugs accumulation in wastewater of Lucknow and Ghaziabad (Singh, Rai, Singh, Verma, & Gupta, 2014).

Drug name	Drug structure	Unit	Range in Lucknow	Range in Ghaziabad
Ibuprofen		$\mu g L^{-1}$	BDL 10.02	BDL 26.45
Ketoprofen		$\mu g L^{-1}$	BDL 1.69	BDL 16.21
Naproxen		$\mu g L^{-1}$	BDL	BDL 1.65
Diclofenac		$\mu g L^{-1}$	BDL 1.43	BDL 25.68
Trimethoprim		$\mu g L^{-1}$	BDL 5.11	BDL 51.82
Carbamazepine		$\mu g L^{-1}$	BDL 12.36	BDL 5.42

BDL, below detection limit.

In southern India, Carbamazepine concentration in the Kaveri River was reported by Ramaswamy et al. to be 28.3 ng/L (Ramaswamy, Shanmugam, Velu, Rengarajan, & Larsson, 2011).

Thalla et al. studied the presence of NSAIDs in municipal wastewaters in southwest India. They conducted the research in a secondary treatment plant situated in Kavoor, Karnataka; the southwest monsoon region. They studied to find out the presence and removal of the NSAIDs diclofenac, naproxen, ibuprofen, ketoprofen, and acetylsalicylic acid. These drugs are usually consumed at a higher rate, are toxic, and belong to the class of prescribed medicine; they also increase environmental hazard and have a negative effect on biodiversity. This is why it is important to determine their level in this region. From their study, the authors observed a significant variation of drug presence in wastewater and its removal rate by the treatment plant in two different seasons. This further implies that both parameters are directly related to the seasonal changes in that region (Thalla & Vannarath, 2020).

4. Analytical techniques used to determine pharmaceutical compounds

Various techniques are used to detect the presence of pharmaceutical products (Table 3) and, with modernization and technological improvement, the precision of detecting samples has also been upgraded.

5. Toxicological effect of pharmaceuticals

The effect of pharmaceuticals on the environment and public health is worrying as it can include severe toxicity that leads to mutagenic changes and genotoxicity (Akintonwa et al., 2009; Bakare, Okunola, Adetunji, & Jenmi, 2009; Nadal, Schuhmacher, & Domingo, 2004). Many pharmaceuticals don't degrade easily and persist in the environment for a long time. Furthermore, the fact that wastewater

Table 3 Detection of pharmaceuticals.

Technique used	Drugs detected	Reference
Gas chromatography-mass spectrometry (GC-MS)	Antiinflammatory Drugs: diclofenac, ibuprofen, ketoprofen, indomethacin, naproxen, etc.	Sacher, Lange, Brauch, and Blankenhorn (2001)
High-performance liquid chromatography/electrospray ionization tandem mass spectrometry (HPLC-ESI-MS-MS)	Beta Blockers: metoprolol, propranolol, atenolol, bisoprolol, sotalol, pindolol, betaxolol	Sacher et al. (2001)
Gas chromatography-mass spectrometry (GC-MS)	Carbamazepine, clofibrate, phenazone, aminopyrine	Lin, Chen, and Ding (2005)
High-performance liquid chromatography-mass spectrometry (HPLC-MS-MS)	Antibiotics: ofloxacin, ciprofloxacin, norfloxacin, sulfamethoxazole, sulfamethazine, lincomycin, oxytetracycline, penicillin G1	Brown, Kulis, Thomson, Chapman, and Mawhinney (2006)

treatment plants are ineffective in removing the compounds completely might impact the environment severely. In several studies, it has been observed that fish are highly exposed to pharmaceutical compounds through the water system. Indirectly, it can impact human health too. It has been shown that compounds like diclofenac and 17α-ethinylestradiol can affect the kidneys and intestines of fish and alter genetic expressions (Jobling, Nolan, Tyler, Brighty, & Sumpter, 1998; Lyssimachou & Arukwe, 2007). Long exposure to diclofenac can harm the reproductive systems of fish and of mammals, and may even affect humans, as suggested by one study (Mehinto, Hill, & Tyler, 2010). In the aquatic environment, pharmaceutical compounds can affect algae, which negatively impacts the ecology. Vannini et al. conducted a study that said that high amounts of diclofenac and carbamazepine impact the chloroplasts of algae. Further investigation suggested that antibiotics damages microorganisms and algae (Vannini et al., 2011), and can even target the photosynthetic pathway of algae, and as a consequence, their survival will be compromised (Liu et al., 2011).

Hormonal tablets can trigger serious environmental damage at very low concentration, as they are biologically active. In 2010, Santos et al. published a work where he discussed that 17α-ethinylestradiol has an estrogenic effect on fish (Santos et al., 2010). Studies have found that it effects the fertilization system of fish and can even change their sex characteristics (Parrott & Blunt, 2005; Pawlowski, Van Aerle, Tyler, & Braunbeck, 2004).

6. Classical methods of pharmaceutical wastewater treatment

An enormous volume of pharmaceutical waste is difficult to treat as it is toxic. These compounds contain volatile, hazardous substances, and residues with high biological oxygen demand and chemical oxygen demand, which make them difficult to handle and harmful for the ecological system. That is why it is important to remove them. There are several classical treatment methods for pharmaceutical wastewater including different physicochemical and biological methods. As the nature of pharmaceutical compounds is varied, the removal of compounds is treatment-specific.

6.1 Physicochemical methods

Physicochemical methods include coagulation-flocculation, precipitation, and others (Cheng et al., 2007). To remove pharmaceutical products and personal care products, certain processes are preferred over others. These include coagulation, volatilization, sorption and sedimentation. There are different filtration methods that have been employed to remove contaminants successfully but the efficiency of these processes solely depends on the size and molecular weight of the targeted molecules (Rajapaksha, Premarathna, Gunarathne, Ahmed, & Vithanage, 2019). Hormones can be removed by nanomembrane, depending on their molecular size, hydrogen bonding with the functional group of the membrane, and hydrophobic interaction with the membrane (Nghiem, Schäfer, & Elimelech, 2004). Coagulation, flocculation, and sedimentation are considered to be primary treatments as the costs are negligible, however, they don't provide much accuracy (Rajapaksha et al., 2019).

6.1.1 Coagulation and flocculation

Coagulation and flocculation are important classical treatment methods that remove suspended particulates and other compounds to increase the biodegradability of wastewater. In the coagulation and flocculation process, pollutants are settled with a change in pH or by adding chemical agents. Coagulation destabilizes a suspension or solution; it changes the factors that provide stability to a solution. In flocculation, the destabilized particles come together and form larger particles or agglomerates, which suspend easily. The chemical agents that cause coagulation and flocculation are known as coagulants and flocculants, respectively (Bratby, 1980). Coagulation and flocculation is the first step in removing any type of waste from municipal and industrial wastewater. Some factors need to be considered before opting for this process, like the high amount of sludge formation and deposition (Shammas, 2005).

6.1.2 Sedimentation

Sedimentation is the most traditional method of separating particles, and depends on the coagulation and flocculation process to produce flocs, which allows particles to settle faster (Henderson, Parsons, & Jefferson, 2008). Sedimentation is the process in which particles come out of the solution and settle, mostly at the base of water system. The settling is due to the motion of the fluid, which is often caused by gravitational force. There are different types of sedimentation. Type 1 sedimentation means that the particles settle individually and do not agglomerate throughout the settling. In type 2 sedimentation, particles settle with flocculation, as the size and the velocity changes. Type 3, or zone, sedimentation involves settling of a high of concentration particles, forming one clear zone along with a sludge zone. These processes are applied for removing organic chemicals (Speight, 2017).

6.2 Biological methods

There have been studies to degrade pharmaceutical wastes biologically. These methods include composting, vermicomposting, and aerobic or anaerobic digestive methods to degrade pharmaceutical compounds and turn them into valuable by-products (Järvis, Ivask, Nei, Kuu, & Luud, 2016).

6.2.1 Constructed wastelands

The major obstacle in treating pharmaceutical wastewater is that it enters the sewage stream with other wastewater streams, so it's difficult to separate the composites. Constructed wetlands are engineered systems that use natural procedures (Vymazal, 2010). Types include free-water surface constructed wastelands, horizontal subsurface flow constructed wetlands, and vertical subsurface flow constructed wetlands.

Free-water surface constructed wastelands are water basins with aquatic plants covering up to 50% of the water, which also contain many microbial species (Kadlec, 1995). This type of constructed wetlands is efficient in removing organic substances by microbial degradation, as particles are suspended and filtered by dense macrophytes (Vymazal, 2010). This technique is mainly used in North America, Australia, and also in Sweden and Denmark (Kadlec & Wallace, 2008; Vymazal, Greenway, Tonderski, Brix, & Mander, 2006).

Horizontal subsurface flow constructed wetlands are blocked and planted by macrophytes. The water enters through the inlet, and flows through the horizontal path until it reaches the outlet. The wastewater is treated with microbial degradation in the filtration bed, coupled with chemical and, physical processes to separate composites (Vymazal, 2007). In vertical subsurface flow constructed wetlands, wastewaters are fed in in large amounts and move through a sand medium; the next batch is only allowed to pass when the previous one is completed. This allows more oxygen in the bed and is very effective in removing organic and solid particles. It doesn't allow denitrification (Brix & Arias, 2005; Cooper, 1999; Molle, Liénard, Boutin, Merlin, & Iwema, 2005; Weedon, Murphy, & Sweaney, 2017). These are often used in treating municipal and domestic wastewater (Cooper, 1999).

6.2.2 Activated sludge treatment

This process is usually conducted in big cities and towns to remove organic compounds. It comprises a dense microbial culture in suspension, which degrades organic compounds and forms biological flocs that get separated in a settling tank (Mara & Horan, 2003). It is an aerobic treatment process, which entirely depends on biodegradation. The mixed microbial cultures break down the complexes for nutritional requirements, which allows the separation of the compounds. The main component of this process is an aeration tank where the wastewater comes through the inlet and gets mixed up with oxygen and activated sludge in a continuous mode. The water flows out through the clarifier tank where the sludge is removed from the purified water. The settled sludge is then recycled back to the aeration tank (Lampinen, Assad, & Curd, 2001).

6.2.3 Vermicomposting

This process is based on earthworms and microorganisms, which jointly degrade and detoxify the organic compounds to a product that can then be used for agronomic purposes. This is an eco-friendly and sustainable biological process (Lim & Wu, 2016). It is a zero-waste process, where organic matter is decomposed completely by earthworms. Organic matter is decomposed and converted into renewable energy sources such as biofertilizers, matching to a circular economy concept (Rorat & Vandenbulcke, 2019). Although this process has been successful in managing solid wastes, regarding treating pharmaceutical wastewater, the efficiency has not yet been measured.

6.2.4 Bioaugmentation

Bioaugmentation is the process of adding selected microbial strains or mixture of different cultures to enhance the catabolism rate, and removing specific compounds. Microbial cultures are added to increase the microbial population and degrade these compounds (Ren & Zhang, 2019). Suitable microbial communities are identified and grown to be used for biodegradation. Here, the added microorganisms help in reducing contamination by converting the complex compounds to less toxic, simple compounds (Herrero & Stuckey, 2015).

6.2.5 Biotransformation

Biotransformation is the process of removing contamination from soil and water systems using microorganisms or enzymes. The process helps to transform unwanted complexes to a desired product. Over the last few years, researchers have been focusing on implementing biocatalysts to enhance the stability and efficiency of enzymes (Moo-Young, 2019). Biotransformation processes can be separated into in situ operations, where the contaminated soil and wastewater is treated at the place it is found, and ex situ operations, where the toxic substances are excavated by mechanical or chemical methods and taken to some other place where biotransformation happens.

In situ operation are cheaper than ex situ operations, but in situ takes more time than ex situ, as the sample quantity is often large (Speight, 2018).

Biotransformation holds an important role in controlling the toxicity of volatile organic compounds (VOCs). Many VOCs are insoluble in water but biotransformation makes them soluble; it further helps to remove the compounds easily from water (Anand, Philip, & Mehendale, 2014).

7. Conclusions

As pharmaceutical and personal care products are an indispensable part of our lives, so we need to be conscientious about their disposal. The harmful and toxic effect of pharmaceuticals can create health hazards, and long-term exposure to pharmaceutical wastes can be fatal. Considering all these effects, pharmaceutical wastewater treatment is an essential. The classical treatment methods are low-cost processes, but often show disadvantages due to inefficiency of removing wastes properly. Even the biological treatments lead to incomplete degradation and formation of toxic products (Kaur, Hippargi, Pophali, & Bansiwal, 2019). This is why treatment methods have been evolving rapidly and advanced treatments are getting increased attention. Membrane filtration, advanced oxidation, ozonization, and application of UV are used nowadays to get better separation. In spite of having such advanced technologies, more efficient and cost-effective methods need to be developed (Kaur et al., 2019).

References

Akintonwa, A., Awodele, O., Olofinnade, A. T., Anyakora, C., Afolayan, G. O., & Coker, H. A. B. (2009). Assessment of the mutagenicity of some pharmaceutical effluents. *American Journal of Pharmacology and Toxicology*, *4*(4), 144–150.

Anand, S. S., Philip, B. K., & Mehendale, H. M. (2014). Volatile organic compounds. In *Encyclopedia of toxicology* (pp. 967–970). Elsevier.

Bakare, A. A., Okunola, A. A., Adetunji, O. A., & Jenmi, H. B. (2009). Genotoxicity assessment of a pharmaceutical effluent using four bioassays. *Genetics and Molecular Biology*, *32*(2), 373–381.

Barnes, K. K., Christenson, S. C., Kolpin, D. W., Focazio, M. J., Furlong, E. T., Zaugg, S. D., et al. (2004). Pharmaceuticals and other organic waste water contaminants within a leachate plume downgradient of a municipal landfill. *Groundwater Monitoring & Remediation*, *24*(2), 119–126.

Batt, A. L., Kostich, M. S., & Lazorchak, J. M. (2008). Analysis of ecologically relevant pharmaceuticals in wastewater and surface water using selective solid-phase extraction and UPLC – MS/MS. *Analytical Chemistry*, *80*(13), 5021–5030.

Bratby, J. (1980). *Coagulation and flocculation*. Croydon, England: Uplands.

Brix, H., & Arias, C. A. (2005). The use of vertical flow constructed wetlands for on-site treatment of domestic wastewater: New Danish guidelines. *Ecological Engineering*, *25*(5), 491–500.

Brown, K. D., Kulis, J., Thomson, B., Chapman, T. H., & Mawhinney, D. B. (2006). Occurrence of antibiotics in hospital, residential, and dairy effluent, municipal wastewater, and the Rio Grande in New Mexico. *Science of the Total Environment*, *366*(2–3), 772–783.

Buszka, P. M., Yeskis, D. J., Kolpin, D. W., Furlong, E. T., Zaugg, S. D., & Meyer, M. T. (2009). Waste-indicator and pharmaceutical compounds in landfill-leachate-affected ground water near Elkhart, Indiana, 2000–2002. *Bulletin of Environmental Contamination and Toxicology*, *82*(6), 653–659.

Calisto, V., & Esteves, V. I. (2009). Psychiatric pharmaceuticals in the environment. *Chemosphere*, *77*(10), 1257–1274.

Campagnolo, E. R., Johnson, K. R., Karpati, A., Rubin, C. S., Kolpin, D. W., Meyer, M. T., et al. (2002). Antimicrobial residues in animal waste and water resources proximal to large-scale swine and poultry feeding operations. *Science of the Total Environment*, *299*(1–3), 89–95.

Cheng, H., Xu, W., Liu, J., Wang, H., He, Y., & Chen, G. (2007). Pretreatment of wastewater from triazine manufacturing by coagulation, electrolysis, and internal microelectrolysis. *Journal of Hazardous Materials*, *146*(1–2), 385–392.

Choi, K., Kim, Y., Park, J., Park, C. K., Kim, M., Kim, H. S., et al. (2008). Seasonal variations of several pharmaceutical residues in surface water and sewage treatment plants of Han River, Korea. *Science of the Total Environment*, *405*(1–3), 120–128.

Cooper, P. (1999). A review of the design and performance of vertical-flow and hybrid reed bed treatment systems. *Water Science and Technology*, *40*(3), 1–9.

Daughton, C. G., & Ternes, T. A. (1999). Pharmaceuticals and personal care products in the environment: Agents of subtle change? *Environmental Health Perspectives*, *107*(suppl 6), 907–938.

Drewes, J. E. (2009). Ground water replenishment with recycled water—Water quality improvements during managed aquifer recharge. *Groundwater*, *47*(4), 502–505.

Ekpeghere, K. I., Kim, U. J., Sung-Hee, O., Kim, H. Y., & Oh, J. E. (2016). Distribution and seasonal occurrence of UV filters in rivers and wastewater treatment plants in Korea. *Science of the Total Environment*, *542*, 121–128.

Fent, K., Weston, A. A., & Caminada, D. (2006). Ecotoxicology of human pharmaceuticals. *Aquatic Toxicology*, *76*(2), 122–159.

Giger, W., Alder, A. C., Golet, E. M., Kohler, H. P. E., McArdell, C. S., Molnar, E., et al. (2003a). Occurrence and fate of antibiotics as trace contaminants in wastewaters, sewage sludges, and surface waters. *Chimia International Journal for Chemistry*, *57*(9), 485–491.

Giger, W., Alder, A. C., Golet, E. M., Kohler, H. P. E., McArdell, C. S., Molnar, E., et al. (2003b). Occurrence and fate of antibiotics as trace contaminants in wastewaters, sewage sludges, and surface waters. *Chimia International Journal for Chemistry*, *57*(9), 485–491.

Glassmeyer, S. T., Furlong, E. T., Kolpin, D. W., Cahill, J. D., Zaugg, S. D., Werner, S. L., et al. (2005). Transport of chemical and microbial compounds from known wastewater discharges: Potential for use as indicators of human fecal contamination. *Environmental Science & Technology*, *39*(14), 5157–5169.

Halling-Sørensen, B. N. N. S., Nielsen, S. N., Lanzky, P. F., Ingerslev, F., Lützhøft, H. H., & Jørgensen, S. E. (1998). Occurrence, fate and effects of pharmaceutical substances in the environment—A review. *Chemosphere*, *36*(2), 357–393.

Harrison, E. Z., Oakes, S. R., Hysell, M., & Hay, A. (2006). Organic chemicals in sewage sludges. *Science of the Total Environment*, *367*(2–3), 481–497.

Henderson, R., Parsons, S. A., & Jefferson, B. (2008). The impact of algal properties and pre-oxidation on solid–liquid separation of algae. *Water Research*, *42*(8–9), 1827–1845.

Herrero, M., & Stuckey, D. C. (2015). Bioaugmentation and its application in wastewater treatment: A review. *Chemosphere*, *140*, 119–128.

Hirsch, R., Ternes, T., Haberer, K., & Kratz, K. L. (1999). Occurrence of antibiotics in the aquatic environment. *Science of the Total Environment, 225*(1–2), 109–118.

Järvis, J.Ü. R. I., Ivask, M., Nei, L., Kuu, A., & Luud, A. (2016). Effect of green waste compost application on afforestation success. *Baltic Forestry, 22*(1), 90–97.

Jobling, S., Nolan, M., Tyler, C. R., Brighty, G., & Sumpter, J. P. (1998). Widespread sexual disruption in wild fish. *Environmental Science & Technology, 32*(17), 2498–2506.

Johnson, A. C., Williams, R. J., & Matthiessen, P. (2006). The potential steroid hormone contribution of farm animals to freshwaters, the United Kingdom as a case study. *Science of the Total Environment, 362*(1–3), 166–178.

Jones, O. A. H., Voulvoulis, N., & Lester, J. N. (2002). Aquatic environmental assessment of the top 25 English prescription pharmaceuticals. *Water Research, 36*(20), 5013–5022.

Kadlec, R. H. (1995). Overview: Surface flow constructed wetlands. *Water Science and Technology, 32*(3), 1–12.

Kadlec, R. H., & Wallace, S. (2008). *Treatment wetlands.* CRC Press.

Kaur, H., Hippargi, G., Pophali, G. R., & Bansiwal, A. K. (2019). Treatment methods for removal of pharmaceuticals and personal care products from domestic wastewater. In *Pharmaceuticals and personal care products: Waste management and treatment technology* (pp. 129–150). Butterworth-Heinemann.

Kay, P., Blackwell, P. A., & Boxall, A. B. (2005). A lysimeter experiment to investigate the leaching of veterinary antibiotics through a clay soil and comparison with field data. *Environmental Pollution, 134*(2), 333–341.

Kolpin, D. W., Furlong, E. T., Meyer, M. T., Thurman, E. M., Zaugg, S. D., Barber, L. B., et al. (2002). Pharmaceuticals, hormones, and other organic wastewater contaminants in US streams, 1999–2000: A national reconnaissance. *Environmental Science & Technology, 36*(6), 1202–1211.

Kümmerer, K. (2009). Antibiotics in the aquatic environment – A review – Part I. *Chemosphere, 75*(4), 417–434.

Lampinen, M. A. R. K. K. U., Assad, M. E. H., & Curd, E. F. (2001). *Physical fundamentals. Industrial ventilation design guidebook* (pp. 41–171).

Lapworth, D. J., Baran, N., Stuart, M. E., & Ward, R. S. (2012). Emerging organic contaminants in groundwater: A review of sources, fate and occurrence. *Environmental Pollution, 163*, 287–303.

Larsson, D. J., de Pedro, C., & Paxeus, N. (2007). Effluent from drug manufactures contains extremely high levels of pharmaceuticals. *Journal of Hazardous Materials, 148*(3), 751–755.

Li, D., Yang, M., Hu, J., Zhang, Y., Chang, H., & Jin, F. (2008). Determination of penicillin G and its degradation products in a penicillin production wastewater treatment plant and the receiving river. *Water Research, 42*(1–2), 307–317.

Li, W. C. (2014). Occurrence, sources, and fate of pharmaceuticals in aquatic environment and soil. *Environmental Pollution, 187*, 193–201.

Lim, S. L., & Wu, T. Y. (2016). Characterization of matured vermicompost derived from valorization of palm oil mill byproduct. *Journal of Agricultural and Food Chemistry, 64*(8), 1761–1769.

Lin, W. C., Chen, H. C., & Ding, W. H. (2005). Determination of pharmaceutical residues in waters by solid-phase extraction and large-volume on-line derivatization with gas chromatography–mass spectrometry. *Journal of Chromatography A, 1065*(2), 279–285.

Lindberg, R., Jarnheimer, P.Å., Olsen, B., Johansson, M., & Tysklind, M. (2004). Determination of antibiotic substances in hospital sewage water using solid phase extraction and liquid chromatography/mass spectrometry and group analogue internal standards. *Chemosphere*, *57*(10), 1479–1488.

Liu, B. Y., Nie, X. P., Liu, W. Q., Snoeijs, P., Guan, C., & Tsui, M. T. (2011). Toxic effects of erythromycin, ciprofloxacin and sulfamethoxazole on photosynthetic apparatus in *Selenastrum capricornutum*. *Ecotoxicology and Environmental Safety*, *74*(4), 1027–1035.

Luo, Y., Mao, D., Rysz, M., Zhou, Q., Zhang, H., Xu, L., et al. (2010). Trends in antibiotic resistance genes occurrence in the Haihe River, China. *Environmental Science & Technology*, *44*(19), 7220–7225.

Lyssimachou, A., & Arukwe, A. (2007). Alteration of brain and interrenal StAR protein, P450 scc, and Cyp11β mRNA levels in Atlantic salmon after nominal waterborne exposure to the synthetic pharmaceutical estrogen ethynylestradiol. *Journal of Toxicology and Environmental Health, Part A*, *70*(7), 606–613.

Mara, D., & Horan, N. J. (Eds.). (2003). *Handbook of water and wastewater microbiology* Elsevier.

Matamoros, V., Hijosa, M., & Bayona, J. M. (2009). Assessment of the pharmaceutical active compounds removal in wastewater treatment systems at enantiomeric level. Ibuprofen and naproxen. *Chemosphere*, *75*(2), 200–205.

Mehinto, A. C., Hill, E. M., & Tyler, C. R. (2010). Uptake and biological effects of environmentally relevant concentrations of the nonsteroidal anti-inflammatory pharmaceutical diclofenac in rainbow trout (*Oncorhynchus mykiss*). *Environmental Science & Technology*, *44*(6), 2176–2182.

Moldovan, Z., Schmutzer, G., & Alder, A. (2006). Assurance of quality control in analytical procedures using mass spectrometry. In *Vol. 2. 2006 IEEE international conference on automation, quality and testing, robotics, May* (pp. 205–207). IEEE.

Molle, P., Liénard, A., Boutin, C., Merlin, G., & Iwema, A. (2005). How to treat raw sewage with constructed wetlands: An overview of the French systems. *Water Science and Technology*, *51*(9), 11–21.

Moo-Young, M. (2019). *Comprehensive biotechnology*. Elsevier.

Nadal, M., Schuhmacher, M., & Domingo, J. L. (2004). Metal pollution of soils and vegetation in an area with petrochemical industry. *Science of the Total Environment*, *321*(1–3), 59–69.

Nghiem, L. D., Schäfer, A. I., & Elimelech, M. (2004). Removal of natural hormones by nanofiltration membranes: Measurement, modeling, and mechanisms. *Environmental Science & Technology*, *38*(6), 1888–1896.

Nikolaou, A., Meric, S., & Fatta, D. (2007). Occurrence patterns of pharmaceuticals in water and wastewater environments. *Analytical and Bioanalytical Chemistry*, *387*(4), 1225–1234.

Pal, A., Gin, K. Y. H., Lin, A. Y. C., & Reinhard, M. (2010). Impacts of emerging organic contaminants on freshwater resources: Review of recent occurrences, sources, fate and effects. *Science of the Total Environment*, *408*(24), 6062–6069.

Parrott, J. L., & Blunt, B. R. (2005). Life-cycle exposure of fathead minnows (*Pimephales promelas*) to an ethinylestradiol concentration below 1 ng/L reduces egg fertilization success and demasculinizes males. *Environmental Toxicology*, *20*(2), 131–141.

Pawlowski, S., Van Aerle, R., Tyler, C. R., & Braunbeck, T. (2004). Effects of 17α-ethinylestradiol in a fathead minnow (*Pimephales promelas*) gonadal recrudescence assay. *Ecotoxicology and Environmental Safety*, *57*(3), 330–345.

Pedrouzo, M., Reverté, S., Borrull, F., Pocurull, E., & Marcé, R. M. (2007). Pharmaceutical determination in surface and wastewaters using high-performance liquid chromatography-(electrospray)-mass spectrometry. *Journal of Separation Science*, *30*(3), 297–303.

Petrovic, M., Gros, M., & Barcelo, D. (2006). Multi-residue analysis of pharmaceuticals in wastewater by ultra-performance liquid chromatography–quadrupole–time-of-flight mass spectrometry. *Journal of Chromatography A*, *1124*(1–2), 68–81.

Rajapaksha, A. U., Premarathna, K. S. D., Gunarathne, V., Ahmed, A., & Vithanage, M. (2019). Sorptive removal of pharmaceutical and personal care products from water and wastewater. In *Pharmaceuticals and personal care products: Waste management and treatment technology* (pp. 213–238). Butterworth-Heinemann.

Ramaswamy, B. R., Shanmugam, G., Velu, G., Rengarajan, B., & Larsson, D. J. (2011). GC–MS analysis and ecotoxicological risk assessment of triclosan, carbamazepine and parabens in Indian rivers. *Journal of Hazardous Materials*, *186*(2–3), 1586–1593.

Ren, H., & Zhang, X. (Eds.). (2019). *High-risk pollutants in wastewater* Elsevier.

Rorat, A., & Vandenbulcke, F. (2019). Earthworms converting domestic and food industry wastes into biofertilizer. In *Industrial and municipal sludge* (pp. 83–106). Butterworth-Heinemann.

Sacher, F., Lange, F. T., Brauch, H. J., & Blankenhorn, I. (2001). Pharmaceuticals in groundwaters: Analytical methods and results of a monitoring program in Baden-Württemberg, Germany. *Journal of Chromatography A*, *938*(1–2), 199–210.

Samaras, V. G., Stasinakis, A. S., Mamais, D., Thomaidis, N. S., & Lekkas, T. D. (2013). Fate of selected pharmaceuticals and synthetic endocrine disrupting compounds during wastewater treatment and sludge anaerobic digestion. *Journal of Hazardous Materials*, *244*, 259–267.

Santos, L. H., Araújo, A. N., Fachini, A., Pena, A., Delerue-Matos, C., & Montenegro, M. C. B. S. M. (2010). Ecotoxicological aspects related to the presence of pharmaceuticals in the aquatic environment. *Journal of Hazardous Materials*, *175*(1–3), 45–95.

Sarmah, A. K., Meyer, M. T., & Boxall, A. B. (2006). A global perspective on the use, sales, exposure pathways, occurrence, fate and effects of veterinary antibiotics (VAs) in the environment. *Chemosphere*, *65*(5), 725–759.

Shammas, N. K. (2005). Coagulation and flocculation. In *Physicochemical treatment processes* (pp. 103–139). Humana Press.

Sim, W. J., Lee, J. W., Lee, E. S., Shin, S. K., Hwang, S. R., & Oh, J. E. (2011). Occurrence and distribution of pharmaceuticals in wastewater from households, livestock farms, hospitals and pharmaceutical manufactures. *Chemosphere*, *82*(2), 179–186.

Singh, K. P., Rai, P., Singh, A. K., Verma, P., & Gupta, S. (2014). Occurrence of pharmaceuticals in urban wastewater of north Indian cities and risk assessment. *Environmental Monitoring and Assessment*, *186*(10), 6663–6682.

Snyder, S. A. (2008). Occurrence, treatment, and toxicological relevance of EDCs and pharmaceuticals in water. *Ozone: Science & Engineering*, *30*(1), 65–69.

Speight, J. G. (2017). Introduction into the environment. In *Environmental organic chemistry for engineers* (pp. 263–303). Butterworth-Heinemann (Chapter 6).

Speight, J. G. (2018). *Reaction mechanisms in environmental engineering: Analysis and prediction*. Butterworth-Heinemann.

Ternes, T. A. (1998). Occurrence of drugs in German sewage treatment plants and rivers. *Water Research*, *32*(11), 3245–3260.

Thalla, A. K., & Vannarath, A. S. (2020). Occurrence and environmental risks of nonsteroidal anti-inflammatory drugs in urban wastewater in the southwest monsoon region of India. *Environmental Monitoring and Assessment*, *192*(3), 1–13.

Vanderford, B. J., Pearson, R. A., Rexing, D. J., & Snyder, S. A. (2003). Analysis of endocrine disruptors, pharmaceuticals, and personal care products in water using liquid chromatography/tandem mass spectrometry. *Analytical Chemistry*, *75*(22), 6265–6274.

Vannini, C., Domingo, G., Marsoni, M., De Mattia, F., Labra, M., Castiglioni, S., et al. (2011). Effects of a complex mixture of therapeutic drugs on unicellular algae Pseudokirchneriella subcapitata. *Aquatic Toxicology*, *101*(2), 459–465.

Vieno, N. M., Härkki, H., Tuhkanen, T., & Kronberg, L. (2007). Occurrence of pharmaceuticals in river water and their elimination in a pilot-scale drinking water treatment plant. *Environmental Science & Technology*, *41*(14), 5077–5084.

Vymazal, J. (2007). Removal of nutrients in various types of constructed wetlands. *Science of the Total Environment*, *380*(1–3), 48–65.

Vymazal, J. (2010). Constructed wetlands for wastewater treatment. *Water*, *2*(3), 530–549.

Vymazal, J., Greenway, M., Tonderski, K., Brix, H., & Mander, Ü. (2006). Constructed wetlands for wastewater treatment. In *Wetlands and natural resource management* (pp. 69–96). Berlin, Heidelberg: Springer.

Weedon, C. M., Murphy, C., & Sweaney, G. (2017). Establishing a design for passive vertical flow constructed wetlands treating small sewage discharges to meet British Standard EN 12566. *Environmental Technology*, *38*(2), 220–229.

Wishart, D. S., Knox, C., Guo, A. C., Shrivastava, S., Hassanali, M., Stothard, P., et al. (2006). DrugBank: A comprehensive resource for in silico drug discovery and exploration. *Nucleic Acids Research*, *34*(Suppl_1), D668–D672.

Wong, H., Mok, K. M., & Fan, X. J. (2007). Natural organic matter and formation of trihalomethanes in two water treatment processes. *Desalination*, *210*(1–3), 44–51.

Zhang, Y., Geißen, S. U., & Gal, C. (2008). Carbamazepine and diclofenac: Removal in wastewater treatment plants and occurrence in water bodies. *Chemosphere*, *73*(8), 1151–1161.

Current situation of pharmaceutical wastewater around the globe

2

Ahmad Saud Masood[a], *Md. Sajid Ali*[b], *Mohammad Saood Manzar*[c], *Nadeem A. Khan*[d], *and Afzal Husain Khan*[e]
[a]Politecnico di Milano, Milan, Italy, [b]Department of Pharmaceutics, College of Pharmacy, Jazan University, Jazan, Saudi Arabia, [c]Department of Environmental Engineering, College of Engineering, Imam Abdulrahman Bin Faisal University, Dammam, Saudi Arabia, [d]Department of Civil Engineering, Mewat Engineering College, Nuh, India, [e]Civil Engineering Department, College of Engineering, Jazan University, Jazan, Saudi Arabia

1. Introduction

Pharmaceuticals are important for human and animal life as they improve the quality of life by preventing disease and curing it. This helps us to live longer and enjoy a more productive and higher quality life. However, when these pharmaceutical products disperse into the environment by various means, they can have a harmful and negative impact on the ecosystem, and human and animal health. Hospital wastewater (HWW) should be treated before being discharged to surface water or municipal wastewater treatment plants, according to WHO rules, medical professionals, and hospital engineers, due to potential toxicity present in HWW before it is transported to these treatment plants. To overcome these challenges, comprehensive emphasis has been given to policies and regulatory systems in assessing pharmaceutical production and management, pretreatment of waste, and initiating dialogue in the public domain with different national and transnational agencies like WHO personnel, UN personnel, medical specialists, national governments, etc. (Chartier et al., 2014; Escher et al., 2011).

Often, the relevant regulations are contradictory in implementation. For example, according to WHO (Chartier et al., 2014) and European Union Directives (EC, 2014), mercury, which is quite commonly used in daily life, has been put in List I, with a maximum discharge of 5% in the effluent for HWW; whereas, more than half of dental amalgams originating from hospitals in Europe and the United Kingdom are tainted with metals such as mercury and zinc. To avoid pathogen activity, countries with high rates of enteric and malignant illnesses, such as China and Japan, have established specific on-site pretreatment before dumping HWWs into the municipal wastewater system (Kumari, Maurya, & Tiwari, 2020).

Before 1990, there was no special legislation in place in European countries to deal with hospital wastewater, because there was no distinction between hospital discharge

The Treatment of Pharmaceutical Wastewater. https://doi.org/10.1016/B978-0-323-99160-5.00013-8

and municipal wastewater (EC, 2014). As a result, each EU nation has its peculiar codes of rules and decrees for managing and treating HWW. Thus, to improve the quality of water released into waterbodies, after the EU-WFD was passed in 1991, industrial waste discharges into household garbage collection systems required prior authorization (EC, 2014). In Europe, there are around 2.6 hospitals per 10,000 people, with an average of 500 beds per hospital (the lowest being around 300 in Spain and the highest being around 800 in Germany). In Europe, the average consumption of antibodies is nearly 10,000 tons/year, a large portion of which is eventually discharged in wastewater in the form of urine and feces. Toxic compounds are burned at the source of creation in Europe's current waste management system. European Pills Project research suggests that (Chonova et al., 2018) pharmaceutical substances can be removed from HWW using membrane bioreactor (MBR) technology along with other technologies. The most frequent treatments are MBR with granular activated carbon (GAC) and ozone treatment (Netherlands), MBR with powdered activated carbon (PAC) (Switzerland), and MBR with ozone/UV/RO/H_2O_2 treatment (Luxembourg) (aus der Beek et al., 2016; Kumari et al., 2020). The hospital effluent discharge limitations and legislation/laws are represented in Table 1.

2. Characteristics of pharmaceutical wastewater

Pharmaceuticals are medications that are used to diagnose, cure, or prevent ailments. Several methods have been developed to incorporate chemical substances and may be used for drugs as well (Huang, Shen, & Shuai, 2021). Curatives, antirheumatic medications, synthetic hormones, cytotoxins, and statin drugs are noted types of human medications used on a large scale, collectively (Giwa et al., 2020). The succeeding criteria separate pharmaceuticals from various types of noxious waste (Putschew & Jekel, 2007): (a) they can be constituted by large and chemically complex molecules with a broad range of molecular weight, framework, functionality, and form; (b) they are polar in nature with many ionizable groups, and the extent of ionization and its characteristics are influenced by the medium's pH; (c) they are lipid-soluble, but some of them seem to be moderately soluble in water; (d) pharmaceuticals such as naproxen, erythromycin, and sulfamethoxazole; and (e) the molecules are swallowed, dispersed, and exposed to metabolic processes in humans that possibly will lead them to modify chemical makeup (Bean et al., 2018; Giwa et al., 2020; Huang et al., 2021).

Across medicinal products as a group, waterborne wastes and residues can be divided into pharmacological, biopharmaceutical, chemical, and herbal production wastewater (Guo, Qi, & Liu, 2017; Yao, Wang, & He, 2013). In addition, large pharmacological firms manufacture on a batch basis, with different raw materials and manufacturing routes, yielding a wide variety of wastes. The composition and properties of pharmaceutical wastewater are summarized in Table 2 (Boroski et al., 2009; Chen et al., 2008, 2011; Cyr et al., 2002; Domínguez et al., 2012; Melero et al., 2009; Peng et al., 2004; Rajkumar & Palanivelu, 2004; Zheng, 2011). Strong quantity fluctuation, small C/N, large SS content, large sulfate concentration, complex structure, high biological toxicity, and high chrominance are all characteristics of

Table 1 The hospital effluent discharge limitations and legislation/law.

Rule or legislations	Region	Made effective since	Onsite treated effluent	Influent received at STP	HWW disposal to surface or groundwater
EU, 1991. Directive 91/271/EEC on hazardous waste (renewed EU, 1998. Dir. 98/15/EC)	EU	1998	Not available	Not available	Not available
Wastewater Treatment law	England and Wales	1994	BOD_5 (mg/dm^3) 24 mg/dm^3 COD (mg/dm^3) $\leq$ 123 mg/dm^3 P total ≤ 2.2 $N \leq 14\,mg/dm^3$ of nitrogen	Not available	Not available
Decreto No. 26,042-S-MINAE discharges management	Spain	1997	Not available	Temp. around 42°C pH 6.6–9.1 BOD $\leq 250\,mg/dm^3$ COD $\leq 900\,mg/dm^3$ SS $\leq 410\,mg/dm^3$ Oil and grease $\leq 200\,mg/dm^3$ Sulfate $\leq 450\,mg/dm^3$ Fluorine $\leq 10\,mg/dm^3$ Organophosphorate $\leq 0.2\,mg/dm^3$ Carbamate $\leq 0.11\,mg/dm^3$ Organochlorine $\leq 0.03\,mg/dm^3$	All categories: Temp. 12–35°C pH 5.5–9.5 Sedimentary solids $\leq 1.4\,mg/dm^3$ Floating material $\leq$ absent Total cyanides $\leq 1.4\,mg/dm^3$ Sulfite $\leq 1.4\,mg/dm^3$ Sulfur $\leq 23\,mg/dm^3$ Clorine $\leq 2.1\,mg/dm^3$ Organophosphorate $\leq 0.2\,mg/dm^3$ Carbamates $\leq 0.2\,mg/dm^3$ Organochlorines $\leq 0.04\,mg/dm^3$

Continued

Table 1 Continued

Rule or legislations	Region	Made effective since	Onsite treated effluent	Influent received at STP	HWW disposal to surface or groundwater
Decreto 57/2005	Spain	2005	Not available	Temp around 40°C pH 6.6–9.2 BOD around 911 mg/dm^3 COD around 1100 mg/dm^3 SS around 800 mg/dm^3 Oil and grease around 75 mg/dm^3 Cyanides around 3 mg/dm^3 Conductivity around 6400 µS/cm^2 Total cleaning around 24 mg/dm^3 Fluorine around 8 mg/dm^3 SO_4 around 750 mg/dm^3 H_2S around 4.2 mg/dm^3 Toxicity (1/CE50) × 100 around 24 AOX around 2 mg/dm^3 P total around 35 mg/dm^3 N total around 110 mg/dm^3	
Federal Ministry for the Environment, Nature Conservation, and Nuclear Safety. Wastewater ordinance	Germany	2004	Sewage category (i.e., hospitals): BOD5 around 14–40 mg/dm^3 COD around 65–120 mg/dm^3 TSS around 20 mg/dm^3 NH_4 around 06 mg/dm^3 N total around 12–14 mg/dm^3 P total around 0.9–1.2 mg/dm^3	Not available	

DLgs No. 152/2006 on environmental protection	Italy	2006	Not available	Not available	pH 5.4–9.1 Temp. $\leq 31°C$ Color not perceptible with the dilution of 12 Material roughness-absent TSS less than 72 mg/dm^3 BODS less than 34 mg/dm^3 COD less than 123 mg/dm^3 COD/BOD less than 1.3 P total less than 8 mg/dm^3 NH_4 less than 9 mg/dm^3 NO_2 less than 0.3 mg/dm^3 NO_3 less than 18 mg/dm^3 Oil and grease less than 15 mg/dm^3 Tension actives less than 1 mg/dm^3 Total hydrocarbons less than 2.2 mg/dm^3 Phenols less than 0.23 mg/dm^3 Aldehydes less than 1.1 mg/dm^3 Organic aromatic solvents less than 0.2 mg/dm^3 Organic nitrogen solvents less than 0.11 mg/dm^3 Pesticides phosphorus less than 0.1 mg/dm^3; Total pesticides less than 0.02 Chlorinated solvents less than 2.1 mg/dm^3; Total cyanides less than 0.2 mg/dm^3 Sulfites less than 1.1 mg/dm^3 Sulfur less than 1.1 mg/dm^3 Sulfates less than 1100 Fluorine less than 1 mg/dm^3 Free chlorine ≤ 0.1 mg/dm^3 *E. coli* UFC/100 mL

Continued

Table 1 Continued

Rule or legislations	Region	Made effective since	Onsite treated effluent	Influent received at STP	HWW disposal to surface or groundwater
Law in China (GB897888)	China	1998	Not available	Industrial effluent: pH around 5.9–9.2 SS around 350 mg/dm^3 BOD_5 around 250 mg/dm^3 COD–450 mg/dm^3 Fluoride around 18 mg/dm^3 Phosphorus around 0.1 mg/dm^3 AOX (as CI) around 5 mg/dm^3 Fecal coliform (individual/L) 850–3500 Total chlorine after disinfection >1 or >6 (contact time ≥ 1.1 h)	pH around 6.5–9.2 SS around 150 mg/dm^3 BOD_5 around 50 mg/dm^3 COD around 140 mg/dm^3 Ammonia nitrogen around 21 mg/dm^3 Fluoride around 9 mg/dm^3 Phosphate around 1.3 mg/dm^3 Phosphorus around 0.11 mg/dm^3 Fecal coliform (individual/L) 500–1000 and 100–400 Total chlorine after disinfection <0.2 to >2.2 (contact time ≥ 1.2 h) and around 7.3 (contact time ≥ 1.6 h)

Table 2 Common pharmaceutical wastewater composition and physicochemical properties (Boroski et al., 2009; Chen et al., 2011; Chen, Ren, Wang, Zhang, & Shi, 2008; Cyr, Suri, & Helmig, 2002; Domínguez, González, & Palo, 2012; Melero, Botas, Molina, Pariente, & Martı, 2009; Peng, Li, Peng, & Wang, 2004; Rajkumar & Palanivelu, 2004; Zheng, 2011).

Parameters	Minimum value	Maximum value	Average composition
COD (mg/L)	375	32,500	8854
BOD_5 (mg/L)	200	6000	2344
D_5/COD ratio	0.1	0.6	0.32
TN (mg/L)	500	1500	1000
TP (mg/L)	50	250	150
SS (mg/L)	200	500	350
TOC (mg/L)	860	4940	2467
TKN (mg/L)	165	770	383
NH_3 (Nmg/L)	148	363	244
Cl^- (mg/L)	760	4200	2820
SO_4^{2-} (mg/L)	890	1500	1260
Chromaticity (times)	500	1000	750
Temp. (°C)	25	80	52.5
pH	3.9	9.2	6.9

biopharmaceutical wastewater. Chemical waste is low in nutrition, difficult to biodegrade, harmful to microbes, as well as having a high salt content.

3. Source and occurrence of pharmaceuticals in water

Many antibiotics have been found in environmental illustrations all over the world. Sewage treatment plant seepages, surface water, oceans, groundwater, soils, sediment, and fish have all been shown to contain them. In general, the regularly used antibiotics noticed in water include antiinflammatory drugs/analgesics, antibiotics, lipid regulators, steroids and related hormones, Beta-blockers, cancer therapeutics, diuretics, antiepileptics, antidepressants, and tranquilizers (Bush, 1997; Nikolaou, Meric, & Fatta, 2007). Prescription medications, over-the-counter medicinal drugs, and veterinary treatments all contain pharmaceuticals, which can be manufactured or natural substances. Pharmaceuticals are made up of active chemicals that are intended to have pharmacological effects and provide considerable societal advantages. The advancements in analytical techniques and instruments are primarily responsible for the rise in detection. Several surveys and investigations have established the existence of antibiotics in municipal wastewater and seepages, and these have been identified as a leading source of antibiotics in drinking water. Continuous monitoring programs for pharmaceuticals in drinking water, as well as prescribed chemical and microbiological criteria, have not been established. Informal surveys or specific investigation programs and training have often yielded data on the occurrence of drugs in drinking water. According to existing research, pharmaceutical concentrations in groundwater, surface water, and moderately treated water are often less than 0.1 g/L (or 100 ng/L), whereas amounts in treated water are typically less than 0.05 g/L (or 50 ng/L). Organized research will aid in our understanding of pharmaceutical movement, occurrence, and fate in the environment, particularly in sources of drinking water. Standardization of collection and analysis techniques for medicines would make data comparability easier.

4. Source, occurrence, and fate of pharmaceuticals in the environment

Thousands of tons of medicines for human and animal use are produced and administered globally each year (Ganiyu, Van Hullebusch, Cretin, Esposito, & Oturan, 2015). Pharmaceuticals are released into the environment as an outcome of their discharge from a variety of places, including hospitals, homes, aquaculture, agriculture, and livestock farms (Kümmerer, 2001). APIs (active pharmaceutical ingredients) can reach the environment via three main pathways. The bulk of APIs found in water systems are present as a consequence of everyday patient and consumer usage, as well as emissions into sewage and wastewater treatment systems. A second route is the inappropriate or improper disposal of unwanted or outdated medications by users, who flush them down toilets or pour them down drains. The wastewater discharged by API manufacturing plants is the third route.

Researcher studied the presence and fate of pharmaceuticals in the environment (Caldwell, 2016). It is seen that, due to low digestion, the drugs and their residues are excreted in one form or another by humans and animals. Pharmaceuticals are primarily received by wastewater treatment plants (WWTPs) through excretion, medical disposals (Bound, Kitsou, & Voulvoulis, 2006), and hospital effluent (Larsson, de Pedro, & Paxeus, 2007). Many WWTPs have excellent removal systems for analgesics and antiinflammatory medications, whereas extremely poor or no removal systems for therapeutic groups such as antiepileptics, and antibiotics (Kosma, Lambropoulou, & Albanis, 2014). The sludge containing sorbed pharmaceuticals is landfilled or used as agricultural manure, while the WWTP effluent containing residual medications is deposited into surface rivers (Kosma et al., 2014). Municipal solid waste dumps also accept pharmaceutical waste as a final destination for its decomposition. The excrement of cattle is utilized as fertilizer in fields for farming. These landfills and farms may release pharmaceuticals into the surface and groundwater, respectively. Numerous studies have discovered their existence in a variety of aquatic ecosystems across the world, including drinking water sources such as surface and groundwater (Robles-Molina, Gilbert-López, García-Reyes, & Molina-Díaz, 2014).

5. Health hazards of discharged pharmaceuticals

Pharmaceutical pollution of surface and groundwater has been identified as an environmental hazard by several countries, prompting the creation of the field of antibiotics in the environment research. Active pharmaceutical ingredients (APIs) are compounds that are biologically active, resistive to decomposition, exceedingly stable in an aqueous medium, and probably capable of causing harmful effects to aquatic species and having a negative impact on the ecosystem. It has been seen that many pharmaceuticals show the tendency of bioactivity, resistance, and accumulation. Even at modest dosages, these characteristics are a leading cause of endocrine disturbance, reproductive abnormalities, bacterial resistance development, and environmental toxicity. Furthermore, their metabolites can be far more dangerous than the parent molecule.

The hazardous behavior of pharmaceutical residues towards living creatures has been described in several researches, including reviews (Gunnarsson et al., 2009; Ortiz de García, Pinto Pinto, García-Encina, & Irusta-Mata, 2014). Globally, over half of all pharmaceutical wastewater is estimated to be released untreated. The existence of endocrine-disrupting chemicals (EDCs) in aquatic systems has prompted significant uneasiness due to the fact that EDCs are believed to damage the human endocrine system. As pharmaceutical residues enter the environment, they have consequences, such as the generation of antibiotic-resistant organisms in aquatic habitats, slower oxidation of nitrite and metagenomics, and the likelihood of heightened toxicities of chemical combinations and metabolites. Despite their widespread use in human and veterinary medicine, antibiotics pose a menace to human health and the environment, including fueling the spread of resistant microorganisms. Resistance to antibiotics is undesirable from the standpoint of human health because environmental bacteria contribute resistance genes to human infections (Martinez, 2009). One of the major difficulties confronting the global healthcare system is the growing rise in resistance (Larsson, 2010).

According to a new study, pharmaceutical residues in water streams have been related to gendering in fish and alligators. Pharmaceutical residues have an impact on salmon behavior and migration patterns. Forty priority pharmaceuticals were selected for a study to cultivate and implement a testing regime for their potential ecotoxicity in a specific region, plus drug, pain reliever/antiinflammatory medications, and their metabolites. According to an analogous research study, thyroid hormone, antihypertensive medications, antibiotics, and analgesics were among the most important pharmaceuticals. Antibiotics, psychiatric drugs, analgesics and antiinflammatories, lipid regulators, and Beta-blockers were shown to represent a high to medium risk (Kosma et al., 2014). However, the number and quantities of antibiotics and pharmaceutical and personnel care products (PPCPs) that are eradicated by the remediation methods are unknown, and several of the complexes discharged are nondecaying and thus ineffectively remediated by traditional (primary, secondary, and tertiary) treatment methods, resulting in a dangerous buildup in aquatic environments. An enormous amount of effluent wastewater is created during batch and multistage pharmaceutical production operations. Furthermore, research indicates that pharmaceutical manufacturing and utilization will increase; growing in tandem with worldwide lifestyle and lifespan advancements (Goossens, Ferech, Coenen, & Stephens, 2004; Kümmerer, 2009; Van der, Kommer, Van Montfoort, & Versteegh, 2011).

6. Pharmaceutical wastewater treatment options

Several research publications on the handling of pharmaceuticals, EDCs, and pharmaceutical and personal care products (PPCPs) have been published over the past few decades, with an emphasis on seepage from tertiary WWTPs. Treatment techniques at the source, on the other hand, might not only minimize costs and environmental effects but also enable the possibility of compound recovery. Despite extensive study in this field and the publication of multiple studies in recent years, they do not give a full picture of pharmaceutical wastewater composition and treatment processes (Klavarioti, Mantzavinos, & Kassinos, 2009; Larsson & Fick, 2009; Lawrence, Wang, Howard, & Constantine, 2005; Michael et al., 2013). The pharmacological trade requires a steady supply of high-grade water for manufacturing and wastewater handling in order to come up to the requirement of stringent governing discharge limits. To overcome the aforementioned difficulties, firms need to question traditional detection methods, as well as investigate new technology and solutions in order to stay competitive. There are countless conventional and advanced approaches utilized for the handling of pharmaceutical residues, as discussed below.

6.1 Conventional methods

6.1.1 Coagulation

Coagulation is a chemical water treatment method that encompasses modifying the electrostatic charges of particles suspended in water to remove solids. Small, highly charged molecules are introduced into water to disrupt the charges on particles, colloids, or greasy materials in suspension. Choosing the correct coagulant for a system improves overall system performance and, in particular, solids elimination efficiency by improving filter and clarifier performance. Coagulation is very effective in removing chemical phosphorus from water.

6.1.2 Flocculation

In wastewater treatment, flocculation and coagulation work in tandem. After coagulation has clumped the waste particles together, flocculating chemicals are employed to eliminate the clumps. Flocculants are polymers that induce destabilized clusters of particles to agglomerate and drop out of the solution, rejecting them from the treated water. They can be lightweight, medium-weight, or heavy. The weight utilized is determined by the type of particle.

6.1.3 Sedimentation

Sedimentation is the method of permitting solid particles that are heavier than water particles to settle underneath gravity, i.e., particles with a particle size of less than 0.2 mm and a specific gravity of 2.65. Sedimentation is employed in wastewater treatment to eliminate both inorganic and organic impurities (Puspalatha & Kalpana, 2016).

6.1.4 Conventional activated sludge

The most popular aerobic treatment, activated sludge, has been shown to be effective for a variety of pharmaceutical wastewaters. The activated sludge process is a common biological process used to reduce organic matter in wastewater. It essentially entails the oxidation of carbonaceous biological matter in order to reduce organic contaminants. Conventional activated sludge (CAS) handling is an economical technology that is primarily governed by two constraints: temperature and hydraulic retention period (HRT). The benefits of this technique include the ability to eliminate the bulk of biological materials. The downsides of this approach include partial degradation, which produces harmful degradation products, and the inability to remove stubborn pollutants. (e.g., carbamazepine) (Al Qarni, Collier, O'Keeffe, & Akunna, 2016; Oulton, Kohn, & Cwiertny, 2010).

6.1.5 Membrane bioreactor

The membrane bioreactor (MBR) has evolved as a small and effective device, in comparison to municipal and industrial untreated water handling (Friha, Karray, Feki, Jlaiel, & Sayadi, 2014; Mutamim, Noor, Hassan, Yuniarto, & Olsson, 2013). An MBR is a system that combines a standard biological treatment system with physical liquid-solid separation by membrane filtration (Chang, Le Clech, Jefferson, & Judd, 2002). The usage of membranes reduces the need for secondary clarifiers in the system. The removal of secondary clarifiers and the application of MBR over a lower HRT result in considerably decreased plant space requirements. Conversely, there are several drawbacks to using MBR technology, such as greater power costs, the necessity to handle membrane fouling issues, and the possible high cost of the periodic replacement of membranes (Tchobanoglous, Stensel, Tsuchihashi, & Burton, 2014). Nevertheless, according to Oulton et al., MBR is a more efficient technique for pharmaceutical elimination than the CAS method (Oulton et al., 2010).

6.1.6 Constructed wetland

Constructed wetlands are one of the most promising municipal wastewater treatment alternatives for decentralized communities, particularly in rural and suburban regions, because they are inexpensive in cost and maintenance needs while providing high performance. Constructed wetlands (CWs) are cheap, simple in use and handling skills, and have been widely researched for the elimination of conventional contamination problems, such as organic matter and nutrients (nitrogen and phosphorus), as well as pharmaceutical residues from wastewater (Ilyas & Masih, 2018; Kadlec, 2009; Vymazal, 2013). There is a possibility that constructed wetlands (CWs) might be used in small communities. Biodegradation, adsorption, and plant uptake are all methods used to remove PPCPs from CW. In all kinds of wetland systems tested, ibuprofen was the most frequently removed PPCP. CWs' ability to remove micropollutants is also influenced by periodic variations, with better rates of elimination in the summer-time than in the winter-time.

6.2 Advanced treatment methods

New procedures involving oxidation, adsorption, and membrane processes have been developed due to the failure of standard treatment methods to effectively remove large amounts of PPCPs. It also helps that advanced oxidation processes (AOPs) are able to completely mineralize organic pollutants, which encourages broad application. In physical techniques, carbonaceous compounds like activated carbon and graphene have been extensively investigated. Increased oxidation, such as ozonation, UV treatment, and photocatalysis using titania or peroxide, has also garnered considerable study.

6.2.1 Activated carbon as adsorbent

Adsorption is a phase transfer procedure that is widely utilized to enrich chemical species from a fluid phase to a solid or liquid surface. Adsorption-related solid surfaces have active and energy-rich regions that can interact with solutes in the adjacent aqueous phase via their spatial and particular electronic characteristics. The solid substance that provides the surface for adsorption is known as an adsorbent, and the species that will be adsorbed is known as an adsorbate. According to research on the use of adsorbents to extract pharmaceuticals from aqueous systems, hydrophobic pharmaceuticals have a higher affinity for the adsorbent surface than their hydrophilic counterparts (Akhtar et al., 2016; Kyzas, Fu, Lazaridis, Bikiaris, & Matis, 2015). The most often utilized ingredient in wastewater treatment is activated carbon. It might be powdered activated carbon (PAC) or granular activated carbon (GAC) (Serrano-Ruiz, Luque, & Sepúlveda-Escribano, 2011). Because of its huge surface area (about $1000\,m^2/g$) and the blend of a sophisticated permeable structure and surface chemistry features, activated carbon (AC) adsorption is ideally adapted to eliminate OCs. Mestre, Pires, Nogueira, and Carvalho (2007) has shown the removal of ibuprofen using waste-derived activated carbon. As a result, the AC procedure has the benefit of having a simple raw substance input for the manufacture of carbon. PAC has a lead over GAC in that it is generally fresh, whereas GAC is frequently regenerated in fixed-bed columns.

6.2.2 Graphene-based adsorption

Graphene is an innovative "wonder substance" that has piqued the interest of scientists in recent years. It is a single-atom-thick, two-dimensional layer of sp2-hybridized carbon. Graphene's exceptional physicochemical features, including its large specific surface area, distinctive shape, chemical structure, and electrical, thermal, and mechanical strength properties, have fueled interest in the material. Because of its massive, delocalized electron system, which may create strong connections with other contaminants, graphene has been employed as an efficient sorbent for numerous toxins (Wang et al., 2013). It can be utilized to diminish contamination weight via sorption, to breakdown organic contaminants, and to decompose tenacious organic contaminants. Nanoplatelets were explored by reporting the elimination of aspirin, acetaminophen, and caffeine via graphene (Al-Khateeb, Almotiry, & Salam, 2014). Adsorption utilizing graphene materials is a straightforward and low-cost technology (Sophia, Bhalambaal, Lima, & Thirunavoukkarasu, 2016), particularly for treating unique wastes like pharmaceuticals, which may only be present in trace amounts (g/L to ng/L).

6.2.3 Carbon nanotubes as adsorbent

Carbon nanotubes (CNTs) have sparked significant interest as a novel form of adsorbent owing to their exceptional capacity to remove numerous inorganic and organic toxins, plus radionuclides, from enormous quantities of wastewater. CNTs are fundamentally different from ACs in that their atomic structure is significantly more well-defined and homogeneous. CNT applications for the elimination of unsafe contaminants from gas streams and large capacities of aqueous solutions have been extensively investigated via theoretical designs, trial measurements, and molecular mockups, due to their extremely permeable and hollow structure, large specific surface area, light mass density, and the strong interface between CNTs and contaminant particles (Ren, Chen, Nagatsu, & Wang, 2011).

6.2.4 Membrane filtration

Nanofiltration (NF) and ultrafiltration (UF) membrane filtration methods have been employed as advanced treatments in water treatment plants (WTPs) and sewage treatment plants (STPs) for the removal of natural organic waste and pesticides (Yoon, Westerhoff, Snyder, Wert, & Yoon, 2007). In terms of PPCP removal, membrane methods such as NF, UF, and RO (reverse osmosis) have been reported to be effective. NF, also known as low-pressure RO, features nanoscale pores and was created to bridge the gap between RO and UF (Alfonso-Muniozguren, Serna-Galvis, Bussemaker, Torres-Palma, & Lee, 2021; Yang, Lohwacharin, & Takizawa, 2014). The advantage of this method is that it consumes little space and eliminates a wide variety of impurities; however, there are some drawbacks, such as the need to replace the membrane every 3–5 years and the formation of a large volume of concentrates.

6.2.5 Advanced oxidation processes

Certain micropollutants may be released into the environment when treated effluents are discharged into receiving waters due to the poor biodegradability of many pharmaceuticals, which makes existing treatment processes inefficient. As a result of this pollution, aquatic species have been demonstrated to be at risk. There are a number of different treatments that may create and employ radical species (such as hydroxyls) to remove contaminants from the environment. Ozone, ozonation, UV irradiation, Fenton, and persulfate are the most prevalent AOPs. Based on the pharmaceutical residues and the treatment aim of either destruction or transformation, AOPs may be utilized alone or in concert with other physicochemical or biological processes.

6.2.5.1 Advanced oxidation processes involving hydroxyl radicals

Hydroxyl radicals are very reactive entities that damage the majority of organic compounds. Because of their oxidation potential, they are extremely oxidizing in nature. Furthermore, due to their nonselectivity, many sensitive organic compounds can be readily removed or destroyed utilizing hydroxyl radical (acids, alcohols, aldehyde, aromatics, amines, ethers, ketones, etc.) (Munter, 2001). Hydroxyl radicals destroy organic contaminants via four fundamental mechanisms: radical addition, hydrogen abstraction, electron transfer, and radical combination. Carbon-centered radicals (R• or R•–OH) are formed as a result of their interactions with organic molecules. These carbon-centered radicals can be converted to organic peroxyl radicals (ROO) by using O_2. All of the radicals continue to react, resulting in the creation of highly reactive species, such as H_2O_2 and superoxide (O_2•), which leads to chemical deterioration and even mineralization of these organic molecules (Deng & Zhao, 2015).

6.2.5.2 Ozonation

Ozone is a potent oxidant that degrades organic molecules in both direct and indirect ways. It decomposes in water to create hydroxyl radicals, which are stronger oxidizing agents than ozone and will subsequently oxidize the contaminants, or it can selectively target particular functional groups of organic molecules via an electrophilic process (Klavarioti et al., 2009). The existence of varied micropollutants in wastewater of varying compositions makes determining appropriate ozone dosages for organic micropollutant removal in wastewater ozonation difficult (Lee, Kovalova, McArdell, & von Gunten, 2014). Selective oxidative reactions occur between unsaturated aromatic and aliphatic molecules and ozone.

6.2.5.3 Photocatalytic oxidation

Photochemical oxidation, commonly known as UV photocatalytic oxidation, is a process that combines UV light with an oxidant. When exposed to UV light, oxidative breakdown of oxidants releases free radicals with greater oxidative capability, allowing them to oxidize more difficult-to-decompose refractory organic contaminants than using just oxidants. Photochemical oxidation can be classified according to the kind of oxidant used, such as UV/O_3, UV/H_2O, UV/H_2O_2/O_3, and so on. The impact of this technology on the treatment of refractory organic materials, on the other hand, is undeniable (Khan et al., 2020).

6.2.5.4 Fenton-based advanced oxidation processes

Fenton is a highly successful advanced treatment method. In Fenton oxidation, hydroxyl radicals are created by a free radical chain reaction involving ferrous or ferric ions and hydrogen peroxide. This is a metal-catalyzed oxidation process in which iron serves as a catalyst (Klavarioti et al., 2009). The primary drawbacks of using the Fenton process are the relatively expensive cost of H_2O_2 and the large volume of ferric sludge produced during the neutralization stage of the treated solution before disposal. To solve the aforementioned limitations, the traditional Fenton method was merged with UV radiation (photo-Fenton process), electrochemical processes (electro-Fenton process), ultrasound (sono-Fenton process), and other wastewater treatment technologies to create the enhanced Fenton process (Khan, Khan, Tiwari, Zubair, & Naushad, 2021).

6.2.5.5 Persulfate-based advanced oxidation processes

Advanced oxidation methods built on persulfate have been fruitfully applied to the handling of wastewater containing drugs and personal care products, colorants, endocrine-disrupting chemicals, chlorinated organic pollutants, and phenolics for the breakdown of refractory carbon-based toxins (Tian et al., 2022). Persulfate ($S_2O_8^{-2}$) is a powerful oxidizing agent when exposed to light or heat, produces extremely reactive sulfate radicals, leading to sulfate radical-mediated AOPs. Furthermore, sulfate radicals can create hydroxyl radicals throughout the oxidation procedure. pH, Fe^{+2}, and Fe^{+3} ions, or the combination of Fe^{+2} and irradiation, are alternative activators of sulfate radical-mediated AOPs. The use of activated persulfate can completely remove organic contaminants (Song, Li, Wang, & Zhang, 2019).

7. Hospital wastewater management in developed countries

7.1 United States

In the recent study, it was found that North American and European Union surface waters were contaminated with more than 80 pharmaceuticals (Jose, Sandra Pinto, Kotian, Mathew Thomas, & Narayana Charyulu, 2020). The water stream of the United States contains various hormones, pharmaceuticals, and organic wastewater contaminants (OWC) (Kolpin et al., 2002). The United States Food and Drug Administration (USFDA) is solely responsible for evaluating the content of drugs that enter the environment under the US National Environmental Policy Act of 1969 (NEPA) (Kar & Roy, 2012). After pharmaceutical residues have been disposed of, it is necessary to calculate the its content in the environment (Shaw, 2013). Special requirements for HWW pollutant treatment, handling, and discharge approval have been implemented as a result of this law. The United States Environmental Protection Agency (USEPA) has documented a list of toxic chemicals, which includes a pharmaceuticals called erythromycin and five synthetic hormones. US Commercial Services has authorized its firms to transmit membrane biofilm reactor (MBR) technology to Chinese hospitals amidst the eruption of severe acute respiratory syndrome (SARS) because MBR's treatment is efficient for the elimination of germs and viruses (SARS) (Verlicchi, Al Aukidy, & Zambello, 2015).

A study was conducted on psychoactive drugs and antihypertensives in two wastewater handling firms in New York State, USA, as detailed by Subedi and Kannan (2015), with the process amenities having 2.5 and 3.5 million gallons per day treatment capacity, respectively. It was found that all the studied 27 psychoactive ingredients were present in untreated water and the sludge of the WWTPs. The concentration of these psychoactive medicinal products was 0.98–1.220 and 0.26–1.490ng/L, respectively; and the removal efficiency of pharmaceuticals in WWTPs after the final treatment ranged from 0.3% to 87%. This represents a low elimination rate of the pharmaceuticals listed as carbamazepine, carboxylic acid, lorazepam, norsertraline, clopidogrel, propranolol, and 2-(diphenylmethoxy) acetic acid. This is owing to the bacteriological metamorphosis of a conjugate drug into its paternal chemical. The average amount of psychoactive drugs from the range of 0.001–31.6mg/day/100 people were discharged from each WWTP. Fig. 1 shows the generation and transmission route of pharmaceuticals to the environment.

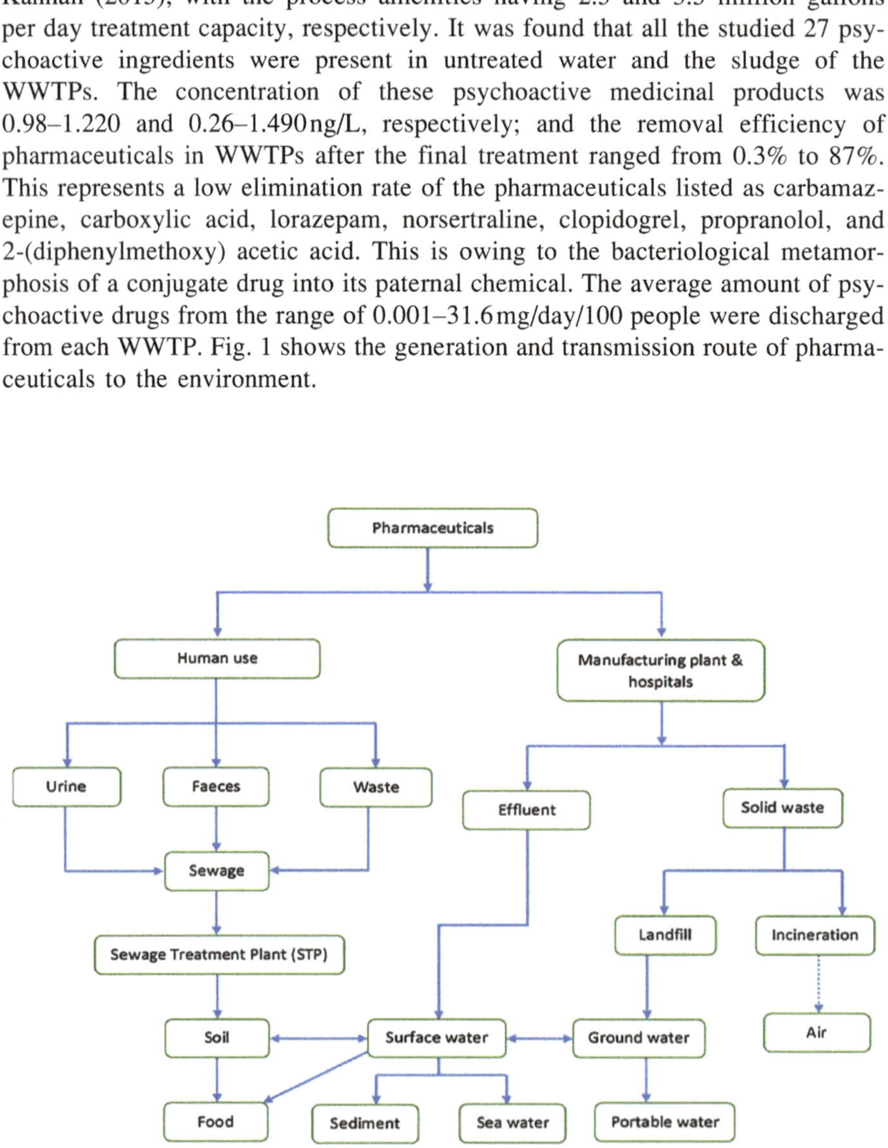

Fig. 1 The route by which pharmaceuticals reach the environment (Jose et al., 2020).

7.2 Italy

In Italy, pharmaceuticals can be categorized into three categories based on the summer and winter seasons. The removal rates in the first category are higher in the summertime as compared to that in winter, the second category has similar exclusion rates in the summer-time and winter-time, and the third category has removal rates near to zero in both seasons. Because pharmaceuticals, notably antibiotics, are used less in the summer, the total loads are nearly half those in the winter. The antibiotics ciprofloxacin, ofloxacin, and sulphamethoxazole had marginally inferior loads in the summertime compared to the winter, whereas atenolol, furosemide, ibuprofen, hydrochlorothiazide, and ranitidine had the same loads in both seasons.

To understand more about drug behavior and elimination during treatment, a new study in Italy used mass balance calculations to examine six sewage treatment plants (STPs). The removal efficiencies of 30 chemicals were evaluated, and the parameters impacting elimination in STPs were discovered (Zuccato et al., 2006). Certain STPs (in Naples, Latina, and Cuneo) were sampled in the winter (January–March), while others (in Cagliari and Varese Lago) were sampled in the summer-time (June–September); only Varese Olona was collected from in both periods. The weight of specific drugs in inlet and exit water at a specific plant was compared to determine removal rates. The removal rates for total loads were often less than 40%, except for the Varese Lago facility, where it reached 64%. In the four STPs studied in the winter (Naples, Latina, Varese Olona, and Cuneo), removal rates were 0%, 16%, 31%, and 39%, whereas, in the summer (Varese Lago, Varese Olona, and Cagliari), removal rates were 0%, 31%, and 64%. Based on this data, comparison could be made for Varese Olona: elimination rates appear to remain higher in the summertime than in the wintertime (31% vs 0%), consistent with a temperature-dependent increase in bacteriological action. In the Varese Olona STP, entire loads in the summer-time remained around half of those in the winter-time (1.5 vs 3 g/day/1000 residents). Seasonal changes in the elimination rates of the individual compounds were also noted. As a result, this study discovered that pharmaceutical removal rates in STPs fluctuate greatly during the year, but that abatement is always incomplete. In the not-too-distant future, tertiary treatments including ozonation, advanced oxidation, membrane seperation (activated carbon adsorption), and photocatalysis may be utilized to improve removal.

7.3 Denmark

In 2013, Denmark built the world's first full-scale wastewater treatment plant (WWTP) for HWW treatment. The hospital WWTP was constructed to treat wastewater from Herlev Hospital in 2012 (Nielsen & Jacob, 2016), which discharges 150,000 m^3 of wastewater per year, and is expected to discharge 200,000 m^3 by 2020. Around 850 pharmaceutical active compounds are used in Herlev Hospital's various departments. The hospital was identified as a large point source, releasing 700 kg of toxic pharmaceuticals into the maritime environment each year. The raw

HWW causes the death of crustaceans and Zebra fish, which die within 96h of exposure, and also contain high levels of antibiotic-resistant bacteria (ARB) and norovirus.

To treat the pharmaceutical residues of Herlev Hospital, the constructed WWTP has a capacity of 500 m^3 per day of wastewater, two biological tanks for secondary treatment, followed by granular activated carbon (GAC) adsorption, ozonation, and UV radiation. The wastewater is screened to remove solid particles as part of the primary treatment. Following the screening process, the wastewater is pumped to a biological process tank with intermittent aeration and coagulant (aluminum) to remove nitrogen and phosphorus simultaneously. The effluent is then injected into a membrane bioreactor, which has a 0.2 μm pore size ceramic membrane. With a TSS (total suspended solids) concentration of 5000 kg, the biological process retained sludge for about 30 days. After the secondary treatment, tertiary treatment such as 3.4 mg O_3/L granular activated carbon (GAC) with three filter columns in series and UV radiation installation consisting of one UV lamp of 220 W is implemented in series for the maximum removal of pharmaceuticals.

One hundred and twenty-two different pharmaceuticals and their metabolic byproducts were subjected to the MBR, which was shown to remove 95% of them without damaging the contrast medium. Ozone, GAC, and UV polishing of MBR permeate removed 99% of all pharmaceuticals, with no pharmaceutical concentrations surpassing the freshwater "no effect" concentration during the polishing procedure. Neither *Daphnia magna* nor zebra fish embryos were harmed in the ecotoxicology test using treated water.

7.4 France

In a study conducted in France, it was found that the effluent from two traditional activated sludge plants (WWTP A and WWTP B) is discharged into the Rivers Doubs and Loue, respectively (Chiffre, Degiorgi, Buleté, Spinner, & Badot, 2016). To establish upper and lower levels of the WWTP A, which handles domestic and industrialized unrefined water for the municipality of Pontarlier and its population of 55,000 people, water from the River Doubs was tested. For a higher and lower level of the WWTP B, which gathers raw water from the town of Ornans, with a population of 5000 people, river water was sampled in the Loue. The samples had the highest quantities of diclofenac (965 and 467 ng/L), sulfamethoxazole (655 and 615 ng/L), and carbamazepine (442 and 566 ng/L). However, the amounts of antiinflammatory pharmaceuticals (diclofenac, naproxen), antibiotics (sulfamethoxazole, trimethoprim, metronidazole), oestrogen (estrone), and Beta-blockers (propanolol) were all lower detected in April 2014 (Chiffre et al., 2016). The antiinflammatory naproxen was one of the highest concentrated pharmaceuticals (254 ng/L). On the other hand, the psychiatric drug oxazepam was detected (167 ng/L). Diclofenac and sulfamethoxazole were identified in reasonably higher amounts in WWTP discharges and surface water, indicating that WWTPs are incapable of effectually eradicate pharmaceuticals.

This research also revealed that high amounts of ofloxacin and sulfamethoxazole in external water can be harmful to aquatic ecosystems, underlining their potential danger. Furthermore, the presence of certain antibiotics upstream of the WWTP and in the Loue River's karst spring suggested other residential and agricultural discharges in the catchment, as well as contamination transfer through the karst system. Owing to the severe threat of transmissions in the karst structure, authorities must view WWTPs as chief sources of contagion and take diffusing the sources in the watershed into account to improve water quality control and limit its bearing on the marine network (Chiffre et al., 2016).

7.5 United Kingdom

Although pharmaceuticals are produced, used, and disposed of in substantial quantities in the UK, the presence and dispersion of pharmaceutical residues in HWW have less of an impact here than they do in other nations. Over 100 tons of the country's top 25 pharmaceutical compounds were utilized annually in 2000. Pharmaceutical residues in the UK were specified under the Controlled Waste Regulations (1992) (Bound & Voulvoulis, 2005). The Special Waste Regulations (1996) also keep records of clinical waste (Bound & Voulvoulis, 2005). Landfill sites were built to accommodate toxic sludge, according to the legislation. HWW sludge was put in landfills, and it undergoes biological decomposition, but it also raises the risk of leachate percolating into the groundwater (Bound & Voulvoulis, 2005).

7.6 Australia

In Australia, hospital wastewater generation is around 750 L/bed/day. The HWW produced there is directly released into the municipal sewer system, where it is treated with municipal wastewater at the MWWTP (municipal wastewater treatment plant). This method is referred to as cotreatment (Al Aukidy, Al Chalabi, & Verlicchi, 2017). However, Australian policies for HWWT recommend on-site treatment to safeguard the environment from toxins that may diffuse into the natural ecosystem (Preliminary, 2020). Despite the fact that pharmaceutical residues have not been added to drinking water quality requirements, Australia has been the greatest user of medications for the past 60 years and contributes a significant amount of pharmaceutical residues to the environment. Nine of the top 10 antibiotics used in Australia were found in WWTP influents (Chartier et al., 2014). Certain clonal populations of *E. coli* resistant to beta-lactam antibiotics were found in both HWW as well as the STP influents, according to the literature (Minguez et al., 2016). In general, HWW bacteria are more resistant than STP strains, which signifies heavy antibiotic usage in HWW. As per recommendation by the Joint Expert Advisory Committee study based on antibiotic resistance, the average use of beta-lactam inhibitors, carbapenems, and cephalosporins was over 40,000 kg per year between 1992 and 1997 (Kumari et al., 2020).

8. Hospital wastewater management in developing countries

8.1 India

In contrast to the industrialized world, India currently lacks formal legislation or policies that particularly oversee the fate or administration of emerging pollutants. India is presently one of the world's top pharmaceutical developing markets, as well as one of the leading worldwide suppliers of medications, accounting for 20% of global exports. In India, proper waste management systems do not exist, and traditional treatment facilities are poor at removing emerging pollutants, with sewage treatment plants discharged into rivers. According to some accounts, 10%–90% of the pharmaceuticals taken is eliminated from the human body in its parent form, while the remainder is metabolized and/or conjugated (Ferronato & Torretta, 2019).

The pharmaceutical industry makes use of a variety of wastewater treatment and disposal technologies. Wastewaters produced by these industries vary not only in composition but also in the amount based on the raw materials and techniques utilized in the manufacture of various medications (Gadipelly et al., 2014). There are several alternative treatment techniques available to cope with the large range of waste produced by these industries, but they are particular to the kind of industry and accompanying wastes. To treat pharmaceutical residues, general procedures such as sedimentation, floatation and advanced oxidation processes are used, as well as aerobic/anaerobic biological treatment in membrane bioreactors or bio-aeration and innovative hybrid technologies particular to the pharmaceutical industry.

On-site treatment of HWWs is now being considered since it can alleviate the problem of pathogen dilution and spread from its discharge into the municipal sewer system. Before discharging to MWWTPs, this onsite treatment may involve sedimentation and/or coagulation, filtration followed by disinfection by $FeCl_3$/UV, and traditional digesting of the settled solids (Kumari et al., 2020). The sewage sludge includes large quantities of helminths and other pathogens that must be treated with extreme caution (Rodriguez-Mozaz, Lucas, & Barceló, 2017; Thirumala, 2013). A few researchers have looked at the existence and dangers of helminth eggs and other biologically active infections in humans.

Despite the high rates of pathogen development, adequate sewage treatment facilities (STPs) are falling short across the country, and most research is primarily focused on parent pharmaceuticals. As a result, it is required to reduce the concentration ranges of metabolites discharged into the environment.

8.2 China

In recent years, the pharmaceutical industry's rapid expansion has become a significant component of China's national economy. In 2014, the output of chemical origin pharmaceuticals reached 3.034 million tons, a 15.2% increase over the previous year. Some features of the Chinese pharmaceutical sector include a plethora of small-scale firms and products, as well as a dispersed general layout. By the end of 2014, there were 7108 pharmaceutical firms in China, which was 8.9% more than in 2013. Only 1609 of these represented major or medium-sized businesses (Guo et al., 2017). However, as the pharmaceutical business grows, environmental degradation is becoming increasingly significant. There are many different types of pharmaceutical wastewater due to the range of pharmaceutical industry products and the differences in manufacturing size and procedure. In recent years, the primary focus of scientific study and technical application has switched to enhanced pharmaceutical wastewater treatment, the primary approach of which is physicochemical technology (Al Aukidy et al., 2017). HWW is successfully treated using hybrid physical and chemical procedures such as coagulation and sedimentation, flotation, activated carbon adsorption, advanced oxidation processes, and membrane separation.

To search for a technology that would improve public health and the environment while also complying with increasingly strict rules, the cost-effective membrane bioreactor approach, known as MBR, is being adopted for HWW treatment. In China, more than 50 commercial MBR facilities for HWW treatment with capacities ranging from 20 to 2000 m/day have already been installed (Liu, Zhou, Chen, & Zheng, 2010). Liu et al. (2010) examined the treatment efficacy of several Chinese HWW treatment plants using the MBR method followed by a chlorination phase for disinfection. The use of MBR in the treatment of HWW decreases the concentration and reaction time of hypochlorite during the disinfection process. According to the study, MBR-treated effluent followed by chlorine disinfection fulfills Chinese wastewater discharge criteria. Yuan, Jiang, Xia, Zhang, and Zheng (2013) studied the incidence and treatment efficiency of two wastewater treatment plants (WWTP-P1 and WWTP-P2) at a Chinese psychiatric hospital. The WWTP of psychiatric institutions uses initial treatment to remove solid particles, followed by a subsequent activated sludge process (Yuan et al., 2013). The comparison of influent and effluent from both WWTPs (WWTP-P1 and WWTP-P2) revealed that secondary treatment of HWW was effective in removing the majority of psychiatric pharmaceuticals (aripiprazole (64%–70%), olanzapine (93%–98%), quetiapine (>73%), and risperidone (72%–95%)). Despite the high removal rate of psychiatric pharmaceuticals in WWTPs, the residual pharmaceutical content in treated effluent was higher than in MWWTP influent (Yuan et al., 2013). To neutralize pharmaceutical chemicals, advanced secondary treatment (such as MBR or sequence batch reactor) or tertiary treatment (such as linked process advance oxidation process or disinfection) was necessary.

8.3 Africa

In South Africa, there is a dearth of baseline research to counteract any impacts that may be produced by the presence of pharmaceuticals in South African water systems. South Africa faces unique issues, such as a high prevalence of HIV/Aids and tuberculosis (TB), with resistant forms of TB widespread in the population. Consequently, antiretroviral (ARV) medicines and antibiotics are widely used, leading to quite large amounts being discharged into aquatic habitats. As a result, it is necessary to measure and establish their fate, as well as estimate their potential long-term consequences on the environment and public health. Despite the fact that South Africa has an estimated 7.7 million individuals living with HIV, and 62% of all persons living with HIV in South Africa are on ARV medication, there have been few investigations on pharmaceutical residues in water and their biodegradation profile (Ngqwala & Muchesa, 2020). As a result, it is projected that the concentration of ARV medications in waste and surface water could be very high. This can be assessed by rigorous quantitative pharmacological risk evaluations tailored to South Africa's specific needs.

Currently, pharmaceuticals removal by WWTP techniques is used in South Africa, which is incapable of totally removing drugs, resulting in their discharge into bodies of water. There is a need for research to discover how effective these sewage treatments are at removing various types of drugs. Most quantitative pharmacological risk estimates have concentrated on metropolitan regions, ignoring rural people, who frequently consume untreated water. As a result, research in South Africa's rural regions will give important information on the prevalence and destiny of pharmaceuticals in the environment that can be compared to studies in metropolitan areas. South Africa is a semiarid country, and pharmaceutical residues are especially detrimental to aquatic life during drought since the concentration increases owing to low water quantities. In South Africa, waste management is an urgent issue, and inappropriate disposal of hazardous healthcare waste is a major concern. Nurses, pharmacists, and other healthcare professionals are to blame for the mismanagement of waste. It also imposes an unacceptably high financial and human resource load on health authorities in order to tackle the problem (Madikizela, Tavengwa, & Chimuka, 2017).

Wesström and Andersson (2014) studied the existence of pharmaceutical residues in agricultural water derived from wastewater at Chirapatre Estates in Kumasi, Ghana. The findings revealed that malaria control medicine and paracetamol were found in irrigation water derived from unrefined water. Olarinmoye, Adekunle, Obih, and Arne (2016) studied the presence of pharmaceuticals such as estrogens, and lipid-lowering agents in surface water released from an unrefined wastewater treatment facility in Lagos, Nigeria; and 37 pharmaceutical residues were detected at concentrations ranging from 8.84 to 560 g/L. It was suggested that expanded environmental monitoring of pharmaceutical residues be performed in Nigerian locations with large population densities. Matongo, Birungi, Moodley, and Ndungu (2015) studied the presence of pharmaceutical residues in the Msunduji Stream's water and sediments in Kwazulu-Natal, South Africa. The study revealed that many pharmacological clusters, including antipyretics, antibiotics, antiepileptic, and antipsychotic, were discovered in the samples. Ibuprofen was the most prevalent residue in unrefined water (117 g/L), surface water (84.60 g/L), and sediments (659 ng/L) across all groups. Agunbiade and Moodley (2016) conducted a similar investigation in which they discovered eight acidic pharmaceuticals in wastewater, surface water, and sediments from the Msunduzi River in Kwazulu-Natal, South Africa. Aspirin and nalidixic acid were found to be the most prevalent pharmaceutical residues, in the ranges of 118 g/L and 25.2–29.9 g/L, respectively.

As per the United Nations Development Program, Africa is inadequately implementing rules for pharmaceutical wastewater disposal. As a result, HWW is released directly into the municipal sewer system, where it is cotreated with urban wastewater. This is a frequent practice in Australia, Iran, Egypt, India, and Japan. Ghana, Lesotho, and Eritrea continue to follow any HWW treatment and disposal guidelines, although Nigeria, Gambia, and Kenya are signatories to the Stockholm Convention 2009 (Verlicchi et al., 2015). South Africa typically follows USEPA laws for the treatment of Class A sludge from hospitals, which is then utilized unrestrictedly in agriculture with no additional monitoring. To fulfill national effluent discharge limits, the resulting sludge is further treated (air-drying, composting, and long-term storage).

8.4 Malaysia

The Center of Excellence for Environmental Forensics at the Faculty of Environmental Studies at the University of Putra Malaysia (UPM) undertook preliminary monitoring to establish whether drug residues are present in Malaysian waterways and in what amounts. The first monitoring was carried out in the Langat Basin, where surface water, STP effluent samples, and a few tap water samples were obtained. Sungai Langat is one of Selangor's four major river systems. It has a catchment area of 2350 km^2 and a length of around 180 km. The river is fed by three major tributaries: Sungai Semenyih, Sungai Beranang, and Sungai Labu. According to reports, the major source of organic pollution in the Langat basin is untreated sewage from STPs (organic pollution and pathogens), animal husbandry waste (organic pollution, metals, and pathogens), and industrial waste (organic and inorganic pollution). Pollution to the river system generated by STPs is regarded as the second most significant source of pollution after industrial sources and is followed by pig farms. The Langat River gets STP wastewater from over 520 STPs throughout the basin, as well as raw sewage. There are eight water treatment facilities (WTPs) along the Langat basin. Except for WTPs located upstream of the Langat River, the remaining WTPs pose a considerable contamination risk to the raw water intake for the plants. Pharmaceutical contaminants were also found in all of the STP effluent samples obtained from the basin's five distinct STPs. This confirms that STPs are the primary source of these contaminants in the river and that current treatment methods are incapable of completely removing them. A few pharmaceutical contaminants were also found in tap water samples, indicating that these pollutants are not eliminated by WTPs and so enter the tap water. These findings are consistent with earlier studies conducted in other regions of the world, which reveal that pharmaceutical pollution poses a danger to a healthy aquatic ecosystem and safe drinking water (Al-Odaini, Zakaria, Yaziz, Surif, & Abdulghani, 2013; Khan et al., 2022).

8.5 Brazil

Drugs in intake water and the environment are not regulated under Brazilian law. Scholars and environmental organizations, like the So Paulo State Environmental Firm, are spearheading the monitoring efforts (CETESB). Brazil still does not have an approved list of drugs to be examined and scrutinized in water. The Brazilian Health Regulatory Agency (Anvisa) controlled procedures of, 2018 in handling health schemes remain in force for when health facilities (hospitals, ambulances, health centers, etc.), drugstores, and medication distributors dispose of pharmaceuticals (Resolution n. 222/2018). According to the resolution, pharmaceuticals that do not lead to radiological, chemical, or biological concerns shall be handled via reverse approaches. Additional countrywide rules that attempt to adopt other approaches for drugs are inefficient.

Contamination of surface water and, as an outcome, the quality of intake water in Brazil is a periodic issue owing to a scarcity of appropriate community strategies aimed at addressing fundamental sanitation issues. In a nation of approximately 200 million people, the number of houses having access to a sewage handling facility does not reach 50%, and much of the accumulated sewage is not properly treated before being released into source waters (IBGE, 2011). According to statistics reviewed in the 2011 Sanitation Atlas published by the Brazilian Institute of Geography and Statistics (IBGE, 2011), approximately 31% of Brazilian cities discharge unprocessed sewage into waterways, lakes, or pools, and those sources of water are utilized by individuals for a variety of applications, including agriculture and supplying drinking water for the people. Precarious sanitation in some areas, along with population increase in urban agglomerations, is now one of the primary drivers of surface water deterioration in Brazil. According to a recent national valuation by the National Water Agency (ANA) on the Conjunction of Water Resources, the quality of water sources in urban and surrounding areas is rated as low or extremely bad. Because surface water treatment produces the great majority of drinking water in Brazilian urban areas, poor water quality needs sophisticated treatment for the appropriate eradication of various uncontrolled chemicals. On the contrary, the majority of the DWTPs use traditional procedures like coagulation, flocculation, sedimentation, filtering, and disinfection with chlorine-based chemicals.

An initial review of the records of the Minas Gerais state Environmental Agency (FEAM) in the pharmaceutical industry licensing process revealed discrepancies in statistics on the monitoring of solid waste production, as well as nonconformity with the requirements to meet seepage release protocols (Deschamps et al., 2012). Inspections of businesses revealed that waste administration is stuck in its embryonic stage and that the wide range of output resulted in an extremely changing seepage configuration. This significantly reduces the efficacy of existing handling techniques. There has been little research in Minas Gerais on the elimination of pharmaceutical residues, particularly in hospital effluents. The hospitals in Minas Gerais utilize traditional treatment, and there is no assessment of elimination efficacy.

Santoro et al. (2015) investigated the antibiotic resistance profile of *Pseudomonas aeruginosa* in a hospital wastewater treatment plant in Rio de Janeiro, Brazil. The hospital WWTP processes 220 m/day of wastewater using an activated sludge technique. The hospital's WWTP features an extended aeration tank, after which sludge settles in a settling tank and the treated wastewater is piped for chlorination. After subsequent chlorine treatment, the effluent is discharged via a rainwater network. The 27 strains of *P. aeruginosa* were identified at various stages of the wastewater handling method, including inlet wastewater, activated sludge, settling, and treated seepage (Kumari et al., 2020).

8.6 Mexico

There are partial statistics related to pharmaceutical concentration levels in wastewater in Mexico owed to an absence of state guidelines targeted at giving resources to organizations and investigators equipped to conduct this sort of research. In Mexico, 54% of untreated water is inadequately handled, posing the utmost serious civic well-being issues because this type of water is reclaimed for farming use and, in some circumstances, human consumption (Ronderos-Lara et al., 2020).

To date, all stratagems and rules for water resource management in Mexico have been futile, owing to the monetary and practical benefits of specific groups in society, which has prevented the utilization of the ideologies recognized in integrated water resources management (IWRM), taking into account the contribution of awareness groups. There have been several studies in Mexico that have assessed the concentration levels of this sort of substance in untreated water, groundwater, and surface water, practically all of which were conducted in the country's center (Guanajuato, Hidalgo, Jalisco, Morelos states, and Mexico City). Estradiol, ethinylestradiol, 4-nonylphenol, bisphenol A, 4-tert-octylphenol, naproxen, acetaminophen, diclofenac, bezafibrate, atenolol, and carbamazepine are among the identified substances.

Innumerable government agencies are accountable for conducting an investigation in administration progressions and the progress of water remediation approaches. Only a few studies in recent years intended for the application of innovative techniques for the elimination and/or breakdown of organic compounds in untreated wastewater. For illustration, the Mexican Institute of Water Technology (IMTA) has been exploring several solutions for the elimination of EC since 2014. They employed biofiltration devices for the biodegradation of two pharmaceuticals, metformin and ciprofloxacin, in these solutions. During 103 days of operation, breakdown efficiencies of 83% and 71% were obtained (Estrada-Arriaga, Mijaylova Nacheva, & García-Sánchez, 2015). In another study, conducted in two wastewater treatment plants in the states of Guanajuato and Mexico, employed a system comprised of oxidation channels and UV light lamps, achieving pharmaceuticals removal efficacies ranging from 20% to 22% (Guanajuato); whereas in the other treatment, which included anaerobic/anoxic/aerobic tanks as well as two disinfection methods, chlorine dioxide and UV lamps, pharmaceuticals elimination was considerable (up to 80%) (Mexico) (Estrada-Arriaga et al., 2015). A submerged membrane bioreactor was also exploited to degrade the substances estrone, estradiol, and 17-ethinylestradiol, with clearance rates close to 96% for all compounds (Estrada-Arriaga et al., 2013). Velázquez and Nacheva (2017) used four aerated submerged attached growth bioreactors to remove three pharmaceuticals (fluoxetine, mefenamic acid, and metoprolol) from municipal unrefined water, with elimination efficacies of 95%, 82%, and 73% for fluoxetine, mefenamic acid, and metoprolol, respectively . In another study, Garca-Espinosa et al. reported 88.7% breakdown percentages of carbamazepine in wastewater utilizing an electrochemical oxidation technique (García-Espinoza, Mijaylova-Nacheva, & Avilés-Flores, 2018).

9. Conclusions

The chapter has gathered statistics on the presence of pharmacological concentrations in countries around the globe. The key conclusions are as follows. First, pharmaceuticals are found in all the environmental samples, including groundwater, oceans, surface water, etc. According to some studies, pharmaceuticals are found in ground water, surface water, and moderately treated water in concentrations of less than 0.1 g/L, while amounts in treated water are normally less than 0.05 g/L. Hence there is a need for a planned investigation in order to understand the occurrence of pharmaceuticals predominantly in sources of drinking water. In developing countries like India, despite the high consumption rates of pharmaceuticals, there is a scarcity of treatment facilities. However, most research is chiefly done on parent pharmaceuticals. In Brazil, the majority of the cities discharge untreated sewage into water streams, which leads to the high contamination of the water matrices. Despite poor sanitation, and agglomeration of unwanted chemicals in the surface water, the majority of the Brazilian cities WWTPs use outdated procedures like coagulation, flocculation, and sedimentation.

The WWTPs dealing with hospital effluents require advanced treatment strategies for the extermination of numerous uncontrolled compounds. Several pharmaceuticals, like estradiol, ethinylestradiol, 4-nonylphenol, bisphenol A, 4-tert-octylphenol, naproxen, acetaminophen, diclofenac, bezafibrate, atenolol, and carbamazepine, have been detected in untreated water, groundwater, and surface water in developing countries like India, Mexico, Pakistan, Bangladesh, etc. A lot of procedures, including oxidation, adsorption, and membrane processes, have been established to effectively remove large amounts of antibiotics. Notably, carbonaceous compounds like activated carbon and graphene have been extensively explored. In order to remove pharmaceuticals, advanced secondary treatment (such as MBR or sequence batch reactor) or tertiary treatment is also being used extensively. MBR is also effective in removing the majority of the pharmaceuticals present in wastewater. Recently, South Africa applied USEPA regulations for the handling of sludge discharged from hospitals, which is then employed in agriculture with no further treatment.

Despite noteworthy advances in understanding the fate of pharmaceuticals, toxicity, removal, or breakdown of drugs in untreated water and its associated hazards over several decades, still, a great deal of effort needs to be applied to minimize the risks, and to develop strict laws to stop the overuse of drugs or discharge of drugs into water matrices.

Conflict of interest

The authors declare no conflict of interest.

References

Agunbiade, F. O., & Moodley, B. (2016). Occurrence and distribution pattern of acidic pharmaceuticals in surface water, wastewater, and sediment of the Msunduzi River, Kwazulu-Natal, South Africa. *Environmental Toxicology and Chemistry*, *35*(1), 36–46.

Akhtar, M. F., Ashraf, M., Javeed, A., Anjum, A. A., Sharif, A., Saleem, A., et al. (2016). Toxicity appraisal of untreated dyeing industry wastewater based on chemical characterization and short term bioassays. *Bulletin of Environmental Contamination and Toxicology*, *96*(4), 502–507.

Al Aukidy, M., Al Chalabi, S., & Verlicchi, P. (2017). *Hospital wastewater treatments adopted in Asia, Africa, and Australia* (pp. 171–188). https://doi.org/10.1007/698.

Al Qarni, H., Collier, P., O'Keeffe, J., & Akunna, J. (2016). Investigating the removal of some pharmaceutical compounds in hospital wastewater treatment plants operating in Saudi Arabia. *Environmental Science and Pollution Research*, *23*(13), 13003–13014.

Alfonso-Muniozguren, P., Serna-Galvis, E. A., Bussemaker, M., Torres-Palma, R. A., & Lee, J. (2021). A review on pharmaceuticals removal from waters by single and combined biological, membrane filtration and ultrasound systems. *Ultrasonics Sonochemistry*, *76*, 105656.

Al-Khateeb, L. A., Almotiry, S., & Salam, M. A. (2014). Adsorption of pharmaceutical pollutants onto graphene nanoplatelets. *Chemical Engineering Journal*, *248*, 191–199.

Al-Odaini, N. A., Zakaria, M. P., Yaziz, M. I., Surif, S., & Abdulghani, M. (2013). The occurrence of human pharmaceuticals in wastewater effluents and surface water of Langat River and its tributaries, Malaysia. *International Journal of Environmental Analytical Chemistry*, *93*, 245–264. https://doi.org/10.1080/03067319.2011.592949.

aus der Beek, T., Weber, F. A., Bergmann, A., Hickmann, S., Ebert, I., Hein, A., & Küster, A. (2016). Pharmaceuticals in the environment—Global occurrences and perspectives. *Environmental Toxicology and Chemistry*, *35*(4), 823–835. https://doi.org/10.1002/etc.3339.

Bean, T. G., Rattner, B. A., Lazarus, R. S., Day, D. D., Burket, S. R., Brooks, B. W., et al. (2018). Pharmaceuticals in water, fish and osprey nestlings in Delaware River and Bay. *Environmental Pollution*, *232*, 533e545.

Boroski, M., Cl, A., Garcia, J. C., Sampaio, L. C., Nozaki, J., & Hioka, N. (2009). Combined electrocoagulation and TiO 2 photoassisted treatment applied to wastewater effluents from pharmaceutical and cosmetic industries. *Journal of Hazardous Materials*, *162*, 448–454.

Bound, J. P., Kitsou, K., & Voulvoulis, N. (2006). Household disposal of pharmaceuticals and perception of risk to the environment. *Environmental Toxicology and Pharmacology*, *21*, 301–307.

Bound, J. P., & Voulvoulis, N. (2005). Household disposal of pharmaceuticals as a pathway for aquatic contamination in the United Kingdom. *Environmental Health Perspectives*, *113* (12), 1705–1711. https://doi.org/10.1289/ehp.8315.

Bush, K. (1997). Antimicrobial agents. *Current Opinion in Chemical Biology*, *1*, 169–175.

Caldwell, D. J. (2016). Sources of pharmaceutical residues in the environment and their control. In R. E. Hester, & R. M. Harrison (Eds.), *Pharmaceuticals in the environment* (pp. 92–119). The Royal Society of Chemistry.

Chang, I. S., Le Clech, P., Jefferson, B., & Judd, S. (2002). Membrane fouling in membrane bioreactors for wastewater treatment. *Journal of Environmental Engineering*, *128*(11), 1018–1029.

Chartier, Y., Emmanuel, J., Pieper, U., Prüss, A., Rushbrook, P., Stringer, R., et al. (2014). *Safe management of wastes from healthcare activities* (2nd. ed.). WHO.

Chen, Z., Ren, N., Wang, A., Zhang, Z., & Shi, Y. (2008). A novel application of TPAD-MBR system to the pilot treatment of chemical synthesis-based pharmaceutical wastewater. *Water Research*, *42*, 3385–3392.

Chen, Z., Wang, H., Ren, N., Cui, M., Nie, S., & Hu, D. (2011). Simultaneous removal and evaluation of organic substrates and NH 3-N by a novel combined process in treating chemical synthesis-based pharmaceutical wastewater. *Journal of Hazardous Materials*, *197*, 49–59.

Chiffre, A., Degiorgi, F., Buleté, A., Spinner, L., & Badot, P. M. (2016). Occurrence of pharmaceuticals in WWTP effluents and their impact in a karstic rural catchment of Eastern France. *Environmental Science and Pollution Research*, *23*(24), 25427–25441. https://doi.org/10.1007/s11356-016-7751-5.

Chonova, T., Lecomte, V., Bertrand-Krajewski, J.-L., Bouchez, A., Labanowski, J., Dagot, C., et al. (2018). The SIPIBEL project: Treatment of hospital and urban wastewater in a conventional urban wastewater treatment plant. *Environmental Science and Pollution Research*, *25*(10), 9197–9206. https://doi.org/10.1007/s11356-017-9302-0.

Cyr, P. J., Suri, R. P. S., & Helmig, E. D. (2002). A pilot scale evaluation of removal of mercury from pharmaceutical wastewater using granular activated carbon. *Water Research*, *36*, 4725–4734.

Deng, Y., & Zhao, R. (2015). Advanced oxidation processes (AOPs) in wastewater treatment. *Current Pollution Reports*, *1*(3), 167–176.

Deschamps, E., Vasconcelos, O., Lange, L., Donnici, C. L., Silva, M. C. D., & Sales, J. A. (2012). Management of effluents and waste from pharmaceutical industry in Minas Gerais, Brazil. *Brazilian Journal of Pharmaceutical Sciences*, *48*, 727–736.

Domínguez, J. R., González, T., & Palo, P. (2012). Electrochemical degradation of a real pharmaceutical effluent. *Water, Air, and Soil Pollution*, *223*, 2685–2694.

EC. (2014). *Council directive 91/271/EEC of 21 May 1991 concerning urban waste-water treatment. EUR-Lex.* Retrieved January 16, 2022, from https://eur-lex.europa.eu/legal-content/EN/LSU/?uri=CELEX:31991L0271.

Escher, B. I., Baumgartner, R., Koller, M., Treyer, K., Lienert, J., & McArdell, C. S. (2011). Environmental toxicology and risk assessment of pharmaceuticals from hospital wastewater. *Water Research*, *45*(1), 75–92. https://doi.org/10.1016/j.watres.2010.08.019.

Estrada-Arriaga, E. B., et al. (2013). Occurrence and treatment of endocrine disrupter chemicals in Mexico City wastewater employing submerged membrane bioreactor. *Ingeniería. Investigación y Tecnología*, *14*(2), 275–284.

Estrada-Arriaga, E. B., Mijaylova Nacheva, P., & García-Sánchez, L. (2015). Effect of mixed liquor volatile suspended solids on membrane fouling during short and long-term operation of membrane bioreactor. *Ingeniería Y. Ciencia*, *11*(21), 137–155. https://doi.org/10.17230/ingciencia.11.21.7.

Ferronato, N., & Torretta, V. (2019). Waste mismanagement in developing countries: A review of global issues. *International Journal of Environmental Research and Public Health*, *16*(6), 1060.

Friha, I., Karray, F., Feki, F., Jlaiel, L., & Sayadi, S. (2014). Treatment of cosmetic industry wastewater by submerged membrane bioreactor with consideration of microbial community dynamics. *International Biodeterioration & Biodegradation*, *88*, 125–133.

Gadipelly, C., Pérez-González, A., Yadav, G. D., Ortiz, I., Ibáñez, R., Rathod, V. K., & Marathe, K. V. (2014). Pharmaceutical industry wastewater: Review of the technologies for water treatment and reuse. *Industrial & Engineering Chemistry Research*, *53*(29), 11571–11592.

Ganiyu, S. O., Van Hullebusch, E. D., Cretin, M., Esposito, G., & Oturan, M. A. (2015). Coupling of membrane filtration and advanced oxidation processes for removal of pharmaceutical residues: A critical review. *Separation and Purification Technology*, *156*, 891–914.

García-Espinoza, J. D., Mijaylova-Nacheva, P., & Avilés-Flores, M. (2018). Electrochemical carbamazepine degradation: Effect of the generated active chlorine, transformation pathways and toxicity. *Chemosphere*, *192*, 142–151.

Giwa, A., Yusuf, A., Balogun, H. A., Sambudi, N. S., Bilad, M. R., Adeyemi, I., et al. (2020). Recent advances in advanced oxidation processes for removal of contaminants from water: A comprehensive review. *Process Safety and Environment Protection*, *146*, 220–256.

Goossens, H., Ferech, M., Coenen, S., & Stephens, P. (2004). Comparison of outpatient systemic antibacterial use in 2004 in the United States and 27 European countries. *Clinical Infectious Diseases*, *44*, 1091–1095.

Gunnarsson, L., Kristiansson, E., Rutgersson, C., Sturve, L., Fick, J. J., Förlin, L., et al. (2009). Pharmaceutical industry effluent diluted 1:500 affects global gene expression, cytochrome P450 1A activity, and plasma phosphate in fish. *Environmental Toxicology and Chemistry*, *28*, 2639–2647.

Guo, Y., Qi, P. S., & Liu, Y. Z. (2017). A review on advanced treatment of pharmaceutical wastewater. *IOP Conference Series: Earth and Environmental Science*, *63*, 012025.

Huang, L., Shen, R., & Shuai, Q. (2021). Adsorptive removal of pharmaceuticals from water using metal-organic frameworks: A review. *Journal of Environmental Management*, *277*, 111389.

IBGE. (2011). *Qualidade e Eficiência dos Serviços de Saneamento* (pp. 28–46). Brasília: MCidades. http://www.ibge.gov.br/home/estatistica/populacao/atlas_saneamento/default_zip.shtm.

Ilyas, H., & Masih, I. (2018). The effects of different aeration strategies on the performance of constructed wetlands for phosphorus removal. *Environmental Science and Pollution Research*, *25*(6), 5318–5335.

Jose, J., Sandra Pinto, J., Kotian, B., Mathew Thomas, A., & Narayana Charyulu, R. (2020). Comparison of the regulatory outline of ecopharmacovigilance of pharmaceuticals in Europe, USA, Japan and Australia. *The Science of the Total Environment*, *709*, 134815. https://doi.org/10.1016/J.SCITOTENV.2019.134815.

Kadlec, R. H. (2009). Comparison of free water and horizontal subsurface treatment wetlands. *Ecological Engineering*, *35*(2), 159–174.

Kar, S., & Roy, K. (2012). Risk assessment for ecotoxicity of pharmaceuticals—An emerging issue. *Expert Opinion on Drug Safety*, *11*(2), 235–274. https://doi.org/10.1517/14740338.2012.644272.

Khan, A. H., Aziz, H. A., Khan, N. A., Hasan, M. A., Ahmed, S., Farooqi, I. H., … Mahtab, M. S. (2022). Impact, disease outbreak and the eco-hazards associated with pharmaceutical residues: A critical review. *International Journal of Environmental Science and Technology*, *19*, 677–688. https://doi.org/10.1007/s13762-021-03158-9.

Khan, A. H., Khan, N. A., Ahmed, S., Dhingra, A., Singh, C. P., Khan, S. U., … Ali, I. (2020). Application of advanced oxidation processes followed by different treatment technologies for hospital wastewater treatment. *Journal of Cleaner Production*, *269*, 122411. https://doi.org/10.1016/j.jclepro.2020.122411.

Khan, N. A., Khan, A. H., Tiwari, P., Zubair, M., & Naushad, M. (2021). New insights into the integrated application of Fenton-based oxidation processes for the treatment of pharmaceutical wastewater. *Journal of Water Process Engineering*, *44*, 102440. https://doi.org/10.1016/j.jwpe.2021.102440.

Klavarioti, M., Mantzavinos, D., & Kassinos, D. (2009). Removal of residual pharmaceuticals from aqueous systems by advanced oxidation processes. *Environment International*, *35*, 402–417.

Kolpin, D. W., Furlong, E. T., Meyer, M. T., Thurman, E. M., Zaugg, S. D., Barber, L. B., et al. (2002). Pharmaceuticals, hormones, and other organic wastewater contaminants in U.S. streams, 1999 – 2000: A national reconnaissance. *Environmental Science & Technology*, *36*(6), 1202–1211. https://doi.org/10.1021/es011055j.

Kosma, C. I., Lambropoulou, D. A., & Albanis, T. A. (2014). Investigation of PPCPs in wastewater treatment plants in Greece: Occurrence, removal and environmental risk assessment. *Science of the Total Environment*, *466–467*, 421–438.

Kumari, A., Maurya, N. S., & Tiwari, B. (2020). Hospital wastewater treatment scenario around the globe. In *Current developments in biotechnology and bioengineering* (pp. 549–570). Elsevier. https://doi.org/10.1016/b978-0-12-819722-6.00015-8.

Kümmerer, K. (2001). Drugs in the environment: Emission of drugs, diagnostic aids and disinfectants into wastewater by hospitals in relation to other sources: A review. *Chemosphere, 45*, 957–969.

Kümmerer, K. (2009). The presence of pharmaceuticals in the environment due to human use-present knowledge and future challenges. *Journal of Environmental Management, 90*(8), 2354–2366.

Kyzas, G. Z., Fu, J., Lazaridis, N. K., Bikiaris, D. N., & Matis, K. A. (2015). New approaches on the removal of pharmaceuticals from wastewaters with adsorbent materials. *Journal of Molecular Liquids, 209*, 87–93.

Larsson, D. G. J. (2010). Release of active pharmaceutical ingredients from manufacturing sites need for new management strategies. *Integrated Environmental Assessment and Management, 6*, 184–186.

Larsson, D. G. J., de Pedro, C., & Paxeus, N. (2007). Effluent from drug manufactures contains extremely high levels of pharmaceuticals. *Journal of Hazardous Materials, 148*(3), 751–755.

Larsson, D. G. J., & Fick, J. (2009). Transparency throughout the production chain—A way to reduce pollution from the manufacturing of pharmaceuticals? *Regulatory Toxicology and Pharmacology, 53*, 161–163.

Lawrence, K., Wang, Y.-T. H., Howard, H. L., & Constantine, Y. (2005). *Waste water treatment in the process industries*. CRS Press. 2005.

Lee, Y., Kovalova, L., McArdell, C. S., & von Gunten, U. (2014). Prediction of micropollutant elimination during ozonation of a hospital wastewater effluent. *Water Research, 64*, 134–148.

Liu, Q., Zhou, Y., Chen, L., & Zheng, X. (2010). Application of MBR for hospital wastewater treatment in China. *Desalination, 250*, 605–608. https://doi.org/10.1016/j.desal.2009.09.033.

Madikizela, L. M., Tavengwa, N. T., & Chimuka, L. (2017). Status of pharmaceuticals in African water bodies: Occurrence, removal and analytical methods. *Journal of Environmental Management, 193*, 211–220.

Martinez, J. L. (2009). Environmental pollution by antibiotics and by antibiotic resistance determinants. *Environmental Pollution, 157*, 2893–2902.

Matongo, S., Birungi, G., Moodley, B., & Ndungu, P. (2015). Pharmaceutical residues in water and sediment of Msunduzi River, Kwazulu-Natal, South Africa. *Chemosphere, 134*, 133–140.

Melero, J. A., Botas, J. A., Molina, R., Pariente, M. I., & Martı, F. (2009). Heterogeneous catalytic wet peroxide oxidation systems for the treatment of an industrial pharmaceutical wastewater. *Water Research, 43*, 4010–4018.

Mestre, A. S., Pires, J., Nogueira, J. M. F., & Carvalho, A. P. (2007). Activated carbons for the adsorption of ibuprofen. *Carbon, 45*(10), 1979–1988.

Michael, I., Rizzo, L., Mcardell, C. S., Manaia, C. M., Merlin, C., Schwartz, T., et al. (2013). Urban wastewater treatment plants as hotspots for the release of antibiotics in the environment: A review. *Water Research, 47*(3), 957–995.

Minguez, L., Pedelucq, J., Farcy, E., Ballandonne, C., Budzinski, H., & Halm-Lemeille, M. P. (2016). Toxicities of 48 pharmaceuticals and their freshwater and marine environmental assessment in northwestern France. *Environmental Science and Pollution Research, 23*(6), 4992–5001. https://doi.org/10.1007/s11356-014-3662-5.

Munter, R. (2001). Advanced oxidation processes—Current status and prospects. *Proceedings of the Estonian Academy of Sciences, Chemistry*, *50*(2), 59–80.

Mutamim, N. S. A., Noor, Z. Z., Hassan, M. A. A., Yuniarto, A., & Olsson, G. (2013). Membrane bioreactor: Applications and limitations in treating high strength industrial wastewater. *Chemical Engineering Journal*, *225*, 109–119.

Ngqwala, N. P., & Muchesa, P. (2020). Occurrence of pharmaceuticals in aquatic environments: A review and potential impacts in South Africa. *South African Journal of Science*, *116*(7–8), 1–7.

Nielsen, U., & Jacob, S. (2016). *Full scale advanced wastewater treatment at Herlev Hospital-Treatment performance and evaluation*. Denmark: Herlev.

Nikolaou, A., Meric, S., & Fatta, D. (2007). Occurrence patterns of pharmaceuticals in water and wastewater environments. *Analytical and Bioanalytical Chemistry*, *387*, 1225–1234.

Olarinmoye, O., Adekunle, B., Obih, U., & Arne, H. (2016). Quantification of pharmaceutical residues in wastewater impacted surface waters and sewage sludge from Lagos, Nigeria. *Journal of Environmental Chemistry and Ecotoxicology*, *8*(3), 14–24.

Ortiz de García, S. A., Pinto Pinto, G., García-Encina, P. A., & Irusta-Mata, R. (2014). Ecotoxicity and environmental risk assessment of pharmaceuticals and personal care products in aquatic environments and wastewater treatment plants. *Ecotoxicology*, *23*, 1517–1533.

Oulton, R. L., Kohn, T., & Cwiertny, D. M. (2010). Pharmaceuticals and personal care products in effluent matrices: A survey of transformation and removal during wastewater treatment and implications for wastewater management. *Journal of Environmental Monitoring*, *12* (11), 1956–1978.

Peng, Y. Z., Li, Y. Z., Peng, C. Y., & Wang, S. Y. (2004). Nitrogen removal from pharmaceutical manufacturing wastewater with high concentration of ammonia and free ammonia via partial nitrification and denitrification. *Water Science and Technology*, *50*, 31–36.

Preliminary, P. (2020). *South Australian public health (Wastewater) regulations 2013* (pp. 1–21). Australia: Legislation Revision and Publication Act 2002.

Puspalatha, P., & Kalpana, P. (2016). Design approach for sewage treatment plant: A case study of Srikakulam Greater Municipality, India. *Environmental Science: An Indian Journal*, *12* (9), 112.

Putschew, A., & Jekel, M. (2007). Analysis, fate and removal of pharmaceuticals in the water cycle. *Comprehensive Analytical Chemistry*, *50*, 61–93.

Rajkumar, D., & Palanivelu, K. (2004). Electrochemical treatment of industrial wastewater. *Journal of Hazardous Materials*, *113*, 123–129.

Ren, X., Chen, C., Nagatsu, M., & Wang, X. (2011). Carbon nanotubes as adsorbents in environmental pollution management: A review. *Chemical Engineering Journal*, *170*(2–3), 395–410.

Robles-Molina, J., Gilbert-López, B., García-Reyes, J. F., & Molina-Díaz, A. (2014). Monitoring of selected priority and emerging contaminants in the Guadalquivir River and other related surface waters in the province of Jaén, South East Spain. *Science of the Total Environment*, *479*(480), 247–257.

Rodriguez-Mozaz, S., Lucas, D., & Barceló, D. (2017). Full-scale plants for dedicated treatment of hospital effluents. *Hospital Wastewaters*, 189–208.

Ronderos-Lara, J. G., Saldarriaga-Noreña, H., Reyes-Romero, P. G., Chávez-Almazán, L. A., Vergara-Sánchez, J., Murillo-Tovar, M. A., & Torres-Segundo, C. (2020). Emerging compounds in Mexico: Challenges for their identification and elimination in wastewater. In *Emerging contaminants* IntechOpen.

Santoro, D. O., Cardoso, A. M., Coutinho, F. H., Pinto, L. H., Vieira, R. P., Albano, R. M., & Clementino, M. M. (2015). Diversity and antibiotic resistance profiles of *Pseudomonads* from a hospital wastewater treatment plant. *Journal of Applied Microbiology*, *119*(6), 1527–1540.

Serrano-Ruiz, J. C., Luque, R., & Sepúlveda-Escribano, A. (2011). Transformations of biomass-derived platform molecules: From high added-value chemicals to fuels via aqueous-phase processing. *Chemical Society Reviews*, *40*(11), 5266–5281.

Shaw, I. C. (2013). Should We legalise recreational drugs? An ethical question for animals and humans. *Pharmaceutical Medicine*, *27*(4), 231–233. https://doi.org/10.1007/s40290-013-0022-6.

Song, W., Li, J., Wang, Z., & Zhang, X. (2019). A mini review of activated methods to persulfate-based advanced oxidation process. *Water Science and Technology*, *79*(3), 573–579.

Sophia, A. C., Bhalambaal, V. M., Lima, E. C., & Thirunavoukkarasu, M. (2016). Microbial desalination cell technology: Contribution to sustainable waste water treatment process, current status and future applications. *Journal of Environmental Chemical Engineering*, *4*(3), 3468–3478.

Subedi, B., & Kannan, K. (2015). Occurrence and fate of select psychoactive pharmaceuticals and antihypertensives in two wastewater treatment plants in New York State, USA. *Science of the Total Environment*, *514*, 273–280. https://doi.org/10.1016/J.SCITOTENV.2015.01.098.

Tchobanoglous, G., Stensel, H. D., Tsuchihashi, R., & Burton, F. L. (2014). *Wastewater engineering: Treatment and resource recovery* (5th ed.). New York, NY: Metcalf and Eddy I AECOM. McGraw-Hill Book Company.

Thirumala, S. (2013). Study of bio-medical waste generation and management in various hospitals in Davangere city of Karnataka, India. *Journal of Health and Allied Sciences NU*, *3*(3), 22–24.

Tian, K., Hu, L., Li, L., Zheng, Q., Xin, Y., & Zhang, G. (2022). Recent advances in persulfate-based advanced oxidation processes for organic wastewater treatment. *Chinese Chemical Letters*, *33*(10), 4461–4477.

Van der, A. N. G. F. M., Kommer, G. J., Van Montfoort, J. E., & Versteegh, J. F. M. (2011). Demographic projections of future pharmaceutical consumption in the Netherlands. *Water Science and Technology*, *63*, 825–831.

Velázquez, Y. F., & Nacheva, P. M. (2017). Biodegradability of fluoxetine, mefenamic acid, and metoprolol using different microbial consortiums. *Environmental Science and Pollution Research*, *24*, 6779–6793. https://doi.org/10.1007/s11356-017-8413-y.

Verlicchi, P., Al Aukidy, M., & Zambello, E. (2015). What have we learned from worldwide experiences on the management and treatment of hospital effluent?—An overview and a discussion on perspectives. *Science of the Total Environment*, *514*, 467–491. https://doi.org/10.1016/j.scitotenv.2015.02.020.

Vymazal, J. (2013). The use of hybrid constructed wetlands for wastewater treatment with special attention to nitrogen removal: A review of a recent development. *Water Research*, *47*(14), 4795–4811.

Wang, H., Yuan, X., Wu, Y., Huang, H., Peng, X., Zeng, G., et al. (2013). Graphene-based materials: Fabrication, characterization and application for the decontamination of wastewater and wastergas and hydrogen storage/generation. *Advances in Colloid and Interface Science*, *195*, 19–40.

Wesström, T., & Andersson, J. (2014). *Pharmaceutical pollution in irrigation water: A minor field study in Chirapatre Estates in Kumasi, Ghana (Master's Thesis)* (pp. 1–40). Uppsala,

Sweden: Department of Earth Sciences, Disciplinary Domain of Science and Technology, Earth Sciences, Uppsala University.
Yang, Y., Lohwacharin, J., & Takizawa, S. (2014). Hybrid ferrihydrite-MF/UF membrane filtration for the simultaneous removal of dissolved organic matter and phosphate. *Water Research*, *65*, 177–185.
Yao, H., Wang, Y. K., & He, Y. S. (2013). Study on antibiotic and starch mixed wastewater treatment by combined two-phase anaerobic, aerobic and anammox process. *Journal of Environmental Engineering and Technology*, *3*(3), 183–187.
Yoon, Y., Westerhoff, P., Snyder, S. A., Wert, E. C., & Yoon, J. (2007). Removal of endocrine disrupting compounds and pharmaceuticals by nanofiltration and ultrafiltration membranes. *Desalination*, *202*(1–3), 16–23.
Yuan, S., Jiang, X., Xia, X., Zhang, H., & Zheng, S. (2013). Detection, occurrence and fate of 22 psychiatric pharmaceuticals in psychiatric hospital and municipal wastewater treatment plants in Beijing, China. *Chemosphere*, *90*(10), 2520–2525.
Zheng, Y. (2011). *Pretreatment of pharmaceutical castewater by catalytic wet air oxidation (CWAO)* (pp. 1316–1318). IEEE.
Zuccato, E., Castiglioni, S., Fanelli, R., Reitano, G., Bagnati, R., Chiabrando, C., et al. (2006). Pharmaceuticals in the environment in Italy: Causes, occurrence, effects and control. In *Vol. 13 (1). Environmental science and pollution research* (pp. 15–21). https://doi.org/10.1065/espr2006.01.004.

Further reading

Besse, J. P., & Garric, J. (2008). Human pharmaceuticals in surface waters. Implementation of a prioritization methodology and application to the French situation. *Toxicology Letters*, *176*, 104–123.
Brillas, E., Garcia-Segura, S., Skoumal, M., & Arias, C. (2012). Electrochemical incineration of diclofenac in neutral aqueous medium by anodic oxidation using Pt and boron-doped diamond anodes. *Chemosphere*, *79*, 605–612.
Chartier, Y., Emmanuel, J., Pieper, U., Pruess, A., Rushbrook, P., Stringer, R., Townend, W., Wilbum, S., & Zghondi, R. (2014). *Safe management of wastes from health-care activities* (2nd ed., xvii, 308 pp.). Geneva, Switzerland: World Health Organization (WHO).
Dong, Z., Senn, D. B., Moran, R. E., & Shine, J. P. (2013). Prioritizing environmental risk of prescription pharmaceuticals. *Regulatory Toxicology and Pharmacology*, *65*, 60–67.
Dos Santos, D. M., Buruaem, L., Gonçalves, R. M., Williams, M., Abessa, D. M. S., Kookana, R., et al. (2018). Multiresidue determination and predicted risk assessment of contaminants of emerging concern in marine sediments from the vicinities of submarine sewage outfalls. *Marine Pollution Bulletin*, *129*, 299–307.
Edwards, Q. A., Sultana, T., Kulikov, S. M., Garner-O'Neale, L. D., Yargeau, V., & Metcalfe, C. D. (2018). Contaminants of emerging concern in wastewaters in Barbados, West Indies. *Bulletin of Environmental Contamination and Toxicology*, *101*, 1–6.
EUR-Lex-32010L0075-EN-EUR-Lex. (n.d.). Retrieved January 16, 2022, from https://eur-lex.europa.eu/legal-content/EN/TXT/?uri=CELEX%3A32010L0075.
Fairbairn, D. J., Elliott, S. M., Kiesling, R. L., Schoenfuss, H. L., Ferrey, M. L., & Westerhoff, B. M. (2018). Contaminants of emerging concern in urban storm water: Spatiotemporal patterns and removal by iron-enhanced sand filters (IESFs). *Water Research*, *145*, 332–345.

Grube, A., Donaldson, D., Kiely, T.A., Wu, L., Pesticides industry sales and usage: 2006 and 2007 market. Estimates. United States Environmental Protection Agency: Washington, DC, 2011; p. 41.

Igwe, P. U., Chukwudi, C. C., Ifenatuorah, F. C., Fagbeja, I. F., & Okeke, C. A. (2017). A review of environmental effects of surface water pollution. *International Journal of Advances in Engineering Research Science*, *4*(12), 128–137.

Kumar, N. A. (2007). View on freshwater environment. *Ecology, Environment and Conservation*, *3*(3), 386–393.

Li, Y., Yi, R., Yi, C., Zhou, B., & Wang, H. (2017). Research on the degradation mechanism of pyridine in drinking water by dielectric barrier discharge. *Journal of Environmental Sciences*, *53*, 238–247.

Luo, X. C., Guo, Y., Ngo, W., Nghiem, H. H., Hai, L. D., Zhang, F. I., et al. (2014). A review on the occurrence of micropollutants in the aquatic environment and their fate and removal during wastewater treatment. *Science of the Total Environment*, *473-474*, 619–641.

Luo, Y., Yang, Y., Lin, Y., Tian, Y., Wu, L., Yang, L., et al. (2018). Low temperature and atmospheric pressure sample digestion using dielectric barrier discharge. *Analytical Chemistry*, *90*, 1547–1553.

Mulbry, W., Kondrad, S., Pizarro, C., & Kebede-Westhead, E. (2008). Treatment of dairy manure effluent using freshwater algae: Algal productivity and recovery of manure nutrients using pilot-scale algal turf scrubbers. *Bioresource Technology*, *99*, 8137–8142.

Reponses Hydrodynamiques des Hauts Bassins du Doubs et de la Loue. (n.d.). Retrieved January 16, 2022, from https://www.documentation.eauetbiodiversite.fr/notice/00000000015df1022d1188c2bc3c954a.

Sophia, A. C., & Lima, E. C. (2018). Removal of emerging contaminants from the environment by adsorption. *Ecotoxicology and Environmental Safety*, *150*, 1–17.

Tijani, J. O., Fatoba, O. O., & Petrik, L. F. (2013). A review of pharmaceuticals and endocrine-disrupting compounds: Sources, effects, removal, and detections. *Water, Air, and Soil Pollution*, *224*, 1770.

Tran, N. H., Reinhard, M., & Gin, K. Y.-H. (2018). Occurrence and fate of emerging contaminants in municipal wastewater treatment plants from different geographical regions—A review. *Water Research*, *133*, 182–207.

Verlicchi, P., Al Aukidy, M., & Zambello, E. (2012). Occurrence of pharmaceutical compounds in urban wastewater: removal, mass load and environmental risk after a secondary treatment—Review. *Science of the Total Environment*, *429*, 123–155.

Yadav, D., Rangabhashiyam, S., Verma, P., Singh, P., Devi, P., Kumar, P., et al. (2021). Environmental and health impacts of contaminants of emerging concerns: Recent treatment challenges and approaches. *Chemosphere*, 129492.

Yuan, F., Hu, C., Hu, X., Qu, J., & Yang, M. (2009). Degradation of selected pharmaceuticals in aqueous solution with UV and UV/H2O2. *Water Research*, *43*, 1766–1774.

Recent trends in graphene-based materials for pharmaceuticals wastewater treatment

3

Hasara Samaraweera[a] *and Afzal Husain Khan*[b]
[a]Department of Chemistry, Mississippi State University, Mississippi State, MS, United States, [b]Civil Engineering Department, College of Engineering, Jazan University, Jazan, Saudi Arabia

1. Introduction

Due to the excessive growth of urbanization, the industrial revolution, and the development of the health sector, the production, and usage of emerging contaminants (ECs) have risen. ECs are mostly polar compounds; therefore, their distribution and dissolution in water are raising a major environmental issue. More than 700 ECs including 20 different groups have been identified by the European aquatic environment (www.norman-network.net) recently. In the absence of adequate control of them, various detrimental effects can be caused to the aquatic community and humans. Pharmaceuticals and related compounds have emerged as common aquatic pollutants due to their persistent nature as they easily accumulated through food chains. Fish tissue analysis indicated that hydrophobic organic compounds are concentrated in highly lipophilic tissues (Barber, Keefe, Antweiler, Taylor, & Wass, 2006). This concentration depends on bioconcentration factors and K_{ow} values (Spyriouni, Krokidis, & Economou, 2011). Accumulation of even trace amounts (ng/L-g/L) of these substances causes acute and chronic toxicity, toxicological effects, endocrine disruption, etc. Therefore, the analysis of occurrence, fate, risks (Geissen et al., 2015), and removal of aqueous ECs are urgently required to monitor existing water pollution and reduce water scarcity.

Pharmaceutical molecules can be observed in common human drugs such as analgesics, antidepressants, antibiotics, hormones, antiinflammatory, antiepileptic, statins, beta-blockers, anticancer agents, antibiotics, lipid-regulators, contrast agents, disinfectants (Deblonde, Cossu-Leguille, & Hartemann, 2011). Prior to analyzing these pharmaceuticals compounds, accurate detection is required (Lekkas, 2013). Complex environmental matrices are screened to detect the presence of trace pharmaceuticals using various qualitative and quantitative analytical techniques such as LC/MS, LC/UV, and GC/MS. Before incorporating LC-MS, ng/L level sample was preconcentrated using solid phase extraction (Afonso-Olivares, Torres-Padrón, Sosa-Ferrera, & Santana-Rodríguez, 2013). However, high detection limits of analytical instruments sometimes can reduce the detection ability and, therefore, the proper risk assessment (Geissen et al., 2015).

The Treatment of Pharmaceutical Wastewater. https://doi.org/10.1016/B978-0-323-99160-5.00001-1

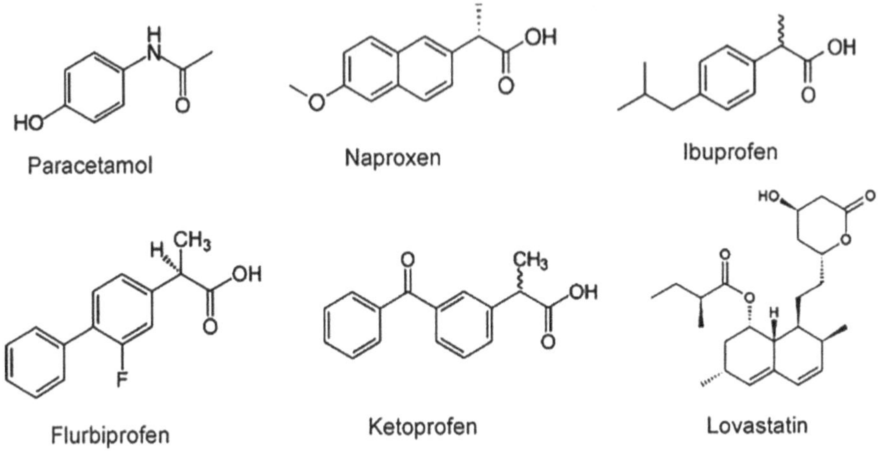

Fig. 1 Structures of some common pharmaceutical drugs.
Reproduced with permission from Spyriouni, T., Krokidis, X., Economou, I. G. (2011). Thermodynamics of pharmaceuticals: Prediction of solubility in pure and mixed solvents with PC-SAFT. *Fluid Phase Equilibria 302*, 331–337. https://doi.org/10.1016/j.fluid.2010.08.029, Elsevier.

Continuous discharge of pharmaceuticals into wastewater is occurred due to their wide use in human medicine, veterinary medicine, and aquaculture. These substances are released from hospitals, wastewater treatments plants with inadequate removal drugs, and agricultural discharges involving animal excretions. Pharmaceutical compounds can also reach the environment when contaminated sludge is used as fertilizer (Carlsson, Johansson, Alvan, Bergman, & Kühler, 2006). Annually, a large number of drugs are developed for human consumption and common drugs are shown in Fig. 1. Recently, world's top markets including Pfizer, Johnson and Johnson, GSK contributes to the massive production of COVID-19 pandemic related vaccines and drugs. However, their long-term effects are still unclear. Further investigations are needed to predict the chronic ecotoxicity precisely, which can be resulted from the novel drugs. This can take up to several decades.

Pharmaceutical products mainly enter the water systems due to their disposal in landfills or wastewater (Nikolaou, Meric, & Fatta, 2007). When discharged, drugs with long half-life (e.g., Erythromycin, carbamazepine, naproxen, sulfamethoxazole, ibuprofen, etc.) can stay longer in the environment than others. Moreno Ríos, Gutierrez-Suarez, Carmona, Ramos, and Silva Oliveira (2021) showed that Naproxen (nonsteroidal antiinflammatory drugs (NSAIDs)) causes oxidative stress, bioaccumulation, and genetic disruption. It also affects the behavior, development, and expression of antioxidant genes in fish species. The presence of even trace amounts of pollutants can be problematic due to their severe side effects. Hospital effluents release most of the pharmaceutical waste, which causes alterations in zebrafish embryos (teratogenic index of 2.45%) (Tenorio-Chávez et al., 2020). However, the chronic toxic effects of some pharmaceuticals and their metabolites are unknown. Some of the pharmaceuticals are more polar when they are in aquatic environments than in the natural environment, due to oxidation of hydrolysis. However, they can be hydrophobic being dynamic through the food chain, and build up in tissues (Vasilachi, Asiminicesei, Fertu, & Gavrilescu, 2021).

Surprisingly, there are no tight regulations to control water pharmaceutical levels in the United States (Finn, Giampietro, Mazyck, & Rodriguez, 2021). Pharmaceuticals in water streams can be effectively treated using advanced filtration or chemical removal methods (Cecconet, Molognoni, Callegari, & Capodaglio, 2017). WWTPs offer a water reuse option after the decontamination. Ozonation, ultraviolet radiation, and adsorption are currently used in drinking water treatment plants. Recent reviews (Ebele, Abou-Elwafa Abdallah, & Harrad, 2017; Khan et al., 2020) have addressed the efficiency and environmental friendliness of different treatment methods. Khan et al. (2020) showed that ozone-assisted oxidation is an effective treatment method for pharmaceutical removal. Low energy consumption and costs need to consider when selecting a proper removal method. Recover of treated water adds an additional advantage that can be achieved in treatment. Highly polar substances can be effectively removed using biological degradation and mineralization with the use of microbes (Vasilachi et al., 2021). Adsorption is simple approach that has multitude of advantages including high pharmaceutical binding capacities, large regeneration potential and energy efficiency. However, finding low-cost adsorbents can be critical. Nowadays, a variety of adsorbents like carbon-based materials (e.g., lignite, biochar), biosorbents, agricultural products, polymeric materials are designed, synthesized, characterized, and evaluated for their different pollutant removal efficiencies (Samaraweera, Sharp, et al., 2021). Activated carbon shows a large organic contaminant remediation potential because of its large surface area and oxygenated functional groups. But it is expensive. Nanocomposites are fabricated by combining nanoparticles with various supporting materials (e.g., activated carbon, biochar) and the resulted hybrid shows a high surface area (Ioniţă et al., 2017) and plenty of surface oxygen functions, which are effective in contaminant adsorption (Samaraweera, Pittman, et al., 2021; Tan et al., 2016). Graphene is a widely used nanomaterial in composite formation because of its outstanding adsorptive and catalytic properties toward pharmaceutical remediation (Vasseghian, Dragoi, Almomani, & Le, 2022).

2. Graphene-based materials

2.1 Graphene, graphene oxide (GO), and reduced graphene oxide (rGO)

Carbon allotropes such as carbon nanotubes (CNTs) and fullerenes have drawn significant interest in various fields for several decades. Their outstanding electronic and mechanical properties have resulted from the basic building block, graphene. Graphene has an atomic diameter of 0.334 nm and is widely used in catalysis, sensor applications, and environmental remediation. It is composed of a single layer carbon atoms arranged in 2D hexagons in a honeycomb structure (Fig. 2). It also features a

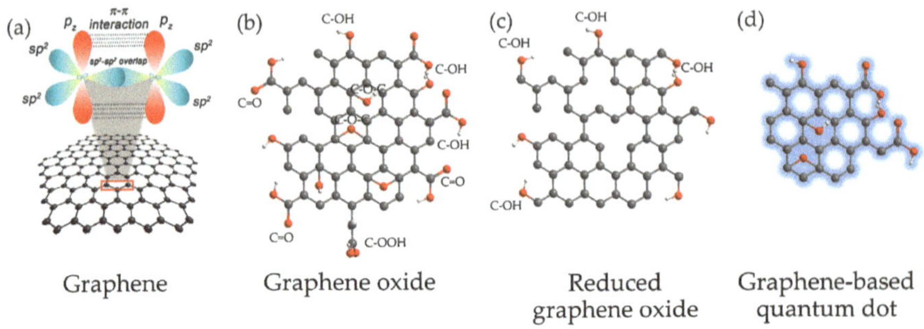

Fig. 2 Structure of graphene (A), graphene oxide (B), reduced graphene oxide (rGO) (C), and graphene-based quantum dot (D).
Reproduced with permission from Suvarnaphaet, P., & Pechprasarn, S. (2017). Graphene-based materials for biosensors: A review. *Sensors 17*. https://doi.org/10.3390/s17102161, MDPI.

high surface area (2630 m^2/g for single-layer graphene) (Zhu et al., 2010), excellent thermal conductivity at room temperature of the order of 5×10^3 W/mK (Balandin et al., 2008), high Young modulus (~1 TP) (Lee, Yoon, & Cheong, 2012) and superior carrier mobility (room-temperature mobilities of 2×10^4 cm^2/V s). In electrically and chemically doped devices, the graphene mobility is high even at high carrier density, indicating occurrence of ballistic transport on the submicrometer scale (Schedin et al., 2007). Further, graphene's charge carriers are massless Dirac fermions (Geim & Novoselov, 2007) and graphene has excited physical properties like the ambipolar field effect (Novoselov et al., 2005), quantum Hall effect (Novoselov et al., 2004), breakdown of the Born-Oppenheimer approximation (Zhang, Tan, Stormer, & Kim, 2005) are among others.

There are many methods to synthesize exfoliated graphene, and the major routes are shown in Fig. 3. In the scotch-tape technique, a graphene platelet was deposited on SiO_2 wafer by exfoliation. Graphenes can be produced using chemical vapor deposition (CVD) on large scale. It also produces high quality graphene. Liquid phase exfoliation is an ideal way to synthesize mono- or several layer graphene at a low-cost (Amiri, Naraghi, Ahmadi, Soleymaniha, & Shanbedi, 2018). However, the selection of the liquid (solvent) is critical here. Organic solvents with surface tension of about 40 mJ m^{-2} are ideal for single layer graphene production (Amiri et al., 2018). Hernandez et al. (2008) synthesized high yields of mono- and few-layer graphenes (12% by mass) by exfoliating 0.01 mg/mL graphite suspensions in *N*-methyl-pyrrolidone. XPS, Raman and FTIR analyses proved that the synthesized layers are also defects-free. However, for commercial applications, search for alternative organic solvents is required to promote the dispersion. Graphenes can also be grown on SiC by sublimation of Si atoms.

Graphene-based materials are widely utilized in environmental remediation. However, pristine graphene has low aqueous dispersibility due to its large hydrophobicity (Yang, Wang, & Chen, 2014). This results in the aggregation of graphene sheets via π-π interactions which makes it difficult to recover. Graphene oxide (GO) is an oxidized version of graphene containing hydroxyls, carboxyls, carbonyl, and epoxides on

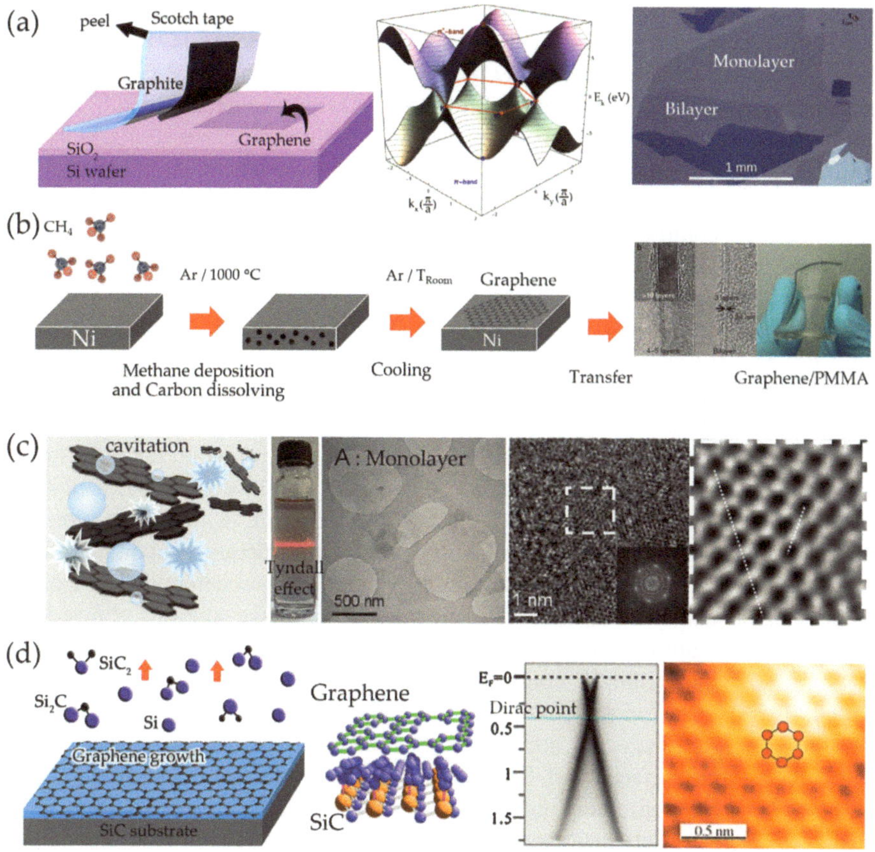

Fig. 3 Synthesis routes of graphene (A) scotch-tape technique, (B) chemical vapor deposition, (C) liquid exfoliation of graphene, and (D) epitaxial graphene growth on a silicon carbide (SiC). Reproduced with permission from Suvarnaphaet, P., & Pechprasarn, S. (2017). Graphene-based materials for biosensors: A review. *Sensors 17*. https://doi.org/10.3390/s17102161, MDPI.

basal planes and edges, which makes the GO surface negatively charged. During water purification, GO reacts with positively charged molecules like heavy metals and cationic dyes (Sabzevari, Cree, & Wilson, 2018). Many GO synthesis methods were reported earlier (Brodie, 1859). Of which, Hummer's method is the most popular (Hummers & Offeman, 1958). Here, graphite oxide is first prepared using the oxidation of graphite (Brodie, 1859) using $KMnO_4$, conc. H_2SO_4 acid, and $NaNO_3$. This method increased the interlayer distance in graphite, 0.34 nm (Franklin, 1951) due to the intercalation of oxygenated surface groups. This will further weaken the Vander Waals forces between the graphene layers and supports with exfoliation of graphite oxide to graphene oxide after ultrasonic or mechanical stirring (Zhu et al., 2010). Improved interlayer space assists with more pollutant adsorption via intercalation. However, GO shows a low conductivity due to oxygenated functional groups, which is problematic for photocatalytic reactions (Ciesielski & Samorì, 2014). This can be avoided by preparing reduced graphene oxide (rGO). rGO is also known as functionalized graphene or chemically reduced graphene (Eda & Chhowalla, 2010), a reduced version of GO. rGO typically has a larger surface area, strength like graphene and a higher water compatibility like GO (Stankovich et al., 2007). rGO's improved conductivity, due to the formation of unsaturated carbon atoms, is used in polymer composite formation. Chemical reduction, electrochemical reduction, high temperature thermal reduction, solvothermal reduction techniques, and microwave reduction are common rGO preparing strategies (Lavin-Lopez, Paton-Carrero, Sanchez-Silva, Valverde, & Romero, 2017). Conventional reductants such as $NaBH_4$, anhydrous hydrazine, and hydrazine monohydrate are highly toxic and explosive, which are not good for bulk rGO production. Therefore, green reductants (e.g., alcohols, water, L-ascorbic acid, etc.) have been employed in rGO preparation. He and Fang (2016) have utilized different alcohols for the reduction and observed the reducibility order as PhOH > EtOH > PrOH > BuOH. Also, PhOH showed an electrical conductivity similar to graphite. rGO has some implications due to the reduction of oxygenated sorption sites (Feng et al., 2017; Park et al., 2019) and is prone to agglomeration. Biocompatible and hydrophilic network systems are fabricated to avoid that issue. Feng et al. (2017) developed a 3D porous network structure, containing reduced graphene oxide-sodium alginate (rGO-SA) hydrogel microspheres via the external emulsification and thermal reduction method. SEM image analysis showed that rGO-SA lacks agglomerating. Also, the utilization of multiple reduction methods was effective in oxidized compound removal such as phenol (Lavin-Lopez et al., 2017).

Graphene-based materials are fascinating materials for adsorption due to their smaller size (large surface area). However, obtaining stable aqueous suspensions will be critical as smaller sized colloidal particles tend to aggregate in the solution phase. The presence of divalent cations (Ca^{2+}, Ni^{2+}, etc.) can also promote the aggregation, due to the decrease of surface charge (Bayati & Fidalgo de Cortalezzi, 2019). Further, divalent cations cross-links the GO sheets via "bridging" of the functional groups at the sheet edges (Wu et al., 2013). This can reduce their practical applicability. Geng, Wang, and Kim (2009) prepared a stable aqueous graphene suspension by oxidating graphene nanoplatelets to graphite oxide nanoplatelets, followed by chemical reduction of graphite oxide nanoplatelets to graphene. Also, graphene surfaces are modified using various metallic additives (Fe^{2+}, Ni^{2+} etc.) to magnetize the surfaces. Magnetic materials are easy to separate using an external magnet, avoiding the filtration.

2.2 Functionalized forms of graphene

Graphene itself shows a large affinity to antibiotics because of its graphitic surface structure which facilitates π-π bonding and hydrophobic interactions. Also, their rapid antibiotic elimination capacity was due to their higher surface area and porous structure supporting diffusion or surface adherence of molecules. Further, low-cost graphenes are more suitable than other carbon allotropes like CNTs for practical applications. However, low surface hydrophilicity and self-aggregation prevent their use on a large scale. In this case, graphene functionalization received much attention

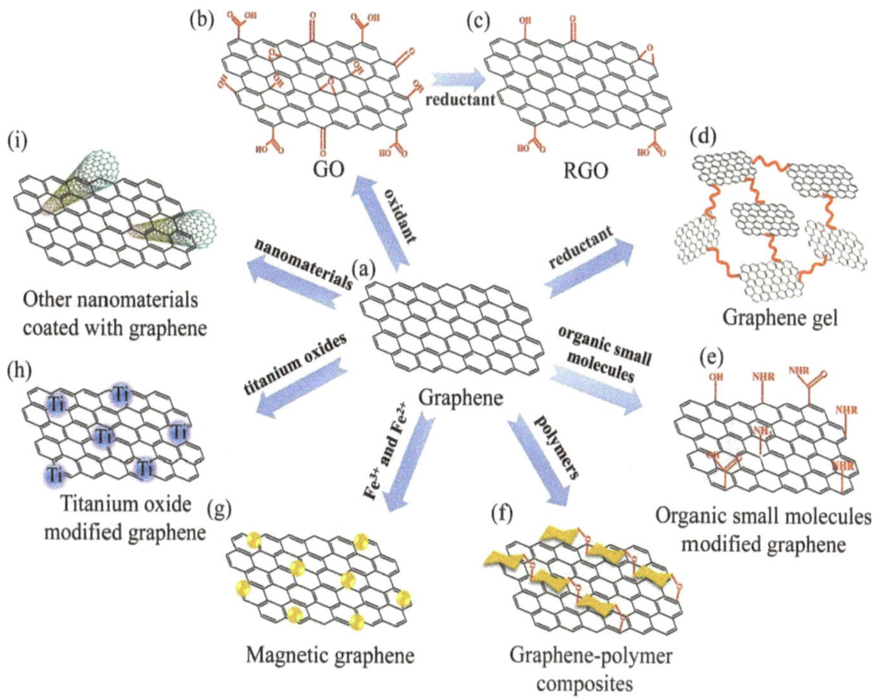

Fig. 4 Functionalization strategies for graphene used in environmental remediation. Reproduced with permission from Li, M., Liu, Y., Zeng, G., Liu, N., & Liu, S. (2019). Graphene and graphene-based nanocomposites used for antibiotics removal in water treatment: A review. *Chemosphere 226*, 360–380. https://doi.org/10.1016/j.chemosphere.2019.03.117, Elsevier.

as an excellent alternative. Graphene can be functionalized to form graphene polymer composites, graphene nanoparticle composites, and graphene CNT composites. Different functionalized graphene composites used in environmental remediation are shown in Fig. 4. These composites showed a better performance than precursor materials (Li, Liu, Zeng, Liu, & Liu, 2019).

2.2.1 Graphene polymer composites

Pristine graphene shows a low polymer intercalation ability due to the aggregation of platelets (Stankovich et al., 2007). This leads to poor homogeneity in solutions. Therefore, graphene surfaces are modified using various agents. Graphene-polymer composites exhibit superior mechanical properties, high thermal stability, excellent electrical conductivity and good barrier properties, etc. They are obtained using nucleophilic addition of organic molecules. Use of cost-effective amine compounds and solvents (e.g., water) under air improves dispersion in organic media. The solvent dispersible properties of surface-modified graphene will assist in the preparation of graphene-based polymer nanocomposites. Compared to conventional fillers, these composites are advantageous as low graphene contents are required (<2%).

The development in stress-strain properties through graphene was reported up to 100%, whereas more than 42% for EG. These developments were primarily depending upon the high aspect ratio of the graphene filler (Kuilla et al., 2010). On the other hand, polymer matrix from nanocomposites containing homogeneous graphene and PANI/graphene nanocomposites also showed excellent barrier properties.

2.2.2 Graphene nanoparticle composites

Nanoparticles (NPs) are particles (size, ~100 nm) and are widely used in many different fields due to their outstanding physicochemical and optical properties. Because NPs with different sizes show different colors, they are used in bioimaging (Khan, Saeed, & Khan, 2019). NPs are three-layered materials containing a surface layer (available for functionalization), a shell layer (chemically different from the core) and the core (central portion of the NP) (Shin, Cho, Kannan, Lee, & Kim, 2016). There are many types of NPs including carbon-based NPs (e.g., fullerenes, carbon nanotubes), metal NPs (e.g., Cu, Pt, Pd Au, Ag), semiconductor NPs, ceramic NPs (SiO_2, Al_2O_3, CeO_2, ZrO_2, and $BaTiO_3$), metal oxide nanoparticles (e.g., Co_3O_4), semiconductor nanoparticles (e.g., ZnO, TiO_2, CdSe) and magnetic nanoparticles (e.g., Fe_2O_3, Fe_3O_4) (Khan et al., 2019). Owning to their electron affinity, electrical conductivity, high strength, flexibility, and localized surface plasmon resonance (LSPR) characteristics, these nanoparticles have introduced in various nanocomposite formations. Graphene composites are prepared using a solvothermal approach (Wu et al., 2016) and chemical coprecipitation using Fe^{2+}/Fe^{3+} salts (Lin, Xu, & Li, 2013). These composites are highly effective and fast in treating tetracycline and sulfonamides. Wu et al. (2016) prepared a recyclable rGO/Fe_3O_4 hybrid using one-pot solvothermal approach. Prepared hybrids were enriched magnetite particles (~35 nm) with higher loading capacity and utilized for real wastewater treatment. Optimum removal of sulfonamides was observed at pH 6.0 and the further increase of pH decreased the removal efficiency. Magnetic adsorbents have attracted wider attentions due to easy and fast removal. Zhang et al. (2014) developed a recyclable supermagnetic Fe_3O_4 functional GO nanoparticles with a large removal toward BPA (39.1 mg/g). The major removal mechanism was π-π interaction. Further, TC removal negligibly depends on the solution pH and ionic strength.

2.2.3 Graphene CNT composites

Carbon nanotube (CNT) is a cylindrical nanomaterial that is widely used in electronics, and composite preparation, coatings, and biomedical device (WTEC, 2007). Single walled CNTs has a large theoretical surface area (~1315 m^2/g) (Peigney, Laurent, Flahaut, Bacsa, & Rousset, 2001). Its low surface functionality and poor dispersibility limits its application in water treatment. Recently, graphene-modified CNTs were prepared to obtain porous materials with improved electrical conductivity, mechanical properties, and thermal stability. Hence, a higher absorptivity of organic and inorganic pollutants can be easily obtained. These composites are used in membrane fabrication for ultrafiltration systems (Hu & Mi, 2013).

3. Removal of pharmaceuticals using graphene-based materials

Metronidazole (MNZ) (brand name Flagyl), a cytotoxic drug which is used to treat bacterial, protozoal, and microaerophilic infections in human and animal infections. Effluents from meat processing industries and hospitals contain large MNZ levels which can be mutagenic or carcinogenic. Typical MNZ concentrations in wastewater are range from 1 ng/L to 9.4 μg/L (Leung et al., 2013). MNZ's amine group is protonated below pH 2.4 (pK_{a1}) and –OH group is deprotonated above pH 14.4 (pK_{a2}) (Fig. 5). Kalhori, Al-Musawi, Ghahramani, Kazemian, and Zarrabi (2017) showed that pH dependence of MNZ onto aggregated adsorbents is negligible from pH 5–9. MNZ can be remediated via both adsorption and photocatalysis in the presence of graphene-based materials (Vasseghian et al., 2022).

However, MNZ's high polarity and poor biodegradability make it difficult to remove. Graphene shows poor MNZ removal efficiency due to its large hydrophobicity and is rarely used. Comparatively GO is highly favorable for MNZ removal due to a large number of oxygenated groups and localized π electron system, which facilitate dispersion in polar media. Segovia-Sandoval et al. (2020) synthesized a series of urea functionalized carbon-xerogel/GO (GO/CX) hybrids and evaluated for their MNZ removal potentials (Fig. 6). Large MNZ removal capacity (166 mg/g) was reported for the hybrid with large GO content (1% GO) at pH 5 and 298 K. Further, this hybrid has the highest BET surface area (816 m^2/g) and larger micropore (0.34 cm^3/g) and mesopore volumes (1.2 cm^3/g). Typical MNZ removal mechanisms involve physisorption, cation exchange, and complexation between the functional groups of MNZ molecule and graphene layers (Vasseghian et al., 2022). For this hybrid, π-π interactions are the major removal route (Fig. 6).

Currently, the development of 3D materials is of great interest due to their fast and efficient removal (Li et al., 2016). He et al. (2020) fabricated a 3D hydrogel using Y-immobilized GO and alginate for tetracycline (TC) sorption. After converting GO into 3D material, separation from water is much facile and, therefore, is ideal for practical water treatment. Further, it showed a large TC sorption (477.9 mg/g) and good reusability without a regeneration.

Utilizing reduced graphene oxide (rGO) in composite formation is necessary as the highly hydrophilic structure of GO is undesirable for aromatic contaminant removal. Feng et al. (2017) prepared a porous rGO-sodium alginate (rGO-SA) hydrogel

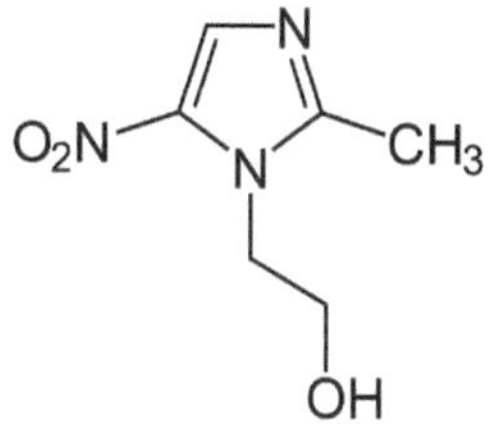

Fig. 5 Structure of metronidazole.

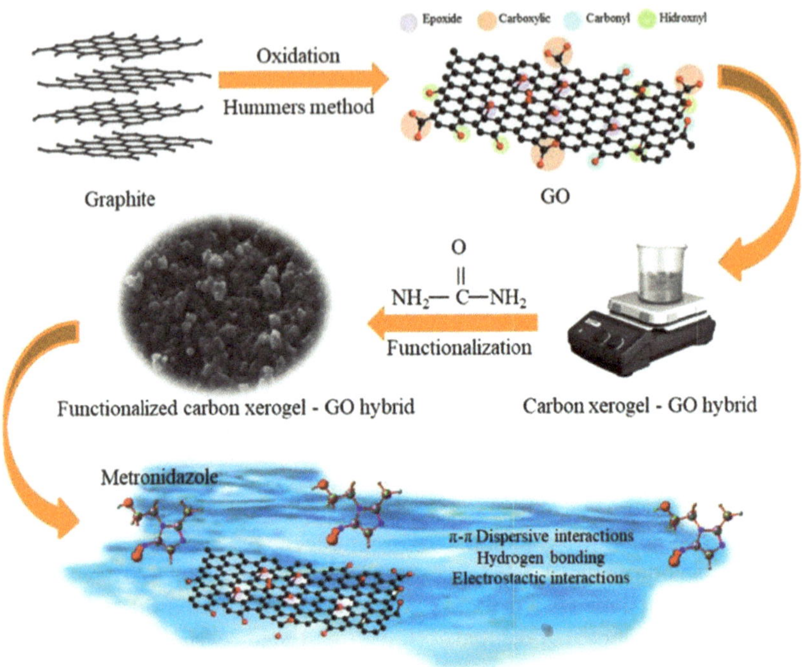

Fig. 6 Formation of functionalized carbon xerogel-GO hybrid from MNZ remediation. Reproduced with permission from Segovia-Sandoval, S. J., Pastrana-Martínez, L. M., Ocampo-Pérez, R., Morales-Torres, S., Berber-Mendoza, M. S., & Carrasco-Marín, F. (2020). Synthesis and characterization of carbon xerogel/graphene hybrids as adsorbents for metronidazole pharmaceutical removal: Effect of operating parameters. *Separation and Purification Technology 237*, 116341. https://doi.org/10.1016/j.seppur.2019.116341, Elsevier.

microspheres (Fig. 7) using external emulsification and thermal reduction method. This synthesis starts from graphite oxide and SA (template) is added followed by ultrasonication to obtain a GO-SA solution. This solution was mixed with $CaCl_2$, in which Ca^{2+} was then exchanged with Na^+ ions. Resulting GO-SA polymers show low dispersion ability, which is then reduced to form rGO-SA hydrogels. These hydrogels lack self-aggregation due to the incorporation of SA and were employed for phenol, BPA, and tetracycline sorption.

Magnetic adsorbents offer high efficiency and ease of separation developed a magnetic graphene oxide/diethylenetriaminepentaacetic acid (DDMGO) nanocomposite for TC and ciprofloxacin removal (Li et al., 2018). DDMGO has a highly oxygenated surface and shows a high regeneration potential as well. For the competitive adsorption on binary compounds, the inhibitory effect of TC was attributed to the simultaneous presence of resistant antibiotics other than CIP, which was more pronounced to the adsorption capacity and the affinity of a particular antibiotic for the adsorbent.

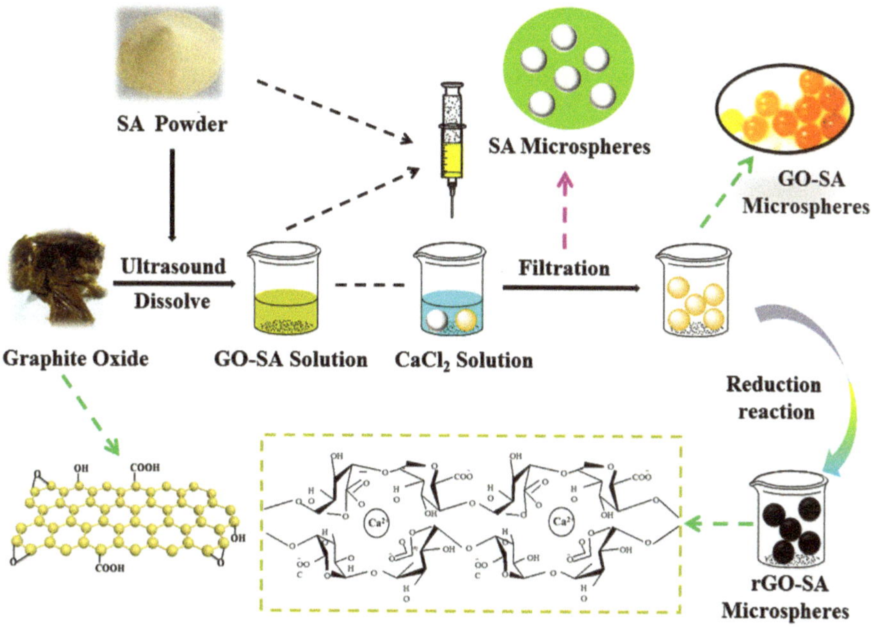

Fig. 7 Synthetic strategy of black-pearl reduced graphene oxide-sodium alginate hydrogel. Reproduced with permission from Feng, J., Ding, H., Yang, G., Wang, R., Li, S., Liao, J., Li, Z., & Chen, D. (2017). Preparation of black-pearl reduced graphene oxide–sodium alginate hydrogel microspheres for adsorbing organic pollutants. *Journal of Colloid and Interface Science 508*, 387–395. https://doi.org/10.1016/j.jcis.2017.07.113, Elsevier.

4. Knowledge gaps

This chapter highlighted graphene materials/composites and their potentials applications for the removal of a wide range of pollutant species from wastewater, including pharmaceuticals. Graphene-based material has been very well established effectively for the removal of a wide range of pollutant species such as inorganic heavy metals, organic contaminants, etc. This opens up a wide range of possibilities and plays a crucial role in hospital wastewater treatment. In the upcoming future, these graphene-based materials will be further developed and tailored for the specificity of hybrid treatment in urban wastewater treatment plants. The discovery of graphene-based materials has opened a new dimension for the production of lightweight, low-cost, and high-performance composite materials for a range of applications in hybrid treatment in urban wastewater treatment plants.

5. Conclusions

Graphene is ideal for organic contaminant remediation as it facilitates donor-acceptor interaction formation and Vander Waals bonds formation. However, the inertness and low aqueous dissolution of raw graphene's limits its practical applicability. Chemically modified graphene's contained improved surface properties (large number of surface functional groups, high electrical and mechanical conductivity, etc.) compared to raw graphene. GO is a derivative of graphene, rich in epoxides, carboxyl, carbonyl groups in its basal and edge planes ideal for adsorption and photocatalytic performance.

References

Afonso-Olivares, C., Torres-Padrón, M. E., Sosa-Ferrera, Z., & Santana-Rodríguez, J. J. (2013). Assessment of the presence of pharmaceutical compounds in seawater samples from coastal area of gran Canaria Island (Spain). *Antibiotics*, *2*, 274–287. https://doi.org/10.3390/antibiotics2020274.

Amiri, A., Naraghi, M., Ahmadi, G., Soleymaniha, M., & Shanbedi, M. (2018). A review on liquid-phase exfoliation for scalable production of pure graphene, wrinkled, crumpled and functionalized graphene and challenges. *FlatChem*, *8*, 40–71. https://doi.org/10.1016/j.flatc.2018.03.004.

Balandin, A. A., Ghosh, S., Bao, W., Calizo, I., Teweldebrhan, D., Miao, F., et al. (2008). Superior thermal conductivity of single-layer graphene. *Nano Letters*, *8*, 902–907. https://doi.org/10.1021/nl0731872.

Barber, L. B., Keefe, S. H., Antweiler, R. C., Taylor, H. E., & Wass, R. D. (2006). Accumulation of contaminants in fish from wastewater treatment wetlands. *Environmental Science & Technology*, *40*, 603–611. https://doi.org/10.1021/es0514287.

Bayati, M., & Fidalgo de Cortalezzi, M. (2019). Aggregation of graphene oxide in natural waters: Role of solution chemistry and specific interactions. *Journal of Environmental Engineering*, *145*, 04019050. https://doi.org/10.1061/(ASCE)EE.1943-7870.0001561.

Brodie, B. C. (1859). XIII. On the atomic weight of graphite. *Philosophical Transactions of the Royal Society of London*, *149*, 249–259. https://doi.org/10.1098/rstl.1859.0013.

Carlsson, C., Johansson, A.-K., Alvan, G., Bergman, K., & Kühler, T. (2006). Are pharmaceuticals potent environmental pollutants?: Part I: Environmental risk assessments of selected active pharmaceutical ingredients. *The Science of the Total Environment*, *364*, 67–87. https://doi.org/10.1016/j.scitotenv.2005.06.035.

Cecconet, D., Molognoni, D., Callegari, A., & Capodaglio, A. G. (2017). Biological combination processes for efficient removal of pharmaceutically active compounds from wastewater: A review and future perspectives. *Journal of Environmental Chemical Engineering*, *5*, 3590–3603. https://doi.org/10.1016/j.jece.2017.07.020.

Ciesielski, A., & Samorì, P. (2014). Graphene via sonication assisted liquid-phase exfoliation. *Chemical Society Reviews*, *43*, 381–398. https://doi.org/10.1039/C3CS60217F.

Deblonde, T., Cossu-Leguille, C., & Hartemann, P. (2011). Emerging pollutants in wastewater: A review of the literature. *International Journal of Hygiene and Environmental Health*, *214*, 442–448. https://doi.org/10.1016/j.ijheh.2011.08.002. The second European PhD students workshop: Water and health? Cannes 2010.

Ebele, A. J., Abou-Elwafa Abdallah, M., & Harrad, S. (2017). Pharmaceuticals and personal care products (PPCPs) in the freshwater aquatic environment. *Emerging Contaminants*, *3*, 1–16. https://doi.org/10.1016/j.emcon.2016.12.004.

Eda, G., & Chhowalla, M. (2010). Chemically derived graphene oxide: Towards large-area thin-film electronics and optoelectronics. *Advanced Materials*, *22*, 2392–2415. https://doi.org/10.1002/adma.200903689.

Feng, J., Ding, H., Yang, G., Wang, R., Li, S., Liao, J., et al. (2017). Preparation of black-pearl reduced graphene oxide–sodium alginate hydrogel microspheres for adsorbing organic pollutants. *Journal of Colloid and Interface Science*, *508*, 387–395. https://doi.org/10.1016/j.jcis.2017.07.113.

Finn, M., Giampietro, G., Mazyck, D., & Rodriguez, R. (2021). Activated carbon for pharmaceutical removal at point-of-entry. *Processes*, *9*, 1091. https://doi.org/10.3390/pr9071091.

Franklin, R. E. (1951). The structure of graphitic carbons. *Acta Crystallographica*, *4*, 253–261. https://doi.org/10.1107/S0365110X51000842.

Geim, A. K., & Novoselov, K. (2007). The rise of graphene. *Nature Materials*. https://doi.org/10.1038/NMAT1849.

Geissen, V., Mol, H., Klumpp, E., Umlauf, G., Nadal, M., van der Ploeg, M., et al. (2015). Emerging pollutants in the environment: A challenge for water resource management. *International Soil and Water Conservation Research*, *3*, 57–65. https://doi.org/10.1016/j.iswcr.2015.03.002.

Geng, Y., Wang, S. J., & Kim, J.-K. (2009). Preparation of graphite nanoplatelets and graphene sheets. *Journal of Colloid and Interface Science*, *336*, 592–598. https://doi.org/10.1016/j.jcis.2009.04.005.

He, J., & Fang, L. (2016). Controllable synthesis of reduced graphene oxide. *Current Applied Physics*, *16*, 1152–1158. https://doi.org/10.1016/j.cap.2016.06.011. Special Section on Nanostructure Physics and Materials Science at Center for Integrated Nanostructure Physics, Institute for Basic Science at Sungkyunkwan University.

He, J., Ni, F., Cui, A., Chen, X., Deng, S., Shen, F., et al. (2020). New insight into adsorption and co-adsorption of arsenic and tetracycline using a Y-immobilized graphene oxide-alginate hydrogel: Adsorption behaviours and mechanisms. *The Science of the Total Environment*, *701*, 134363. https://doi.org/10.1016/j.scitotenv.2019.134363.

Hernandez, Y., Nicolosi, V., Lotya, M., Blighe, F. M., Sun, Z., De, S., et al. (2008). High-yield production of graphene by liquid-phase exfoliation of graphite. *Nature Nanotechnology*, *3*, 563–568. https://doi.org/10.1038/nnano.2008.215.

Hu, M., & Mi, B. (2013). Enabling graphene oxide nanosheets as water separation membranes. *Environmental Science & Technology*, *47*, 3715–3723. https://doi.org/10.1021/es400571g.

Hummers, W. S., & Offeman, R. E. (1958). Preparation of graphitic oxide. *Journal of the American Chemical Society*, *80*, 1339. https://doi.org/10.1021/ja01539a017.

Ioniţă, M., Vlăsceanu, G. M., Watzlawek, A. A., Voicu, S. I., Burns, J. S., & Iovu, H. (2017). Graphene and functionalized graphene: Extraordinary prospects for nanobiocomposite materials. *Composites Part B: Engineering, Bioinspired Nanoengineered Materials*, *121*, 34–57. https://doi.org/10.1016/j.compositesb.2017.03.031.

Kalhori, E. M., Al-Musawi, T. J., Ghahramani, E., Kazemian, H., & Zarrabi, M. (2017). Enhancement of the adsorption capacity of the light-weight expanded clay aggregate surface for the metronidazole antibiotic by coating with MgO nanoparticles: Studies on the kinetic, isotherm, and effects of environmental parameters. *Chemosphere*, *175*, 8–20. https://doi.org/10.1016/j.chemosphere.2017.02.043.

Khan, N. A., Khan, S. U., Ahmed, S., Farooqi, I. H., Yousefi, M., Mohammadi, A. A., et al. (2020). Recent trends in disposal and treatment technologies of emerging-pollutants—A critical review. *Trends in Analytical Chemistry*, *122*, 115744. https://doi.org/10.1016/j.trac.2019.115744.

Khan, I., Saeed, K., & Khan, I. (2019). Nanoparticles: Properties, applications and toxicities. *Arabian Journal of Chemistry*, *12*, 908–931. https://doi.org/10.1016/j.arabjc.2017.05.011.

Kuilla, T., Bhadra, S., Yao, D., Kim, N. H., Bose, S., & Lee, J. H. (2010). Recent advances in graphene based polymer composites. *Progress in Polymer Science*, *35*, 1350–1375. https://doi.org/10.1016/j.progpolymsci.2010.07.005.

Lavin-Lopez, M. P., Paton-Carrero, A., Sanchez-Silva, L., Valverde, J. L., & Romero, A. (2017). Influence of the reduction strategy in the synthesis of reduced graphene oxide. *Advanced Powder Technology*, *28*, 3195–3203. https://doi.org/10.1016/j.apt.2017.09.032.

Lee, J.-U., Yoon, D., & Cheong, H. (2012). Estimation of Young's modulus of graphene by Raman spectroscopy. *Nano Letters*, *12*, 4444–4448. https://doi.org/10.1021/nl301073q.

Lekkas, P. T. (2013). *Pharmaceuticals and related compounds as emerging pollutants in water: Analytical aspects [WWW document]*. https://journal.gnest.org/publication/969. accessed 11.19.21.

Leung, H. W., Jin, L., Wei, S., Tsui, M. M. P., Zhou, B., Jiao, L., et al. (2013). Pharmaceuticals in tap water: Human health risk assessment and proposed monitoring framework in China. *Environmental Health Perspectives*, *121*, 839–846. https://doi.org/10.1289/ehp.1206244.

Li, Y., Cui, W., Liu, L., Zong, R., Yao, W., Liang, Y., et al. (2016). Removal of Cr(VI) by 3D TiO2-graphene hydrogel via adsorption enriched with photocatalytic reduction. *Applied Catalysis B: Environmental*, *199*, 412–423. https://doi.org/10.1016/j.apcatb.2016.06.053.

Li, M., Liu, Y., Liu, S., Zeng, G., Hu, X., Tan, X., et al. (2018). Performance of magnetic graphene oxide/diethylenetriaminepentaacetic acid nanocomposite for the tetracycline and ciprofloxacin adsorption in single and binary systems. *Journal of Colloid and Interface Science*, *521*, 150–159. https://doi.org/10.1016/j.jcis.2018.03.003.

Li, M., Liu, Y., Zeng, G., Liu, N., & Liu, S. (2019). Graphene and graphene-based nanocomposites used for antibiotics removal in water treatment: A review. *Chemosphere*, *226*, 360–380. https://doi.org/10.1016/j.chemosphere.2019.03.117.

Lin, Y., Xu, S., & Li, J. (2013). Fast and highly efficient tetracyclines removal from environmental waters by graphene oxide functionalized magnetic particles. *Chemical Engineering Journal*, *225*, 679–685. https://doi.org/10.1016/j.cej.2013.03.104.

Moreno Ríos, A. L., Gutierrez-Suarez, K., Carmona, Z., Ramos, C. G., & Silva Oliveira, L. F. (2021). Pharmaceuticals as emerging pollutants: Case naproxen an overview. *Chemosphere*, 132822. https://doi.org/10.1016/j.chemosphere.2021.132822.

Nikolaou, A., Meric, S., & Fatta, D. (2007). Occurrence patterns of pharmaceuticals in water and wastewater environments. *Analytical and Bioanalytical Chemistry*, *387*, 1225–1234. https://doi.org/10.1007/s00216-006-1035-8.

Novoselov, K. S., Geim, A. K., Morozov, S. V., Jiang, D., Katsnelson, M. I., Grigorieva, I. V., et al. (2005). Two-dimensional gas of massless Dirac fermions in graphene. *Nature*, *438*, 197–200. https://doi.org/10.1038/nature04233.

Novoselov, K. S., Geim, A. K., Morozov, S. V., Jiang, D., Zhang, Y., Dubonos, S. V., et al. (2004). Electric field effect in atomically thin carbon films. *Science*, *306*, 666–669. https://doi.org/10.1126/science.1102896.

Park, C. M., Kim, Y. M., Kim, K.-H., Wang, D., Su, C., & Yoon, Y. (2019). Potential utility of graphene-based nano spinel ferrites as adsorbent and photocatalyst for removing organic/inorganic contaminants from aqueous solutions: A mini review. *Chemosphere*, *221*, 392–402. https://doi.org/10.1016/j.chemosphere.2019.01.063.

Peigney, A., Laurent, C., Flahaut, E., Bacsa, R. R., & Rousset, A. (2001). Specific surface area of carbon nanotubes and bundles of carbon nanotubes. *Carbon*, *39*, 507–514. https://doi.org/10.1016/S0008-6223(00)00155-X.

Sabzevari, M., Cree, D. E., & Wilson, L. D. (2018). Graphene oxide–chitosan composite material for treatment of a model dye effluent. *ACS Omega*, *3*, 13045–13054. https://doi.org/10.1021/acsomega.8b01871.

Samaraweera, H., Pittman, C. U., Thirumalai, R. V. K. G., Hassan, E. B., Perez, F., & Mlsna, T. (2021). Characterization of graphene/pine wood biochar hybrids: Potential to remove aqueous Cu2+. *Environmental Research*, *192*, 110283. https://doi.org/10.1016/j.envres.2020.110283.

Samaraweera, H., Sharp, A., Edwards, J., Pittman, C. U., Jr., Zhang, X., Hassan, E. B., et al. (2021). Lignite, thermally-modified and Ca/Mg-modified lignite for phosphate remediation. *The Science of the Total Environment*, *145631*. https://doi.org/10.1016/j.scitotenv.2021.145631.

Schedin, F., Geim, A. K., Morozov, S. V., Hill, E. W., Blake, P., Katsnelson, M. I., & Novoselov, K. S. (2007). Detection of individual gas molecules adsorbed on graphene. *Nature Materials*, *6*, 652–655. https://doi.org/10.1038/nmat1967.

Segovia-Sandoval, S. J., Pastrana-Martínez, L. M., Ocampo-Pérez, R., Morales-Torres, S., Berber-Mendoza, M. S., & Carrasco-Marín, F. (2020). Synthesis and characterization of carbon xerogel/graphene hybrids as adsorbents for metronidazole pharmaceutical removal: Effect of operating parameters. *Separation and Purification Technology*, *237*, 116341. https://doi.org/10.1016/j.seppur.2019.116341.

Shin, W.-K., Cho, J., Kannan, A. G., Lee, Y.-S., & Kim, D.-W. (2016). Cross-linked composite gel polymer electrolyte using mesoporous methacrylate-functionalized SiO2 nanoparticles for Lithium-ion polymer batteries. *Scientific Reports*, *6*, 26332. https://doi.org/10.1038/srep26332.

Spyriouni, T., Krokidis, X., & Economou, I. G. (2011). Thermodynamics of pharmaceuticals: Prediction of solubility in pure and mixed solvents with PC-SAFT. *Fluid Phase Equilibria*, *302*, 331–337. https://doi.org/10.1016/j.fluid.2010.08.029.

Stankovich, S., Dikin, D. A., Piner, R. D., Kohlhaas, K. A., Kleinhammes, A., Jia, Y., et al. (2007). Synthesis of graphene-based nanosheets via chemical reduction of exfoliated graphite oxide. *Carbon*, *45*, 1558–1565. https://doi.org/10.1016/j.carbon.2007.02.034.

Tan, X., Liu, Y., Gu, Y., Xu, Y., Zeng, G., Hu, X., et al. (2016). Biochar-based nano-composites for the decontamination of wastewater: A review. *Bioresource Technology*, *212*, 318–333. https://doi.org/10.1016/j.biortech.2016.04.093.

Tenorio-Chávez, P., Cerro-López, M., Castro-Pastrana, L. I., Ramírez-Rodrigues, M. M., Orozco-Hernández, J. M., & Gómez-Oliván, L. M. (2020). Effects of effluent from a hospital in Mexico on the embryonic development of zebrafish, *Danio rerio*. *Science of the Total Environment*, *727*, 138716. https://doi.org/10.1016/j.scitotenv.2020.138716.

Vasilachi, I. C., Asiminicesei, D. M., Fertu, D. I., & Gavrilescu, M. (2021). Occurrence and fate of emerging pollutants in water environment and options for their removal. *Water*, *13*, 181. https://doi.org/10.3390/w13020181.

Vasseghian, Y., Dragoi, E.-N., Almomani, F., & Le, V. T. (2022). Graphene-based materials for metronidazole degradation: A comprehensive review. *Chemosphere*, *286*, 131727. https://doi.org/10.1016/j.chemosphere.2021.131727.

WTEC. (2007). *Final Reports [WWW Document]*. Baltim: MD World Technol. Eval. Cent. WTEC. http://scienceus.org/wtec/ (accessed 11.23.21).

Wu, L., Liu, L., Gao, B., Muñoz-Carpena, R., Zhang, M., Chen, H., et al. (2013). Aggregation kinetics of graphene oxides in aqueous solutions: Experiments, mechanisms, and modeling. *Langmuir*, *29*, 15174–15181. https://doi.org/10.1021/la404134x.

Wu, J., Zhao, H., Chen, R., Pham-Huy, C., Hui, X., & He, H. (2016). Adsorptive removal of trace sulfonamide antibiotics by water-dispersible magnetic reduced graphene oxide-ferrite hybrids from wastewater. *Journal of Chromatography B*, *1029–1030*, 106–112. https://doi.org/10.1016/j.jchromb.2016.07.018.

Yang, K., Wang, J., & Chen, B. (2014). Facile fabrication of stable monolayer and few-layer graphene nanosheets as superior sorbents for persistent aromatic pollutant management in water. *Journal of Materials Chemistry A*, *2*, 18219–18224. https://doi.org/10.1039/C4TA04300F.

Zhang, Y., Cheng, Y., Chen, N., Zhou, Y., Li, B., Gu, W., et al. (2014). Recyclable removal of bisphenol A from aqueous solution by reduced graphene oxide–magnetic nanoparticles: Adsorption and desorption. *Journal of Colloid and Interface Science*, *421*, 85–92. https://doi.org/10.1016/j.jcis.2014.01.022.

Zhang, Y., Tan, Y.-W., Stormer, H. L., & Kim, P. (2005). Experimental observation of the quantum hall effect and Berry's phase in graphene. *Nature*, *438*, 201–204. https://doi.org/10.1038/nature04235.

Zhu, Y., Murali, S., Cai, W., Li, X., Suk, J. W., Potts, J. R., et al. (2010). Graphene and graphene oxide: Synthesis, properties, and applications. *Advanced Materials*, *22*, 3906–3924. https://doi.org/10.1002/adma.201001068.

Fate of common pharmaceuticals in the environment

4

Omar Fawzi Suleiman Khasawneh, Puganeshwary Palaniandy, and Hamidi Abdul Aziz
School of Civil Engineering, Engineering Campus, Universiti Sains Malaysia, Nibong Tebal, Pulau Pinang, Malaysia

1. Introduction

Over the last decades, pharmaceutical compounds (PhACs) have been recognized as a new class of micropollutants and/or organic pollutants. Pharmaceuticals are considered as contaminants of emerging concern (CEC) because of their physiochemical properties, rendering them persistent in the environment and potentially toxic to the aquatics and human health. PhACs can be defined as chemical substances that are used for the treatment, prevention, and diagnosis of illnesses in humans and fauna (Khasawneh & Palaniandy, 2020; Lee, Palaniandy, & Dahlan, 2017; Yang, Ok, Kim, Kwon, & Tsang, 2017). Pharmaceuticals are not limited to prescription drugs (medicines) (i.e., diabetic, lipid regulators, etc.), they can as well include over-the-counter drugs (i.e., aspirin, acetaminophen, ibuprofen, and naproxen), and veterinary drugs. In addition, they involve personal care products (PCPs) that are mostly used to improve the standard of everyday life, such as the active ingredients present in bleaches, soaps, deodorants, shampoos, detergents, dyes, musk, dental and skin care products are also considered pharmaceutically active compounds (PACs). PhACs consist of chemical compounds that are biologically energetic, hormone disruptive, and potentially toxic. They are also designed to be bioavailable and to have a biological effect. Fig. 1 shows that possible pathways and fate of common PhACs in the environment.

The topic of PhACs occurrence and fate in the environment has gained significant attention in the recent years due to the potential eco-toxicological effects (K'oreje et al., 2018; Niemuth & Klaper, 2015; Parrott & Metcalfe, 2019) and frequent detection of these micro-organic pollutants in different environmental compartments (i.e., surface water, groundwater, drinking water wastewater treatment plants, and soil) (Gao, Ding, Li, & Xagoraraki, 2012; K'oreje et al., 2018; Tran, Reinhard, & Gin, 2018; Yang et al., 2017). The ever-growing worldwide population and the emergence of new diseases rendered PhACs vital to improve the life of humans and fauna. Meanwhile, the incomplete ingestion of pharmaceuticals in the human/animal body and inadequate/limited removal after excretion, PhACs have developed as serious environmental contaminants in the last decades. Over the past decades, more than 100 PhACs have been detected and identified in rivers (Gopal et al., 2021; Khan et al., 2021), wastewater treatment plants (Bartelt-Hunt, Snow, Damon, Shockley, &

The Treatment of Pharmaceutical Wastewater. https://doi.org/10.1016/B978-0-323-99160-5.00011-4

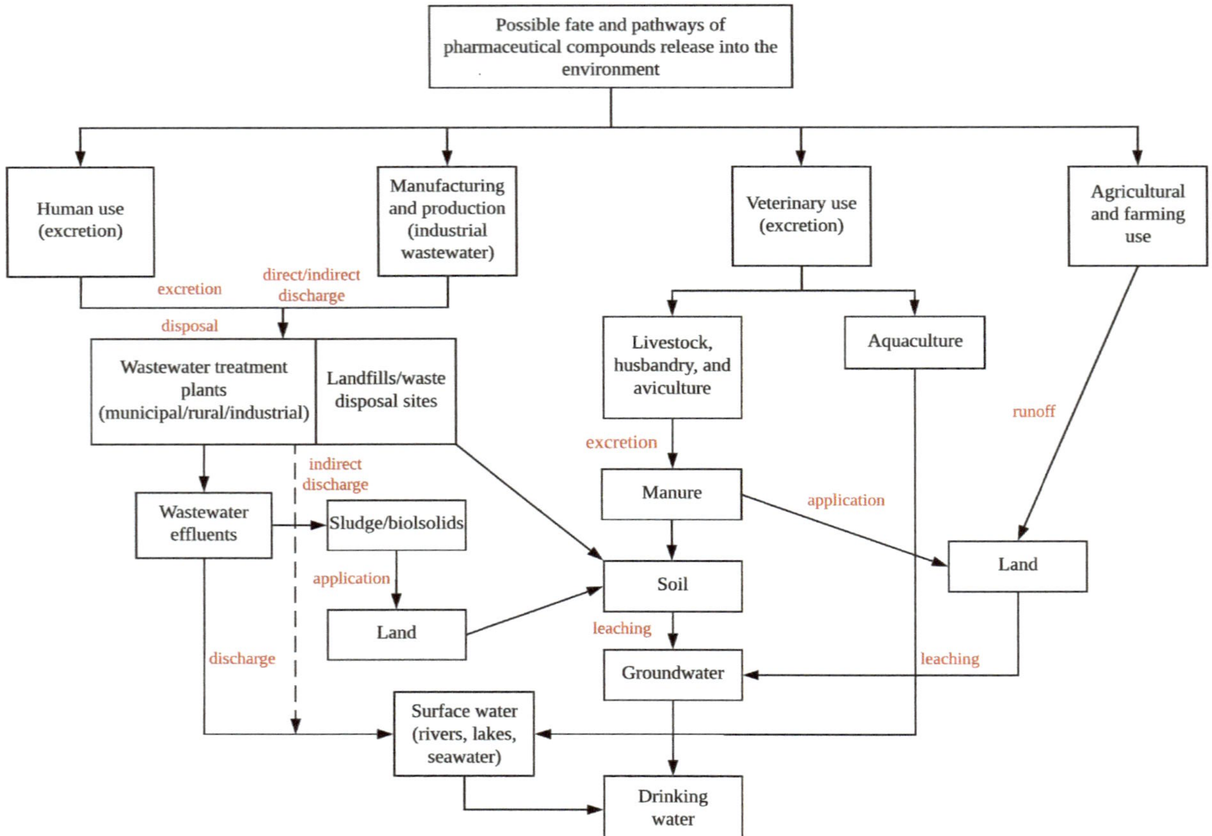

Fig. 1 Possible fate and pathways of PhACs release into the environment.

Hoagland, 2009; D'Alessio, Onanong, Snow, & Ray, 2018), seawaters (Biel-Maeso, Baena-Nogueras, Corada-Fernández, & Lara-Martín, 2018; Zhang et al., 2021), groundwater (Ebele, Oluseyi, Drage, Harrad, & Abou-Elwafa Abdallah, 2020; K'oreje et al., 2016), and soil (Malvar, Santos, Martín, Aparicio, & Alonso, 2021; Mejías, Martín, Santos, Aparicio, & Alonso, 2021).

Numerous and different PhACs, such as synthetic hormones (e.g., estrone), nonsteroidal antiinflammatories (e.g., ibuprofen, acetaminophen), antibiotics (e.g., sulfamethoxazole, ciprofloxacin, tetracycline), stimulants (e.g., caffeine), lipid regulators (e.g., bezafibrate, atenolol), insect repellents (e.g., DEET), and UV blockers (e.g., methyl benzylidene camphor) are constantly produced and consumed in large quantities worldwide, with annual amounts reaching thousands of tons (Khasawneh & Palaniandy, 2019; Quesada et al., 2019). Moreover, it has been reported that more than 3000 different chemicals/substances are being used as ingredients for the manufacturing of active PhACs (Rodriguez-Narvaez, Peralta-Hernandez, Goonetilleke, & Bandala, 2017). Several studies have reported the occurrence of different PhACs in the environment. For instance, the highest concentrations of the following pharmaceuticals: ibuprofen (223.4 ng/L), carbamazepine (183.1 ng/L), diazepam (457.8 ng/L), erythromycin (178.1 ng/L), furosemide (312.2 ng/L), and simvastatin (414.9 ng/L) were detected at the Yamuna River in New Delhi, India (Khan et al., 2021). Since WWTPs are considered the main source of pharmaceuticals release into the environment (Khasawneh & Palaniandy, 2021). Therefore, previous research have also investigated the occurrence of PhACs in the influents and effluents of WWTPs in different regions of the world. For example, several pharmaceuticals belonging to different classes, such caffeine, DEET, methamphetamine, and carbamazepine have detected in the effluent of a WWTP located in Omaha, Nebraska, USA, with mean concentrations of 535.3, 773.9, 350.1, and 119.5 ng/L, respectively (Bartelt-Hunt et al., 2009). Meanwhile, some psychoactive pharmaceuticals, such as diazepam, lorazepam, oxazepam, clopidogrel, and diphenhydramine have been detected in the influents and effluents of two WWTPs in the Albany area, New York, USA, with concentrations ranging from 1.28 to 4.75, 1.8 to 114, 2.3 to 16.9, 7.39 to 92.8, and 14.4 to 764 ng/L, respectively (Subedi & Kannan, 2015).

2. Common PhACs in the environment and their physiochemical properties

PhACs as a class of emerging contaminants, they generally possess dissimilar characteristics that distinctive them from other traditional pollutants (i.e., industrial chemical pollutants). Therefore, it is imperative to take these characteristics into consideration when evaluating their environmental fate in the environment. The following section will discuss the unique characteristics of PhACs that make them different than conventional environmental contaminants. In addition, the physiochemical properties of the common detected PhACs, belonging to different classes (i.e., antibiotics, NSAIDs, β-blocker) is summarized. Some of the exceptional characteristics of PhACs can be summarized as follows:

- The parent PhACs and its related salts have a proclivity for forming polymorphic solid states.
- Metabolism: PhACs and their metabolites are administered into the environment after human digestion.
- The chemical structure of PhACs is generally complex and large.
- Multiple ionization sites are distributed throughout the molecule, making it generally ionizable.

Previously mentioned characteristics of the PhACs shall be considered whenever investigating their fate, interpreting analysis results, establishing, and conducting environmental risk assessments, and modeling and predicting environmental fate.

2.1 Antibiotics

Antibiotics are considered one of the most significant groups of PhACs, and they have been extensively used in both veterinary and human medicine. Antibiotics refer to any synthetic, semisynthetic, or natural compounds that have selective toxic properties that allow them to kill or inhibit the growth of metabolic activity of microorganisms, including bacteria, fungi, and/or viruses. Antibiotics are biologically active molecules with antifungal, antibacterial, and antiparasitic activities intentionally designed as medicines to treat the bacterial infections in human and fauna. The first ever invented antibiotic was derived from natural origins, i.e., penicillin obtained from *Penicillium fungi*. However, nowadays, antibiotics are manufactured by the chemical modification of natural compounds or by chemical synthesis (Kovalakova et al., 2020; Monteiro & Boxall, 2010). Antibiotics can be classified into different groups, either by their mode (mechanism) of action, including bacteriostatic (inhibiting the growth of bacteria) and bactericidal (killing the bacteria), or by their chemical structures, such as sulfonamides, penicillins, tetracyclines, macrolides, β-lactams, and fluoroquinolones, as illustrated in Table 1 (Kovalakova et al., 2020; Ying, Zhao, Zhou, & Liu, 2013). It has been reported that the annual total global production of antibiotics reached 100,000 tons, with more than 250 different antibiotic drugs (Burch, Han, Pichtel, & Zubkov, 2019; Kumar, Lee, & Cho, 2012).

The global consumption of antibiotics have increased significantly in the past decades, and it is estimated that the consumption of antibiotics will raise by 200% by the end of the year 2030, in case if there has been no change in the policies regarding their consumption and removal of from WWTPs (Burch et al., 2019). Several reasons contributed to the increase in the global consumption of antibiotics, such as (i) increased prosperity, (ii) increased human population raised the worldwide use of antibiotics, (iii) the easy access to medicines increased the utilization of antibiotics, and (iv) increased usage of antibiotics and growth promoters for animals due the surging demand of animal protein (Kovalakova et al., 2020).

The occurrence of antibiotics in the aquatic environment threatens the public health due to their chronic toxicity and hazardous effects on the aquatics. Their occurrence has gained significant attention among the scientific communities worldwide due to the potential development of antibiotics resistance genes (*ARGs*) and antibiotic resistance bacteria (ARB), ranking them as priority risk groups (Gao et al., 2012;

Table 1 Physiochemical properties of common antibiotics.

Class/chemical structure	Compound/side chain or chemical structure	Formula	Water solubility (mg/L)	log *P*	*p*K$_a$	CAS no.
Sulfonamides	Sulfamethoxazole	$C_{10}H_{11}N_3O_3S$	610	0.89	–	723-46-6
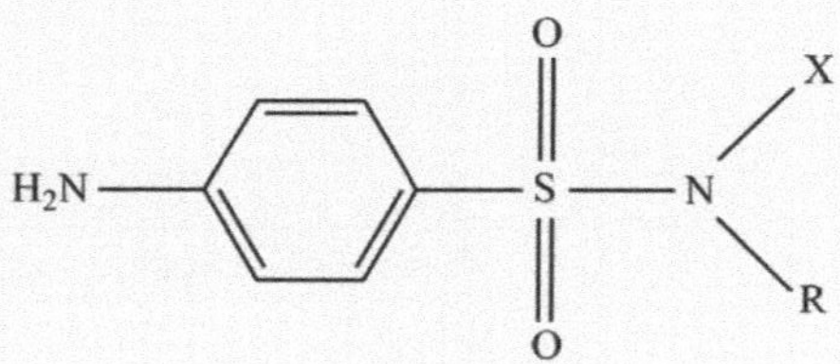	Sulfapyridine	$C_{11}H_{11}N_3O_2S$	270	0.35	2.58, 8.43, 8.4	144-83-2
	Sulfadimethoxine	$C_{12}H_{14}N_4O_4S$	343	1.63	6.3, 2.13, 6.08	122-11-2
	Sulfadoxine	$C_{12}H_{14}N_4O_4S$	2700	0.7	3.15, 6.16	2447-57-6
	Sulfameter	$C_{11}H_{12}N_4O_3S$	730	0.41	1.48, 6.49	651-06-9
	Sulfadiazine	$C_{10}H_{10}N_4O_2S$	77	−0.09	1.6, 6.4	68-35-9
	Sulfaguanidine	$C_7H_{10}N_4O_2S$	2200	−1.22	11.3	57-67-0
Penicillins	Amoxicillin	$C_{16}H_{19}N_3O_5S$	3430 (at 25°C, est)	0.87	–	26787-78-0
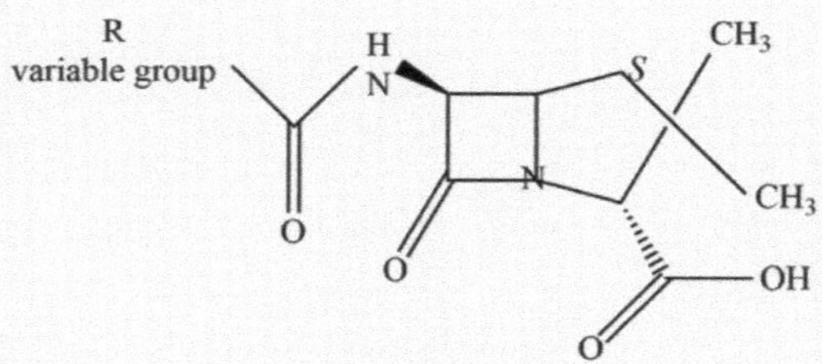	Penicillin G	$C_{16}H_{18}N_2O_4S$	210	1.83	2.74	61-33-6
	Penicillin V	$C_{16}H_{18}N_2O_5S$	101	2.09	2.79	101

Continued

Table 1 Continued

Class/chemical structure	Compound/side chain or chemical structure	Formula	Water solubility (mg/L)	log *P*	*p*K$_a$	CAS no.
Tetracyclines 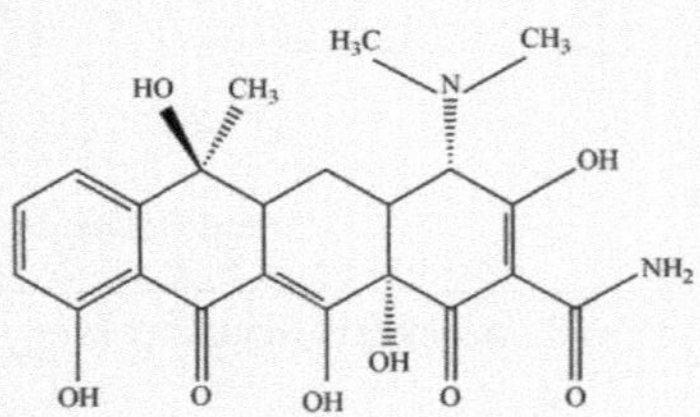	Methacycline	$C_{22}H_{22}N_2O_8$	7550	−1.37	4.05, 6.87, 9.59	914-00-1
	Tetracycline	$C_{22}H_{24}N_2O_8$	1700	−1.19	3.32, 7.78, 9.58	60-54-8
	Chlortetracycline	$C_{22}H_{23}ClN_2O_8$	630	−0.62	3.33, 7.55, 9.33	57-62-5
	Oxytetracycline	$C_{22}H_{24}N_2O_8$	1000	−1.22	3.22, 7.46, 8.94	79-57-2
	Doxycycline	$C_{22}H_{24}N_2O_8$	630	−0.02	3.02, 7.97, 9.15	564-25-0

Macrolides 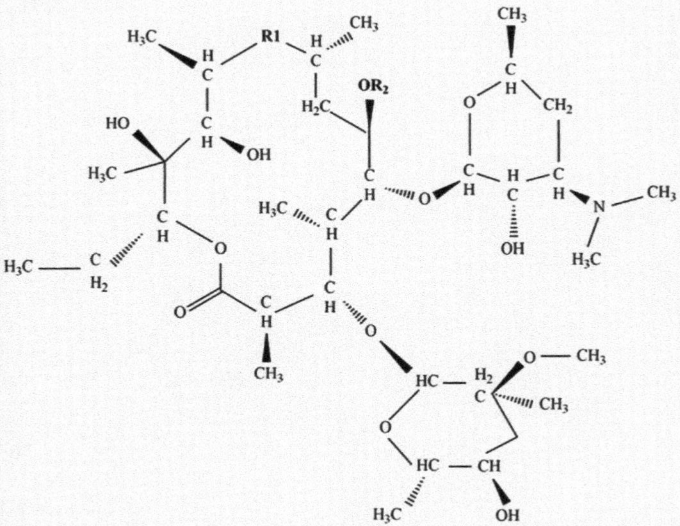	Leucomycin Erythromycin Clarithromycin Tylosin Azithromycin Roxithromycin	$C_{39}H_{65}NO_{14}$ $C_{37}H_{67}NO_{13}$ $C_{38}H_{69}NO_{13}$ $C_{46}H_{77}NO_{17}$ $C_{38}H_{72}N_2O_{12}$ $C_{41}H_{76}N_2O_{15}$	– 2000 0.342 5 7.09 870	– 3.06 3.16 1.63 4.02 2.75	– 8.9 8.99 7.73 8.74 9.17	1392-21-8 114-07-8 81103-11-9 1401-69-0 83905-01-5 80214-83-1

Continued

Table 1 Continued

Class/chemical structure	Compound/side chain or chemical structure	Formula	Water solubility (mg/L)	log*P*	*p*K$_a$	CAS no.
Lincosamides	Lincomycin	$C_{18}H_{34}N_2O_6S$	~927 at 25°C	0.56	7.97, 12.37	154-21-2
β-Lactams 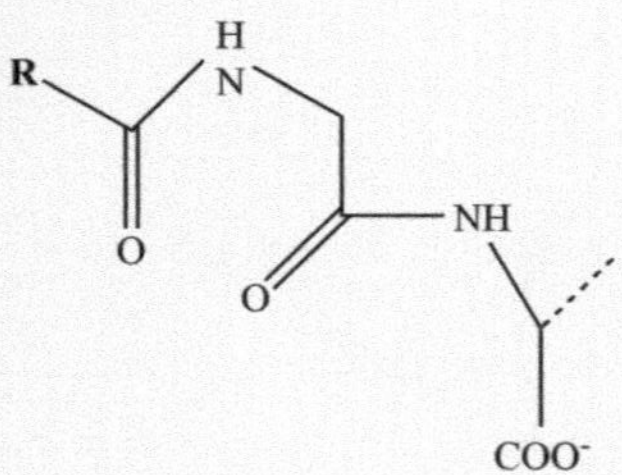	Cloxacillin	$C_{19}H_{18}ClN_3O_5S$	53.2 (drugbank)	2.48	−0.41, 3.75	61-72-3
	Ceftiofur	$C_{19}H_{17}N_5O_7S_3$	105 (drugbank)	1.22	2.83, 4.19	80370-57-6

Fluoroquinolones 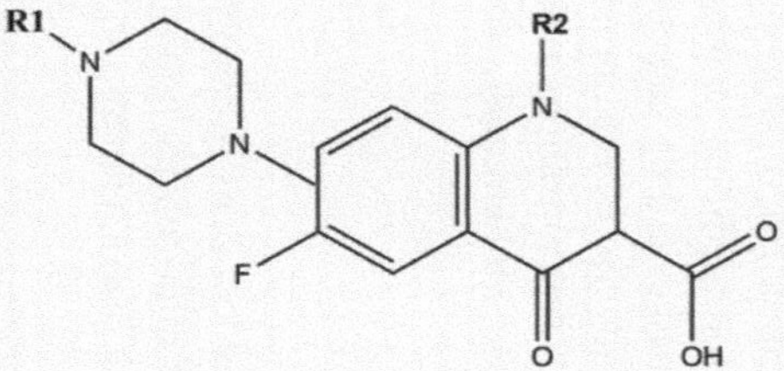	Difloxacin	$C_{21}H_{19}F_2N_3O_3$	1330	0.89	5.64, 6.45	98106-17-3
	Norfloxacin	$C_{16}H_{18}FN_3O_3$	17,800	−1.03	3.11, 6.10, 8.6, 10.56	70458-96-7
	Fleroxacin	$C_{17}H_{18}F_3N_3O_3$	7320	0.24	5.44, 6.06	79660-72-3
	Enrofloxacin	$C_{19}H_{22}FN_3O_3$	1,30,000	1.1	3.85, 6.19, 7.59, 9.86	93106-60-6
	Ciprofloxacin	$C_{17}H_{18}FN_3O_3$	30,000	0.4	3.01, 6.14, 8.70, 10.58	85721-33-1
	Ofloxacin	$C_{18}H_{20}FN_3O_4$	2830	0.36	5.97, 8.28	82419-36-1
	Sarafloxacin	$C_{20}H_{17}F_2N_3O_3$	100	1.07	6.0, 8.6	98105-99-8

Data acquired from drugbank.com.

Khasawneh & Palaniandy, 2021; Mirzaei et al., 2018) and designated as "emerging environmental contaminants" by the US Environmental Protection Agency (Burch et al., 2019). The developed *ARGs* and ARB can pose potential major negative effects in human health by promoting resistant bacteria and altering the human microbiome (Khasawneh & Palaniandy, 2021). *ARGs* and ARB have been frequently detected in several WWTPs worldwide (Pazda, Kumirska, Stepnowski, & Mulkiewicz, 2019). Due to the continuous release of antibiotics and permanent/persistent presence in the aquatics and WWTPs, they are regarded as "pseudo persistent" pollutants (Mirzaei et al., 2018). Therefore, it is imperative to investigate the fate of antibiotics in the different compartments of the environment (i.e., WWTPs, surface water, groundwater, and soil), which will be discussed further in Section 6.

2.2 Nonsteroidal antiinflammatory drugs

Nonsteroidal antiinflammatory drugs (NSAIDs) are chemical compounds which are utilized as medications to reduce fever, relieve pain, treatment and suppression of inflammation in a way similar to steroids, but without their unwanted side effects, and they are considered as "cure-all-pains" medications, such as toothache, cold symptoms, headache, and labor pain (Madikizela & Ncube, 2021; Ying et al., 2013). NSAIDs are regarded as one of the most therapeutic class of pharmaceuticals detected in the ecosystems (Parolini, 2020). These chemicals act by hindering the release and synthesis of prostaglandins from arachidonic acid, through working as "nonselective inhibitors" of the COX (cyclooxygenase) enzymes, including the both isoforms of cyclooxygenase; COX-1 and COX-2 (Gierse et al., 1995).

NSAIDs are deemed as acidic compounds with various hydrophobicities, and since their molecular structures are combined with acid group, they present high hydrophilic property. Meanwhile, their pK_a (experimental/predicted), $\log P$, and water solubility values vary between different compounds, as shown in Table 2. NSAIDs are commonly prescribed medicines or commercialized over-the-counter in pharmacies or stores worldwide. It has been reported that with considering the over-the-counter consumption of NSAIDs, more than 30 billion dosages of NSAIDs are consumed annually in the United States alone (Parolini, 2020). Among the NSAIDs, acetaminophen is considered the most widely used medicine, and one of the top 200 drugs prescribed worldwide. Moreover, the analgesic drug, acetaminophen has attracted significant attention due to its potential adverse effects, persistent in the environment, and frequent detection in WWTPs with concentrations reaching up to 9822 ng/L (Khasawneh, Palaniandy, Ahmadipour, Mohammadi, & Bin Hamdan, 2020).

2.3 Lipid regulators

Lipid regulators are PhACs that are used to reduce the levels of low-density lipoproteins (LDL) and triglycerides, as well as to increase the levels of high-density lipoproteins (HDL) in the blood. Such compounds are utilized by patients at risk of heart attack. Lipid

Table 2 Physiochemical properties of common NSAIDs.

Compound	Formula	Chemical structure	Solubility in water (mg/L)	log*P*	*p*K$_a$	CAS no.
Acetaminophen	$C_8H_9NO_2$	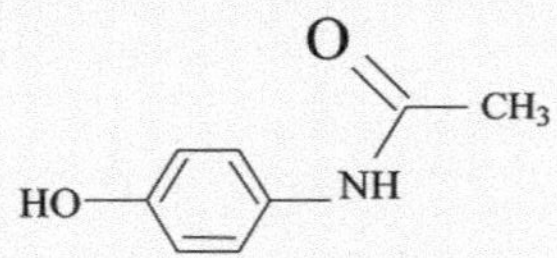	14,000 (exp) 4150 (pred)	0.91 (exp) 0.41, 0.91 (pred)	−4.4 (exp) 9.46 (pred)	103-90-2
Ibuprofen	$C_{13}H_{18}O_2$	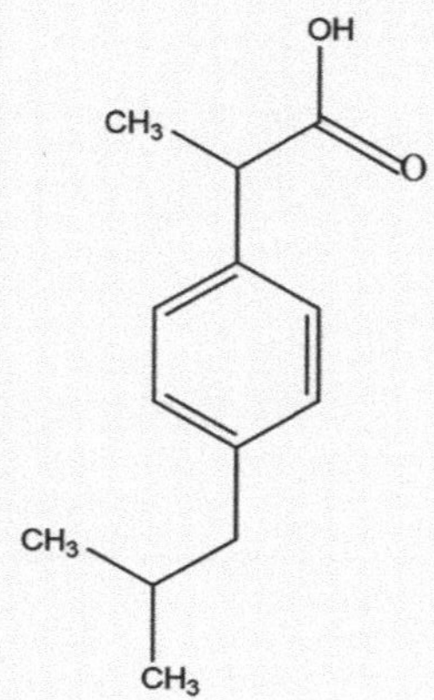	21 (exp)	3.97 (exp)	5.3 (exp)	15687-27-1

Continued

Table 2 Continued

Compound	Formula	Chemical structure	Solubility in water (mg/L)	log*P*	*p*K_a	CAS no.
Fenoprofen	$C_{15}H_{14}O_3$	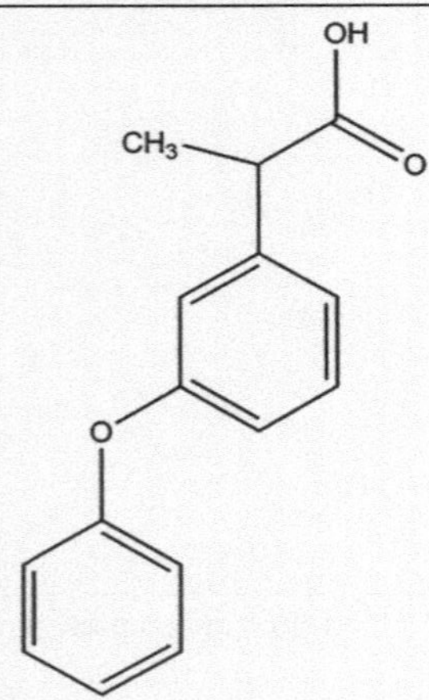	81.1 (est)	3.1 (exp)	4.5 (exp)	29679-58-1

Ketoprofen	$C_{16}H_{14}O_3$	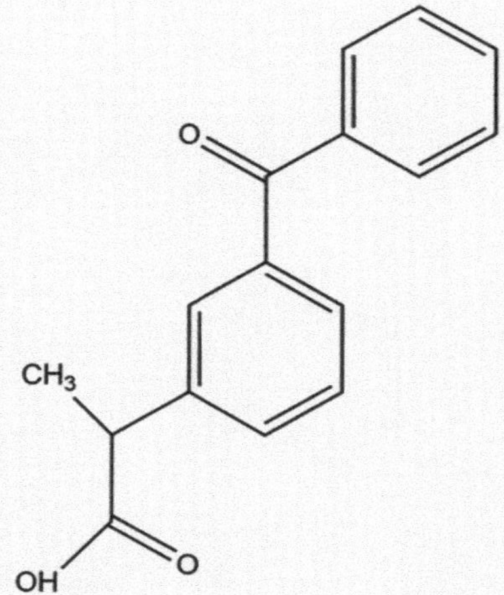	51 (exp)	3.12 (exp)	4.45 (exp)	22071-15-4
Naproxen	$C_{14}H_{14}O_3$	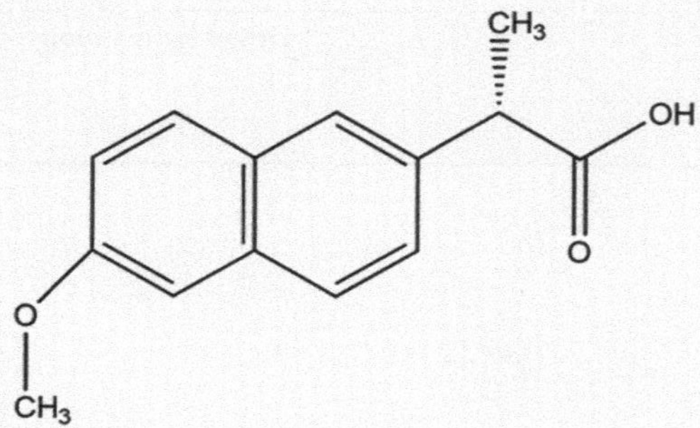	15.9 (exp)	3.18 (exp)	4.15 (exp)	22204-53-1

Continued

Table 2 Continued

Compound	Formula	Chemical structure	Solubility in water (mg/L)	log*P*	*p*K_a	CAS no.
Diclofenac	$C_{14}H_{11}Cl_2NO_2$	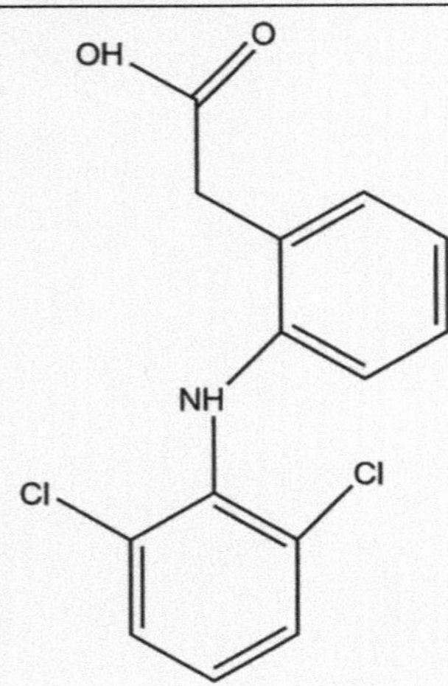	2.37 (exp)	4.51 (exp)	5.15 (exp)	15307-86-5

Meclofenamic acid	$C_{14}H_{11}Cl_2NO_2$	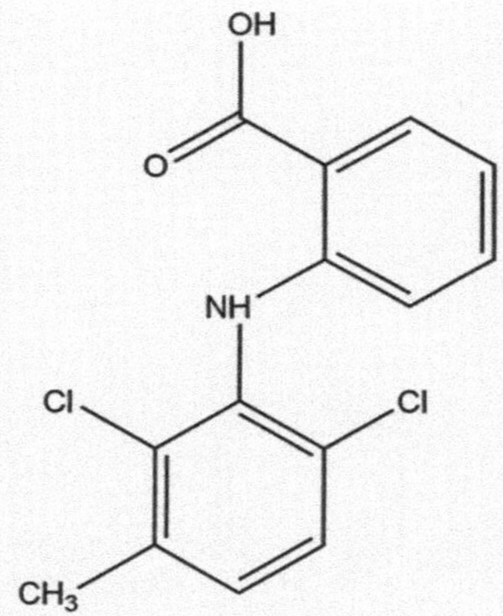	30 (exp)	5 (exp)	3.79 (est)	644-62-2
Mefenamic acid	$C_{15}H_{15}NO_2$	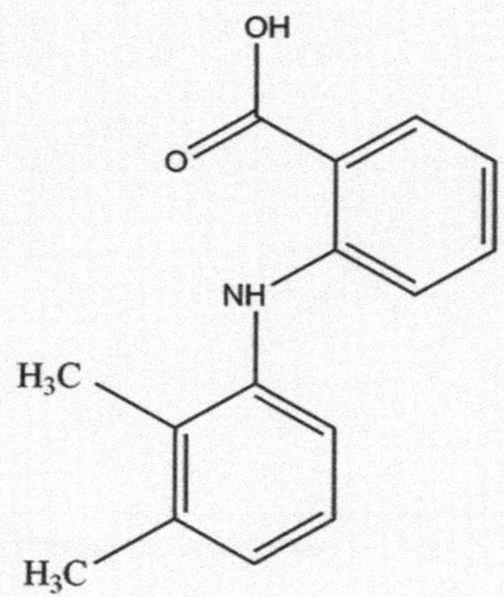	20 (exp)	5.12 (exp)	4.2 (exp)	61-68-7

Continued

Table 2 Continued

Compound	Formula	Chemical structure	Solubility in water (mg/L)	log *P*	*p*K$_a$	CAS no.
Acetylsalicylic acid	$C_9H_8O_4$	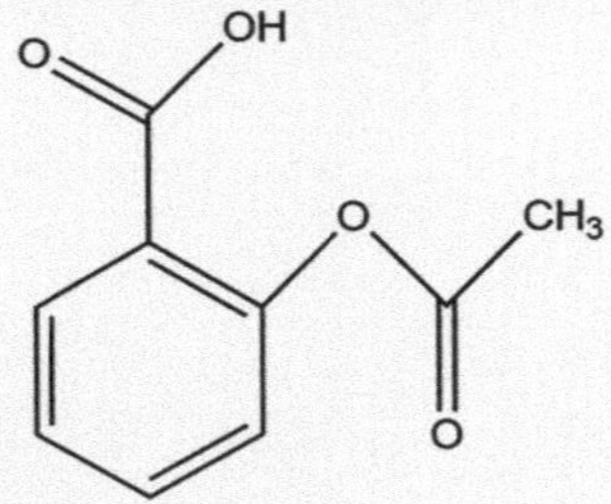	3000 (exp)	1.18 (exp)	3.5 (exp)	50-78-2
Salicylic acid	$C_7H_6O_3$	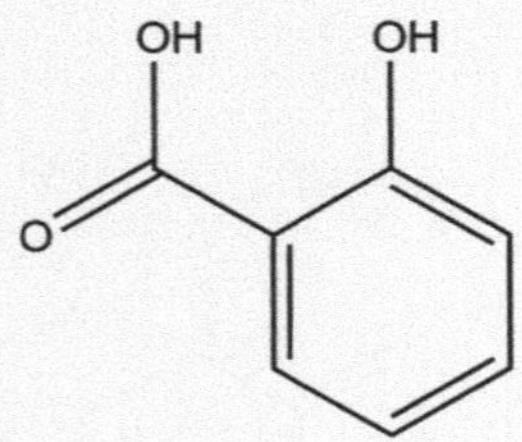	2240 (exp)	2.26 (exp)	2.97 (exp)	69-72-7

Tolfenamic acid	$C_{14}H_{12}ClNO_2$	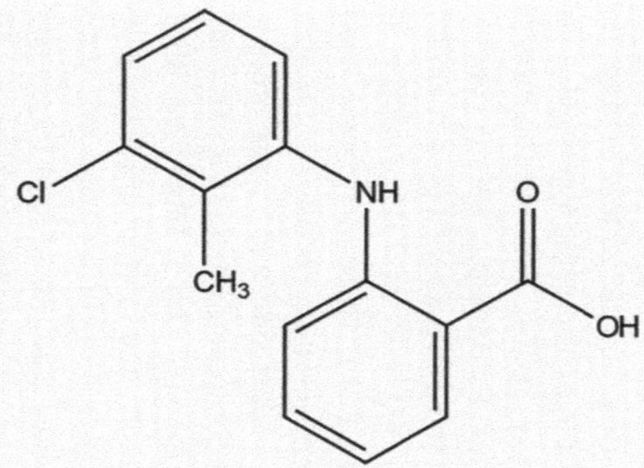	1000 (exp)	5.17 (exp)	5.11 (exp)	13710-19-5
Indomethacin	$C_{19}H_{16}ClNO_4$	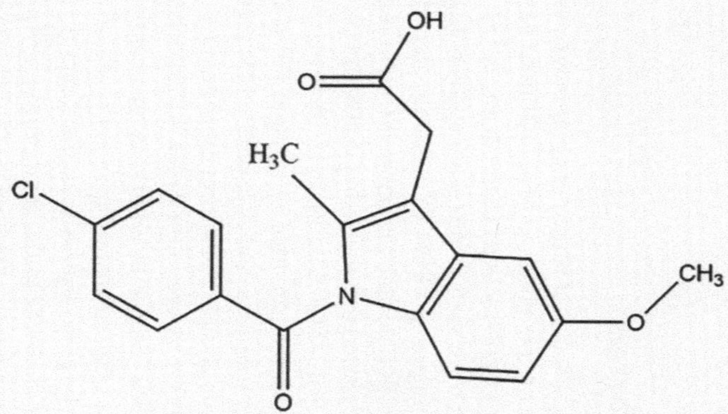	0.937 (exp)	4.27 (exp)	4.5 (exp)	53-86-1

Data acquired from drugbank.com.

regulators are classified into three different groups, including fibric acid derivatives (fibrates), 3-hydroxy-3-methylglutaryl Coenzyme A (HMG-CoA) reductase inhibitors (statins), and nicotinic acid (niacin). Lipid regulators are considered hydrophobic compounds and some lipid regulators such asl, bezafibrate, clofibric acid, and gemfibrozil have similar properties to that in NSAIDs, because they also own "acid group" in their molecular structure (Monteiro & Boxall, 2010; Ying et al., 2013). Among the three types of lipid regulators, fibrates are commonly more targeted analytically compounds in the environment (i.e., aquatics, wastewater, soil, etc.) than the other types.

The lipid regulator class: fibrates, are descended from fibric acid, which is commonly utilized to lower plasma triglycerides and to increase the levels of HDL cholesterol, however, they unlike statins, they are less effective in reducing the LDL cholesterol levels from the blood (Monteiro & Boxall, 2010; Rosal et al., 2010). Examples of fibrates include clofibric acid, bezafibrate, and fenofibrate, as shown in Table 3. The occurrence of fibrates and other lipid regulators in the environment have been frequently reported in the literature (Rosal et al., 2010).

2.4 Beta-blockers

The β-adrenergic receptor blocking compounds, commonly known as β-blockers, are PhACs that act on the blood vessels, inhibiting vasodilatation and lowering the speed and force of heart contractions (Ying et al., 2013). β-Blockers are widely used for the treatment of cardiovascular diseases, such as high blood pressure, abnormal heart rhythms, and angina pectoris (Yi, Sheng, Sui, & Lu, 2020). The consumption β-blockers differs in type and amount in different regions, for example, a study conduction in China reported that the yearly consumption of the β-blockers drug, metoprolol, has increased from 27,961 kg in 2011 to more than 60,000 kg in 2015, which is 10 times higher than that reported in Finland (Yi et al., 2020). These compounds work by blocking the stimulation of beta-adrenergic receptors by noradrenaline in the sympathetic nervous system, resulting in reducing the heart rate and blood pressure (Kümmerer, 2008). A shown in Table 4, β-blockers compounds include an aromatic or heteroaromatic ring with an aryloxypropanolamine side chain. Moreover, all β-blockers compounds consist of at least one chiral center, and their enantioselectivity in binding to β-adrenoceptors and enantiomeric distribution differ from one compound to another. Meanwhile, their solubility in water reach to several thousands' mg/L, with excretion rate from 1% to 75% in the urine (Bekhradnia & Ebrahimzadeh, 2012; Ruan, Wu, Lam, Zhang, & Lam, 2019; Yi et al., 2020). The β1 receptors, which are largely found in the heart muscle, and the β2 receptors, which are found in the blood arteries, are the two types of beta-receptors. Meanwhile, the selective β-adrenergic antagonists act only on the β1 receptors, while the nonselective β-adrenergic antagonists interrelate with the both types of β1 and β2 receptors (Monteiro & Boxall, 2010).

Table 3 Physiochemical properties of common lipid regulators.

Compound	Formula	Chemical structure	Solubility in water (mg/L)	log *P*	*p*K$_a$	CAS no.
Bezafibrate	$C_{19}H_{20}C_lNO_4$	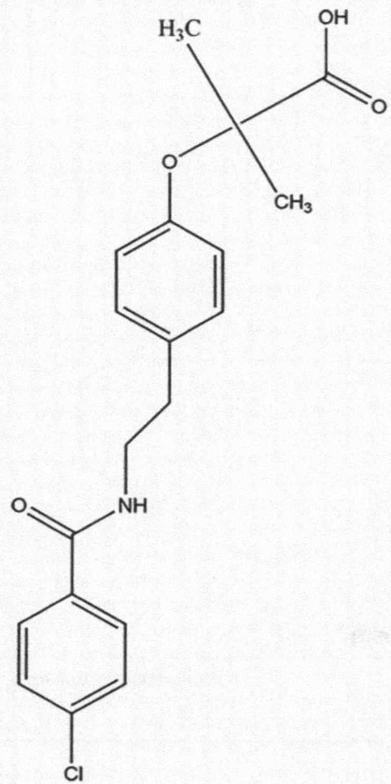	1.55 (pred)	3.97 (pred)	3.83 (pred)	41859-67-0

Continued

Table 3 Continued

Compound	Formula	Chemical structure	Solubility in water (mg/L)	log *P*	*p*K$_a$	CAS no.
Clofibric acid	$C_{10}H_{11}ClO_3$	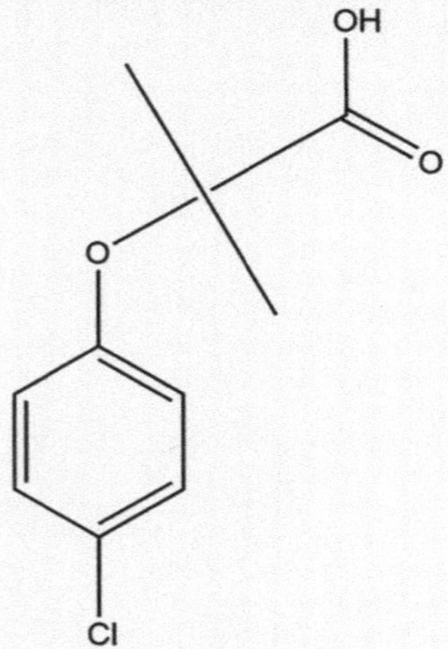	583	2.57	3.2	882-09-7

Gemfibrozil	$C_{15}H_{22}O_3$	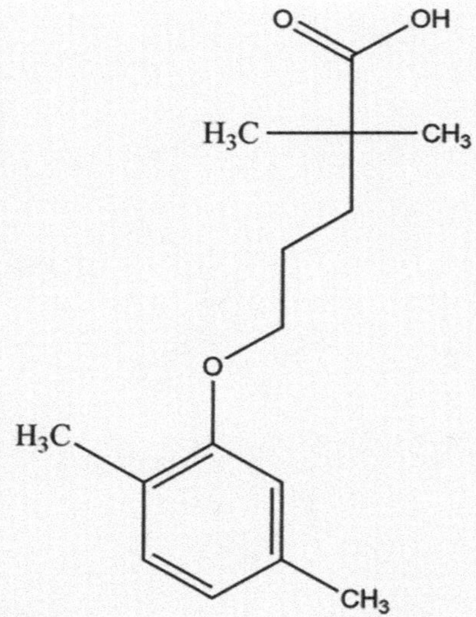	27.8 (pred) 11 (est) at 25°C	4.387 (exp)	4.7	25812- 30-0

Continued

Table 3 Continued

Compound	Formula	Chemical structure	Solubility in water (mg/L)	log *P*	*p*K$_a$	CAS no.
Fenofibrate	$C_{20}H_{21}ClO_4$	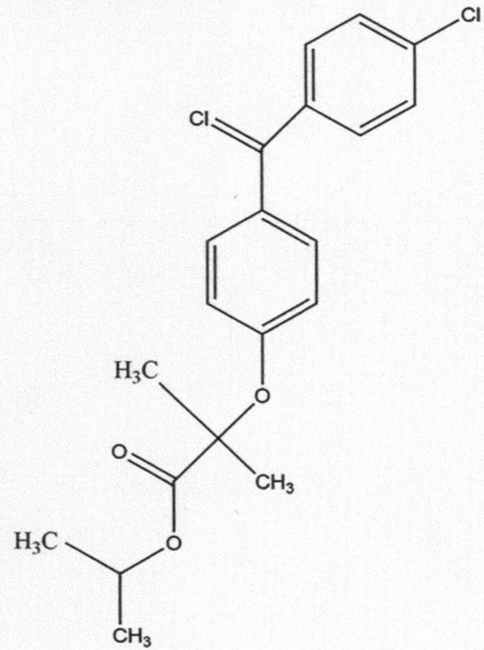	0.707 (pred) 0.42 (est) at 25°C	5.2 (exp)	5.3	49562-28-9

Simvastatin	$C_{25}H_{38}O_5$	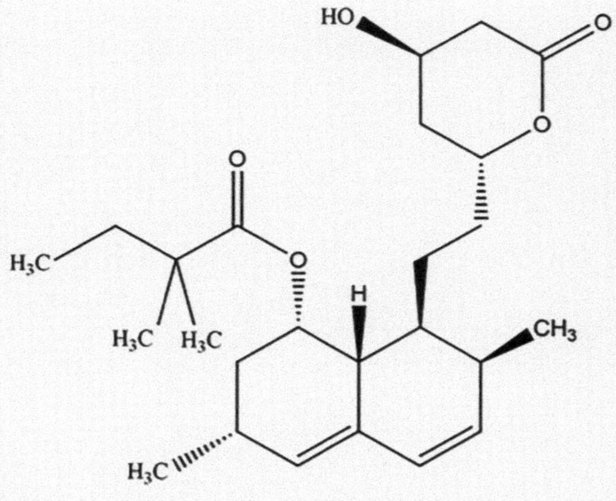	12.2 (pred) insoluble in water (exp)	4.68 (exp)	14.91 (pred)	79902-63-9

Data acquired from drugbank.com.

Table 4 Physiochemical properties of common β-blockers.

Compound	Formula	Chemical structure	Solubility in water (mg/L)	log*P*	*p*K$_a$	CAS no.
Atenolol	$C_{14}H_{22}N_2O_3$	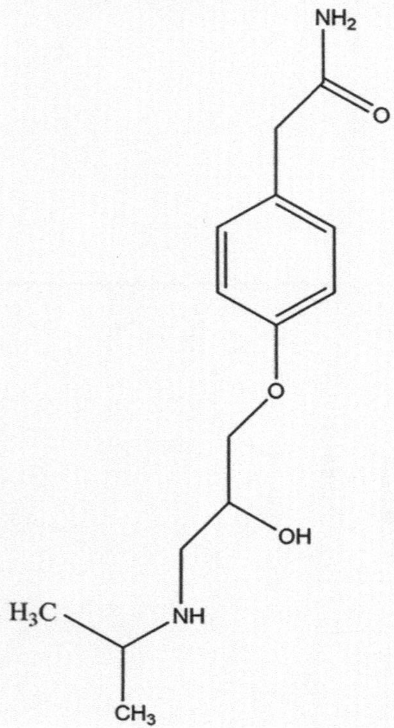	13,300 (exp)	0.16 (exp)	9.6 (exp)	29122-68-7

Metoprolol	$C_{15}H_{25}NO_3$	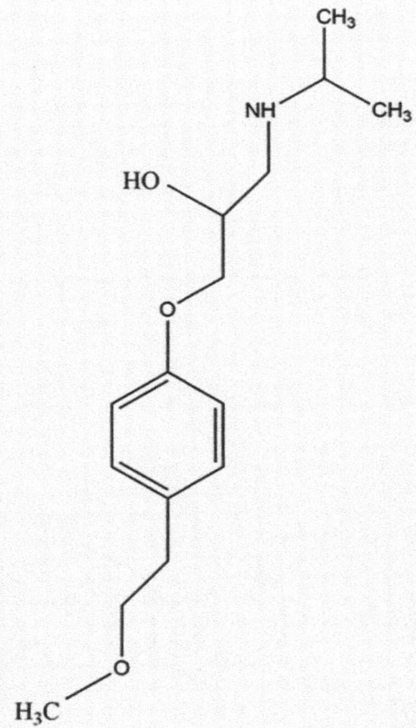	16,900 (exp)	2.15 (exp)	9.7	51384-51-1

Continued

Table 4 Continued

Compound	Formula	Chemical structure	Solubility in water (mg/L)	log *P*	*p*K$_a$	CAS no.
Propranolol	$C_{16}H_{21}NO_2$	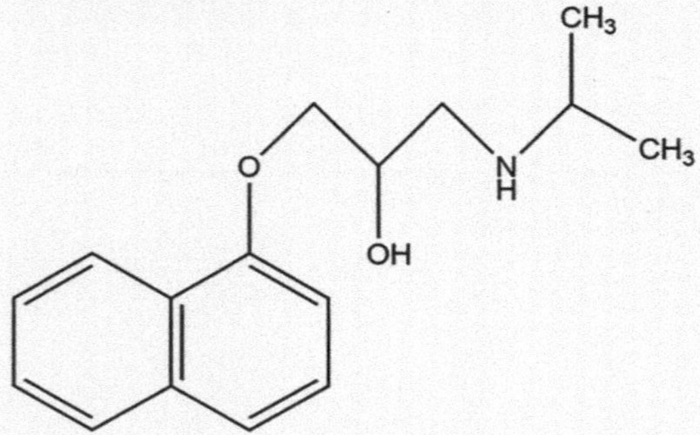	61.7 (exp)	3.48 (exp) 3.03 (pred)	9.42 (exp)	525-66-6

Sotalol	$C_{12}H_{20}N_2O_3S$	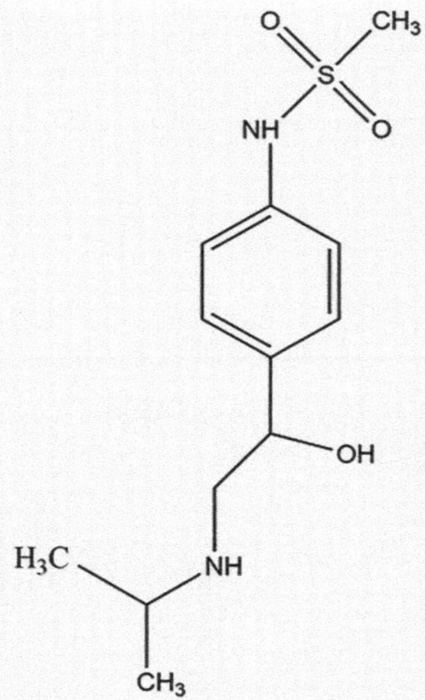	782 (pred)	0.85 (pred)	10.07 (pred)	3930-20-9

Continued

Table 4 Continued

Compound	Formula	Chemical structure	Solubility in water (mg/L)	log*P*	*p*K_a	CAS no.
Amlodipine	$C_{20}H_{25}ClN_2O_5$	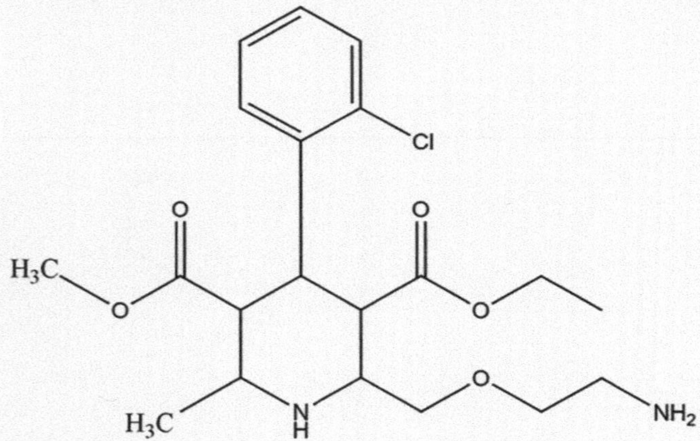	7.4 (pred)	3 (exp)	8.6	88150-42-9

Data acquired from drugbank.com.

2.5 Endocrine disrupting compounds

According to the US Environmental Agency (USPEA), endocrine disrupting chemical (EDC) is defined as "an exogenous agent that interferes with the synthesis, secretion, transport, binding, action, or elimination of natural hormones in the body that are responsible for the maintenance of homeostasis, reproduction, development, and/or behavior" (Crisp et al., 1997). EDCs have emerged as contaminants of major public health concern, due to their properties, that allow them to interact with the endocrine system, thus, potentially producing harmful effects on the human health, fauna, and aquatics, even at low concentrations. To date, more than 400 different substances were classified by the European Union (EU) commission as candidate EDCs, among those, 94 compounds have been classified EDCs with proven endocrine effects (Čelić et al., 2020). The most abundant EDCs that are commonly found in the environment include, natural hormones (estradiol, estrone, and estriol), synthetic hormones (ethinylestradiol and diethylstilbestrol), bisphenol A (used as plasticizer), and alkylphenols (octylphenol and nonylphenol), which are used in various agricultural, industrial, and domestic applications, and are formed/produced by the degradation of nonionic surfactants (Čelić et al., 2020; Ying, Williams, & Kookana, 2002). Table 5 summarizes the physiochemical properties of the most common detected EDCs in the environment.

2.6 Other pharmaceuticals

In this chapter, other classes of PhACs including opiate drugs, stimulants, antidepressants, anticonvulsants, and antidiabetic were grouped as "other pharmaceuticals." The physiochemical properties of other pharmaceuticals are summarized in Table 6. Opiate drugs, commonly referred to as "illicit drugs" (e.g., cocaine and morphine) have been detected in water environments at concentrations ranging between 0.2 and 183 ng/L (Causanilles et al., 2017; Hu et al., 2019). Once administrated into the body, they are excreted in feces and urine, both parent and altered (metabolite) forms. Different forms of illicit drugs enter WWTPs through the sewage system, and eventually released into the environment, due to their inefficient removal in WWTPs (Aberg, Chaplin, Freeman, Paizs, & Dunn, 2022). The worldwide production and consumption of illicit drugs have increased conspicuously in the past decades. Data revealed that 5% of the world population (between 15 and 64 years old) consumed illicit drugs in 2014, and since these types of PhACs are often nonprescribed drugs, it is challenging to measure their released volume into the environment (Cosenza et al., 2018). However, several strategies have been adopted such as "Wastewater-Based-Epidemiology" to back-calculate the volume of drug use based on the amount of the target drug trace concentration in wastewater (Causanilles et al., 2017). Illicit drugs (e.g., opiate drugs) can have significant potential effect on the environment due to their property of being persistent (e.g., cocaine and its metabolites). In addition, the metabolites of illicit drugs have the same serious potential effects of the parent compound, rendering toxic toward nontarget microorganisms (Causanilles et al., 2017; Cosenza et al., 2018). The fate of PhACs in organisms is influenced by its uptake, accumulation

Table 5 Physiochemical properties of common EDCs.

Compound	Formula	Chemical structure	Solubility in water (mg/L)	log *P*	*p*K$_a$	CAS no.
Estrone	$C_{18}H_{22}O_2$	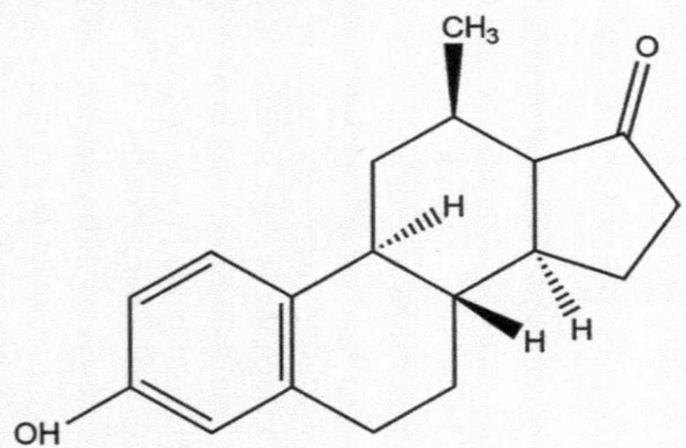	760 (exp)	2.6 (exp)	10.33 (pred)	53-16-7
Estradiol	$C_{18}H_{24}O_2$	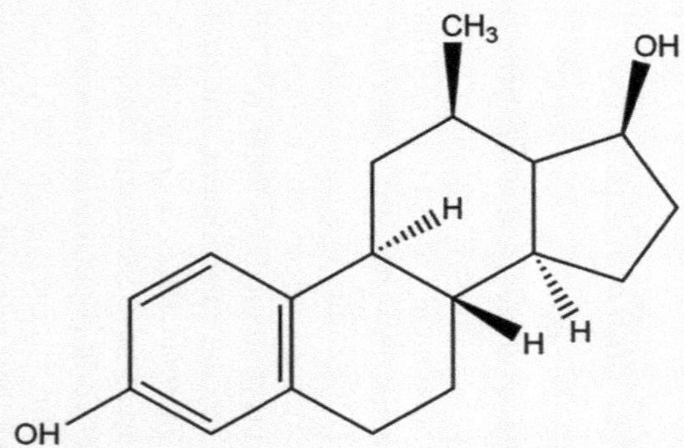	3.6 (exp) at 27°C	4.01 (exp)	10.46±0.03 (exp)	50-28-2

Table 5 Continued

Compound	Formula	Chemical structure	Solubility in water (mg/L)	log*P*	*p*K$_a$	CAS no.
Estriol	$C_{18}H_{24}O_3$	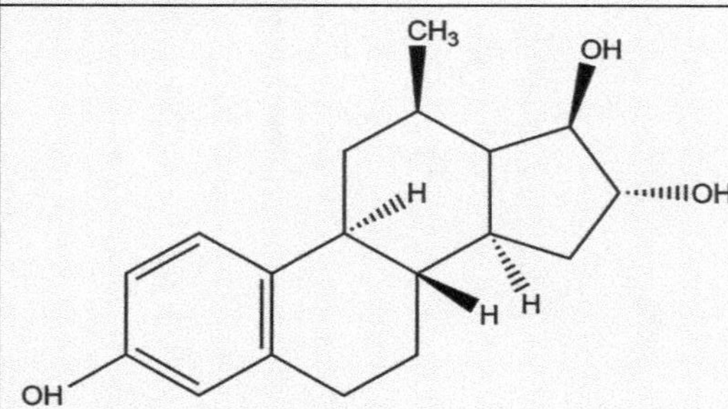	119 (pred)	2.45 (exp)	10.33 (pred)	50-27-1
Ethinylestradiol	$C_{20}H_{24}O_2$	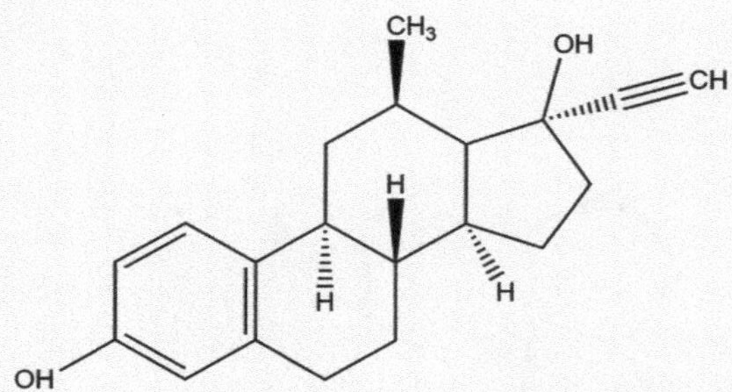	11.3 (exp)	3.67 (exp)	10.33 (pred)	57-63-6

Continued

Table 5 Continued

Compound	Formula	Chemical structure	Solubility in water (mg/L)	log*P*	*p*K$_a$	CAS no.
Testosterone	$C_{19}H_{28}O_2$	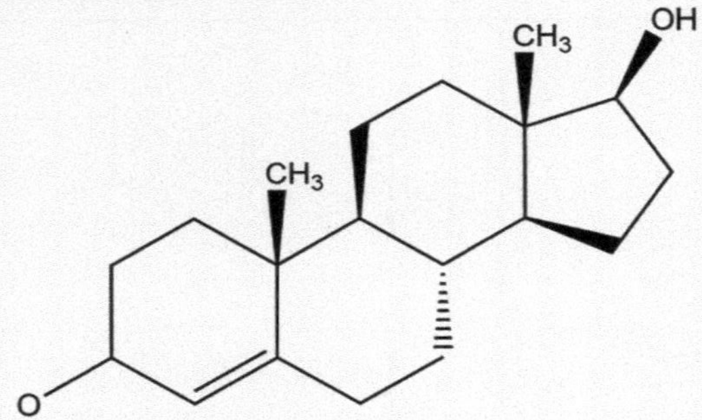	23.4 (exp)	3.32 (exp)	18.52 (pred)	58-22-0
Progesterone	$C_{21}H_{30}O_2$	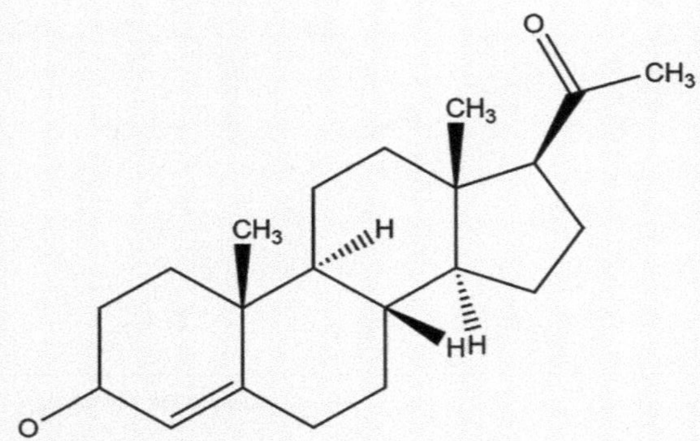	<100 (exp) at 19°C	3.87 (exp)	18.92 (exp)	57-83-0

Table 5 Continued

Compound	Formula	Chemical structure	Solubility in water (mg/L)	log*P*	*p*K$_a$	CAS no.
Cortisol	$C_{21}H_{30}O_5$	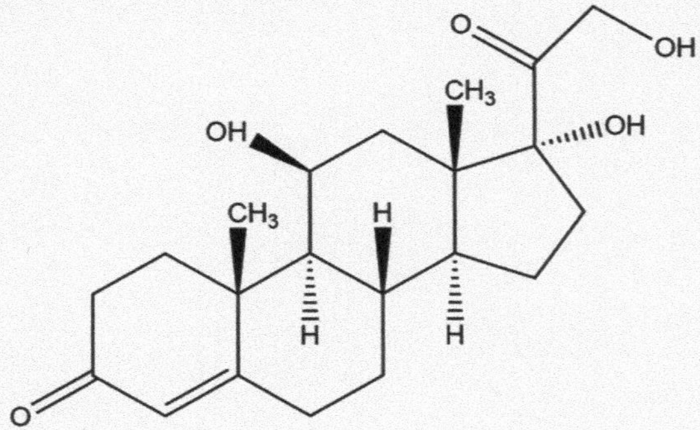	320 (exp) 199 (pred)	1.61 (exp)	12.59 (pred)	50-23-7
Cortisone	$C_{21}H_{28}O_5$	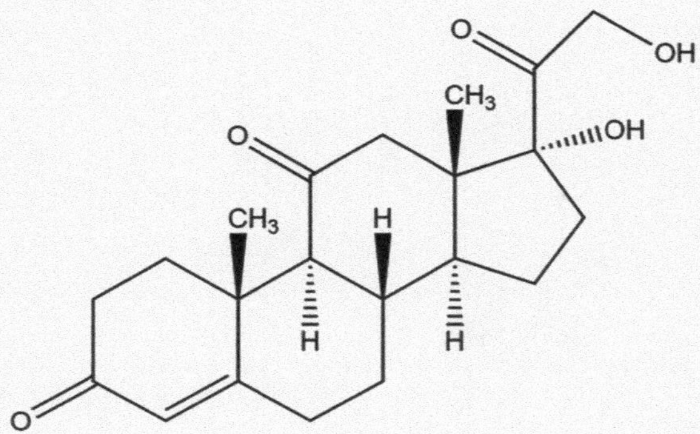	136 (pred)	1.98 (pred)	12.58 (pred)	53-06-5

Continued

Table 5 Continued

Compound	Formula	Chemical structure	Solubility in water (mg/L)	log*P*	*p*K_a	CAS no.
Dexamethasone	$C_{22}H_{29}FO_5$	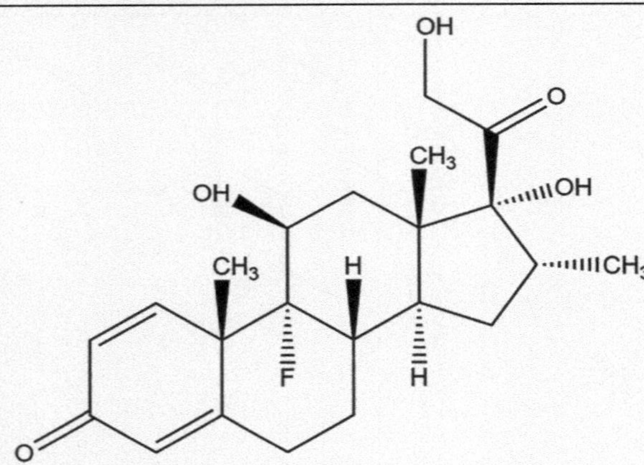	89 (exp) at 25°C	1.83 (exp)	12.42 (pred)	50-02-2
Bisphenol A	$C_{15}H_{16}O_2$	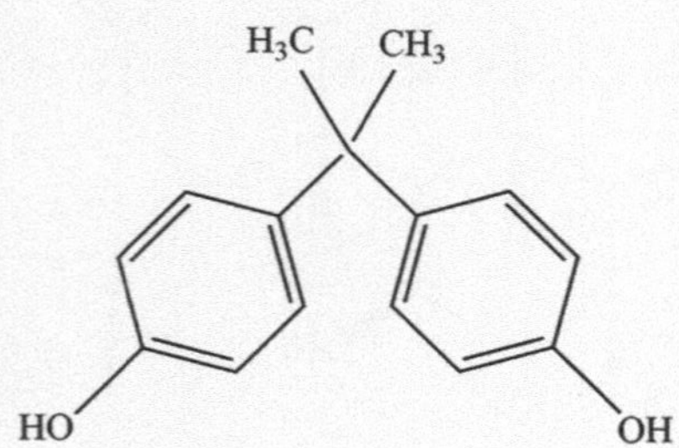	86.5 (pred)	3.81 (pred)	9.78 (pred)	80-05-7
Nonylphenol	$C_{15}H_{24}O$	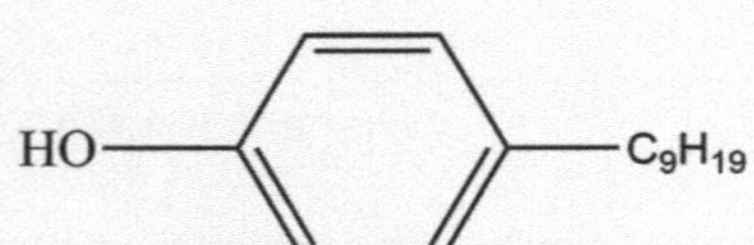	6.4	5.71	10.3	25154-52-3

Data acquired from drugbank.com.

Table 6 Physiochemical properties of other common classes of pharmaceuticals.

Class/ Compound	Formula	Chemical structure	Solubility in water (mg/L)	log*P*	*p*K$_a$	CAS no.
Opiate drugs						
Codeine	$C_{18}H_{21}NO_3$	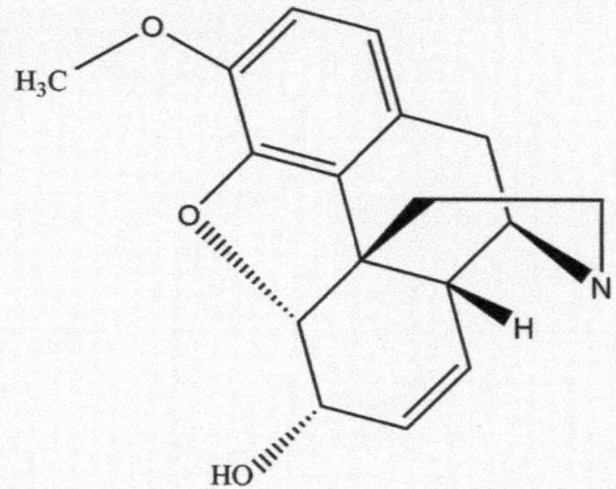	<1000 at 21°C	1.39 (exp)	8.2 (exp)	76-57-3
Morphine	$C_{17}H_{19}NO_3$	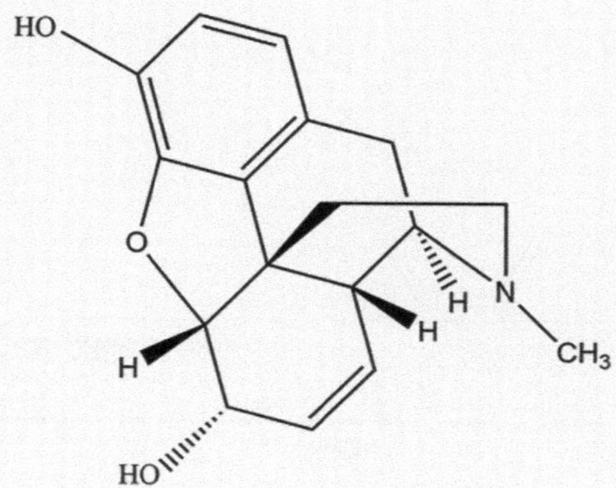	149 (exp)	0.87 (exp)	8.21 (exp)	57-27-2

Continued

Table 6 Continued

Class/ Compound	Formula	Chemical structure	Solubility in water (mg/L)	log *P*	*p*K$_a$	CAS no.
Tramadol	$C_{16}H_{25}NO_2$	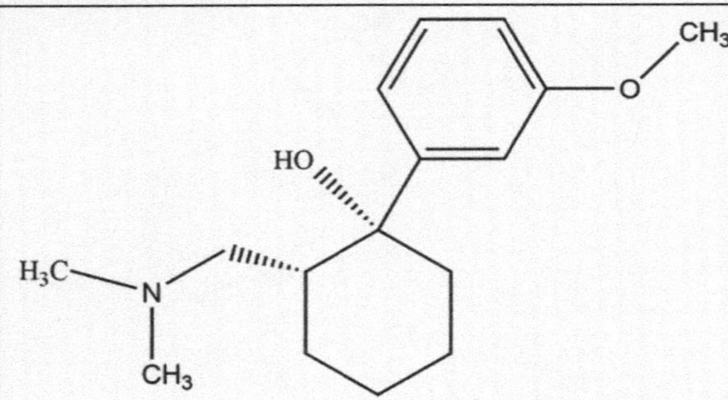	750 (pred)	1.34 (exp)	9.41 (exp)	27203-92-5
Stimulants						
Caffeine	$C_8H_{10}N_4O_2$	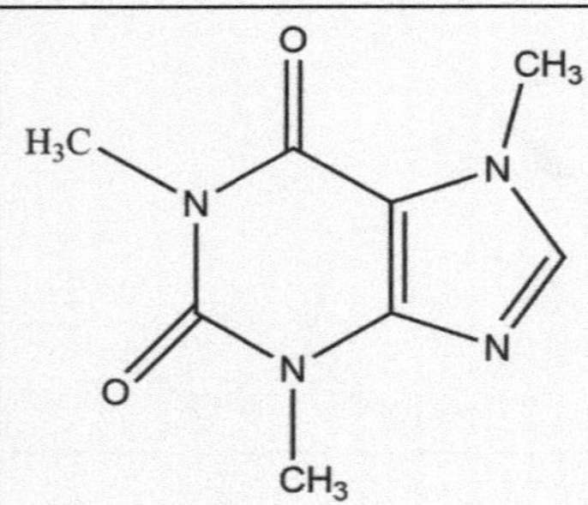	10,000–50,000 (exp) at 25°C	−0.07 (exp)	14 (exp)	58-08-2

Nicotine	$C_{10}H_{14}N_2$	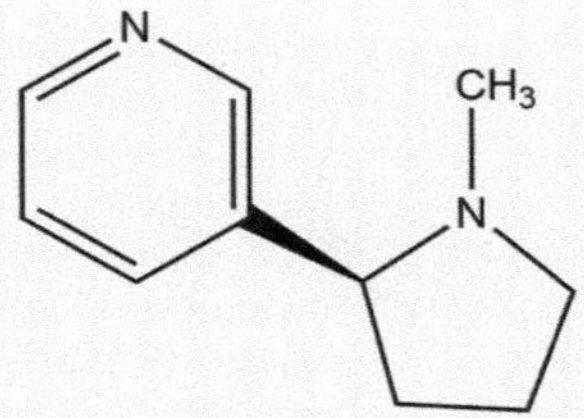	1000,000 (exp)	1.17 (exp)	8.5 (exp)	54-11-5
Antidepressants						
Fluoxetine	$C_{17}H_{18}F_3NO$	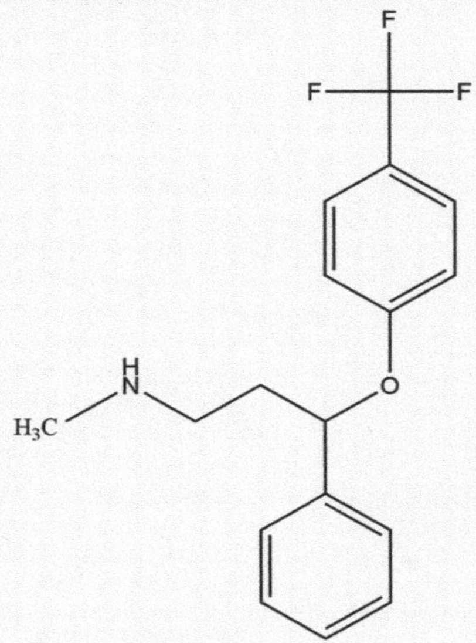	Insoluble	4.05 (exp)	9.8 (pred)	54910-89-3

Continued

Table 6 Continued

Class/ Compound	Formula	Chemical structure	Solubility in water (mg/L)	log *P*	*p*K$_a$	CAS no.
Venlafaxine	$C_{17}H_{27}NO_2$	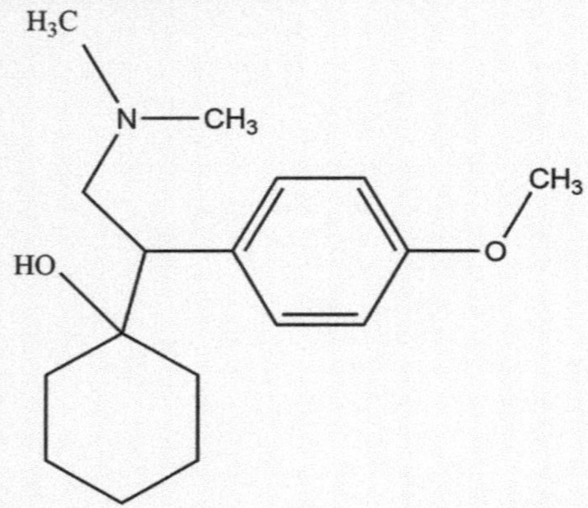	230 (pred)	2.69 (pred)	14.42 (pred)	93413-69-5
Diazepam	$C_{16}H_{13}ClN_2O$	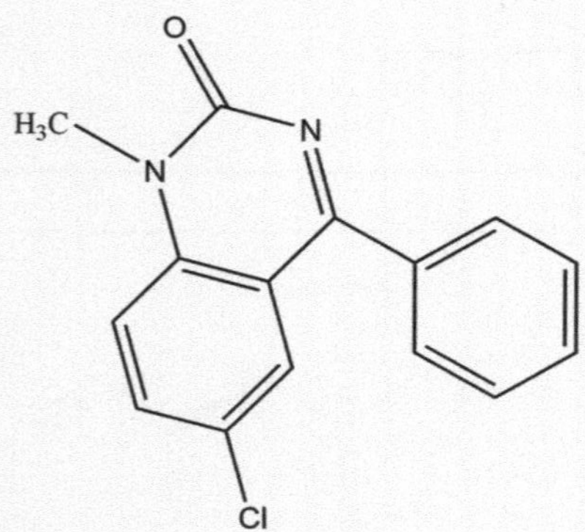	50 (exp)	2.82 (exp)	3.4 (exp)	439-14-5

Anticonvulsants/Antiepileptic						
Gabapentin	$C_9H_{17}NO_2$	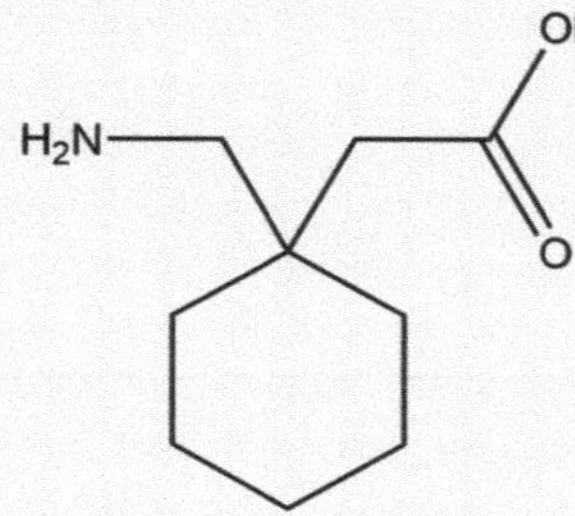	>100,000 (exp)	1.25 (exp)	3.7 (exp)	60142-96-3
Pregabalin	$C_8H_{17}NO_2$	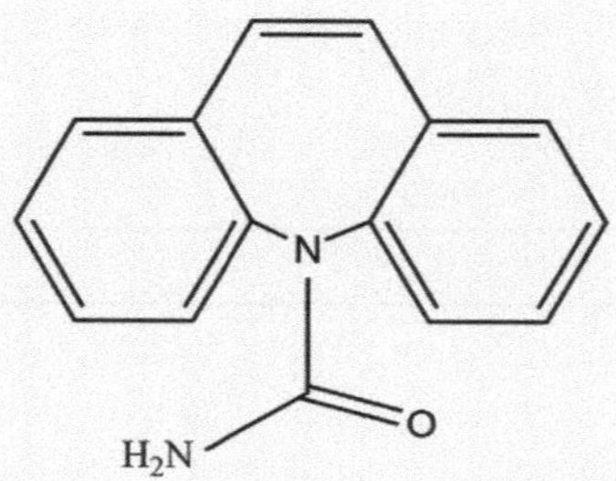	Freely soluble in water 11,300 (pred)	−1.4 (pred)	4.8 (pred)	148553-50-8

Continued

Table 6 Continued

Class/ Compound	Formula	Chemical structure	Solubility in water (mg/L)	log*P*	*p*K_a	CAS no.
Carbamazepine	$C_{15}H_{12}N_2O$	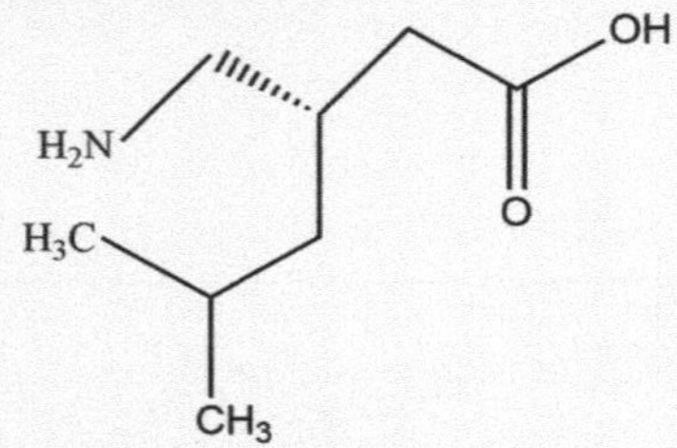	152 (pred)	2.77 (exp)	15.96, −3.8 (exp)	298-46-4
Antidiabetic						
Metformin	$C_4H_{11}N_5$	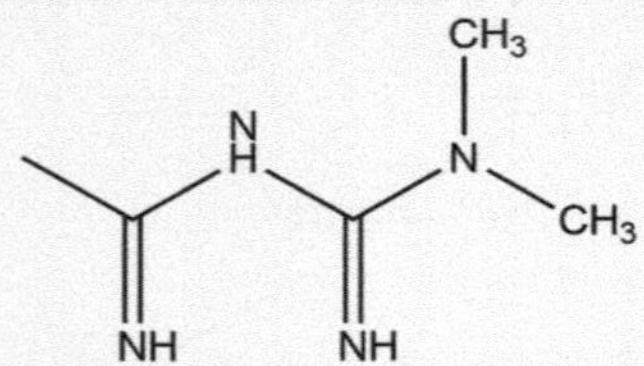	1380 (pred)	−2.6 (exp)	12.4 (exp)	657-24-9

Data acquired from drugbank.com.

potential, and biotransformation enzyme activities in the organisms. However, few studies have investigated the accumulation of illicit drugs in aquatic organisms, while their bioaccumulation and simultaneous biotransformation in organisms have been recently investigated (Yin et al., 2021).

The physiochemical properties of other classes such as stimulants are also reported in Table 2. Among these, caffeine is considered the most common metabolic stimulant and the most common central nervous system. Caffeine is an alkaloid belonging to the family of methylxanthine, is often found in our daily products (e.g., chocolate, coffee, sodas, and tea) and some pharmaceuticals. Caffeine is used widely to reduce the physical fatigue, restore alertness, and due to its stimulating effect on the central nervous system (Khasawneh & Palaniandy, 2021; Li, 2014).

Antidepressants or psychiatric drugs are prescription medications that are designed to treat anxiety and depression. The intake of antidepressants worldwide is increasing alone with the number of diagnosed patients, and its consumption doubled from the year 2000 to 2017 in 28 countries (Guo et al., 2022; Khasawneh & Palaniandy, 2021). This class of PhACs was ranked as the third most frequently prescribed medicine taken by Americans (Khasawneh & Palaniandy, 2021). Antidepressants are often used with anticonvulsants to control and manage depression in patients with epilepsy (Tran et al., 2018). Antidepressants act as synaptic junction in nerve clefts to inhibit the reuptake of noradrenaline, serotonin and other neurotransmitters (e.g., dopamine) (Parrott & Metcalfe, 2019; Wang et al., 2021). Antidepressants are administrated to WWTPs in their parent or metabolized form via human urine. For instance, fluoxetine, which one of the most commonly used antidepressants, is metabolized in the human body around 20%–30% to produce the metabolites: fluoxetine glucuronide and norfluoxetine, while the remaining is discharged to WWTPs as unchanged drug (parent compound) (Munari, Marin, & Matozzo, 2014). The antidepressant drug, fluoxetine, has genotoxic and possible immunotoxic effects on nontarget aquatic organisms (Wang et al., 2021). In addition, other antidepressant drugs have shown to damage organisms even at trace levels (Guo et al., 2022).

Anticonvulsants, also commonly referred to as antiepileptic drugs, are a class of PhACs designed to treat patients with, seizure disorders, epilepsy, neuralgia, and other mental disorders (Nkoom, Lu, Liu, Yang, & Dong, 2019). This class of PhACs act in the central nervous system by (i) reducing the overall neuronal activity, (ii) improving the inhabitancy of aminobutyric neurotransmitters, or (iii) by blocking voltage-dependent sodium channels (Wang et al., 2021). Among the presented compounds in Table 6, carbamazepine is considered the most consumed antiepileptic drug. Nkoom et al. reported that approximately 13% of carbamazepine that is used is excreted as unchanged parent compound through human urine and feces and being primarily excreted as hydroxylated and conjugated metabolites. Due to its high consumption and low biodegradability and removal, it is frequently detected in WWTPs (Nkoom et al., 2019). The removal rate of carbamazepine in WWTPs is very low, accounting for only about 7%, due to its hard photodegradation and biodegradation at trace concentrations (Wang et al., 2021). Moreover, due to its continuous release, resistance to photochemical and biological transformations in the environment, carbamazepine is frequently found in different compartments of the environment (e.g., freshwater, marine water, etc.) (Almeida, Soares, Esteves, & Freitas, 2021; Biel-Maeso et al., 2018; Nödler, Voutsa, & Licha, 2014). At environmentally relevant concentrations (up to 1 μg/L), carbamazepine is able change the biological responses (i.e., cellular, physiological, molecular) in marine bivalves (Almeida et al., 2021).

Metformin is an antidiabetic drug used to control type 2 diabetes and is considered one of the most prescribed drugs worldwide (Ghoshdastidar, Fox, & Tong, 2015). Metformin is administered orally, with an average dose of 2 g/day (Scheurer, Michel, Brauch, Ruck, & Sacher, 2012). The antidiabetic drug, metformin, is not metabolized in human bodies and about 70% is excreted unaltered (as parent compound) via urine, and the rest through feces (Martin & Weiß, 2008; Pentikäinen, Neuvonen, & Penttilä, 1979). In the year 2015, 415 million people worldwide suffered from diabetes, with 60 million patients were in Europe alone. Type 2 diabetes is greater than ever, it is predicted that 642 million people will suffer from in 2040 (IDF, 2013; Tisler & Zwiener, 2018). Studies and data on the ecotoxicological effects of metformin and its transformation products are still lacking. However, some studies reported that metformin had an LC_{50} value >982 mg/L for *Lepomis macrochirus* and an EC_{50} value of 130 mg/L for *Daphnia magna* (Scheurer et al., 2012). Both metformin and its transformation products are extremely polar compounds, making them very mobile in the aquatic environment (Scheurer et al., 2012).

3. Metabolism and transformation processes

In recent years, the risk proposed by PhACs in the environment have begun to be recognized, resulting in significant increase in the number of studies that are targeted at improved the understanding of their impacts on the ecosystems. Currently, it is well known that PhACs are present in different compartments of the environment (i.e., surface water, groundwater, soil, and WWTPs), while researchers and scientists are becoming more aware of and familiar with their mechanism and modes of action on nontarget organism, resulting in better understanding of this phenomenon (Maculewicz et al., 2022). Although it may seem that the occurrence of the PhACs (in their parent form) is the major issue, however, several studies have showed that there is another hidden threat, in the form of transformation products (TPs). According to the literature, TPs refer to the metabolites (which are excreted by human and animal bodies via changes in the chemical structure of the parent compound(s)) and transformation products resulting from biotic and abiotic transformation (which occur in wastewater treatment plants, water treatment plants, and the environment after their excretion) (Kümmerer, 2008; Maculewicz et al., 2022).

After the administration of PhACs into the human/animal body, the absorption process must take place before the medicine reaches the interior of the body. Generally, the absorption of PhACs occurs by simple diffusion, however, the absorption process could be influenced by some physiochemical properties (i.e., degree of ionization, molecular size and shape, and relative lipid solubility). Following the absorption, the administered drug enters the "circulation process," and after performing its action, it may be metabolized to a more hydrophilic compound(s) for excretion. Nevertheless, if the drug stays lipophilic, it will once again be reabsorbed and remain in the body for extended period (Galbraith, Bullock, Manias, Hunt, & Richards, 2004; Monteiro & Boxall, 2010). Usually, the metabolism of PhACs produces more polar metabolites (with lower activity) that are excreted easily from the body. Toxic or biologically active metabolites may be produced in some cases. As demonstrated in Table 7, there are two successive routes (pathways) that are implicated in the metabolism process of PhACs; Phase 1 and Phase 2. The first phase comprises of oxidative reactions (i.e., N-dealkylation, aliphatic hydroxylation, N-oxidation, and deamination) and hydrolysis reactions. However, the second phase consists of conjugation reactions, including the addition of glucuronic acid, acetate, amino acids, or sulfate (Monteiro & Boxall, 2010).

The structural changes of PhACs in the human and animal bodies could be due to the presence of the microorganisms in the gut or by the enzymes, resulting in what is known as "metabolites." Although many pharmaceuticals are metabolized before they are even excreted, however, other drugs can be poorly or moderately metabolized. Metabolites possess different toxicological and pharmacological properties compared to their parent compounds (pharmaceutical). While some factors such as the administration procedure (orally or intravenously) can influence the metabolism and the structural changes that occur. Nevertheless, the metabolism process is often incomplete, and the process of metabolism may improve water solubility and/or lower the activity. Fig. 2 illustrates the involved reactions in the metabolism of PhACs. It is worth noted that further and different molecules could be formed after the release of parent compounds and their metabolites into the environment. As shown in Fig. 2, transformation processes are a result of other processes, which could be biotic, hydrolysis, or photooxidation. While the developed structural changes could also be a result of the treated wastewater effluent (Kümmerer, 2008).

As a result of the transformation processes occurring during wastewater treatment processes and in the environment, transformation products (TPs) are formed. Several studies have investigated and reported the occurrence of TPs in different environmental compartments (Ibáñez, Bijlsma, Pitarch, López, & Hernández, 2021; Maculewicz et al., 2022; Meffe et al., 2021; Zind et al., 2021). Table 8 summarizes the TPs formed during their degradation under different treatment processes and their respective ecotoxic effects toward various species. Generally, the nature of TPs is still indetermined. In other words, the toxic effects of TPs vary, as some TPs are more toxic and persistent than their parent compound(s) or other TPs. Among these TPs, acridine and 9(10H)-acridone, which occur during the degradation of the anticonvulsant drug, carbamazepine (CBZ), under UV photolysis, are considered considerably more toxic than their parent compound across multiple trophic levels (Donner et al., 2013). Results have showed the chronic toxicity of CBZ (at concentrations of 200 μg/L) testing using the *Daphina pulex* reduced the rates of population growth due to the retardation of juvenile somatic growth and the consequent delays in maturation and time to first production (Lürling, Sargant, & Roessink, 2006). It is worth noting that some TPs and human metabolites (HMs) could convert back to their parent compounds under suitable conditions. For instance, presence of the antibiotic sulfamethoxazole in water-sediments environment make it exposed to phototransformation during daylight

Table 7 Involved reactions in PhACs metabolism (Monteiro & Boxall, 2010).

		Reaction	Example
Phase 1	*i. Oxidative reactions*		
	N-dealkylation	$RNHCH_3 \longrightarrow RNH_2 + CH_2O$	Caffeine, codeine, erythromycin, diazepam tamoxifen
	Aliphatic hydroxylation	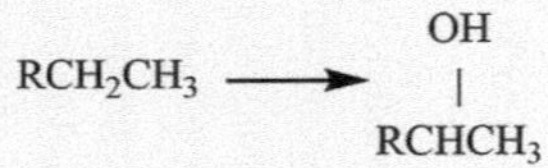	Meprobamate, ibuprofen
	N-oxidation	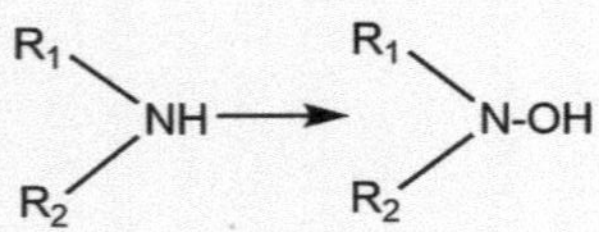	Acetaminophen, quinidine
	Deamination	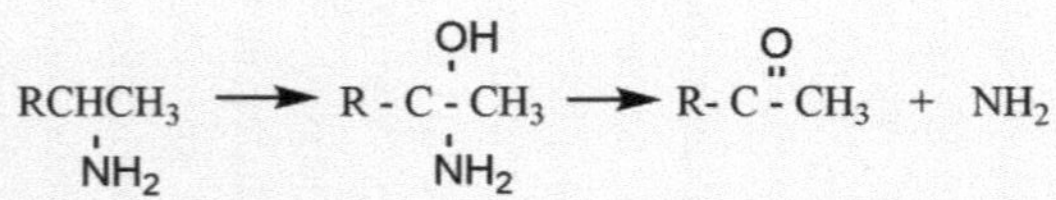	Diazepam
	ii. Hydrolysis reactions		
		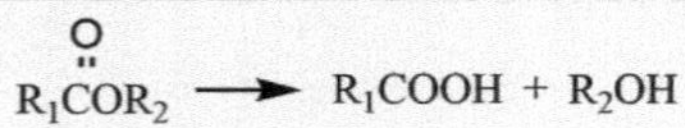	Cocaine, clofibrate, aspirin, enalapril
		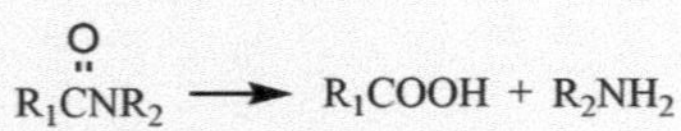	Indomethacin, lidocaine

Phase 2	iii. *Conjugation reactions*		
	Glucuronidation	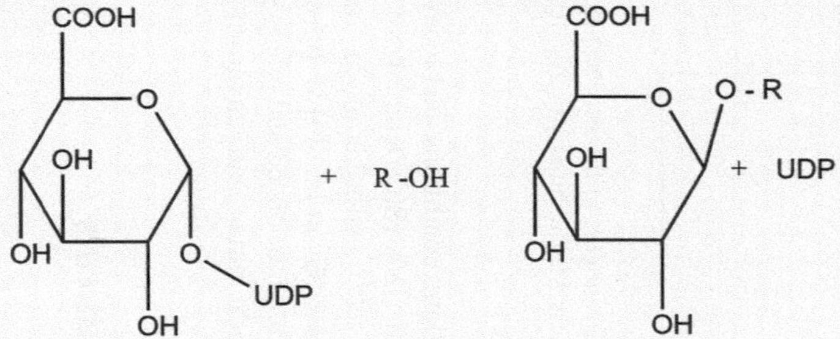 UDP - glucuronic acid	Oxazepam, acetaminophen, morphine
	Sulfation	ROH + 3' - phosphoadenosine-5'-phosphosulfate (PAPS) 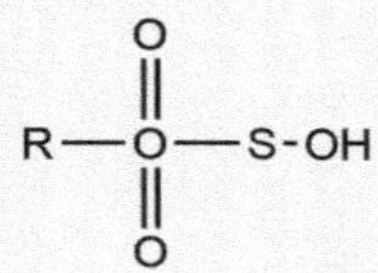 + 3' - phosphoadenosine-5'- phosphate	Sulfonamides

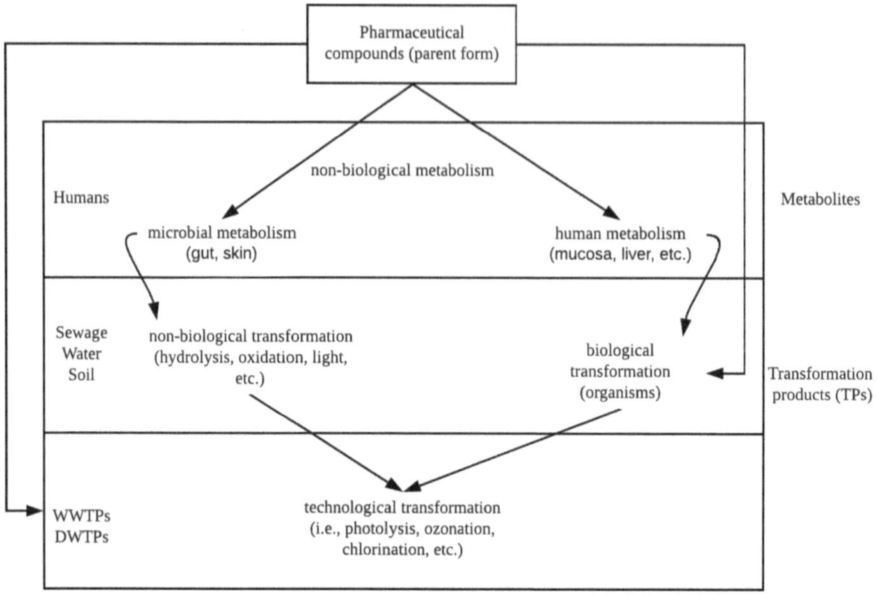

Fig. 2 Metabolism and transformation products.
Adapted and modified from Kümmerer, K. (2008). *Pharmaceuticals in the environment: Sources, fate, effects and risks*. Springer Science \& Business Media. https://doi.org/10.1007/978-3-540-74664-5.

and reformation throughout the night under certain abiotic conditions via biodegradation, resulting in back-transformation of its TPs and HMs (Su, Deng, Benskin, & Radke, 2016). Qu et al. (2013) reported similar back-transformation of the TPs (formed due to photolysis) of steroid trenbolone acetate (Qu et al., 2013). In conclusion, TPs are deemed as secondary contaminants that are introduced and released into the environment from their parent compounds with varies and unknown large ecotoxicity toward different species.

4. Occurrence of common pharmaceuticals in the environment

Numerous studies have investigated the occurrence of PhACs in different environmental matrices (Burns, Carter, Kolpin, Thomas-Oates, & Boxall, 2018; Khasawneh & Palaniandy, 2021; Matongo, Birungi, Moodley, & Ndungu, 2015; Szymczycha, Borecka, Białk-Bielińska, Siedlewicz, & Pazdro, 2020). The continuous development in analytical techniques have allowed the identification and quantification of low traces of PhACs in different and complex environmental matrices more feasible (Khasawneh & Palaniandy, 2021). Over the last years, more than 100 different PhACs have been detected as environmental pollutants in the surface water, groundwater, sediments, influents and effluents of WWTPs, sludge, and even drinking

Table 8 Formed TPs during degradation of pharmaceuticals and their potential toxic effects.

Parent compounds	Transformation process	TPs	Ecotoxic effects	References
Acyclovir	Ozonation	Carboxy-acyclovir, N-(4-carbamoyl-2-imino-5-oxoimidazolidin)-formamido-*n*-methoxy-acid	Enhanced reduction in reproduction levels of *Daphnia magna*	Schlüter-Vorberg, Prasse, Ternes, Mückter, and Coors (2015)
Cefazolin	Photolysis	5-Methyl-1,3,4-thiadiazole-2-thiol	Luminescence inhibition in *V. fischeri*	Wang and Lin (2012)
Acetaminophen	Photocatalysis	1,4-Benzoquinone	Hepatotoxicity, genotoxic and mutagenic	Chang, Wang, Ouyang, Zhang, and Jing (2015)
Ibuprofen	Ozonation	Hydroxy-ibuprofen	Improved growth inhibition in *S. capricornium* alga	Quero-Pastor, Garrido-Perez, Acevedo, and Quiroga (2014)
Carbamazepine	Biodegradation	2-Hydroxy carbamazepine, 3-hydroxy carbamazepine	Luminescence inhibition in *V. fischeri*	Kaiser, Prasse, Wagner, Bröder, and Ternes (2014)
		Carbamazepine-10,11-epoxide	Decrease in population of the nonbiting midge *Chironomus riparius*	Heye et al. (2016)
	UV photolysis	Acridine and 9(10H)-acridone	The study indicated the TPs of CBZ are significantly more toxic than the parent compound and resulted in higher: bioluminescence inhabitation in *V. fischeri*; growth inhibition in *Pseudokirchneriella subcapitata* alga; and immobilization in *Daphnia magna*	Donner et al. (2013)
Sulfamethoxazole	Biodegradation	N4acetylsulfamethoxazole	Growth inhibition in duckweed and algae	Grabarczyk et al. (2020)
Phenazone	Ozonation	Aniline	Enhanced growth inhibition in *Daphnia magna*	Miao et al. (2015)

Continued

Table 8 Continued

Parent compounds	Transformation process	TPs	Ecotoxic effects	References
Clotrimazole	VUV/UVC photolysis	(2-Chlorophenyl) bisphenylmethanol	Relatively low inhibitory effect towards green microalgae *R. subcapitata*, the crustacean *D. magna* and the fungus *F. graminearum*	Gonçalves, del Puerto, Medana, Calza, and Roslev (2021)
Gemfibrozil	Chlorination	Chloro- and bromo-gemfibrozil	Increased antiandrogenic activity in Japanese medaka	Bulloch et al. (2012)
Diclofenac	Biodegradation	4-Hydroxydiclofenac	Luminescence inhibition in *V. fischeri*	Grabarczyk et al. (2020)
Metoprolol	Biodegradation	Metoprolol acid	Growth inhibition in *Ganoderma lucidum*	Jaén-Gil et al. (2019)
Propranolol	Photolysis	4-Hydroxy-propranolol, 5-hydroxy-propranolol	Cytotoxicity toward *V. fischeri*	Menz et al. (2017)

water (Monteiro & Boxall, 2010). In this section, an overview and summary of the most common pharmaceuticals (belonging to different pharmaceutical classes) that are detected in various environmental compartments is provided.

4.1 Occurrence in wastewater treatment plants

The presence of PhACs in WWTPs has been reported in several studies (D'Alessio et al., 2018; Khasawneh & Palaniandy, 2021; Palli et al., 2019; Tran et al., 2018). Antibiotics, NSAIDs, β-blockers, lipid regulators, antidepressants, and endocrine disrupting compounds (EDCs), are examples of the common detected classes of pharmaceuticals in WWTPs. Table 9 summarized the acquired data from different studies on the occurrence of different classes of PhACs in WWTPs. Among these classes, antibiotics are the most valuable and commonly used drugs among people, and

Table 9 Concentration of commonly detected PhACs (μg/L) in the influents and effluents of WWTPs around the world.

Class/ Pharmaceutical	Country	Concentration in IN/EF	References
Antibiotics			
Amoxicillin	Italy, Qatar	0.066–3.55/ 0.01–4.6	Al-Maadheed, Goktepe, Latiff, and Shomar (2019), Palli et al. (2019), and Spataro, Ademollo, Pescatore, Rauseo, and Patrolecco (2019)
Erythromycin	USA, Poland, Czech	0.016–2.3/ 0.011–2.7	de Jesus Gaffney et al. (2017), Kot-Wasik, Jakimska, and Śliwka-Kaszyńska (2016), and Phonsiri et al. (2019)
Norfloxacin	Columbia	0.633–2.01/ 0.35–0.606	Botero-Coy et al. (2018)
Ofloxacin	Tunisia, Span, USA	0.138–2.28/ 0.1–0.64	Afonso-Olivares, Sosa-Ferrera, and Santana-Rodríguez (2017), Meador, Yeh, Young, and Gallagher (2016), and Moslah, Hapeshi, Jrad, Fatta-Kassinos, and Hedhili (2018)
Sulfamethoxazole	Germany, Czech, China	0.043–3.93/ 0.0018–0.71	Ben et al. (2018), Golovko, Kumar, Fedorova, Randak, and Grabic (2014), and Gurke et al. (2015)
Tetracycline	Tunisia, China	0.002–0.319/ 0.0004–0.319	Al-Maadheed et al. (2019) and Ben et al. (2018)

Continued

Table 9 Continued

Class/ Pharmaceutical	Country	Concentration in IN/EF	References
NSAIDs			
Ibuprofen	USA	8.6–96.5/ 0.013–0092	D'Alessio et al. (2018) and Yu, Wu, and Chang (2013)
Ketoprofen	Colombia	0.519–3.72/ 0.432–3.16	Botero-Coy et al. (2018)
Acetaminophen	Greece	0062–52.5/ 0.5–1.7	Kosma, Lambropoulou, and Albanis (2010) and Papageorgiou, Kosma, and Lambropoulou (2016)
Naproxen	Spain	0.25–5.83/ 0.03–4.02	Fernández-López, Guillén-Navarro, Padilla, and Parsons (2016) and Gracia-Lor, Sancho, Serrano, and Hernández (2012)
Diclofenac	Spain	0.26–15.63/ 0.06–0.91	Fernández-López et al. (2016) and Gracia-Lor et al. (2012)
Lipid regulators			
Bezafibrate	Mexico	0.57–3.44/ 0.265–0.95	Rivera-Jaimes et al. (2018)
Gemfibrozil	Mexico, USA	0.02–6.9/ 0.012–0.38	D'Alessio et al. (2018) and Rivera-Jaimes et al. (2018)
Atorvastatin	Czech	0.070–0.75/ 0.004–0.24	Golovko et al. (2014)
Simvastatin	China	0.039–0.39/ 0.014–0.0495	Yan et al. (2014)
β-Blockers			
Atenolol	Germany, Italy	0.216–0.540/ 0.061–0.341	Gurke et al. (2015) and Palli et al. (2019)
Propranolol	Germany	0.039–0.096/ 0.048–0.103	Gurke et al. (2015)
Metoprolol	Germany, France	0.277–5.48/ 1.75–5.76	Gurke et al. (2015) and Thiebault, Boussafir, and Le Milbeau (2017)
EDCs			
Estrone	South Korea	0.017–0.224/ 0.003–0.023	Ekpeghere, Sim, Lee, and Oh (2018)
Estradiol	Malaysia	0.088–0.093/ 0.0351–0.085	Yien Fang, Praveena, Aris, Syed Ismail, and Rasdi (2019)
Estriol	South Korea	0.083–3.66/ 0.017–0.039	Ekpeghere et al. (2018)

Table 9 Continued

Class/ Pharmaceutical	Country	Concentration in IN/EF	References
Ethinylestradiol	Malaysia	0.0002–0.0049/ 0.00002–0.0001	Yien Fang et al. (2019)
Antidiabetic			
Metformin	Poland, Portugal	3.81–325/0.05–62.9	de Jesus Gaffney et al. (2017) and Kot-Wasik et al. (2016)
Stimulant			
Caffeine	Canada, USA	>15–63.8/ 0.013–115.141	Basiuk, Brown, Cartwright, Davison, and Wallis (2017), D'Alessio et al. (2018), and Ghoshdastidar et al. (2015)
Opiate drugs			
Morphine	Costa Rica	0.067–0.077/ 0.015–0.061	Causanilles et al. (2017)
Codeine	Costa Rica, France	0.143–2.11/ 0.011–0.669	Causanilles et al. (2017) and Thiebault et al. (2017)
Tramadol	France, Germany	0.381–9.86/ 0.072–1.19	Gurke et al. (2015) and Thiebault et al. (2017)
Antidepressant			
Diazepam	USA	0.0029–0.00978/ 0.00128–0.0040	Subedi and Kannan (2015)
Fluoxetine	Canada	0.009–0.026/ 0.006–0.042	Lajeunesse, Smyth, Barclay, Sauvé, and Gagnon (2012)
Venlafaxine	Colombia, Kenya	0.003–0.074/ 0.003–0.049	Botero-Coy et al. (2018) and K'oreje et al. (2016)
Anticonvulsant			
Carbamazepine	USA, Australia	0.034–0.685/ 0.062–0.7	Roberts et al. (2016) and Yu et al. (2013)
Gabapentin	Singapore	4.825–15.3588/ 0.213–0.3654	Tran and Gin (2017)
Pregabalin	Germany	1.288–3.363/ 0.972–1.242	Gurke et al. (2015)

they are considered a significant class of pharmaceuticals that are commonly utilized for the prevention of treatment of diseases in human and veterinary medicine, development of livestock growth rate, and enhancing feed production. Antibiotics have been recognized as a significant class of contaminants, and in the past decades, this class was regarded as a new class of "contaminants of emerging concerns." This is because the presence of antibiotics in the environment poses potential effects on human health, fauna, and the aquatics (Tran et al., 2018). Similar to antibiotics, NSAIDs are deemed the most investigated group of PhACs. NSAIDs consist of pain-killing (analgesic) and fever-reducing (antipyretic) compounds, such as ibuprofen, ketoprofen, naproxen, acetaminophen, and diclofenac. NSAIDs pharmaceuticals are regularly detected in high concentration levels up to nearly 100 μg/L in different regions. It is also worth noted that with the development and improvements of analytical methods, it is possible that more PhACs will be identified in the future.

As shown in Table 8, the concentrations of the detected PhACs in the influents/effluents of WWTPs vary between different regions/countries. This variation could be attributed to several reasons, for example, the difference in the concentration of pharmaceuticals in the influents of WWTPs is mostly influenced by the consumption rate and pattern in that region/country/area, water consumption per capita per day, land use, weather conditions, population size and density, type of the sewer system, and environmental persistence. On the other hand, variation in the effluent of WWTPs is mainly caused and affected by the efficiency of wastewater treatment processes, and whether the treatment plant applies any tertiary or advanced treatment processes for the removal of persistent pollutants, physiochemical properties of these pharmaceutical substances, operating conditions (pH, temperature, sludge retention time SRT, hydraulic retention time HRT, biomass concentration, and age of the activated sludge applied in the WWTP) (Khasawneh & Palaniandy, 2021).

4.2 Occurrence in aquatic environments

The first report on the occurrence of pharmaceuticals in river water was published in 1970 (Madikizela et al., 2020). Ever since, thorough research has been devoted to the monitoring of PhACs in aquatic environments (i.e., surface water and groundwater). Although the major studies are dedicated to investigating the occurrence of PhACs in WWTPs and rivers, however, only before 2 decades, specifically in the early 2000s, the occurrence of PhACs was discovered in seawater (Weigel, Bester, & Hühnerfuss, 2001). This resulted in diverse and abundant scientific research dedicated to investigating the occurrence of pharmaceuticals in the marine environment (Arpin-Pont, Bueno, Gomez, & Fenet, 2016; Ojemaye & Petrik, 2019).

In this section, data on the occurrence of common PhACs in surface waters (SW), including rivers, lakes, and seawater, and groundwater (GW), which is obtained from several studies was summarized in Table 10. From Table 10, it can be noted that although the concentration of the detected pharmaceuticals in SW and GW is relatively high, however, it is still lower than that detected in the influents and effluents of WWTPs (Table 9). This could be attributed to the treatment processes applied in WWTPs as well as the natural processes occurring in the environmental, such as

Table 10 Concentration of commonly detected PhACs (μg/L) in the aquatic environments.

Class/ Pharmaceutical	Country	Concentration	References
Antibiotics			
Amoxicillin	Greece	<0.128 (SW)	Alygizakis et al. (2016)
Ampicillin	India	13.8 (SW)	Balakrishna, Rath, Praveenkumarreddy, Guruge, and Subedi (2017)
Ciprofloxacin	India	1.4 (SW)	Balakrishna et al. (2017)
Erythromycin	India	0.023–0.067 (GW)	Fick et al. (2009)
	USA	0.217 (SW)	Nödler et al. (2014)
Enrofloxacin	India	0.021–0.031 (GW)	Fick et al. (2009)
Norfloxacin	India	0.026–0.480 (GW)	Fick et al. (2009)
	Spain	n.d.–3.551 (SW)	Afonso-Olivares, Torres-Padrón, Sosa-Ferrera, and Santana-Rodríguez (2013)
Ofloxacin	Korea	n.d.–0.0124 (SW)	Kim, Lee, and Oh (2017)
Roxithromycin	Germany	0.016 (SW)	Nödler et al. (2014)
Sulfamethoxazole	Greece, Germany	<0.0001–0.042 (SW)	Alygizakis et al. (2016) and Nödler et al. (2014)
	Poland	0.0023–0.016 (GW)	Szymczycha et al. (2020)
Tetracycline	Spain	n.d.–0.063 (SW)	Biel-Maeso et al. (2018)
NSAIDs			
Ibuprofen	India	2.32 (SW)	Balakrishna et al. (2017)
	Singapore	0.041–0.121 (SW)	Wu, Qian, Yang, and Zhang (2010)
Ketoprofen	Spain, Portugal	n.d.–0.106 (SW)	Afonso-Olivares et al. (2013) and Lolić et al. (2015)
	Poland	0.027–0.7318 (GW)	Szymczycha et al. (2020)
Acetaminophen	Spain, Germany, Portugal	n.d.–0.584 (SW)	Afonso-Olivares et al. (2013), Lolić et al. (2015), and Nödler et al. (2014)
	USA	1.890	Fram and Belitz (2011)
Naproxen	Spain	n.d.–0.0958 (SW)	Alygizakis et al. (2016) and Biel-Maeso et al. (2018)
	South Africa	<LOD–0.160 (SW)	Ngubane, Naicker, Ncube, Chimuka, and Madikizela (2019)

Continued

Table 10 Continued

Class/ Pharmaceutical	Country	Concentration	References
Diclofenac	Spain, Greece, Taiwan	n.d.–0.344 (SW)	Afonso-Olivares et al. (2013), Alygizakis et al. (2016), Biel-Maeso et al. (2018), and Fang, Nan, Chin, and Feng (2012)
	Poland	0.0013–2.77 (GW)	Kapelewska et al. (2018) and Szymczycha et al. (2020)
Lipid regulators			
Bezafibrate	Spain, Turkey	n.d.–0.0035 (SW)	Biel-Maeso et al. (2018) and Nödler et al. (2014)
Fenofibrate	Spain, Tunisia	n.d.–0.014 (SW)	Afsa, Hamden, Lara Martin, and Mansour (2020) and Biel-Maeso et al. (2018)
Gemfibrozil	Singapore, USA, Spain, Taiwan, Turkey	<0.001–0.043 (SW)	Biel-Maeso et al. (2018), Jiang, Lee, and Fang (2014), Nödler et al. (2014), and Wu et al. (2010)
Atorvastatin	USA	n.d.–0.0004 (SW)	Vidal-Dorsch et al. (2012)
β-Blockers			
Atenolol	India	3.18 (SW)	Balakrishna et al. (2017)
	Spain, Turkey	0.0004–0.194 (SW)	Biel-Maeso et al. (2018) and Nödler et al. (2014)
Propranolol	Korea, Spain	n.d.–0.00119 (SW)	Biel-Maeso et al. (2018) and Kim et al. (2017)
Metoprolol	India	0.090 (GW)	Fick et al. (2009)
	Germany	0.158 (SW)	Nödler et al. (2014)
EDCs			
Estrone	Ireland, Singapore, USA, Germany	0.00008–0.011 (SW)	Bayen, Zhang, Desai, Ooi, and Kelly (2013), Beck, Bruhn, Gandrass, and Ruck (2005), Ronan and McHugh (2013), and Singh et al. (2010)
	Poland	<0.01–0.31 (GW)	Kapelewska et al. (2018)
Estradiol	Canada, USA	n.d.–0.018 (SW)	Robinson, Hui, Soo, and Hellou (2009) and Singh et al. (2010)
	Poland	<0.43 (GW)	Kapelewska et al. (2018)
Ethinylestradiol	Canada, Germany	n.d.–0.017 (SW)	Beck et al. (2005) and Robinson et al. (2009)
Stimulant			

Table 10 Continued

Class/ Pharmaceutical	Country	Concentration	References
Caffeine	Poland	0.024–1.029 (GW)	Szymczycha et al. (2020)
	USA	0.290	Fram and Belitz (2011)
Opiate drugs			
Codeine	Spain	0.0018, 0.0068 (SW)	Moreno-González, Rodriguez-Mozaz, Gros, Barceló, and León (2015)
Tramadol	Greece	<0.0001–0.001 (SW)	Alygizakis et al. (2016)
Amphetamine	India	0.984 (SW)	Balakrishna et al. (2017)
Antidepressant			
Fluoxetine	USA, Spain	n.d.–0.090 (SW)	Biel-Maeso et al. (2018), Fernández-Rubio et al. (2019), and Nödler et al. (2014)
Venlafaxine	Spain	n.d.–0.291 (SW)	Fernández-Rubio et al. (2019)
Anticonvulsant			
Carbamazepine	India	0.013 (SW)	Balakrishna et al. (2017)
	Greece, France, Germany, Taiwan, Turkey	0.00005–0.157 (SW)	Alygizakis et al. (2016), Jiang et al. (2014), Martínez Bueno et al. (2016), and Nödler et al. (2014)
	Poland	<0.50 (GW)	Kapelewska et al. (2018) and Szymczycha et al. (2020)
Disinfectant			
Triclocarban	India	6.18 (SW)	Balakrishna et al. (2017)

photolysis, biotransformation, sorption, volatilization, and dispersion, which can diminish and lower the concentration of pharmaceuticals in SW and GW environments (Li, 2014). Hence, the concentration of pharmaceuticals in WWTPs is generally higher.

4.3 Occurrence in soil

Soil is considered a major exposure path of toxic contaminants; therefore, it is imperative to emphasize the need to understand the chemical behavior of natural and man-made substances that could inhabit and reside in terrestrial environments. This knowledge will help in the identification of priority contaminants for evidence-based evaluation of environmental risk. Generally, emerging contaminants, in particular, PhACs, can reach the soil's surface via three major routes, these include (Zhang, Barron, & Sturzenbaum, 2021):

(i) direct dumping and disposal of waste,
(ii) deposition of contaminated sediments,
(iii) deposition of suspended particles from the atmosphere, and/or
(iv) through the circulation of groundwater.

Although it has been recognized that PhACs may accumulate in plants (Tanoue et al., 2012; Wu, Dodgen, Conkle, & Gan, 2015), however, investigations on the presence and fate of PhACs in soil compartments have been much less studied. This could be due to the limited sensitive and accurate analysis methods for the assessment of trace pharmaceuticals in complex matrices such as soils (Białk-Bielińska et al., 2016; Vazquez-Roig, Andreu, Blasco, & Picó, 2012). Nevertheless, to date, several analytical methods have been successfully employed for the identification and quantification of different pharmaceutical classes, including antibiotics, NSAIDs, estrogens, β-blockers, antiepileptic, and antidepressants in soil samples, as presented in Table 11. These studies allowed the detection and measurement of pharmaceutical drugs from different classes, but none of them allowed the detection of all several drugs belonging to all classes in a single run (Kumirska et al., 2019). Meanwhile,

Table 11 Concentration of commonly detected pharmaceuticals in soil compartments (μg/kg).

Class/ Pharmaceutical	Country	Concentration (μg/kg)	References
Antibiotics			
Norfloxacin	Spain	1.7, 8.4	Vazquez-Roig et al. (2012)
Ofloxacin	Spain	1.4, 3.3	Vazquez-Roig et al. (2012)
Sulfamethoxazole	Spain	<LOD	Vazquez-Roig et al. (2012)
Tetracycline	Spain	<LOD	Vazquez-Roig et al. (2012)
Triclosan	UK	774–949	Butler, Whelan, Ritz, Sakrabani, and van Egmond (2011)
Trimethoprim	USA	0.12–0.26	Franklin, Williams, and Watson (2018)
	Spain	0.01, 0.2	Vazquez-Roig et al. (2012)
NSAIDs			
Ibuprofen	Mexico	0.21–0.29	Durán-Alvarez, Becerril-Bravo, Castro, Jiménez, and Gibson (2009)
	Poland	<1	Kumirska et al. (2019)
	Spain	<LOD	Vazquez-Roig et al. (2012)
Acetaminophen	Poland	<1	Kumirska et al. (2019)
	Spain	0.9, 1.8	Vazquez-Roig et al. (2012)
Naproxen	Mexico	0.54–0.56	Durán-Alvarez et al. (2009)
	Poland	3	Kumirska et al. (2019)

Table 11 Continued

Class/ Pharmaceutical	Country	Concentration (μg/kg)	References
Diclofenac	Pakistan	101–257	Ashfaq et al. (2017)
	Poland	<3	Kumirska et al. (2019)
	Spain	<LOD	Vazquez-Roig et al. (2012)
Lipid regulators			
Clofibric acid	Spain	<LOD	Vazquez-Roig et al. (2012)
β-Blockers			
Propranolol	Spain	0.02, 0.4	Vazquez-Roig et al. (2012)
Metoprolol	Spain	0.1, 0.3	Vazquez-Roig et al. (2012)
EDCs			
Estrone	Poland	5.4, 7.2	Kumirska et al. (2019)
Estriol	Poland	1.9	Kumirska et al. (2019)
Ethinylestradiol	Poland	<1	Kumirska et al. (2019)
Opiate drugs			
Codeine	Spain	0.1, 1.7	Vazquez-Roig et al. (2012)
Antidepressant			
Diazepam	Spain	0.02, 0.3	Vazquez-Roig et al. (2012)
Anticonvulsant			
Carbamazepine	USA	0.19–0.55	Kinney, Furlong, Werner, and Cahill (2006)
	Mexico	5.89–7.07	Durán-Alvarez et al. (2009)
	Spain	0.3, 1.5	Vazquez-Roig et al. (2012)

as summarized in Table 11, considerable number of studies have uncovered the occurrence of PhACs in soil compartments located in different countries (Ashfaq et al., 2017; Franklin et al., 2018; Vazquez-Roig et al., 2012). It is worth noting from Table 11 that the majority of the commonly detected pharmaceuticals in soil samples reported to be from the following classes: antibiotics, NSAIDs, estrogen hormones, β-blockers, antidepressants, anticonvulsants, lipid regulators, and opiate drugs.

5. Fate of common pharmaceuticals in the environment

After the administration of PhACs in the human/animal body, they are introduced into sewage treatment plants through excretion (i.e., urine and/or feces). The shape of these pharmaceuticals is influenced by the processes applied at the sewage treatment plant.

Pharmaceuticals may be removed from wastewater via microbial breakdown or sorption to solids, which are then removed with sludge. PhACs that are more persistent are more likely to escape WWTPs, where are they finally released, transported, and distributed into different environmental compartments. Once they are administrated in the environment, the fate of pharmaceutical is governed by key processes; biotic and abiotic, which include biodegradation, photodegradation, and sorption (Fig. 3). While specific key factors, such as the physiochemical properties of pharmaceuticals (e.g., lipophilicity, vapor pressure, and water solubility), the microbial community, and environmental conditions (e.g., temperature, pH, redox conditions, and irradiation time and intensity), can affect the fate and transport of these pharmaceuticals into water-sediment environments (Khan et al., 2020; Monteiro & Boxall, 2010). Nevertheless, beside the previously mentioned key factors, the soil type, partitioning soil properties, and degradation in soil matrices can also influence their transport, distribution, and behavior in soil environment. Meanwhile, the potential risks posed by PhACs in the environment is reliant their persistence, biological activity of the TPs, toxicity, and bioaccumulation properties in the environment (Ebele, Abou-Elwafa Abdallah, & Harrad, 2017; Monteiro & Boxall, 2010).

5.1 Fate in wastewater treatment plants

Researchers have reported that WWTPs that meet the regulatory requirement for the treatment of wastewater are moderately efficient for the degradation and removal of PhACs, except for WWTPs with applied tertiary treatment system (e.g., UV disinfection, chlorination, ozonation, etc.) (Li, 2014). In fact, complete mineralization and/or

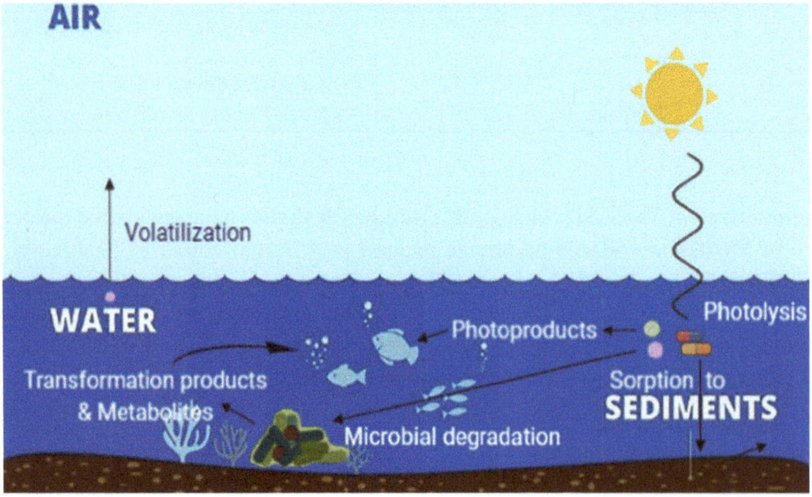

Fig. 3 Natural reduction processes of pharmaceuticals in the aquatic environment. Adapted from Khan, H. K., Rehman, M. Y. A., & Malik, R. N. (2020). Fate and toxicity of pharmaceuticals in water environment: An insight on their occurrence in South Asia. *Journal of Environmental Management, 271*, 111030. https://doi.org/10.1016/j.jenvman.2020.111030.

removal of pharmaceuticals are difficult to attain in most WWTPs. Therefore, large number of PhACs belonging to different classes (i.e., antibiotics, NSAIDs, lipid regulators, estrogens) can still be detected in the effluents of WWTPs and sludge samples (Khasawneh & Palaniandy, 2021; Tran et al., 2018). Examples on the occurrence of different pharmaceuticals have been previously summarized in Table 9. Some of these detected pharmaceutical pollutants exceeded their corresponding Predicted No Effect Concentration (PNEC) values, rendering their fate in the environment to draw attention of researchers and the public regarding concerns on their potential risks (Li, 2014). Water/wastewater treatment system that has been investigated for the removal and degradation of PhACs includes conventional treatment systems, advanced treatment systems, and/or combination of both. The conventional/traditional technologies applied in WWTPs include biological and physical-chemical systems, such as coagulation, sedimentation, activated sludge, filtration, adsorption/bioadsorption, and chemical disinfection (e.g., chlorination). Nonetheless, advanced treatment systems include physical and/or chemical techniques, such as advanced oxidation processes (e.g., UV photolysis, photocatalysis, ozonation, photo-Fenton reactions) and membrane filtration. Table 12 summarizes the removal efficiencies of some common PhACs in via different treatment techniques. From Table 12, it can be noted that there is a large discrepancy in the removal efficiencies of pharmaceuticals depending on the

Table 12 Removal efficiencies of some common PhACs via different treatment techniques in different matrices.

PhACs	Treatment applied	Removal rate (%)	References
Ampicillin	Conventional activated sludge	72	Yang et al. (2017)
Salicylic acid	Ozonation	95	Hu, Zhang, and Hu (2016)
Ciprofloxacin	Biodegradation by algae (*Chlamydomonas mexicana*)	13–56	Wang and Wang (2016)
Levofloxacin	Conventional activated sludge	50–80	Yang et al. (2017)
Sulfamethoxazole	Algae-mediated photolysis	32	Wang and Wang (2016)
	Biocatalyst laccase with mediators	100	Sharma, Ahmad, and Flora (2018)
	UV-photolysis	59	Lee et al. (2019)
Tetracycline	Conventional activated sludge	44	Yang et al. (2017)
Oxytetracycline	Conventional activated sludge	54.6–96.3	Tran et al. (2018)
Carbamazepine	Ozonation + photocatalysis	80	Jankunaite et al. (2017)
	Membrane bioreactor	<10	Sui et al. (2015)
	Conventional activated sludge	90	

Continued

Table 12 Continued

PhACs	Treatment applied	Removal rate (%)	References
Acetaminophen	Reverse osmosis	99	Taheran et al. (2016)
	Fe_2O_3-TiO_2 photocatalysis	98.85	Khasawneh et al. (2020)
	Activated carbon	90	Luján-Facundo, Iborra-Clar, Mendoza-Roca, and Alcaina-Miranda (2019)
Efavirenz	Conventional activated sludge	19–92	Madikizela, Ncube, and Chimuka (2018)
Atenolol	Conventional activated sludge	20	Madikizela et al. (2018)
Metoprolol	Conventional activated sludge	25	Sui et al. (2015)
Propranolol	Conventional activated sludge	60	Sui et al. (2015)
	Membrane bioreactor	66–78	
Codeine	Uptake by plant (*Iris pseudacorus*)	87	Madikizela et al. (2018)
Gemfibrozil	Nanofiltration	86.5	Taheran et al. (2016)
Estradiol	Biodegradation by algae (*Chlamydomonas reinhardtii*)	100	Wang and Wang (2016)
Naproxen	Reverse osmosis	95	Taheran et al. (2016)
Ibuprofen	Nanofiltration	86.5	Taheran et al. (2016)
	Activated carbon	25	Ek, Baresel, Magnér, Bergström, and Harding (2014)
	UV-photolysis	30	Lee et al. (2019)
Diclofenac	Ozonation	60	Jankunaite et al. (2017)
	Uptake by plant (*Ceratophyllum demersum*)	99	Madikizela et al. (2018)
	Membrane bioreactor	63–65	Sui et al. (2015)
	UV-photolysis	85	Lee et al. (2019)
Ketoprofen	Conventional activated sludge	51.5–91.9	Tran et al. (2018)
Ofloxacin	Conventional activated sludge	8	Yang et al. (2017)
Cyclophosphamide	$UV/O_3/H_2O_2$	99	Sharma et al. (2018)

applied treatment technology. These observed variations in the removal efficiencies are influenced by several factors, such as the physiochemical properties of pharmaceutical pollutants, operation conditions (e.g., pH, temperature, biomass concentration, hydraulic retention time HRT, solids retention time SRT, and age of the activated sludge in the WWTP), properties and type of the treatment process, and microbial community, climate conditions, sampling method (i.e., grab sampling composite sampling, pool sampling), and seasonal variations (Khan et al., 2020; Khasawneh & Palaniandy, 2021).

5.1.1 Conventional wastewater treatment

Conventional wastewater treatment systems consist of primary treatment, secondary treatment, and occasionally, tertiary, or advanced technologies. Primary treatment, which is a physical process that is common almost for all urbane WWTPs, may include aerated grift chamber and/or sedimentation tank to separate the suspended solids (SS), such as grease, fats, oils, sands, and grits through filtration or sedimentation from influent wastewater. This treatment step can be assessed based on the reduction in the number of SS and/or biological oxygen demand (BOD) of the primary treated effluent (Gogoi et al., 2018; Khasawneh & Palaniandy, 2021). Removal efficacy of 28% was reported for some pharmaceuticals, e.g., diclofenac and estriol in the primary treatment stage (Luo et al., 2014). Nevertheless, findings revealed that primary treatment is not efficient nor designed for the removal of PhACs due to their hydrophilic nature (i.e., log $D_{ow} < 1.0$) (Verlicchi, Al Aukidy, & Zambello, 2012). Secondary treatment step, such as conventional activated sludge CAS, trickling filter TF, fixed bed bioreactors FBR, moving bed biofilm reactor MBFR, and membrane bioreactors MBR, normally employ biological processes (aerobic and/or anaerobic) for the degradation of organic substances and nutrients. The biological removal process of PhACs occurs through different steps: dilution, partition, adsorption, biodegradation, and biotransformation, which results in complete or partial mineralization/degradation of the parent compounds (i.e., transformation products) (Khasawneh & Palaniandy, 2021). While organic substances and nutrients are removed through the formation of floc under certain conditions by using AS and dissolved oxygen. Lastly, tertiary treatment processes, such as UV disinfection, chlorination, ozonation, membrane filtration, activated carbon, etc., are utilized to improve the whole treatment process and discharge higher quality effluents. Other organics, e.g., phosphorus can be effectively removed (Batt, Kim, & Aga, 2007). Although they can also be effective for the removal of some PhACs, however, findings showed that even after applying tertiary treatment, considerable number of pharmaceuticals can still occur in relatively high concentrations in the final effluent. This could be attributed to the complex nature of pharmaceuticals, their physiochemical properties and persistence, and operating conditions of the tertiary treatment system(s) (Khasawneh & Palaniandy, 2021). It is also worth mentioning that tertiary treatment systems are not extensively used due to their high energy consumption and operating cost (Gracia-Lor et al., 2012).

Only small fraction of PhACs is removed by sorbing to sludge. For example, tetracyclines and fluoroquinolones are more likely to be absorbed to the sewage sludge than biodegraded, due to their sorption properties (Monteiro & Boxall, 2010). While the majority of other pharmaceuticals are discharged with the final effluent. Furthermore, biological, chemical, and photolysis process may transfer PhACs into other forms (i.e., transformation products) via partial oxidation, that can be even more toxic than their parent compounds, e.g., epoxy-carbamazepine, di-OH-carbamazepine, *N*-desmethyl tramadol, hydroxy-diclofenac, *N*-(4-carbamoyl-2-imino-5-oxoimidazolidin)-formamido-*N*-methoxyacetic acid (COFA) and carboxy-acyclovir (Kharel et al., 2021; Schlüter-Vorberg et al., 2015). Another type of TP that can be formed during tertiary treatment is disinfection by products (DBPs). DBPs are formed when organic substances (i.e., pharmaceuticals) present with water and react with disinfection agents (e.g., chlorine, ozone, etc.). To date, more than 600 DBPs have been reported in the literature (Richardson, 2011; Yang et al., 2017). Ozonation (to some extent) is one of the effective technologies for the removal of pharmaceuticals from the effluents of WWTPs. However, the formation of ozonation by-products (products resulted from reactions between ozone and WW matrix) and ozonation TPs (products resulted from reactions between ozone and to be removed contaminants) is inevitable. Nonetheless, the potential adverse effects from such technology can be reduced and mitigated by using a posttreatment; either by biological systems (e.g., polishing bonds, sand-filter, and moving bed biofilm reactors), or adsorptive/biological systems that are based on granular activated carbon (GAC). These techniques are recommended to mitigate the occurrence of TPs and the biologically degradable by-products (Kharel et al., 2021).

5.2 Fate in the aquatic environment

After the final treatment of wastewater effluent, it is discharged into the aquatic environment (i.e., rivers, marine environments, etc.). As a result, persistent pollutants such as PhACs are introduced into these environments. Once these pharmaceutical pollutants enter the aquatic environments, e.g., surface water, they undergo several processes which mainly include sedimentation, adsorption, bioaccumulation, biodegradation, photolysis, and hydrolysis (Evgenidou, Konstantinou, & Lambropoulou, 2015; Wang et al., 2021). The environmental mobility of these pharmaceuticals is influenced by their physiochemical properties, water matrix (i.e., presence of heavy metals), pH, redox conditions (i.e., aerobic, anaerobic, and anoxic), and soil permeability. Moreover, degradation and photolysis rate of pharmaceuticals in surface waters is dependent on several factors, such as organic composition, solar irradiation intensity and duration, and eutrophication conditions (Wang et al., 2021). In the meanwhile, photodegradation is considered the predominant removal method for some pharmaceuticals, e.g., diclofenac, with half-life <1 h, and resulting in three TPs (via photolysis), which are 2-chloroaniline, 2,6-dichlorophenol, and 2,6-dichloroaniline (Fent, Weston, & Caminada, 2006).

5.2.1 Partitioning

Partitioning is a process of significant importance for determining the fate of PhACs in the aquatic environment. Generally, two main processes are accountable for the partitioning of PhACs between the water and sediment phase, namely: sorption and desorption. These two processes can also influence the bioavailability of pharmaceuticals to the aquatic biota (Ying et al., 2013). Sorption can also influence other processes in WWTPs or surface waters, e.g., biodegradation and photolysis (Tran et al., 2018). Nonetheless, given that PhACs can be either weak bases, weak acids, or zwitterions, sorption is not only driven by hydrophobic interactions, but also motivated by another binding methods (e.g., cation bridging, cation exchange, hydrogen bonding, and surface complexation). Furthermore, sorption of PhACs onto sediments or particles could trigger a loss in bioactivity and detectability, which is caused by the irreversible bonding. Several factors can affect the sorption of pharmaceuticals, such as water sediment properties (i.e., content of the organic matter and pH value) and nature and amount of the SS particles in the water matrix (Ying et al., 2013).

In general, neutral PhACs and carboxylic acids tend to have lesser sorption coefficients compared to amines due to the low content of organics. Although solubility and/or log K_{ow} (log$P$) of pharmaceuticals are not adequate for the assessing the sorption rate and behavior of pharmaceuticals (i.e., because they contain polar groups), however, it has been suggested that compounds with higher $\log K_{ow} > 5$ and molecular weight are more likely to be absorbed onto sediments (Pal, Gin, Lin, & Reinhard, 2010). On the other hand, pharmaceuticals with $\log K_{ow} < 2.5$ have lower sorption properties and are more likely to persist in the water surface on in soil zones (Mompelat, Le Bot, & Thomas, 2009).

To better assess the sorption behavior of pharmaceuticals, sorption coefficient (K_d) is commonly used. For instance, it is worth to mention that the pharmaceutical group, antibiotics, has different classes within the same group with various physiochemical properties. It has been reported that K_d values for sulfonamides fall in the range 0.6–7.4 L/kg in soil or sediments, while for tetracyclines, the K_d reported values were from 417 to 2386 L/kg for soil and sediments. Generally, according to the literature, the strength of the K_d for different antibiotics classes followed the following order: fluoroquinolones $\sim$ tetracyclines > macrolides > sulfonamides. On the contrast, NSAIDs have low sorption coefficient values (<1 to >100 L/kg) and low K_{oc} values (100 to >1000 L/kg) due to their high hydrophilicity, implying that they are only partitioned in water phase in the aquatic environments, rather than being absorbed into soil or sediments (Yamamoto et al., 2009; Ying et al., 2013).

5.2.2 Hydrolysis and photolysis

The abiotic degradation of PhACs in surface waters mainly occurs through two processes, which include hydrolysis and photolysis. Since PhACs are taken orally and are generally resistant to hydrolysis process, this mostly render photolysis the prevailing process for the abiotic transformation of pharmaceuticals in the aquatic environments (Monteiro & Boxall, 2010). For example, it has been reported that photodegradation is the predominant process for the removal of diclofenac in surface waters (Fent et al., 2006). Although hydrolysis is a significant process for the degradation of some PhACs (e.g., antibiotics) in the aquatic environment (Wang et al., 2021), however, not all PhACs can hydrolyze (undergo hydrolysis) in water compartments (e.g., acidic drugs and steroids). On the contrary, antibiotics, e.g., sulfonamides and tetracyclines, can hydrolyze in water, and in fact, hydrolysis is considered an important and sensitive process that can affect the fate of such antibiotics in the environment. The optimum conditions for the hydrolysis of sulfonamides are under acidic pH solutions, followed by neutral and alkaline solutions, while an increase in the temperature of the solution can boost the hydrolysis of such compounds (Białk-Bielińska et al., 2012; Ying et al., 2013). Other pharmaceutical groups such as fluoroquinolones, are not influenced by hydrolysis. For example, studies showed that some fluoroquinolones, e.g., flumequine and oxolinic acid were found not to hydrolyze in three different types of water matrices (DI, seawater, and freshwater) (Pouliquen, Delépée, Larhantec-Verdier, Morvan, & Le Bris, 2007). Similarly, the following pharmaceuticals: florfenicol, trimethoprim, and lincomycin were also found to be insensitive to hydrolysis in aquatic environments (Loftin, Adams, Meyer, & Surampalli, 2008).

Apart from hydrolysis, which is commonly negligible (Ebele et al., 2017), Photolysis process, also known as photodegradation, is considered the most substantial process for the degradation of pharmaceuticals (e.g., antibiotics) in the natural aquatic systems. For instance, studies reported that photolysis is the major process responsible for the degradation of sulfonamides in natural waters (Li et al., 2018). Photolysis/photodegradation can be defined as "the direct or indirect chemical transformation of a compound by light-induced mechanism" (Boreen, Arnold, & McNeill, 2003). The process of photodegradation of PhACs in surface water is dependent on several factors, including sensitivity of target pharmaceuticals to sunlight, solar irradiation intensity, latitude, composition of organic substances/dissolved organic matter (DOM) in water, inorganic ions, absorption spectra of the present pharmaceuticals, water pH, chemical properties and structure of target pharmaceuticals, water depth, turbidity, and climate conditions (Scaria, Anupama, & Nidheesh, 2021; Wang et al., 2021; Ying et al., 2013). Studies have showed that some pharmaceuticals (mefenamic acid and acetaminophen) have slow degradation rate under sunlight radiations, and even poor for other compounds, such as carbamazepine and ibuprofen (Yamamoto et al., 2009).

The direct photodegradation of the PhACs group, sulfonamides can vary considerably depending on the source of irradiation (i.e., artificial light or natural sunlight), water matrix, and pH value (Boreen et al., 2003). Various investigations reported the photodegradation of sulfonamides under natural sunlight radiations, and found that the rate constant can range from 10^{-7} to $1.3 \times 10^{-4}\ s^{-1}$ (Ge et al., 2019; Li, Zhao, Feng, Huang, & Sun, 2021). Meanwhile, the half-life ($t_{1/2}$) value of sulfonamides in the winter season was reported to be 31.06–1925.41 h, which is about 1–2 order of magnitude higher than that in the summer season (3.0–44.78 h), indicating that seasonal variation can influence the photodegradation of sulfonamides (Ge et al., 2019). Nevertheless, the photodegradation rates of sulfonamides TPs are considerably distinct compared with their parent compounds. This is because of the effect of substituents on the aniline ring (Li et al., 2021).

5.3 Fate in soils

PhACs are known for their persistence under biological treatment processes. The fact that the majority of WWTPs are not designed to eliminate such compounds, render WWTPs the major source of these persistent pollutants into the environment. As shown in Fig. 1, PhACs can reach soils via different paths. Pharmaceuticals that are absorbed onto the sludge can reach soil compartments when the sludge is utilized as a fertilizer (biosolids) in farmlands to increase the nutrients of the local soil and improve agricultural yields (Zhang, Barron, & Sturzenbaum, 2021). In addition, disposal of animal manure, animal grazing, accidental spills during the manufacturing/distribution of PhACs, and the dumping of unused/expired pharmaceutical drugs are considered other significant routes that contribute to the release and occurrence of pharmaceuticals in soil environments (Li, 2014; Luque-Muñoz, Vílchez, & Zafra-Gómez, 2017). The occurrence and persistence of the pharmaceutical pollutants in solid matrices for long periods could cause potential toxic effects toward the terrestrial organisms (Goeppert, Dror, & Berkowitz, 2014; Li, 2014). For instance, studies have detected estrogenic activity at all sampling locations when biosolids where applied to land (Langdon, Warne, Smernik, Shareef, & Kookana, 2014). Moreover, pharmaceuticals present in soils can as well be transferred and accumulated into plants, as reported in several studies (Kumirska et al., 2019; Puckowski et al., 2016).

The identification and detection of PhACs in biosolids and soils has triggered concerns worldwide due to their potential to be bioactive and/or toxicity toward organisms even at trace levels. Certain compounds can trigger the development of "microbial resistance" (e.g., antibiotics), while other compounds can exhibit acute toxicity (e.g., diclofenac), resulting in endocrine disruption. The major pharmaceuticals that are administrated into soil environments include NSAIDs, antibiotics, β-blockers, lipid regulators, psychostimulants, hormones, and estrogens. Nevertheless, NASAIDs (diclofenac and ibuprofen), antibiotics (sulfamethoxazole, trimethoprim, and triclosan), and the antiepileptic (carbamazepine) are the most common detected pharmaceuticals in soil compartments (Table 11). Meanwhile, additional attention was dedicated to PhACs entering agricultural areas through the application of biosolids and wastewater due to their significant potential risk on the food chain. For instance, a study detected the presence of carbamazepine and its metabolites (i.e., transformation products) in human urine after the consumption of fresh crop that was irrigated with treated wastewater (Paltiel et al., 2016). The fate of PhACs in soil, including degradation, transformation, transportation, and uptake by plants and organisms is influenced by several factors: initial concentration of pharmaceuticals, physiochemical properties of pharmaceuticals, and soil characteristics (e.g., soil pH and ionic strength). The physiochemical properties of PhACs, specifically octanol-water partition coefficient ($\log K_{ow}$), solid-water distribution coefficient ($\log K_d$), and polarity, affect the fate of pharmaceuticals within the soil matrix. Other properties, such as cation-bridging, ionization, and retention in stagnant pore water are also factors that can influence the fate, movement, and behavior of pharmaceutical pollutants in soil environments (Carter et al., 2014; Zhang, Barron, & Sturzenbaum, 2021). The incidence of positive structures, such as amino groups and triazine rings, results in upsurging the sorption (both adsorption and absorption) affinity of lamotrigine, when related to carbamazepine, ensuing in the improved capability of the molecules to generate hydrogen bonds with functional groups on the surfaces of soil, and particularly polar soil organic matter. Other factors, such as soil components, can influence the fate, movement, and behavior of pharmaceuticals in soils. The presence of soil organic carbon (OC) and dissolved organic matter (DOM) in pore water can inhibit or improve the movement of pharmaceuticals in soil (Monteiro & Boxall, 2010; Zhang, Barron, & Sturzenbaum, 2021).

6. Conclusions

This chapter has introduced and discussed the problem of PhACs occurrence, sources, and fate in the environment. PhACs belonging to several classes: antibiotics, NSAIDs, lipid regulators, beta-blockers, EDCs, antidepressants, anticonvulsants, stimulants, opiate drugs, and antidiabetic drugs were included and reviewed in this chapter. Information on the PhACs that are commonly found in WWTPs, and the environment were provided and reviewed, along with their physiochemical properties that can significantly affect their fate and transport in WWTPs and the environment. In addition, reactions involved in the metabolism and transformation processes and mechanisms were introduced. TPs that are formed during wastewater treatment processes or in the environment and their toxic effects toward the environment were discussed. Metabolism and transformation processes were found to influence the fate of PhACs in the environment. The occurrence of common PhACs in WWTPs, aquatic environments (i.e., rivers and marine water), and soils in different regions were summarized and reviewed. Finally, the fate of common PhACs in different compartments of the environment (WWTPs, aquatic environment, and soils) was discussed. The factors that can potentially influence the fate of PhACs were also pointed out and debated.

References

Aberg, D., Chaplin, D., Freeman, C., Paizs, B., & Dunn, C. (2022). The environmental release and ecosystem risks of illicit drugs during Glastonbury festival. *Environmental Research, 204*, 112061. https://doi.org/10.1016/j.envres.2021.112061.

Afonso-Olivares, C., Sosa-Ferrera, Z., & Santana-Rodríguez, J. J. (2017). Occurrence and environmental impact of pharmaceutical residues from conventional and natural wastewater treatment plants in Gran Canaria (Spain). *Science of the Total Environment, 599–600*, 934–943. https://doi.org/10.1016/J.SCITOTENV.2017.05.058.

Afonso-Olivares, C., Torres-Padrón, M., Sosa-Ferrera, Z., & Santana-Rodríguez, J. (2013). Assessment of the presence of pharmaceutical compounds in seawater samples from coastal area of Gran Canaria Island (Spain). *Antibiotics, 2*(2), 274–287. https://doi.org/10.3390/antibiotics2020274.

Afsa, S., Hamden, K., Lara Martin, P. A., & Mansour, H. B. (2020). Occurrence of 40 pharmaceutically active compounds in hospital and urban wastewaters and their contribution to Mahdia coastal seawater contamination. *Environmental Science and Pollution Research, 27*(2), 1941–1955. https://doi.org/10.1007/s11356-019-06866-5.

Al-Maadheed, S., Goktepe, I., Latiff, A. B. A., & Shomar, B. (2019). Antibiotics in hospital effluent and domestic wastewater treatment plants in Doha, Qatar. *Journal of Water Process Engineering, 28*, 60–68. https://doi.org/10.1016/J.JWPE.2019.01.005.

Almeida, Â., Calisto, V., Esteves, V. I., Schneider, R. J., Figueira, E., Soares, A. M. V. M., et al. (2021). Can ocean warming alter sub-lethal effects of antiepileptic and antihistaminic pharmaceuticals in marine bivalves? *Aquatic Toxicology*, *230*, 105673. https://doi.org/10.1016/j.aquatox.2020.105673.

Almeida, Â., Soares, A. M. V. M., Esteves, V. I., & Freitas, R. (2021). Occurrence of the antiepileptic carbamazepine in water and bivalves from marine environments: A review. *Environmental Toxicology and Pharmacology*, *86*, 103661. https://doi.org/10.1016/j.etap.2021.103661.

Alygizakis, N. A., Gago-Ferrero, P., Borova, V. L., Pavlidou, A., Hatzianestis, I., & Thomaidis, N. S. (2016). Occurrence and spatial distribution of 158 pharmaceuticals, drugs of abuse and related metabolites in offshore seawater. *Science of the Total Environment*, *541*, 1097–1105. https://doi.org/10.1016/j.scitotenv.2015.09.145.

Arpin-Pont, L., Bueno, M. J. M., Gomez, E., & Fenet, H. (2016). Occurrence of PPCPs in the marine environment: A review. *Environmental Science and Pollution Research*, *23*(6), 4978–4991. https://doi.org/10.1007/s11356-014-3617-x.

Ashfaq, M., Nawaz Khan, K., Saif Ur Rehman, M., Mustafa, G., Faizan Nazar, M., Sun, Q., et al. (2017). Ecological risk assessment of pharmaceuticals in the receiving environment of pharmaceutical wastewater in Pakistan. *Ecotoxicology and Environmental Safety*, *136*, 31–39. https://doi.org/10.1016/j.ecoenv.2016.10.029.

Balakrishna, K., Rath, A., Praveenkumarreddy, Y., Guruge, K. S., & Subedi, B. (2017). A review of the occurrence of pharmaceuticals and personal care products in Indian water bodies. *Ecotoxicology and Environmental Safety*, *137*, 113–120. https://doi.org/10.1016/j.ecoenv.2016.11.014.

Bartelt-Hunt, S. L., Snow, D. D., Damon, T., Shockley, J., & Hoagland, K. (2009). The occurrence of illicit and therapeutic pharmaceuticals in wastewater effluent and surface waters in Nebraska. *Environmental Pollution*, *157*(3), 786–791. https://doi.org/10.1016/J.ENVPOL.2008.11.025.

Basiuk, M., Brown, R. A., Cartwright, D., Davison, R., & Wallis, P. M. (2017). Trace organic compounds in rivers, streams, and wastewater in southeastern Alberta, Canada. *Inland Waters*, *7*(3), 283–296. https://doi.org/10.1080/20442041.2017.1329908.

Batt, A. L., Kim, S., & Aga, D. S. (2007). Comparison of the occurrence of antibiotics in four full-scale wastewater treatment plants with varying designs and operations. *Chemosphere*, *68*(3), 428–435. https://doi.org/10.1016/j.chemosphere.2007.01.008.

Bayen, S., Zhang, H., Desai, M. M., Ooi, S. K., & Kelly, B. C. (2013). Occurrence and distribution of pharmaceutically active and endocrine disrupting compounds in Singapore's marine environment: Influence of hydrodynamics and physical–chemical properties. *Environmental Pollution*, *182*, 1–8. https://doi.org/10.1016/j.envpol.2013.06.028.

Beck, I.-C., Bruhn, R., Gandrass, J., & Ruck, W. (2005). Liquid chromatography–tandem mass spectrometry analysis of estrogenic compounds in coastal surface water of the Baltic Sea. *Journal of Chromatography A*, *1090*(1), 98–106. https://doi.org/10.1016/j.chroma.2005.07.013.

Bekhradnia, A. R., & Ebrahimzadeh, M. A. (2012). Theoretical study on some non-selective beta-adrenergic antagonists and correlation to their biologically active configurations. *Medicinal Chemistry Research*, *21*(9), 2571–2578. https://doi.org/10.1007/s00044-011-9781-3.

Ben, W., Zhu, B., Yuan, X., Zhang, Y., Yang, M., & Qiang, Z. (2018). Occurrence, removal and risk of organic micropollutants in wastewater treatment plants across China: Comparison of wastewater treatment processes. *Water Research*, *130*, 38–46. https://doi.org/10.1016/J.WATRES.2017.11.057.

Białk-Bielińska, A., Kumirska, J., Borecka, M., Caban, M., Paszkiewicz, M., Pazdro, K., et al. (2016). Selected analytical challenges in the determination of pharmaceuticals in drinking/

marine waters and soil/sediment samples. *Journal of Pharmaceutical and Biomedical Analysis*, *121*, 271–296. https://doi.org/10.1016/j.jpba.2016.01.016.

Białk-Bielińska, A., Stolte, S., Matzke, M., Fabiańska, A., Maszkowska, J., Kołodziejska, M., et al. (2012). Hydrolysis of sulphonamides in aqueous solutions. *Journal of Hazardous Materials*, *221–222*, 264–274. https://doi.org/10.1016/j.jhazmat.2012.04.044.

Biel-Maeso, M., Baena-Nogueras, R. M., Corada-Fernández, C., & Lara-Martín, P. A. (2018). Occurrence, distribution and environmental risk of pharmaceutically active compounds (PhACs) in coastal and ocean waters from the Gulf of Cadiz (SW Spain). *Science of the Total Environment*, *612*, 649–659. https://doi.org/10.1016/j.scitotenv.2017.08.279.

Boreen, A. L., Arnold, W. A., & McNeill, K. (2003). Photodegradation of pharmaceuticals in the aquatic environment: A review. *Aquatic Sciences*, *65*(4), 320–341. https://doi.org/10.1007/s00027-003-0672-7.

Botero-Coy, A. M., Martínez-Pachón, D., Boix, C., Rincón, R. J., Castillo, N., Arias-Marín, L. P., et al. (2018). An investigation into the occurrence and removal of pharmaceuticals in Colombian wastewater. *Science of the Total Environment*, *642*, 842–853. https://doi.org/10.1016/J.SCITOTENV.2018.06.088.

Bulloch, D. N., Lavado, R., Forsgren, K. L., Beni, S., Schlenk, D., & Larive, C. K. (2012). Analytical and biological characterization of halogenated gemfibrozil produced through chlorination of wastewater. *Environmental Science & Technology*, *46*(10), 5583–5589. https://doi.org/10.1021/es3006173.

Burch, K. D., Han, B., Pichtel, J., & Zubkov, T. (2019). Removal efficiency of commonly prescribed antibiotics via tertiary wastewater treatment. *Environmental Science and Pollution Research*, *26*(7), 6301–6310. https://doi.org/10.1007/s11356-019-04170-w.

Burns, E. E., Carter, L. J., Kolpin, D. W., Thomas-Oates, J., & Boxall, A. B. A. (2018). Temporal and spatial variation in pharmaceutical concentrations in an urban river system. *Water Research*, *137*, 72–85. https://doi.org/10.1016/J.WATRES.2018.02.066.

Butler, E., Whelan, M. J., Ritz, K., Sakrabani, R., & van Egmond, R. (2011). Effects of triclosan on soil microbial respiration. *Environmental Toxicology and Chemistry*, *30*(2), 360–366. https://doi.org/10.1002/etc.405.

Carter, L. J., Harris, E., Williams, M., Ryan, J. J., Kookana, R. S., & Boxall, A. B. A. (2014). Fate and uptake of pharmaceuticals in soil–plant systems. *Journal of Agricultural and Food Chemistry*, *62*(4), 816–825. https://doi.org/10.1021/jf404282y.

Causanilles, A., Ruepert, C., Ibáñez, M., Emke, E., Hernández, F., & de Voogt, P. (2017). Occurrence and fate of illicit drugs and pharmaceuticals in wastewater from two wastewater treatment plants in Costa Rica. *Science of the Total Environment*, *599–600*, 98–107. https://doi.org/10.1016/J.SCITOTENV.2017.04.202.

Čelić, M., Škrbić, B. D., Insa, S., Živančev, J., Gros, M., & Petrović, M. (2020). Occurrence and assessment of environmental risks of endocrine disrupting compounds in drinking, surface and wastewaters in Serbia. *Environmental Pollution*, *262*, 114344. https://doi.org/10.1016/j.envpol.2020.114344.

Chang, C.-T., Wang, J.-J., Ouyang, T., Zhang, Q., & Jing, Y.-H. (2015). Photocatalytic degradation of acetaminophen in aqueous solutions by TiO2/ZSM-5 zeolite with low energy irradiation. *Materials Science and Engineering B*, *196*, 53–60. https://doi.org/10.1016/j.mseb.2014.12.025.

Cosenza, A., Maida, C. M., Piscionieri, D., Fanara, S., Di Gaudio, F., & Viviani, G. (2018). Occurrence of illicit drugs in two wastewater treatment plants in the South of Italy. *Chemosphere*, *198*, 377–385. https://doi.org/10.1016/j.chemosphere.2018.01.158.

Crisp, T. M., Clegg, E. D., Cooper, R. L., Anderson, D. G., Baetcke, K. P., Hoffmann, J. L., et al. (1997). *Special report on environmental endocrine disruption: An effects assessment and analysis*. Washington, DC: US Environmental Protection Agency.

D'Alessio, M., Onanong, S., Snow, D. D., & Ray, C. (2018). Occurrence and removal of pharmaceutical compounds and steroids at four wastewater treatment plants in Hawai'i and their environmental fate. *Science of the Total Environment*, *631–632*, 1360–1370. https://doi.org/10.1016/J.SCITOTENV.2018.03.100.

de Jesus Gaffney, V., Cardoso, V. V., Cardoso, E., Teixeira, A. P., Martins, J., Benoliel, M. J., et al. (2017). Occurrence and behaviour of pharmaceutical compounds in a Portuguese wastewater treatment plant: Removal efficiency through conventional treatment processes. *Environmental Science and Pollution Research*, *24*(17), 14717–14734. https://doi.org/10.1007/s11356-017-9012-7.

Donner, E., Kosjek, T., Qualmann, S., Kusk, K. O., Heath, E., Revitt, D. M., et al. (2013). Ecotoxicity of carbamazepine and its UV photolysis transformation products. *Science of the Total Environment*, *443*, 870–876. https://doi.org/10.1016/j.scitotenv.2012.11.059.

Durán-Alvarez, J. C., Becerril-Bravo, E., Castro, V. S., Jiménez, B., & Gibson, R. (2009). The analysis of a group of acidic pharmaceuticals, carbamazepine, and potential endocrine disrupting compounds in wastewater irrigated soils by gas chromatography–mass spectrometry. *Talanta*, *78*(3), 1159–1166. https://doi.org/10.1016/j.talanta.2009.01.035.

Ebele, A. J., Abou-Elwafa Abdallah, M., & Harrad, S. (2017). Pharmaceuticals and personal care products (PPCPs) in the freshwater aquatic environment. *Emerging Contaminants*, *3*(1), 1–16. https://doi.org/10.1016/J.EMCON.2016.12.004.

Ebele, A. J., Oluseyi, T., Drage, D. S., Harrad, S., & Abou-Elwafa Abdallah, M. (2020). Occurrence, seasonal variation and human exposure to pharmaceuticals and personal care products in surface water, groundwater and drinking water in Lagos State, Nigeria. *Emerging Contaminants*, *6*, 124–132. https://doi.org/10.1016/j.emcon.2020.02.004.

Ek, M., Baresel, C., Magnér, J., Bergström, R., & Harding, M. (2014). Activated carbon for the removal of pharmaceutical residues from treated wastewater. *Water Science and Technology*, *69*(11), 2372–2380. https://doi.org/10.2166/wst.2014.172.

Ekpeghere, K. I., Sim, W.-J., Lee, H.-J., & Oh, J.-E. (2018). Occurrence and distribution of carbamazepine, nicotine, estrogenic compounds, and their transformation products in wastewater from various treatment plants and the aquatic environment. *Science of the Total Environment*, *640–641*, 1015–1023. https://doi.org/10.1016/J.SCITOTENV.2018.05.218.

Evgenidou, E. N., Konstantinou, I. K., & Lambropoulou, D. A. (2015). Occurrence and removal of transformation products of PPCPs and illicit drugs in wastewaters: A review. *Science of the Total Environment*, *505*, 905–926. https://doi.org/10.1016/j.scitotenv.2014.10.021.

Fang, T.-H., Nan, F.-H., Chin, T.-S., & Feng, H.-M. (2012). The occurrence and distribution of pharmaceutical compounds in the effluents of a major sewage treatment plant in Northern Taiwan and the receiving coastal waters. *Marine Pollution Bulletin*, *64*(7), 1435–1444. https://doi.org/10.1016/J.MARPOLBUL.2012.04.008.

Fent, K., Weston, A. A., & Caminada, D. (2006). Ecotoxicology of human pharmaceuticals. *Aquatic Toxicology*, *76*(2), 122–159. https://doi.org/10.1016/j.aquatox.2005.09.009.

Fernández-López, C., Guillén-Navarro, J. M., Padilla, J. J., & Parsons, J. R. (2016). Comparison of the removal efficiencies of selected pharmaceuticals in wastewater treatment plants in the region of Murcia, Spain. *Ecological Engineering*, *95*, 811–816. https://doi.org/10.1016/J.ECOLENG.2016.06.093.

Fernández-Rubio, J., Rodríguez-Gil, J. L., Postigo, C., Mastroianni, N., López de Alda, M., Barceló, D., et al. (2019). Psychoactive pharmaceuticals and illicit drugs in coastal waters of North-Western Spain: Environmental exposure and risk assessment. *Chemosphere*, *224*, 379–389. https://doi.org/10.1016/j.chemosphere.2019.02.041.

Fick, J., Söderström, H., Lindberg, R. H., Phan, C., Tysklind, M., & Larsson, D. G. J. (2009). Contamination of surface, ground, and drinking water from pharmaceutical production.

Environmental Toxicology and Chemistry, *28*(12), 2522–2527. https://doi.org/10.1897/09-073.1.

Fram, M. S., & Belitz, K. (2011). Occurrence and concentrations of pharmaceutical compounds in groundwater used for public drinking-water supply in California. *Science of the Total Environment*, *409*(18), 3409–3417. https://doi.org/10.1016/j.scitotenv.2011.05.053.

Franklin, A. M., Williams, C. F., & Watson, J. E. (2018). Assessment of soil to mitigate antibiotics in the environment due to release of wastewater treatment plant effluent. *Journal of Environmental Quality*, *47*(6), 1347–1355. https://doi.org/10.2134/jeq2018.02.0076.

Galbraith, A., Bullock, S., Manias, S., Hunt, B., & Richards, A. (2004). Pharmacokinetics: Absorption and distribution. In *Fundamentals of pharmacology: A text for nurses and health professionals* (4th ed., pp. 109–114). Essex, England: Pearson Education Limited.

Gao, P., Ding, Y., Li, H., & Xagoraraki, I. (2012). Occurrence of pharmaceuticals in a municipal wastewater treatment plant: Mass balance and removal processes. *Chemosphere*, *88*(1), 17–24. https://doi.org/10.1016/J.CHEMOSPHERE.2012.02.017.

Ge, L., Zhang, P., Halsall, C., Li, Y., Chen, C.-E., Li, J., et al. (2019). The importance of reactive oxygen species on the aqueous phototransformation of sulfonamide antibiotics: Kinetics, pathways, and comparisons with direct photolysis. *Water Research*, *149*, 243–250. https://doi.org/10.1016/j.watres.2018.11.009.

Ghoshdastidar, A. J., Fox, S., & Tong, A. Z. (2015). The presence of the top prescribed pharmaceuticals in treated sewage effluents and receiving waters in Southwest Nova Scotia, Canada. *Environmental Science and Pollution Research*, *22*(1), 689–700. https://doi.org/10.1007/s11356-014-3400-z.

Gierse, J. K., Hauser, S. D., Creely, D. P., Koboldt, C., Rangwala, S. H., Isakson, P. C., et al. (1995). Expression and selective inhibition of the constitutive and inducible forms of human cyclo-oxygenase. *Biochemical Journal*, *305*(2), 479–484. https://doi.org/10.1042/bj3050479.

Goeppert, N., Dror, I., & Berkowitz, B. (2014). Detection, fate and transport of estrogen family hormones in soil. *Chemosphere*, *95*, 336–345. https://doi.org/10.1016/j.chemosphere.2013.09.039.

Gogoi, A., Mazumder, P., Tyagi, V. K., Tushara Chaminda, G. G., An, A. K., & Kumar, M. (2018). Occurrence and fate of emerging contaminants in water environment: A review. *Groundwater for Sustainable Development*, *6*, 169–180. https://doi.org/10.1016/j.gsd.2017.12.009.

Golovko, O., Kumar, V., Fedorova, G., Randak, T., & Grabic, R. (2014). Seasonal changes in antibiotics, antidepressants/psychiatric drugs, antihistamines and lipid regulators in a wastewater treatment plant. *Chemosphere*, *111*, 418–426. https://doi.org/10.1016/J.CHEMOSPHERE.2014.03.132.

Gonçalves, N. P. F., del Puerto, O., Medana, C., Calza, P., & Roslev, P. (2021). Degradation of the antifungal pharmaceutical clotrimazole by UVC and vacuum-UV irradiation: Kinetics, transformation products and attenuation of toxicity. *Journal of Environmental Chemical Engineering*, *9*(5), 106275. https://doi.org/10.1016/j.jece.2021.106275.

Gopal, C. M., Bhat, K., Ramaswamy, B. R., Kumar, V., Singhal, R. K., Basu, H., et al. (2021). Seasonal occurrence and risk assessment of pharmaceutical and personal care products in Bengaluru rivers and lakes, India. *Journal of Environmental Chemical Engineering*, *9*(4), 105610. https://doi.org/10.1016/j.jece.2021.105610.

Grabarczyk, Ł., Mulkiewicz, E., Stolte, S., Puckowski, A., Pazda, M., Stepnowski, P., et al. (2020). Ecotoxicity screening evaluation of selected pharmaceuticals and their transformation products towards various organisms. *Environmental Science and Pollution Research*, *27*(21), 26103–26114. https://doi.org/10.1007/s11356-020-08881-3.

Gracia-Lor, E., Sancho, J. V., Serrano, R., & Hernández, F. (2012). Occurrence and removal of pharmaceuticals in wastewater treatment plants at the Spanish Mediterranean area of Valencia. *Chemosphere*, *87*(5), 453–462. https://doi.org/10.1016/J.CHEMOSPHERE.2011.12.025.

Guo, Y., Guo, Z., Wang, J., Ye, Z., Zhang, L., & Niu, J. (2022). Photodegradation of three antidepressants in natural waters: Important roles of dissolved organic matter and nitrate. *Science of the Total Environment*, *802*, 149825. https://doi.org/10.1016/j.scitotenv.2021.149825.

Gurke, R., Rößler, M., Marx, C., Diamond, S., Schubert, S., Oertel, R., et al. (2015). Occurrence and removal of frequently prescribed pharmaceuticals and corresponding metabolites in wastewater of a sewage treatment plant. *Science of the Total Environment*, *532*, 762–770. https://doi.org/10.1016/J.SCITOTENV.2015.06.067.

Heye, K., Becker, D., Lütke Eversloh, C., Durmaz, V., Ternes, T. A., Oetken, M., et al. (2016). Effects of carbamazepine and two of its metabolites on the non-biting midge Chironomus riparius in a sediment full life cycle toxicity test. *Water Research*, *98*, 19–27. https://doi.org/10.1016/j.watres.2016.03.071.

Hu, P., Guo, C., Zhang, Y., Lv, J., Zhang, Y., & Xu, J. (2019). Occurrence, distribution and risk assessment of abused drugs and their metabolites in a typical urban river in North China. *Frontiers of Environmental Science & Engineering*, *13*(4), 56. https://doi.org/10.1007/s11783-019-1140-5.

Hu, R., Zhang, L., & Hu, J. (2016). Study on the kinetics and transformation products of salicylic acid in water via ozonation. *Chemosphere*, *153*, 394–404. https://doi.org/10.1016/j.chemosphere.2016.03.074.

Ibáñez, M., Bijlsma, L., Pitarch, E., López, F. J., & Hernández, F. (2021). Occurrence of pharmaceutical metabolites and transformation products in the aquatic environment of the Mediterranean area. *Trends in Environmental Analytical Chemistry*, *29*, e00118. https://doi.org/10.1016/j.teac.2021.e00118.

IDF. (2013). *Diabetes atlas*. International Diabetes Federation.

Jaén-Gil, A., Castellet-Rovira, F., Llorca, M., Villagrasa, M., Sarrà, M., Rodríguez-Mozaz, S., et al. (2019). Fungal treatment of metoprolol and its recalcitrant metabolite metoprolol acid in hospital wastewater: Biotransformation, sorption and ecotoxicological impact. *Water Research*, *152*, 171–180. https://doi.org/10.1016/j.watres.2018.12.054.

Jankunaite, D., Tichonovas, M., Buivydiene, D., Radziuniene, I., Racys, V., & Krugly, E. (2017). Removal of diclofenac, ketoprofen, and carbamazepine from simulated drinking water by advanced oxidation in a model reactor. *Water, Air, & Soil Pollution*, *228*(9), 353. https://doi.org/10.1007/s11270-017-3517-z.

Jiang, J.-J., Lee, C.-L., & Fang, M.-D. (2014). Emerging organic contaminants in coastal waters: Anthropogenic impact, environmental release and ecological risk. *Marine Pollution Bulletin*, *85*(2), 391–399. https://doi.org/10.1016/j.marpolbul.2013.12.045.

K'oreje, K. O., Kandie, F. J., Vergeynst, L., Abira, M. A., Van Langenhove, H., Okoth, M., et al. (2018). Occurrence, fate and removal of pharmaceuticals, personal care products and pesticides in wastewater stabilization ponds and receiving rivers in the Nzoia Basin, Kenya. *Science of the Total Environment*, *637–638*, 336–348. https://doi.org/10.1016/J.SCITOTENV.2018.04.331.

K'oreje, K. O., Vergeynst, L., Ombaka, D., De Wispelaere, P., Okoth, M., Van Langenhove, H., et al. (2016). Occurrence patterns of pharmaceutical residues in wastewater, surface water and groundwater of Nairobi and Kisumu city, Kenya. *Chemosphere*, *149*, 238–244. https://doi.org/10.1016/J.CHEMOSPHERE.2016.01.095.

Kaiser, E., Prasse, C., Wagner, M., Bröder, K., & Ternes, T. A. (2014). Transformation of oxcarbazepine and human metabolites of carbamazepine and oxcarbazepine in wastewater treatment and sand filters. *Environmental Science & Technology*, *48*(17), 10208–10216. https://doi.org/10.1021/es5024493.

Kapelewska, J., Kotowska, U., Karpińska, J., Kowalczuk, D., Arciszewska, A., & Świrydo, A. (2018). Occurrence, removal, mass loading and environmental risk assessment of emerging organic contaminants in leachates, groundwaters and wastewaters. *Microchemical Journal*, *137*, 292–301. https://doi.org/10.1016/j.microc.2017.11.008.

Khan, A. H., Aziz, H. A., Khan, N. A., Dhingra, A., Ahmed, S., & Naushad, M. (2021). Effect of seasonal variation on the occurrences of high-risk pharmaceutical in drain-laden surface water: A risk analysis of Yamuna River. *Science of the Total Environment*, *794*, 148484. https://doi.org/10.1016/j.scitotenv.2021.148484.

Khan, H. K., Rehman, M. Y. A., & Malik, R. N. (2020). Fate and toxicity of pharmaceuticals in water environment: An insight on their occurrence in South Asia. *Journal of Environmental Management*, *271*, 111030. https://doi.org/10.1016/j.jenvman.2020.111030.

Kharel, S., Stapf, M., Miehe, U., Ekblad, M., Cimbritz, M., Falås, P., et al. (2021). Removal of pharmaceutical metabolites in wastewater ozonation including their fate in different post-treatments. *Science of the Total Environment*, *759*, 143989. https://doi.org/10.1016/j.scitotenv.2020.143989.

Khasawneh, O. F. S., & Palaniandy, P. (2019). Photocatalytic degradation of pharmaceuticals using TiO2 based nanocomposite catalyst—Review. *Civil and Environmental Engineering Reports*, *3*(30), 1–33. https://doi.org/10.2478/ceer-2019-0021.

Khasawneh, O. F. S., & Palaniandy, P. (2020). Removal of organic pollutants from water by Fe2O3/TiO2 based photocatalytic degradation: A review. *Environmental Technology and Innovation*, 101230. https://doi.org/10.1016/j.eti.2020.101230.

Khasawneh, O. F. S., & Palaniandy, P. (2021). Occurrence and removal of pharmaceuticals in wastewater treatment plants. *Process Safety and Environmental Protection*, *150*, 532–556. https://doi.org/10.1016/j.psep.2021.04.045.

Khasawneh, O. F. S., Palaniandy, P., Ahmadipour, M., Mohammadi, H., & Bin Hamdan, M. R. (2020). Removal of acetaminophen using Fe2O3-TiO2 nanocomposites by photocatalysis under simulated solar irradiation: Optimization study. *Journal of Environmental Chemical Engineering*, 104921. https://doi.org/10.1016/j.jece.2020.104921.

Kim, H.-Y., Lee, I.-S., & Oh, J.-E. (2017). Human and veterinary pharmaceuticals in the marine environment including fish farms in Korea. *Science of the Total Environment*, *579*, 940–949. https://doi.org/10.1016/j.scitotenv.2016.10.039.

Kinney, C. A., Furlong, E. T., Werner, S. L., & Cahill, J. D. (2006). Presence and distribution of wastewater-derived pharmaceuticals in soil irrigated with reclaimed water. *Environmental Toxicology and Chemistry*, *25*(2), 317–326. https://doi.org/10.1897/05-187R.1.

Kosma, C. I., Lambropoulou, D. A., & Albanis, T. A. (2010). Occurrence and removal of PPCPs in municipal and hospital wastewaters in Greece. *Journal of Hazardous Materials*, *179*(1–3), 804–817. https://doi.org/10.1016/J.JHAZMAT.2010.03.075.

Kot-Wasik, A., Jakimska, A., & Śliwka-Kaszyńska, M. (2016). Occurrence and seasonal variations of 25 pharmaceutical residues in wastewater and drinking water treatment plants. *Environmental Monitoring and Assessment*, *188*(12), 661. https://doi.org/10.1007/s10661-016-5637-0.

Kovalakova, P., Cizmas, L., McDonald, T. J., Marsalek, B., Feng, M., & Sharma, V. K. (2020). Occurrence and toxicity of antibiotics in the aquatic environment: A review. *Chemosphere*, *251*, 126351. https://doi.org/10.1016/j.chemosphere.2020.126351.

Kumar, R. R., Lee, J. T., & Cho, J. Y. (2012). Fate, occurrence, and toxicity of veterinary antibiotics in environment. *Journal of Korean Society for Applied Biological Chemistry*, *55*(6), 701–709. https://doi.org/10.1007/s13765-012-2220-4.

Kumirska, J., Łukaszewicz, P., Caban, M., Migowska, N., Plenis, A., Białk-Bielińska, A., et al. (2019). Determination of twenty pharmaceutical contaminants in soil using ultrasound-assisted extraction with gas chromatography-mass spectrometric detection. *Chemosphere*, *232*, 232–242. https://doi.org/10.1016/j.chemosphere.2019.05.164.

Kümmerer, K. (2008). *Pharmaceuticals in the environment: Sources, fate, effects and risks.* Springer Science \& Business Media. https://doi.org/10.1007/978-3-540-74664-5.

Lajeunesse, A., Smyth, S. A., Barclay, K., Sauvé, S., & Gagnon, C. (2012). Distribution of antidepressant residues in wastewater and biosolids following different treatment processes by municipal wastewater treatment plants in Canada. *Water Research*, *46*(17), 5600–5612. https://doi.org/10.1016/J.WATRES.2012.07.042.

Langdon, K. A., Warne, M. S.t. J., Smernik, R. J., Shareef, A., & Kookana, R. S. (2014). Persistence of estrogenic activity in soils following land application of biosolids. *Environmental Toxicology and Chemistry*, *33*(1), 26–28. https://doi.org/10.1002/etc.2395.

Lee, G., Lee, S., Ha, N., Kho, Y., Park, K., Kim, P., et al. (2019). Effects of gemfibrozil on sex hormones and reproduction related performances of Oryzias latipes following long-term (155 d) and short-term (21 d) exposure. *Ecotoxicology and Environmental Safety*, *173*, 174–181. https://doi.org/10.1016/j.ecoenv.2019.02.015.

Lee, C., Palaniandy, P., & Dahlan, I. (2017). Pharmaceutical residues in aquatic environment and water remediation by TiO2 heterogeneous photocatalysis: A review. *Environmental Earth Sciences*, *76*(17), 611. https://doi.org/10.1007/s12665-017-6924-y.

Li, W. C. (2014). Occurrence, sources, and fate of pharmaceuticals in aquatic environment and soil. *Environmental Pollution*, *187*, 193–201. https://doi.org/10.1016/j.envpol.2014.01.015.

Li, Y., Liu, X., Zhang, B., Zhao, Q., Ning, P., & Tian, S. (2018). Aquatic photochemistry of sulfamethazine: Multivariate effects of main water constituents and mechanisms. *Environmental Science: Processes & Impacts*, *20*(3), 513–522. https://doi.org/10.1039/C7EM00548B.

Li, J., Zhao, L., Feng, M., Huang, C.-H., & Sun, P. (2021). Abiotic transformation and ecotoxicity change of sulfonamide antibiotics in environmental and water treatment processes: A critical review. *Water Research*, *202*, 117463. https://doi.org/10.1016/j.watres.2021.117463.

Loftin, K. A., Adams, C. D., Meyer, M. T., & Surampalli, R. (2008). Effects of ionic strength, temperature, and pH on degradation of selected antibiotics. *Journal of Environmental Quality*, *37*(2), 378–386. https://doi.org/10.2134/jeq2007.0230.

Lolić, A., Paíga, P., Santos, L. H. M. L. M., Ramos, S., Correia, M., & Delerue-Matos, C. (2015). Assessment of non-steroidal anti-inflammatory and analgesic pharmaceuticals in seawaters of North of Portugal: Occurrence and environmental risk. *Science of the Total Environment*, *508*, 240–250. https://doi.org/10.1016/j.scitotenv.2014.11.097.

Luján-Facundo, M. J., Iborra-Clar, M. I., Mendoza-Roca, J. A., & Alcaina-Miranda, M. I. (2019). Pharmaceutical compounds removal by adsorption with commercial and reused carbon coming from a drinking water treatment plant. *Journal of Cleaner Production*, *238*, 117866. https://doi.org/10.1016/j.jclepro.2019.117866.

Luo, Y., Guo, W., Ngo, H. H., Nghiem, L. D., Hai, F. I., Zhang, J., et al. (2014). A review on the occurrence of micropollutants in the aquatic environment and their fate and removal during wastewater treatment. *Science of the Total Environment*, *473–474*, 619–641. https://doi.org/10.1016/j.scitotenv.2013.12.065.

Luque-Muñoz, A., Vílchez, J. L., & Zafra-Gómez, A. (2017). Multiclass method for the determination of pharmaceuticals and personal care products in compost from sewage sludge using ultrasound and salt-assisted liquid–liquid extraction followed by ultrahigh performance liquid chromatography-tandem mass spectrometr. *Journal of Chromatography A*, *1507*, 72–83. https://doi.org/10.1016/j.chroma.2017.05.051.

Lürling, M., Sargant, E., & Roessink, I. (2006). Life-history consequences for Daphnia pulex exposed to pharmaceutical carbamazepine. *Environmental Toxicology*, *21*(2), 172–180. https://doi.org/10.1002/tox.20171.

Maculewicz, J., Kowalska, D., Świacka, K., Toński, M., Stepnowski, P., Białk-Bielińska, A., et al. (2022). Transformation products of pharmaceuticals in the environment: Their fate, (eco)toxicity and bioaccumulation potential. *Science of the Total Environment*, *802*, 149916. https://doi.org/10.1016/j.scitotenv.2021.149916.

Madikizela, L. M., & Ncube, S. (2021). Occurrence and ecotoxicological risk assessment of non-steroidal anti-inflammatory drugs in South African aquatic environment: What is known and the missing information? *Chemosphere*, *280*, 130688. https://doi.org/10.1016/j.chemosphere.2021.130688.

Madikizela, L. M., Ncube, S., & Chimuka, L. (2018). Uptake of pharmaceuticals by plants grown under hydroponic conditions and natural occurring plant species: A review. *Science of the Total Environment*, *636*, 477–486. https://doi.org/10.1016/j.scitotenv.2018.04.297.

Madikizela, L. M., Ncube, S., Tutu, H., Richards, H., Newman, B., Ndungu, K., et al. (2020). Pharmaceuticals and their metabolites in the marine environment: Sources, analytical methods and occurrence. *Trends in Environmental Analytical Chemistry*, *28*, e00104. https://doi.org/10.1016/j.teac.2020.e00104.

Malvar, J. L., Santos, J. L., Martín, J., Aparicio, I., & Alonso, E. (2021). Occurrence of the main metabolites of the most recurrent pharmaceuticals and personal care products in Mediterranean soils. *Journal of Environmental Management*, *278*, 111584. https://doi.org/10.1016/j.jenvman.2020.111584.

Martin, S., & Weiß, J. (2008). Diabetes mellitus Typ 2. *Der Diabetologe*, *4*(1), 30–40. https://doi.org/10.1007/s11428-007-0155-x.

Martínez Bueno, M. J., Herrera, S., Munaron, D., Boillot, C., Fenet, H., Chiron, S., et al. (2016). POCIS passive samplers as a monitoring tool for pharmaceutical residues and their transformation products in marine environment. *Environmental Science and Pollution Research*, *23*(6), 5019–5029. https://doi.org/10.1007/s11356-014-3796-5.

Matongo, S., Birungi, G., Moodley, B., & Ndungu, P. (2015). Pharmaceutical residues in water and sediment of Msunduzi River, KwaZulu-Natal, South Africa. *Chemosphere*, *134*, 133–140. https://doi.org/10.1016/J.CHEMOSPHERE.2015.03.093.

Meador, J. P., Yeh, A., Young, G., & Gallagher, E. P. (2016). Contaminants of emerging concern in a large temperate estuary. *Environmental Pollution*, *213*, 254–267. https://doi.org/10.1016/J.ENVPOL.2016.01.088.

Meffe, R., de Santiago-Martín, A., Teijón, G., Martínez Hernández, V., López-Heras, I., Nozal, L., et al. (2021). Pharmaceutical and transformation products during unplanned water reuse: Insights into natural attenuation, plant uptake and human health impact under field conditions. *Environment International*, *157*, 106835. https://doi.org/10.1016/j.envint.2021.106835.

Mejías, C., Martín, J., Santos, J. L., Aparicio, I., & Alonso, E. (2021). Occurrence of pharmaceuticals and their metabolites in sewage sludge and soil: A review on their distribution and environmental risk assessment. *Trends in Environmental Analytical Chemistry*, *30*, e00125. https://doi.org/10.1016/j.teac.2021.e00125.

Menz, J., Toolaram, A. P., Rastogi, T., Leder, C., Olsson, O., Kümmerer, K., et al. (2017). Transformation products in the water cycle and the unsolved problem of their proactive assessment: A combined in vitro/in silico approach. *Environment International*, *98*, 171–180. https://doi.org/10.1016/j.envint.2016.11.003.

Miao, H.-F., Cao, M., Xu, D.-Y., Ren, H.-Y., Zhao, M.-X., Huang, Z.-X., et al. (2015). Degradation of phenazone in aqueous solution with ozone: Influencing factors and degradation pathways. *Chemosphere*, *119*, 326–333. https://doi.org/10.1016/j.chemosphere.2014.06.082.

Mirzaei, R., Yunesian, M., Nasseri, S., Gholami, M., Jalilzadeh, E., Shoeibi, S., et al. (2018). Occurrence and fate of most prescribed antibiotics in different water environments of Tehran, Iran. *Science of the Total Environment*, *619–620*, 446–459. https://doi.org/10.1016/j.scitotenv.2017.07.272.

Mompelat, S., Le Bot, B., & Thomas, O. (2009). Occurrence and fate of pharmaceutical products and by-products, from resource to drinking water. *Environment International*, *35*(5), 803–814. https://doi.org/10.1016/j.envint.2008.10.008.

Monteiro, S. C., & Boxall, A. B. A. (2010). Occurrence and fate of human pharmaceuticals in the environment. In *Reviews of environmental contamination and toxicology* (pp. 53–154). New York: Springer. https://doi.org/10.1007/978-1-4419-1157-5_2.

Moreno-González, R., Rodriguez-Mozaz, S., Gros, M., Barceló, D., & León, V. M. (2015). Seasonal distribution of pharmaceuticals in marine water and sediment from a Mediterranean coastal lagoon (SE Spain). *Environmental Research*, *138*, 326–344. https://doi.org/10.1016/j.envres.2015.02.016.

Moslah, B., Hapeshi, E., Jrad, A., Fatta-Kassinos, D., & Hedhili, A. (2018). Pharmaceuticals and illicit drugs in wastewater samples in north-eastern Tunisia. *Environmental Science and Pollution Research*, *25*(19), 18226–18241. https://doi.org/10.1007/s11356-017-8902-z.

Munari, M., Marin, M. G., & Matozzo, V. (2014). Effects of the antidepressant fluoxetine on the immune parameters and acetylcholinesterase activity of the clam Venerupis philippinarum. *Marine Environmental Research*, *94*, 32–37. https://doi.org/10.1016/J.MARENVRES.2013.11.007.

Ngubane, N. P., Naicker, D., Ncube, S., Chimuka, L., & Madikizela, L. M. (2019). Determination of naproxen, diclofenac and ibuprofen in Umgeni estuary and seawater: A case of northern Durban in KwaZulu–Natal Province of South Africa. *Regional Studies in Marine Science*, *29*, 100675. https://doi.org/10.1016/j.rsma.2019.100675.

Niemuth, N. J., & Klaper, R. D. (2015). Emerging wastewater contaminant metformin causes intersex and reduced fecundity in fish. *Chemosphere*, *135*, 38–45. https://doi.org/10.1016/J.CHEMOSPHERE.2015.03.060.

Nkoom, M., Lu, G., Liu, J., Yang, H., & Dong, H. (2019). Bioconcentration of the antiepileptic drug carbamazepine and its physiological and biochemical effects on Daphnia magna. *Ecotoxicology and Environmental Safety*, *172*, 11–18. https://doi.org/10.1016/J.ECOENV.2019.01.061.

Nödler, K., Voutsa, D., & Licha, T. (2014). Polar organic micropollutants in the coastal environment of different marine systems. *Marine Pollution Bulletin*, *85*(1), 50–59. https://doi.org/10.1016/j.marpolbul.2014.06.024.

Ojemaye, C. Y., & Petrik, L. (2019). Pharmaceuticals in the marine environment: A review. *Environmental Reviews*, *27*(2), 151–165. https://doi.org/10.1139/er-2018-0054.

Pal, A., Gin, K. Y.-H., Lin, A. Y.-C., & Reinhard, M. (2010). Impacts of emerging organic contaminants on freshwater resources: Review of recent occurrences, sources, fate and effects. *Science of the Total Environment*, *408*(24), 6062–6069. https://doi.org/10.1016/j.scitotenv.2010.09.026.

Palli, L., Spina, F., Varese, G. C., Vincenzi, M., Aragno, M., Arcangeli, G., et al. (2019). Occurrence of selected pharmaceuticals in wastewater treatment plants of Tuscany: An effect-based approach to evaluate the potential environmental impact. *International Journal of*

Hygiene and Environmental Health, *222*(4), 717–725. https://doi.org/10.1016/J.IJHEH.2019.05.006.

Paltiel, O., Fedorova, G., Tadmor, G., Kleinstern, G., Maor, Y., & Chefetz, B. (2016). Human exposure to wastewater-derived pharmaceuticals in fresh produce: A randomized controlled trial focusing on carbamazepine. *Environmental Science & Technology*, *50*(8), 4476–4482. https://doi.org/10.1021/acs.est.5b06256.

Papageorgiou, M., Kosma, C., & Lambropoulou, D. (2016). Seasonal occurrence, removal, mass loading and environmental risk assessment of 55 pharmaceuticals and personal care products in a municipal wastewater treatment plant in Central Greece. *Science of the Total Environment*, *543*, 547–569. https://doi.org/10.1016/J.SCITOTENV.2015.11.047.

Parolini, M. (2020). Toxicity of the non-steroidal anti-inflammatory drugs (NSAIDs) acetylsalicylic acid, paracetamol, diclofenac, ibuprofen and naproxen towards freshwater invertebrates: A review. *Science of the Total Environment*, *740*, 140043. https://doi.org/10.1016/j.scitotenv.2020.140043.

Parrott, J. L., & Metcalfe, C. D. (2019). Assessing the effects of environmentally relevant concentrations of antidepressant mixtures to fathead minnows exposed over a full life cycle. *Science of the Total Environment*, *648*, 1227–1236. https://doi.org/10.1016/J.SCITOTENV.2018.08.237.

Pazda, M., Kumirska, J., Stepnowski, P., & Mulkiewicz, E. (2019). Antibiotic resistance genes identified in wastewater treatment plant systems—A review. *Science of the Total Environment*, *697*, 134023. https://doi.org/10.1016/j.scitotenv.2019.134023.

Pentikäinen, P. J., Neuvonen, P. J., & Penttilä, A. (1979). Pharmacokinetics of metformin after intravenous and oral administration to man. *European Journal of Clinical Pharmacology*, *16*(3), 195–202. https://doi.org/10.1007/BF00562061.

Phonsiri, V., Choi, S., Nguyen, C., Tsai, Y.-L., Coss, R., & Kurwadkar, S. (2019). Monitoring occurrence and removal of selected pharmaceuticals in two different wastewater treatment plants. *SN Applied Sciences*, *1*(7), 798. https://doi.org/10.1007/s42452-019-0774-z.

Pouliquen, H., Delépée, R., Larhantec-Verdier, M., Morvan, M.-L., & Le Bris, H. (2007). Comparative hydrolysis and photolysis of four antibacterial agents (oxytetracycline oxolinic acid, flumequine and florfenicol) in deionised water, freshwater and seawater under abiotic conditions. *Aquaculture*, *262*(1), 23–28. https://doi.org/10.1016/j.aquaculture.2006.10.014.

Puckowski, A., Mioduszewska, K., Łukaszewicz, P., Borecka, M., Caban, M., Maszkowska, J., et al. (2016). Bioaccumulation and analytics of pharmaceutical residues in the environment: A review. *Journal of Pharmaceutical and Biomedical Analysis*, *127*, 232–255. https://doi.org/10.1016/j.jpba.2016.02.049.

Qu, S., Kolodziej, E. P., Long, S. A., Gloer, J. B., Patterson, E. V., Baltrusaitis, J., et al. (2013). Product-to-parent reversion of Trenbolone: Unrecognized risks for endocrine disruption. *Science*, *342*(6156), 347–351. https://doi.org/10.1126/science.1243192.

Quero-Pastor, M. J., Garrido-Perez, M. C., Acevedo, A., & Quiroga, J. M. (2014). Ozonation of ibuprofen: A degradation and toxicity study. *Science of the Total Environment*, *466–467*, 957–964. https://doi.org/10.1016/j.scitotenv.2013.07.067.

Quesada, H. B., Baptista, A. T. A., Cusioli, L. F., Seibert, D., de Oliveira Bezerra, C., & Bergamasco, R. (2019). Surface water pollution by pharmaceuticals and an alternative of removal by low-cost adsorbents: A review. *Chemosphere*, *222*, 766–780. https://doi.org/10.1016/J.CHEMOSPHERE.2019.02.009.

Richardson, S. D. (2011). Disinfection by-products: Formation and occurrence in drinking water. In J. O. Nriagu (Ed.), *Encyclopedia of environmental health* (pp. 110–136). Burlington: Elsevier Science Inc.

Rivera-Jaimes, J. A., Postigo, C., Melgoza-Alemán, R. M., Aceña, J., Barceló, D., & López de Alda, M. (2018). Study of pharmaceuticals in surface and wastewater from Cuernavaca, Morelos, Mexico: Occurrence and environmental risk assessment. *Science of the Total Environment*, *613–614*, 1263–1274. https://doi.org/10.1016/J.SCITOTENV.2017.09.134.

Roberts, J., Kumar, A., Du, J., Hepplewhite, C., Ellis, D. J., Christy, A. G., et al. (2016). Pharmaceuticals and personal care products (PPCPs) in Australia's largest inland sewage treatment plant, and its contribution to a major Australian river during high and low flow. *Science of the Total Environment*, *541*, 1625–1637. https://doi.org/10.1016/J.SCITOTENV.2015.03.145.

Robinson, B. J., Hui, J. P. M., Soo, E. C., & Hellou, J. (2009). Estrogenic compounds in seawater and sediment from Halifax Harbour, Nova Scotia, Canada. *Environmental Toxicology and Chemistry*, *28*(1), 18–25. https://doi.org/10.1897/08-203.1.

Rodriguez-Narvaez, O. M., Peralta-Hernandez, J. M., Goonetilleke, A., & Bandala, E. R. (2017). Treatment technologies for emerging contaminants in water: A review. *Chemical Engineering Journal*, *323*, 361–380. https://doi.org/10.1016/J.CEJ.2017.04.106.

Ronan, J. M., & McHugh, B. (2013). A sensitive liquid chromatography/tandem mass spectrometry method for the determination of natural and synthetic steroid estrogens in seawater and marine biota, with a focus on proposed water framework directive environmental quality standards. *Rapid Communications in Mass Spectrometry*, *27*(7), 738–746. https://doi.org/10.1002/rcm.6505.

Rosal, R., Rodea-Palomares, I., Boltes, K., Fernández-Piñas, F., Leganés, F., Gonzalo, S., et al. (2010). Ecotoxicity assessment of lipid regulators in water and biologically treated wastewater using three aquatic organisms. *Environmental Science and Pollution Research*, *17* (1), 135–144. https://doi.org/10.1007/s11356-009-0137-1.

Ruan, Y., Wu, R., Lam, J. C. W., Zhang, K., & Lam, P. K. S. (2019). Seasonal occurrence and fate of chiral pharmaceuticals in different sewage treatment systems in Hong Kong: Mass balance, enantiomeric profiling, and risk assessment. *Water Research*, *149*, 607–616. https://doi.org/10.1016/j.watres.2018.11.010.

Scaria, J., Anupama, K. V., & Nidheesh, P. V. (2021). Tetracyclines in the environment: An overview on the occurrence, fate, toxicity, detection, removal methods, and sludge management. *Science of the Total Environment*, *771*, 145291. https://doi.org/10.1016/j.scitotenv.2021.145291.

Scheurer, M., Michel, A., Brauch, H.-J., Ruck, W., & Sacher, F. (2012). Occurrence and fate of the antidiabetic drug metformin and its metabolite guanylurea in the environment and during drinking water treatment. *Water Research*, *46*(15), 4790–4802. https://doi.org/10.1016/j.watres.2012.06.019.

Schlüter-Vorberg, L., Prasse, C., Ternes, T. A., Mückter, H., & Coors, A. (2015). Toxification by transformation in conventional and advanced wastewater treatment: The antiviral drug acyclovir. *Environmental Science & Technology Letters*, *2*(12), 342–346. https://doi.org/10.1021/acs.estlett.5b00291.

Sharma, A., Ahmad, J., & Flora, S. J. S. (2018). Application of advanced oxidation processes and toxicity assessment of transformation products. *Environmental Research*, *167*, 223–233. https://doi.org/10.1016/j.envres.2018.07.010.

Singh, S. P., Azua, A., Chaudhary, A., Khan, S., Willett, K. L., & Gardinali, P. R. (2010). Occurrence and distribution of steroids, hormones and selected pharmaceuticals in South Florida coastal environments. *Ecotoxicology*, *19*(2), 338–350. https://doi.org/10.1007/s10646-009-0416-0.

Spataro, F., Ademollo, N., Pescatore, T., Rauseo, J., & Patrolecco, L. (2019). Antibiotic residues and endocrine disrupting compounds in municipal wastewater treatment plants in Rome,

Italy. *Microchemical Journal, 148*, 634–642. https://doi.org/10.1016/J.MICROC.2019.05.053.

Su, T., Deng, H., Benskin, J. P., & Radke, M. (2016). Biodegradation of sulfamethoxazole photo-transformation products in a water/sediment test. *Chemosphere, 148*, 518–525. https://doi.org/10.1016/j.chemosphere.2016.01.049.

Subedi, B., & Kannan, K. (2015). Occurrence and fate of select psychoactive pharmaceuticals and antihypertensives in two wastewater treatment plants in New York State, USA. *Science of the Total Environment, 514*, 273–280. https://doi.org/10.1016/J.SCITOTENV.2015.01.098.

Sui, Q., Cao, X., Lu, S., Zhao, W., Qiu, Z., & Yu, G. (2015). Occurrence, sources and fate of pharmaceuticals and personal care products in the groundwater: A review. *Emerging Contaminants, 1*(1), 14–24. https://doi.org/10.1016/j.emcon.2015.07.001.

Szymczycha, B., Borecka, M., Białk-Bielińska, A., Siedlewicz, G., & Pazdro, K. (2020). Submarine groundwater discharge as a source of pharmaceutical and caffeine residues in coastal ecosystem: Bay of Puck, southern Baltic Sea case study. *Science of the Total Environment, 713*, 136522. https://doi.org/10.1016/j.scitotenv.2020.136522.

Taheran, M., Brar, S. K., Verma, M., Surampalli, R. Y., Zhang, T. C., & Valero, J. R. (2016). Membrane processes for removal of pharmaceutically active compounds (PhACs) from water and wastewaters. *Science of the Total Environment, 547*, 60–77. https://doi.org/10.1016/j.scitotenv.2015.12.139.

Tanoue, R., Sato, Y., Motoyama, M., Nakagawa, S., Shinohara, R., & Nomiyama, K. (2012). Plant uptake of pharmaceutical chemicals detected in recycled organic manure and reclaimed wastewater. *Journal of Agricultural and Food Chemistry, 60*(41), 10203–10211. https://doi.org/10.1021/jf303142t.

Thiebault, T., Boussafir, M., & Le Milbeau, C. (2017). Occurrence and removal efficiency of pharmaceuticals in an urban wastewater treatment plant: Mass balance, fate and consumption assessment. *Journal of Environmental Chemical Engineering, 5*(3), 2894–2902. https://doi.org/10.1016/J.JECE.2017.05.039.

Tisler, S., & Zwiener, C. (2018). Formation and occurrence of transformation products of metformin in wastewater and surface water. *Science of the Total Environment, 628–629*, 1121–1129. https://doi.org/10.1016/J.SCITOTENV.2018.02.105.

Tran, N. H., & Gin, K. Y.-H. (2017). Occurrence and removal of pharmaceuticals, hormones, personal care products, and endocrine disrupters in a full-scale water reclamation plant. *Science of the Total Environment, 599–600*, 1503–1516. https://doi.org/10.1016/J.SCITOTENV.2017.05.097.

Tran, N. H., Reinhard, M., & Gin, K. Y.-H. (2018). Occurrence and fate of emerging contaminants in municipal wastewater treatment plants from different geographical regions-a review. *Water Research, 133*, 182–207. https://doi.org/10.1016/J.WATRES.2017.12.029.

Vazquez-Roig, P., Andreu, V., Blasco, C., & Picó, Y. (2012). Risk assessment on the presence of pharmaceuticals in sediments, soils and waters of the Pego–Oliva Marshlands (Valencia, eastern Spain). *Science of the Total Environment, 440*, 24–32. https://doi.org/10.1016/j.scitotenv.2012.08.036.

Verlicchi, P., Al Aukidy, M., & Zambello, E. (2012). Occurrence of pharmaceutical compounds in urban wastewater: Removal, mass load and environmental risk after a secondary treatment—A review. *Science of the Total Environment, 429*, 123–155. https://doi.org/10.1016/J.SCITOTENV.2012.04.028.

Vidal-Dorsch, D. E., Bay, S. M., Maruya, K., Snyder, S. A., Trenholm, R. A., & Vanderford, B. J. (2012). Contaminants of emerging concern in municipal wastewater effluents and marine receiving water. *Environmental Toxicology and Chemistry*, *31*(12), 2674–2682. https://doi.org/10.1002/etc.2004.

Wang, X.-H., & Lin, A. Y.-C. (2012). Phototransformation of cephalosporin antibiotics in an aqueous environment results in higher toxicity. *Environmental Science & Technology*, *46* (22), 12417–12426. https://doi.org/10.1021/es301929e.

Wang, J., & Wang, S. (2016). Removal of pharmaceuticals and personal care products (PPCPs) from wastewater: A review. *Journal of Environmental Management*, *182*, 620–640. https://doi.org/10.1016/J.JENVMAN.2016.07.049.

Wang, H., Xi, H., Xu, L., Jin, M., Zhao, W., & Liu, H. (2021). Ecotoxicological effects, environmental fate and risks of pharmaceutical and personal care products in the water environment: A review. *Science of the Total Environment*, *788*, 147819. https://doi.org/10.1016/j.scitotenv.2021.147819.

Weigel, S., Bester, K., & Hühnerfuss, H. (2001). New method for rapid solid-phase extraction of large-volume water samples and its application to non-target screening of North Sea water for organic contaminants by gas chromatography–mass spectrometry. *Journal of Chromatography A*, *912*(1), 151–161. https://doi.org/10.1016/S0021-9673(01)00529-5.

Wu, X., Dodgen, L. K., Conkle, J. L., & Gan, J. (2015). Plant uptake of pharmaceutical and personal care products from recycled water and biosolids: A review. *Science of the Total Environment*, *536*, 655–666. https://doi.org/10.1016/j.scitotenv.2015.07.129.

Wu, J., Qian, X., Yang, Z., & Zhang, L. (2010). Study on the matrix effect in the determination of selected pharmaceutical residues in seawater by solid-phase extraction and ultra-high-performance liquid chromatography–electrospray ionization low-energy collision-induced dissociation tandem mass spectrometry. *Journal of Chromatography A*, *1217*(9), 1471–1475. https://doi.org/10.1016/j.chroma.2009.12.074.

Yamamoto, H., Nakamura, Y., Moriguchi, S., Nakamura, Y., Honda, Y., Tamura, I., et al. (2009). Persistence and partitioning of eight selected pharmaceuticals in the aquatic environment: Laboratory photolysis, biodegradation, and sorption experiments. *Water Research*, *43*(2), 351–362. https://doi.org/10.1016/j.watres.2008.10.039.

Yan, Q., Gao, X., Chen, Y.-P., Peng, X.-Y., Zhang, Y.-X., Gan, X.-M., et al. (2014). Occurrence, fate and ecotoxicological assessment of pharmaceutically active compounds in wastewater and sludge from wastewater treatment plants in Chongqing, the three gorges reservoir area. *Science of the Total Environment*, *470–471*, 618–630. https://doi.org/10.1016/J.SCITOTENV.2013.09.032.

Yang, Y., Ok, Y. S., Kim, K.-H., Kwon, E. E., & Tsang, Y. F. (2017). Occurrences and removal of pharmaceuticals and personal care products (PPCPs) in drinking water and water/sewage treatment plants: A review. *Science of the Total Environment*, *596–597*, 303–320. https://doi.org/10.1016/J.SCITOTENV.2017.04.102.

Yi, M., Sheng, Q., Sui, Q., & Lu, H. (2020). β-Blockers in the environment: Distribution, transformation, and ecotoxicity. *Environmental Pollution*, *266*, 115269. https://doi.org/10.1016/j.envpol.2020.115269.

Yien Fang, T., Praveena, S. M., Aris, A. Z., Syed Ismail, S. N., & Rasdi, I. (2019). Quantification of selected steroid hormones (17β-estradiol and 17α-ethynylestradiol) in wastewater treatment plants in Klang Valley (Malaysia). *Chemosphere*, *215*, 153–162. https://doi.org/10.1016/J.CHEMOSPHERE.2018.10.032.

Yin, X., Guo, C., Deng, Y., Jin, X., Teng, Y., Xu, J., et al. (2021). Tissue-specific accumulation, elimination, and toxicokinetics of illicit drugs in adult zebrafish (Danio rerio). *Science of the Total Environment*, *792*, 148153. https://doi.org/10.1016/j.scitotenv.2021.148153.

Ying, G.-G., Williams, B., & Kookana, R. (2002). Environmental fate of alkylphenols and alkylphenol ethoxylates—A review. *Environment International*, *28*(3), 215–226. https://doi.org/10.1016/S0160-4120(02)00017-X.

Ying, G.-G., Zhao, J.-L., Zhou, L.-J., & Liu, S. (2013). Chapter 14—Fate and occurrence of pharmaceuticals in the aquatic environment (surface water and sediment). In M. Petrovic, D. Barcelo, & S. Pérez (Eds.), *Vol. 62. Analysis, removal, effects and risk of pharmaceuticals in the water cycle* (pp. 453–557). Elsevier. https://doi.org/10.1016/B978-0-444-62657-8.00014-8.

Yu, Y., Wu, L., & Chang, A. C. (2013). Seasonal variation of endocrine disrupting compounds, pharmaceuticals and personal care products in wastewater treatment plants. *Science of the Total Environment*, *442*, 310–316. https://doi.org/10.1016/J.SCITOTENV.2012.10.001.

Zhang, C., Barron, L., & Sturzenbaum, S. (2021). The transportation, transformation and (bio) accumulation of pharmaceuticals in the terrestrial ecosystem. *Science of the Total Environment*, *781*, 146684. https://doi.org/10.1016/j.scitotenv.2021.146684.

Zhang, R., Du, J., Dong, X., Huang, Y., Xie, H., Chen, J., et al. (2021). Occurrence and ecological risks of 156 pharmaceuticals and 296 pesticides in seawater from mariculture areas of Northeast China. *Science of the Total Environment*, *792*, 148375. https://doi.org/10.1016/j.scitotenv.2021.148375.

Zind, H., Mondamert, L., Remaury, Q. B., Cleon, A., Leitner, N. K. V. E. L., & Labanowski, J. (2021). Occurrence of carbamazepine, diclofenac, and their related metabolites and transformation products in a French aquatic environment and preliminary risk assessment. *Water Research*, *196*, 117052. https://doi.org/10.1016/j.watres.2021.117052.

Analytical techniques for the detection of pharmaceuticals in the environment

5

Mohd Sharizal Mohd Sapingi, Omar Fawzi Suleiman Khasawneh, Puganeshwary Palaniandy, and Hamidi Abdul Aziz
School of Civil Engineering, Engineering Campus, Universiti Sains Malaysia, Nibong Tebal, Pulau Pinang, Malaysia

1. Introduction

Pharmaceuticals contaminants (PCs) are considered as emerging contaminants (ECs) or emerging pollutants (EPs), whereas ECs can be described as synthesized or natural compounds that are not incorporated in common standard operating procedure for environmental monitoring but possess the prospective to contaminate the nature and cause an acknowledged or suspected adverse ecological and (or) human health consequences (Daverey, Dutta, & Sarkar, 2019). There are many adverse effects of PhACs to human. The greatest concern on the existence of PhACs in environment is the establishment of antibiotic resistance genes (ARGs) and bacteria (ARB) that is influenced by antibiotics (Schwermer, Krzeminski, Wennberg, Vogelsang, & Uhl, 2018). At certain concentration, pharmaceutical contaminants are able to affect phytoplankton and algal growth as well as reproductive defects in aquatic wildlife (Mohapatra, Huang, Mukherji, & Padhye, 2016).

The majority of pharmaceutical substances are complex molecules having a variety of functions, physical characteristics, and biological qualities. These substances can be basic or acidic with consist of 300–1000 range of molecular weight (Fatta-Kassinos, Nikolaou, & Ioannou-Ttofa, 2019). These substances can be divided by their purpose (Fig. 1).

Humans release pharmaceutical contaminant to the environment mainly through urine and feces where partially the drug is release as its parents metabolites (Veiga-Gómez et al., 2017). Thus, the PCs enter the aquatic system through water treatment system as conventional water treatment system is not able to remove PCs effectively (K'oreje et al., 2016; Krakkó, Licul-Kucera, Záray, & Mihucz, 2019). Low concentration and complex matrices in environment are major factors that limits an appropriate detection approaches of PhACs (Fatta-Kassinos et al., 2019). Multiple sources of PhACs also play a significant part in the elevation of PhACs in environment. This is due to inadequate regulation and executions by agencies and legal

The Treatment of Pharmaceutical Wastewater. https://doi.org/10.1016/B978-0-323-99160-5.00003-5

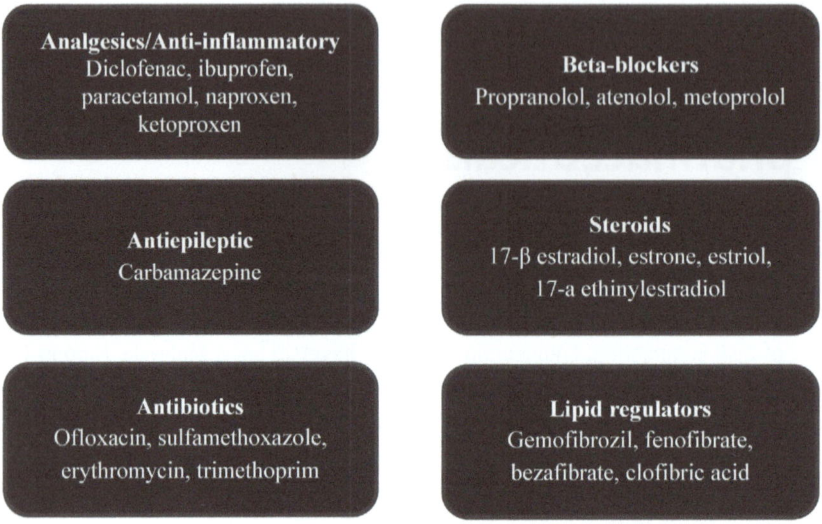

Fig. 1 Frequently detected pharmaceuticals in aqueous matrices.
Adapted and modified from Fatta-Kassinos, D., Nikolaou, A., & Ioannou-Ttofa, L. (2019). Advances in analytical methods for the determination of pharmaceutical residues in waters and wastewaters. *Encyclopedia of environmental health (2nd ed., Vol. 1)*. Elsevier. https://doi.org/10.1016/B978-0-12-409548-9.11247-3.

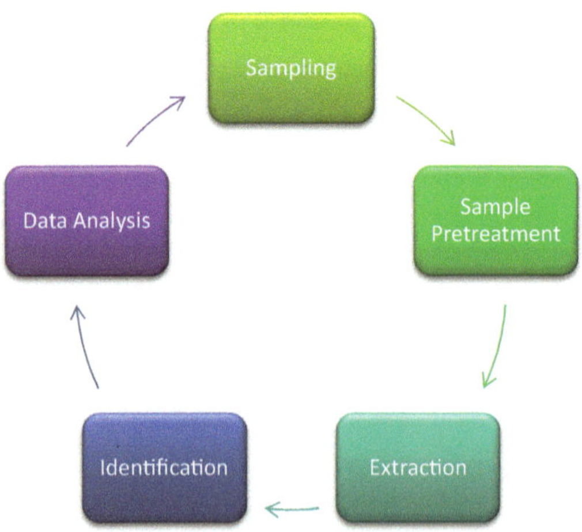

Fig. 2 General approach for pharmaceutical contaminant in environment.

authorities (Wu & Janssen, 2011). Other factors that influence the increment concentrations of PhACs is the nonexistence of routine monitoring standard procedure and some of them has no regulatory limit for its concentration in environment. However, owing to latest technology approaches, the concentrations of PhACs can be quantified up to part per trillion, which gives higher possibilities to detect newer PhACs that are lower than detection limit in previous technologies. Fig. 2 illustrates the common strategies in PhACs detection in environment. Each phase plays an important part in determining the accuracy of data obtained.

2. Preanalysis protocols

2.1 Sampling

Sampling method can be varied based on the character of contaminant of interest. However, it is very common that to use the amber bottle, as the degradation of PCs can occur with the penetration of sunlight to the sample, resulting in a non-representative sample. In certain cases such as LID (legal and illegal drugs) as discussed by Vazquez-Roig, Blasco, and Picó (2013), the type of sampling container used has to be a plastic type to minimize the adsorption of analyte onto glass surface. This is due to the nature of LID that mostly polar and present in the environment in ionized form.

Prior to storage and preservation, certain type of sample requires removal of turbidity and suspended solid (SS) from the sample via filtration (Moret, Hidalgo, & Sanchez, 2012). Generally, two types of glass-fiber filter with pore size of 0.45 μm or 0.2 μm is used to filter water samples, since the presence of natural organic particles may reduce the extraction process's effectiveness (Montaseri & Forbes, 2018). Acidification of samples collected is required in case of PCs having acidic groups in their structure and occurring mostly in their ionized state at neutral pH (Fatta-Kassinos et al., 2019).

2.2 Preservation

The preservation of samples is strongly reliant on the target ingredients as well as the sample composition used in the experiment. Recent findings show that the preservation of aromatic compounds is very challenging due to its rapid depletion and its possibility to happen at low temperature. The avoidance of dehydrohalogenation reaction in chlorinated samples as instance cannot be stopped even by acidification of samples. Nonetheless, common approaches are to store the sample in a cold storage with controlled temperature at 4°C and the sample has to be analyzed as soon as possible. In most cases, the preservation is important due to delay between sampling and analyzing process (Baird, Eaton, & Rice, 2017).

2.3 Pretreatment (clean-up)

Pretreatment is necessary to eliminate the interference from other compounds in complex matrix sample particularly. This step also helps to ensure that the sample suit with the analyzing method without damaging the equipment. Thus, the major purposes of this pretreatment process are for sample clean-up, preconcentration of analyte, and medium transfer from the sample matrix to a different solvent (Cantwell & Losier, 2002; Płotka-Wasylka, Szczepańska, de la Guardia, & Namieśnik, 2016; Poole, 2003; Silvestre, Santos, Lima, & Zagatto, 2009). Sample clean-up demands an excellent selectivity for separating the analyte component from potential interferents, whereas the analyte preconcentration benefits from a high distribution ratio, allowing the analyte of interest to be extracted from a massive volume of sample solution into a small volume of extractant. In pretreatment process, the analyte of interest is isolated

Table 1 Separation method and it's working fundamentals .

Method		**Working fundamentals**
Mechanical separation	Precipitation and filtration	Dissimilarity in solubility
	Distillation	Distinctive volatility
	Extraction	Difference in solubility in two immiscible liquids
	Ion exchange	Dissimilarity in interaction of reactants with ion-exchange resin
Chromatography	Difference in rate of movement of a solute through a stationary phase	
Electrophoresis	Difference in migration rate of charge species in an electric field	
Field-flow fractionation	Difference in interaction with a field or gradient applied in perpendicular to transport direction	

from the matrix. Generally, there are two common approaches in the extraction of analyte of interest which are solid-phase extraction (SPE) and liquid-liquid extraction (LLE).

Table 1 shows several types of separation method that can be applied in sample pretreatment processes as well as in analyzing process. It should be understood that all these approaches can be applied during the preparation of the sample and the while analyzing the samples.

Mechanical separation via precipitation and filtration is mostly applied in the separation of metal contaminant due to difference in solubility of metal salts as examples in different pH. Other than pH adjustment, there are also precipitation via sulfide-salt formation and organic precipitant.

2.3.1 *Liquid-liquid extraction*

LLE is also known as solvent extraction where it is involves the distribution of sample components between two immiscible liquid phases. Due to practicability issue, conventional LLE has been replaced by others method such as SPE (Cantwell & Losier, 2002). LLE's main flaw is that it's labor-intensive, automation mode is challenging, and often beset by practical issues like emulsion formation. It also uses a lot of high-purity solvents, which may be costly to dispose of, and time consuming (Hashemi et al., 2017; Poole, 2003). Table 2 summarized the advantages and disadvantages of LLEs.

2.3.2 *Dispersive liquid-liquid microextraction*

Dispersive liquid-liquid microextraction (DLLME) has received a significant research recognition for sample preparation due to the advantages such as high enrichment ability, minimal reagent/sample consumption, and broad analyte/matrix application, as well as robustness, simplicity, speed, and cheap cost (Hashemi et al., 2017; Wang et al., 2021). DLLME has been known as a preferable extraction approach and this

Table 2 Advantages and drawbacks in liquid-liquid extraction (Hashemi et al., 2017; Montaseri & Forbes, 2018; Płotka-Wasylka et al., 2016).

Advantages	**Drawbacks**
– Allow the processing of heat-sensitive products – Simple techniques	– Require separation phase – Consumption of toxic solvents – expensive equipment – Time consuming

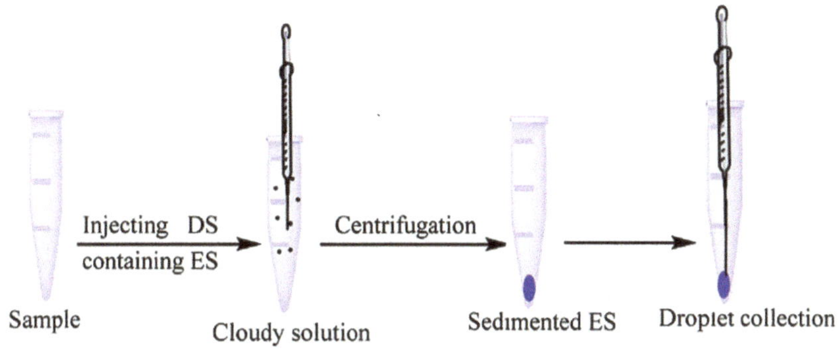

Fig. 3 Dispersive liquid-liquid microextraction.
Adapted from Hashemi, B., Zohrabi, P., Kim, K. H., Shamsipur, M., Deep, A., & Hong, J. (2017). Recent advances in liquid-phase microextraction techniques for the analysis of environmental pollutants. *Trends in Analytical Chemistry, 97*, 83–95. https://doi.org/10.1016/j.trac.2017.08.014.

method is divided into four major groups based on the type of extractant phase: dispersive liquid-liquid microextraction (DLLME), ferrofluid-based microextraction, supramolecular-based liquid-phase microextraction, and vortex-assisted liquid-liquid microextraction. This method creates a cloudy condition by injecting a binary solvent combination (extraction and disperser solvents) into an aqueous sample, which is caused by the formation of solvent droplets. Small droplet enhances the extraction of the compound of interest with due to high surface area. Simplified process of DLLME is illustrated in Fig. 3, where the sample is injected with small volume of mixture of extraction solvent (ES) and disperser solvent (DS). After centrifugation process, the extracted compound is collected at the bottom of the conical tube, resulting in a sedimentation of the mixture. Lastly, syringe is used to collect the analytes at the bottom of conical tube and analyzed with preferable equipment.

2.3.3 Solid-phase extraction

SPE is considered the most widely used "clean-up" ad "pretreatment" method. The working fundamentals of can be summarized as follows (Poole, 2003):

i. The analytes are transported to the solid phase and remain there for the rest of the sampling procedure.
ii. The solid phase of the sample is then separated from it.
iii. The analytes are recovered by elution with other liquid or by thermal desorption into the gas phase.

SPE's major advantages are generally lower in cost with shorter processing duration, lower solvent consumption and less complicated procedure. Additionally, SPEs due to less complex procedure, it is more possible to have the process automated. On the other hands, the uncomplicated procedure also allowed the SPE to be applied during field sampling, thus minimize the errors or degradation of compound of interest during the transportation and eradicate the necessity to store large volume of samples.

Despite of all the advantages of SPE techniques, SPE however suffers a few drawbacks. To start, solvent characteristics are more reproducible than the surface chemistry of solid phases, and hence their sorption properties are less predictable than solvent properties. Retention mechanisms of silica-based, siloxane-bonded sorbent are mixed between low-selectivity adsorption piloted by the bonded phase and ion-exchange interactions which is made available by dissociated silanol group of the silica substrate. Interferences due to this mechanism include less efficiency in analyte recovery due to the character of solvents. As instance, neutral solvent has less efficacy in displacing ionically bound analytes and the retention strength may differ due influence ionic binding into the sorbents. As it is manufactured in large quantity, the quality of production is also triggered a concern. SPE may be contaminated during multiple manufacturing process thus affected the analytes extraction. SPE sample processing issues linked to sorbents' low sorption capacity and analyte displacement or blockage of sorbent pores by matrix components sometimes go unreported, resulting in variations in analyte recovery. Due to the restricted sorption capacity of sorbents in comparison to the solvation characteristics of liquids, sample overload and displacement are more critical for SPE than LLE (Poole, 2003).

In the earliest stage, SPE setup consists of big iron cylinder, and activated carbon is used as sorbent mass. During this first introduction of SPE, 1.5 kg of the sorbent's mass was needed while the sample size contain analyte measured is in liter. Nowadays, as the discoveries of more new sorbent material which has better selectivity and reproducibility, the sorbent mass is reduced to only in microgram (μg) and the sample lessened to within microliter (μL) range (Płotka-Wasylka et al., 2016).

The typical method of SPE is to load a solution onto the SPE solid phase, rinse the unwanted components, and then wash off the desirable analytes into a collecting tube with another solvent (Zwir-Ferenc & Biziuk, 2006). The interaction between the sorbent and analyte is essential to safeguard the extraction efficiency. Thus the understanding of the sorption's mechanism between the analyte and the sorbent plays an important part in the selection where the conventional retention mechanisms include van der Waals forces ("nonpolar interactions"), hydrogen bonding, dipole-dipole forces ("polar" interactions) and cation-anion interactions ("ionic" interactions) (Table 3) (Zwir-Ferenc & Biziuk, 2006).

Table 3 The sample, analyte and mechanism for different type of phase .

Phase	Reverse phase	Normal phase	Ion-exchange
Sample	The mobile phase (sample matrix) is slightly polar or slightly nonpolar stationary phase	Sample composes of polar substance. Polar stationary phase with a mid to nonpolar matrix	– Dissolved substances – Silica surface with an aliphatic quaternary amine group on the surface separates negatively charged—While silica surface that coated with aliphatic sulfonic acid groups separates cationic (positively charged) substances
Analyte	Normally consist of mid- to nonpolar	Most of the time consist of mid- to nonpolar matrix	
Mechanism	The designated functional groups on the surface of stationary adsorbent immobilize the substances with the carbon-hydrogen bond in the analyte	The different polar charge between functional groups in analytes and stationary phase	Mainly the interaction between different charge between the substance in analyte and the stationary phase that is enhance with the functional group with opposite charge

SPE sorbents must provide good separation, mechanical stability, solvent resistance, sample cleaning, and fast analysis plus environmentally friendly. Among numerous of sorbent that accessible, silica-based SPE is the most common especially for polar compound (Ma et al., 2016). Nonetheless, there are many disadvantages to making a silica-based SPE cartridge that must be considered particularly during the pressing into the cartridge (Table 4). An important factor that contributes to the efficiency of a sorbent in SPE is particle size that is indirectly affect extraction efficiency and flow characteristic inside the column/cartridge. In order to have the balance between adsorption and flow efficiency, focus must be given to particle size distribution of sorbent inside the column.

Table 4 Advantages and drawbacks of SPE (Hashemi et al., 2017; Ma et al., 2016; Montaseri & Forbes, 2018; Płotka-Wasylka et al., 2016).

Advantages	Drawbacks
Low solvent consumption Low intrinsic costs Reduction of processing time Possible to automated whole process No separation of phase needed, eliminate errors associated with variable/inaccurately measured extract volumes Environmentally friendly Greater reproducibility Good sensitivity Lower solvent consumption Various sorption characteristics	Requiring several steps of treatment involving washing, conditioning, sample loading, eluting, and drying the solid sorbent

A cartridge-loaded SPE is selected by most researchers to be used in analytical work specifically in isolating and concentrating analytes in different type of samples with varying degrees of content.

SPE cartridge consists of an open-ended tube of syringe preferably made of glass but also available in polypropylene barrels loaded with various types of adsorbents. Glass cartridges are preferred over plastic cartridges owing to superior physicochemical characteristics, particularly higher durability to solvent that minimizes the threat of sample contamination by low molecular weight components and additives used to manufacture plastic tubes. Nonetheless, due to factors such as higher production cost of glass material, plastic cartridge is selected in most research.

The selection of an ideal stationary phase, which offers the chance to collect all analytes as well as the selection of an acceptable column volume, are both critical in achieving the greatest extraction efficiency possible. In order to accommodate this, a very broad range of various kinds of columns in a variety of solid phases and sizes are currently commercially available. Clearly, the lowest capacity cartridges are favored in today's analytical chemistry. Choosing a solution with less sorption material lowers the quantity of organic solvents used for conditioning, washing, and eluting, as well as the time required for the whole analysis. Table 5 shows the comparison of a few approaches in solid phase extraction.

Both columns and discs use the same basic working fundamentals and similar sorbent material. The discrepancies between these two methodological solutions are mostly attributable to the varied packing of the bed and the variable structure of the particles used in each of the solutions. The discs contained a much smaller sorbent particles (8 μm in average) than those utilized in the cartridge filling (40 μmin average) (Płotka-Wasylka et al., 2016).

As listed in Table 5, the disk has the potential to operate at higher flow rate, thus lessen the duration of process. In addition, the increment of the contact surface, homogeneity, and density of the bed, decrease the loss of the analyte. Disks used in various

Table 5 Comparison of formats and devices used in SPE (Płotka-Wasylka et al., 2016).

Format	Advantages	Disadvantages
Cartridge	• In lab preparation is possible • Multiple column merging for desirable results • Inexpensive	• Small cross-sectional area • Time consuming due to slow flow rate • Undesired channeling • High void volume
Disk	• Small amount of elution solvent • No channeling issue thus faster flow rate and faster process • Smaller void volume • Large specific area • Filtration stage may be dismissed • Chances to merge with other sample processing steps	• Volume of breakthrough is small • Costly compared to cartridge
Pipette tips	• Less complex procedure • Only small amount of sample and elution needed • Facilitating multiple sample treatment via multichannel micropipette • Lesser duration of extraction • High recovery factor • Easy automation	• Plugging • Considerable high plastic waste production
Sorbent mix with sample	• Solvent quantity requirement for elution is minimized • Lesser sorbent mass	• Highly reliant on the suitable sorbent for a good extraction

kinds of studies have a diameter ranging from 4 to 96 mm, depending on the kind of analysis. Standard techniques for water sample analysis especially in environmental analysis utilize discs with a diameter of 47 mm, which are ideal for water samples ranging in volume from 0.5 to 1 L. Tran et al. (2016) have successfully developed an approach that applies tandem SPE coupled with UHPLC-MS/MS and isotope dilution where the tandem SPE for simultaneous extraction is managed to obtain acceptable recoveries for all the target analytes.

2.3.4 Methodology of solid-phase extraction and clean-up

There are many and various procedures which can be used to extract environmental samples for the purpose of "clean-up" prior to analysis. Generally, these methods depend on the target PhACs, physiochemical properties of the compounds, type of the environmental sample (liquid or solid), and organic matter content in the collected sample. In this chapter, we have summarized the common procedures reported in the literature and presented in Table 6.

Table 6 Summary of SPE extraction and clean-up processes in different studies.

pH	Cartridge	PhACs	Conditioning	Loading of sample	Rinse	Elution	Residues (reconstituted)	Analytical technique	References
pH 3.6 (acetic acid)	Supelclean LC-18 SPE tubes (6 mL, 1 g)	AMX, TYL, SMX	5 mL MeOH 5 mL pure water (5 mL/min)	500 mL (ww) 10 mL/min	0.5 mL ACN: ultrapure water acidified with formic acid at pH 3 (1:1, v/v)	ACN: MeOH: acetic acid (40:40:20, v/v)	0.5 mL ACN: ultrapure water acidified with formic acid at pH 3 (1:1, v/v)	HPLC-MS	Spataro, Ademollo, Pescatore, Rauseo, and Patrolecco (2019)
pH 2	HLB 60 mg MCX 60 mg	ACT, KET, IBF, KFN, CBZ, SMX	5 mL MeOH 5 mL pure water (5 mL/min)	500 mL IN 1000 mL EF 15 mL/min	3 mL (MeOH: water-5:95, v/v) 3 mL hexane (1 mL/min)	Three aliquots of 1 mL acetone (1 mL/min)	Elute evaporate to dry Residues dissolved in 0.2 mL MeOH, filtered through 0.45 nylon filter	HPLC	Camacho-Muñoz, Martín, Santos, Aparicio, and Alonso (2009)
pH 2.5 (HCl 2 M)	C-18 cart	IBF, NPX, DFC, KFN	6 mL ethyl acetate 6 mL MeOH 6 mL water (5 mL/min)	5 mL/min	2 mL acidified water	Three washes of 2 mL ethyl acetate	Elutes evaporated to dry Residues derivatized with 50 μL BSTFA +1 TMCS and 10 μL pyridine	GC/MS	Samaras, Stasinakis, Mamais, Thomaidis, and Lekkas (2013)
pH 7	HLB (6 mL, 500 mg)	Antibiotics, beta blockers, NSAIDs, etc.	Eliminated	5 mL/mi	5 mL Milli-Q water	Dried (15 min), then eluted 5 mL MeOH 1 mL/min	Eluted dried, recons. with 1 mL of internal standard (mixture of 3 IS in Milli-Q water at 50 μg/L) Final residue filtered 0.2 μm syringe filter	LC/MS	Afonso-Olivares, Sosa-Ferrera, and Santana-Rodríguez (2017)

n.a.	HLB (20 and 200 mg)	ACT, IBF, KTF, DCF	3 mL MeOH 3 mL HPLC-water	100 mL of water sample was spiked with mix of IS working solutions Sample was passed through gravity (3 mL/min)	Eliminated	After dried. Analytes were eluted with 5 mL MeOH	Extract was evaporated to dryness under gentle nitrogen steam (40°C) Residues were reconstituted with 1 mL MeOH-water (10:90, v/v)	UHPLC	Gracia-Lor, Sancho, and Hernández (2010)
n.a.	HLB (200 mg, 6 mL)	SMX, CBZ, ACT	3 mL MeOH 1 mL 0.1 M HCl 5 mL Milli-Q water	100 mL (WW) 3 mL/min	5 mL Milli-Q water	5 mL MeOH Water-150 mg/L EDTA (75:25, v/v)	Elute was collected and reduced to 1 mL	HPLC-MS/MS	Gao, Ding, Li, and Xagoraraki (2012)
n.a.	HLB (500 mg, 6 mL)	CBZ, ACT, IBF, DCF, KTF	2 mL MeOH 2 mL DI water	200 mL IN, 500 EF 5 mL/min	Eliminated	After dried, eluted with 4 × 1 mL MeOH	Eluted were dried under gentle stream of nitrogen Re-dissolved in 900 μL of ethyl acetate	GC/MS	Yu, Wu, and Chang (2013)
n.a.	HLB (200 mg, 6 mL)	CBZ, KTF, ATN, antibiotics	5 mL MeOH 5 mL Milli-Q water 5 mL/min	250/200 mL sample 10 mL/min	5 mL Milli-Q water, then dried under vacuum for 10 min	2 mL MeOH 1 mL/min	Elute was reconstituted with 1 mL HPLC water Analyzed by LC/MS	HPLC	Guedes-Alonso, Afonso-Olivares, Montesdeoca-Esponda, Sosa-Ferrera, and Santana-Rodríguez (2013)

2.4 Pretreatment and extraction of solid samples

Beside liquid samples, PhACs can also be present in sediment and soils of the aquatic environment. Generally, soils have a lesser amount of organic matter content compared to water, which could offer vaguely better efficiency of extraction due to the low matrix effect. Even for polar substances, chemical interactions (such as cation bridging, cation exchange, and surface complexation) bond them chemically to organic materials and clays, making extraction more difficult (e.g., FQs). The distribution-ratio K_d of the chemicals might change significantly depending on the pH of the extraction solvent. In a pH range around the pK_a, the $\log Kow$ of ionizing chemicals changes dramatically (e.g., the adsorption of amphoteric sulfonamides to the soil increases when the pH decreases, relating to their ionization) (Hu et al., 2012; Mutavdžić Pavlović, Pinušić, Periša, & Babić, 2012; Vazquez-Roig, Segarra, Blasco, Andreu, & Picó, 2010).

Solid samples from soils and sediments are generally extracted using one of these techniques (Jelić, Petrović, & Barceló, 2009; Montesdeoca-Esponda, Sosa-Ferrera, & Santana-Rodríguez, 2012; Pérez-Carrera et al., 2010; Vazquez-Roig et al., 2010):

- pressurized liquid extraction (PLE),
- matrix solid-phase dispersion (MSPD),
- microwave-assisted extraction (MAE),
- ultrasonic extraction (US), or
- microwave-assisted micellar extraction (MAME).

Normally, the off-line SPE is applied for purification and preconcentration purposes. Prior to SPE, the content of the organic-solvent of the extract has to be decreased to less than 5% to avert early breakthrough of analytes from the cartridges (Vazquez-Roig et al., 2013). Nevertheless, MAE yields greater recoveries in general, although specific attention to temperature and irradiation time is essential to minimize analyte deterioration. The degradation of PhACs is accelerated by high microwave energies combined with lengthy periods, as documented for clofibric acid, propranolol, and metoprolol.

3. Analytical techniques

Most of common parameter in environment contaminant is gauged as bulk measurement rather than single analyte. Basic theory of analytical separation process is focusing on the treatment of other contaminant that disrupting the indication signifying the analyte of interest. Separations keep the analyte free of any potentially interfering substances. To counteract the effects of interferents, methods including matrix modification, masking, dilution, and saturation are commonly utilized. The separation process is important to ensure only the component of interest is separated from the mixture. Separation of analyte can be occurred in two ways which are partial and complete separation process. Fig. 4 shows the illustration of a complete separation process. The separation of analyte can be preparative or analytical.

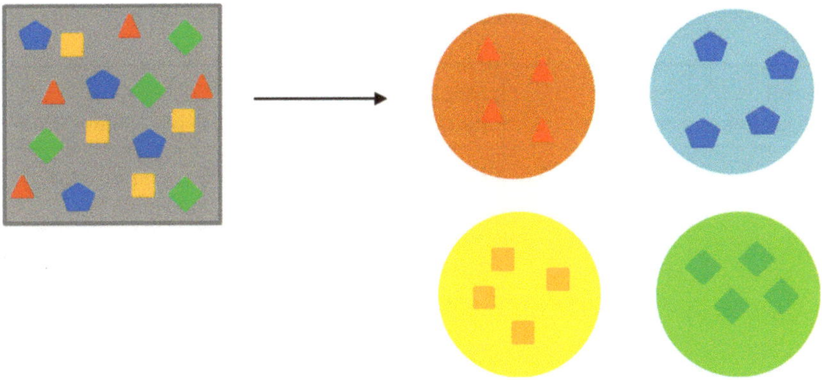

Fig. 4 Complete separation process where all species are completely separated so that each component is separated based on their interaction between stationary and mobile phase (Meyer, 2013; Skoog et al., 2014).

3.1 Chromatography

Chromatographic techniques are categorized as follows: Column chromatography utilizes a thin tube to contain the stationary phase, and the mobile phase is moved through the tube under pressure or gravity. The stationary phase is supported on a flat plate or in the pores of paper in planar chromatography, while the mobile phase passes through the stationary phase through capillary action or gravity. Generally, chromatography is a method of separating mixtures based on variations in the speeds at which a gaseous or liquid mobile phase moves through a fixed or stationary phase. Separation process involves two main phases which are a stationary phase and a mobile phase. In chromatography, the stationary phase is a phase that remains stationary, while the mobile phase is a phase that travels over or through the stationary phase, taking the analyte mixture with it (Skoog et al., 2014).

Most common approaches in chromatography are gas chromatography (GC) and liquid chromatography (LC) at listed in Table 6. In GC, the methods can be divided into two groups which are gas-liquid chromatography (GLC) and gas solid chromatography. Both types are applying gas phase as the mobile phase where the analyte has to be vaporized. However, the difference between these two GCs is the stationary phases, which are in liquid form and solid form for GLC and gas solid chromatography respectively.

LC however consists of more specific approaches as listed in Table 7. Liquid-liquid method is applying the similar concept as liquid-liquid extraction between two immiscible liquid phases. In liquid-solid chromatography, the stationary phase is a solid material, and adsorption between the chemicals and the stationary phase results in various retention times in the column. Table 8 shows the comparison of GC and LC in general.

Table 7 General specification of chromatography system (Ismail & Susan, 2010; Skoog et al., 2014).

General specification	Specific method	Mobile/stationary phase	Type of equilibrium
GC	Gas-liquid	Liquid that supported by solid material	Partition between gas and liquid
	Gas-solid	Solid	Adsorption
LC	Liquid-liquid, or partition	Liquid adsorbed or bonded to a solid surface	Partition between immiscible liquids
	Liquid-solid, or adsorption	Solid	Adsorption
	Hydrophobic interaction	Polar liquid/nonpolar liquid or solid	Molecular size/polarity
	Normal phase	Less polar liquid/more polar liquid or solid	Molecular size/polarity
	Reverse phase	Polar liquid/nonpolar liquid or solid	Molecular size/polarity
	Ion exchange	Ion-exchange resin. Polar liquid/ionic solid	Ion exchange
	Size exclusion	Liquid in interstices of a polymeric solid	Partition/sieving/ molecular size
	Affinity	Group specific liquid bonded to a solid surface, water/binding sites	Partition between surface liquid and mobile liquid. Specific structure
SFC	Supercritical fluid/solid	Supercritical fluid/solid	Molecular size/polarity

Table 8 GC and LC comparison (Fatta, Achilleos, Nikolaou, & Meriç, 2007; Kruve et al., 2015; Montaseri & Forbes, 2018).

LC	GC
Nonvolatile chemical	Volatile chemical
Suitable for heat-sensitive	Not suitable for heat-sensitive
No derivation	Need derivation
large number of parameters have to be at or near optimal values in order to get the desired performance	

3.1.1 Liquid chromatography

LC is selected in the separation process of nonvolatile compounds. It is more preferable for the detection of polar and heat sensitive compound as the derivatization is not necessary (Daverey et al., 2019). As conventional LC is time consuming, the LC system has been upgraded to ultra-high-performance liquid chromatography (UHPLC) which is an upgraded version of the high-performance liquid chromatography (HPLC). UHPLC is superior compared to HPLC due to several factor, such as lesser amount of solvents required for sample elution, much narrower and concentrated bands are obtained, and separation speed increases up to ninefold (Daverey et al., 2019).

Separation mode in LC can be divided into several groups based on the operational fundamentals. The separation mode of LC includes adsorption, ion exchange, size exclusion, bonded phase, and others. For bonded phase chromatography, it can be divided to normal phase and reverse phase (Stout & Dorsey, 2002). Adsorption chromatography is the earliest chromatographic techniques with solid material such as silica gel, porous carbon, and alumina as the stationary phase. Adsorption chromatography however encounters major drawbacks including the sites of adsorption on adsorbent and are not energetically homogeneous, and reproducibility of retention times is often poor and variation in content of mobile phase influence the retention significantly (Stout & Dorsey, 2002).

Ion-exchange chromatography is the process by which ions are exchanged reversibly between mobile and stationary phases that possessed a functional group. The analyte will compete the adsorption site with the mobile phase (Stout & Dorsey, 2002). The size exclusion chromatography (SEC) has a few other names such as gel filtration, gel permeation chromatography and steric exclusion (Stout & Dorsey, 2002; Voedisch & Thie, 1990). The solutes are separated based on the differences in hydrodynamic volume. This particular method is fundamental tool in protein biochemistry and less used in detection of pharmaceutical compounds in environment (Burgess, 2018).

Normal phase chromatography is defined as chromatographic condition that retention of solutes decreases with increment of solvent polarity where the stationary phase normally is more polar compare with mobile phase. Reverse phase chromatography uses a C18 ligand, an 18-carbon hydrocarbon chain as stationary phase where this chain provides hydrophobic retention of the reversed-phase packing (Neue et al., 2007). In the occasion of lesser retention needed, C8 which has shorter hydrocarbon chain is used. The selection of these type of stationary phase is dependent on the peak appearance or lesser retention needed. The difference between normal phase and reversed-phase is summarized in Table 9.

Table 9 Liquid chromatography phases based on type.

LC type	Stationary phase	Mobile phase
Normal phase	Polar	Nonpolar to more polar
Reversed-phase	Nonpolar	Polar to less polar

3.1.2 Gas chromatography

Conventional GC applies single GC column (1D GC) has a few shortcomings that can be improved by the application of GC × GC (2D GC). In 2D GC, two columns (often nonpolar primary and polar secondary) are linked consecutively, enhancing peak capacity and separation efficiency (Daverey et al., 2019). The limitations of 1D GC such as limited type of pollutants content analysis and disability to separate the hydrocarbon compound (>C_{10}) has been addressed in 2D GC. General component for 2D GC is shown in Fig. 5. The modulator that connected between 2 GCs is to trap, focus, and inject small fractions from the first column into the second column (Murray, 2012). Basically, 2D GC possesses the ability to cater greater peak capacity and signal enhancement up to 70 times compared to 1D GC.

Introduction of sample to GC system plays a significant part in order to ensure the separation efficiency and maintain the sample composition (Basil, 2017). There are several types of sample injection method to GC depending on the physical state, chemical stability, and thermal degradation of the sample. Thermal desorption (TD) implicates the elimination of volatile or semivolatile analytes from solid sample by heated gas flow. Then the extracted analytes are transmitted to the GC system by an inert gas in a small volume of concentrated gas. TD process is low in costs and required minimum manual labor but the sensitivity is reduced as the number of samples increases (Basil, 2017).

As conventional GC only allows the injection of 1–2 μL of samples, large volume injection (LVI) (Fig. 6) approaches boost method sensitivity and/or eliminate the requirement for extract concentration stages by injecting greater amounts of final sample extracts than typical (Hoh & Mastovska, 2008; Teske & Engewald, 2002).

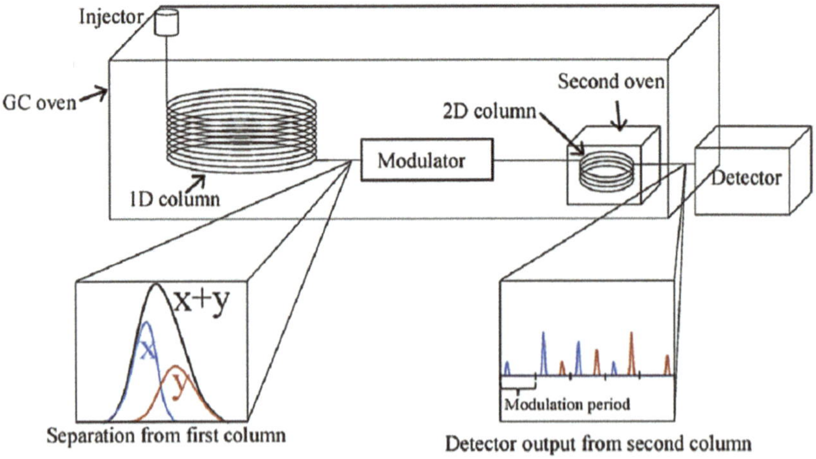

Fig. 5 2D GC components.
Adapted from Murray, J. A. (2012). Qualitative and quantitative approaches in comprehensive two-dimensional gas chromatography. *Journal of Chromatography A, 1261*, 58–68. https://doi.org/10.1016/j.chroma.2012.05.012.

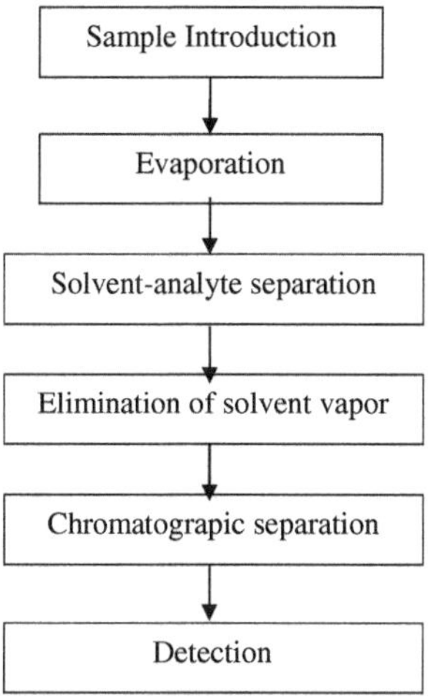

Fig. 6 Events of large-volume injection in gas chromatographic analysis.
Adapted and modified from Teske, J., & Engewald, W. (2002). Methods for, and applications of, large-volume injection in capillary gas chromatography. *Trends in Analytical Chemistry, 21* (9–10), 584–593. https://doi.org/10.1016/S0165-9936(02)00801-4.

With LVI approaches, the sample volume can be increased to several hundred microliters, with the goal of improving the analytical method's performance (by decreasing the detection limits) and incorporating sample pretreatment into instrumental analysis. In addition, an LVI approach may be used to link GC to a sample preparation process, such as solid-phase extraction, on-line (SPE). There are four categories of injection techniques including isothermal (hot) split, isothermal splitless, on-column, and programmed temperature vaporization (PTV).

Certain compounds are not suitable to be used in GC due to several reasons. As instance, compounds with functional group namely –OH, –COOH, –NH, and –SH are less volatile due to its polar character. This attribute are less suitable for the application of GC due to its affinity to the active sites of the columns. Thus, it will cause the appearance of ghost or asymmetric peaks, even worst cause a significant damage to the column. Organometals and halo-acetic acid, for example, are analytes that is susceptible to high temperature make it unsuitable for vaporization at high temperatures (Wang, Ma, Yin, & Xu, 2013). Fig. 7 shows the hydrophobicity and hydrophilicity of groups of pharmaceutical contaminants.

Due to the reasons stated, a process called derivation is necessary to alter the character namely increase the volatility and stability, thus increase the GC separation ability and intensify the detectability. As example, the derivatization of calystegines has to take place due to low molecular weight, and high number of hydroxyl groups which directly affect its polarity (Romera-Torres, Romero-González, Martínez Vidal, & Garrido Frenich, 2018). However, the derivation process has a few significant shortcomings including possibility that several reactions will take place, making the operation difficult, unpleasant, and time-consuming.

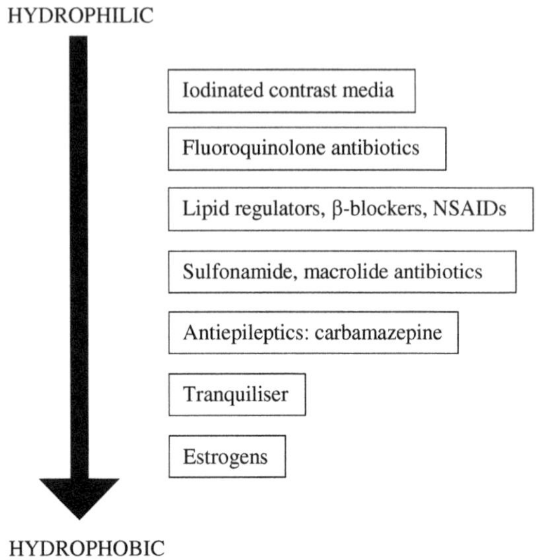

Fig. 7 Level of hydrophilicity and hydrophobicity of pharmaceutical compounds.
Adapted and modified from Fatta, D., Achilleos, A., Nikolaou, A., & Meriç, S. (2007). Analytical methods for tracing pharmaceutical residues in water and wastewater. *Trends in Analytical Chemistry, 26* (6), 515–533. https://doi.org/10.1016/j.trac.2007.02.001.

Additionally, it is conceivable that adverse effects will arise as a result of possible contact to moisture. Moreover, the derivation agents that has been used in the process is high in toxicity and harmful where the fact that it is normally performed manually by operators, the exposure is difficult to be terminated (Fatta et al., 2007). While off-line derivatization is reasonably efficient, it has a low reaction efficiency and thus has poor analytical sensitivity (Wang et al., 2013).

All the drawbacks listed above can be neutralized by On-Line Derivation or known as Injection Port Derivation (IPD) where the derivation occurred in GC system, specifically in GC injection port. IPD help to minimize the solvent quantity used, shorten derivation time which assisting in reducing the degradation of derivatives due to exposure to environment. Moreover, because on-line derivation has a greater reaction efficiency than off-line derivation, detection sensitivity is increased, and more precise quantification is achieved (Wang et al., 2013).

In general, IPD derivatization can be classified into several groups based on the displacement of active hydrogen from polar functional groups (–OH, –SH, –NH, –COOH, –CONH) to alkyl, acyl, or silyl groups, which shield the polar groups and increase the analytes' volatility and stability as well as make the GC column less affected by their lessened affinity.

The derivatization is influenced by a few factors which are: ion-pairing reagent, injection port temperature, and the concentration of ion-pairing reagent (Marinas et al., 2010). Sialylation is the most derivatization process selected in comparison to the parent chemicals, and the derivatives have a higher volatility and thermostability, where in some case, the derivatives can withstand high temperature up to 300°C that is appropriate in the application of GC. In addition, most sialylation consume a short process duration plus it has the ability to enhance detectability, especially to mass spectrometer detector (Wang et al., 2013).

Alkylation corresponds to the substitution of alkyl groups for active hydrogen in polar groups in analytes to produce more volatile ethers, thioethers, N-alkyl amines or esters, and N-alkyl amides.

Tetraalkylammonium (TAA) salts has always been a preference compare with other alkylation reagents (diazomethane, alcohols (acid as catalyst) and alkyl halides) due to its reaction rapidness, low solvent waste, high derivatization efficiency, avoidance of the use of hazardous reagents and operation convenience (Marinas et al., 2010). Acylation is primarily used to derivatize NH, OH, and SH groups utilizing carboxylic acid, acid anhydride, acyl halide, and activated acyl amide as derivatization reagents. By far, acid anhydride has been the most widely employed because it generates volatile amide, ester, or thioester without the presence of water. Aside from three derivatization approaches that have been briefly discussed, there are several other approaches with similar purpose, which is to enhance the applicability of GC for the samples mainly via increasing their volatility and stability in high temperature.

Both GC and LC are coupled with detector which has several types with different working fundamentals. Recently, most advanced chromatography system is combined with mass spectrometer that high sensitivity and accuracy. However, these sets of system do not fit to be used on-site or continuous analysis. Despite of the advantages offered this system is also very expensive and has to be operated by high-skilled labor. As it is costly, this equipment can only be afforded by high-end laboratories. With all the drawbacks, this cutting-edge technology is less accessible to researchers.

3.2 Chromatography detectors

There are several types of sensors available to analyze the separates compound in chromatography system, these may include ultraviolet-visible spectrophotometry (UV-vis), fluorescence, and mass spectrometry (MS) (Kruve et al., 2015). UV-vis detector is one of the earliest detectors in a chromatography system. UV-vis detector can be divided into three groups, which are fixed-wavelength detectors, variable-wavelength detectors, and rapid-scanning detectors (Stout & Dorsey, 2002). Fixed-wavelength detectors generally have lower limit of detection compare with variable-wavelength detector even both were set at similar wavelength. Fixed-wavelength detectors. Photo-emissive and photomultiplier tubes, photovoltaic cells, and photodiode array detectors are types of detectors that applying the UV-vis fundamentals in their detection approaches.

For certain solutes that display molecular fluorescence, the fluorescence detector is a near-ideal detector. This is due to near-zero signal in the absence of a fluorescing solute and it is highly selective (Stout & Dorsey, 2002). The derivatization process available also adds a fluorophore to a nonfluorescent solute thus assisting in detection. Fluorophore to a nonfluorescent solute fluorophore to a nonfluorescent solute. Fluorescence detector can be as simple as consist of a low-pressure mercury lamp as the source to excite compound of interest and a sensor to collect the luminescence. This system is known as the cheapest system of fluorescence detector. However, the system with a better control of excitation and emission of wavelength offers a better selectivity of detection. The use of light emitting diodes (LEDs) as a light source is one of the most significant technological advances that has enabled the growth of micro sized detectors (Lynch, Chen, & Liu, 2018). Laser induced fluorescent (LIF) detection is a very sensitive technology for assessing native fluorescent and fluorescently tagged compounds and has been acknowledge an approach to miniaturize chromatographic system. The factors that contribute to the possibility of miniaturization include low sample consumption, short testing time, high sensitivity, a relative low complexity of their components and are easily integrated into microfluidic chip-based systems (Lynch et al., 2018).

Mass spectrometry (MS) is a powerful and flexible analytical technique for determining the identification, molecular mass, elemental composition, and, in many circumstances, chemical structure of an unknown substance (Skoog et al., 2014). The detection limits of MS are frequently several orders of magnitude better than those of optical techniques. The application of MS system in chromatography system has proven to be high in sensitivity and possess the ability to confirm compounds (Fatta et al., 2007). MS component can be divided into a few steps as illustrated in Fig. 8.

MS ionizes the analytes of interest by any suitable method, separates them according to their mass-to-charge ratio (m/z), and directs them to a transducer that turns the number of ions (abundance) into an electrical signal to obtain the qualitative and quantitative data. The ion abundance plotted against mass-to-charge ratio is called a mass spectrum. While most of other method of detection allow samples recovery, MS however is a destructive analysis (Gross, 2017). On the other hand, as the sample needed for analysis is very low, MS system is still a preferable approach in analytical research. A performance of MS can be gauged by several parameters including sensitivity, limit of detection and signal-to-noise ratio (Gross, 2017).

Generally, there are several types of mass analyzer with various approach of working fundamentals namely single quadrupole mass analyzer, ion trap mass analyzer, triple quadrupole mass analyzer, time-of-flight mass analyzer and others as listed in Table 10.

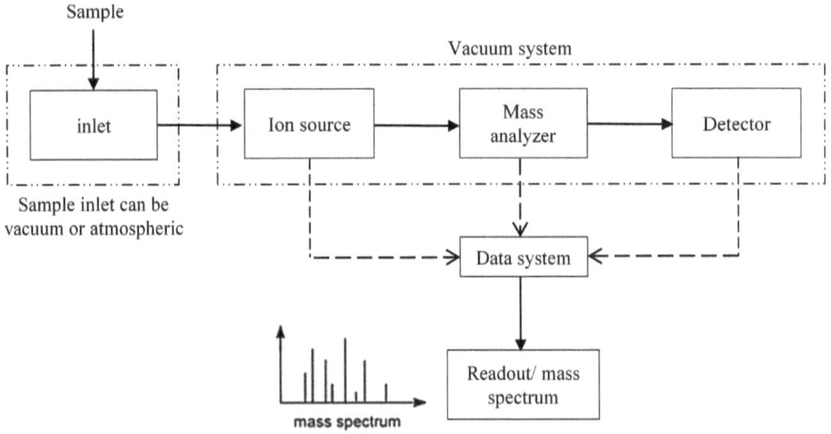

Fig. 8 Components of a mass spectrometer (Gross, 2017; Skoog et al., 2014).

Table 10 Common type of mass analyzer (Skoog et al., 2014).

Type	Analysis principles
Magnetic sector	Application of magnetic field to deflect ions beam. *m/z* value affects trajectories route due to momentum difference (Lorentz Force)
Double-focusing	Application of two mass analyzer. First analyzer assist in focusing via electrostatic force, then deflection by magnetic force *m/z* value affects trajectories route
Linear quadrupole	Ion motion in dc and radio-frequency fields. Only *m/z* with certain values are allowed to pass
Linear quadrupole ion trap	Continuous ion beam delivers ions for trapping; storage, and eventually separation in linear radio frequency quadrupole field by resonant excitation
Ion trap	Encapsulate ions in a space with caps and ring electrodes Ions ejected by electric field force based on *m/z* values
Ion cyclotron resonance	Ion trapping in a cubic cell under the effect of a trapping voltage and a magnetic field The orbital frequency is inversely proportional to the *m/z* value. Separation by cyclotron frequency,image current detection and Fourier transformation of a transient signal
Time-of-flight	The drift tube is filled with ions of equal kinetic energy. Detection time is influenced by drift velocity which is related to mass
Orbitrap	Axial oscillation in inhomogeneous electric field; detection of frequency after Fourier transformation of a transient signal

In order to have a higher resolution, a double focusing mass analyzer combines two approaches which are electrostatic and magnetic sectors. The magnetic sector helps by giving a magnetic force perpendicularly toward the ion, thus reroute the ion based on their momentum. In theory, the ion with higher mass will be deflected more compare to ion with lower mass. This technique however needs a high skilled operator, costly in maintenance for favorable performance, a spacious chamber, and complementary accessories thus it is less preferred by researcher.

Quadrupole mass analyzer only allow particular value of mass to charge ratio to go through the system. A quadrupole system consists of four cylindrical rod that act like a filter and it can be set up via adjustment of the voltages. The basis of separation between compound of interest and unwanted compound via application of ion motion in electric fields. A proper adjusted quadrupole system creates a stable route a certain *m/z* ratio to pass through the analyzer to the transducer thus make the quadrupole mass analyzer is the only analyzer with the ability to operate in mass-selective mode (Basil, 2017; Gross, 2017). MS tandem involved more than one MS analyzer and detector where each system assists to focus and filter out unwanted analytes. MS tandem system can be divided into two groups which are tandem-in-space and tandem-in-time mode. Tandem-in-space consist of multiple MS system that operated sequentially where the first one aimed to select the ion of interest and then transmitted to the second analyzer. Whereas, the tandem-in-time is the application of the similar MS system sequentially (Gross, 2017).

Time of flight (TOF) mass analyzer applies the theory that the time required for the ions to travel a fixed distance to the detector is thus inversely related to the ion mass as theoretically written in Eq. (1). The ions with lower value of m/z will move to the detector faster compared to the ion with high value of m/z. TOF system is suitable for GC systems due to the ability to separate overlapping peaks by using automated mass spectral deconvolution of overlapping signals.

$$v = \sqrt{\frac{2\text{KE}}{m}} \tag{1}$$

where v is ion velocity, KE is kinetic energy, and m is mass.

TOF offers numerous promising advantages. Theoretically, the m/z range of a TOF analyzer is unlimited and possess high ion transmission thus contribute to high sensitivity. Moreover, the design of TOF MS is straightforward compared to another analyzer.

Despite of all the advantages offered by all the detectors, there are restriction or limitation for each type of detector approaches as summarized in Table 11. These limitations should be notice earlier prior to the analysis to ensure the validity of data obtained.

Fatta et al. (2007) have developed a list of pharmaceutical contaminant and suitability of detection method as illustrated in Fig. 9. It can be seen that more pharmaceutical compounds can be analyzed by LC system. This is due to versatility and less complicated sample preparation nature of LC system (Fatta et al., 2007). However, this list is not an absolute reference as there are factors such as the concentrations of contaminants and the accuracy required in the measurement.

3.3 *Electrochemical sensor*

Compared to conventional instrumental analysis techniques, electrochemical detectors (ECDs) built from modified electrodes have shown superior detection sensitivity, repeatability, compactness and enables real-time detection of environmental contaminants (Kumar, Bhadwal, Mizaikoff, Singh, & Kranz, 2019; Mazloum-Ardakani, Sheikh-Mohseni, Mirjalili, & Zamani, 2012). The electrochemical sensor is produced via modification of electrodes using redox modifiers. Electroactive chemicals known as redox modifiers efficiently shuttle electrons between the analyte and the electrode. Regardless of all the benefits listed above, electrochemical sensors are in a development phase.

Torrinha et al. (2020) have summarized the application of nanostructured carbon-based material as the electrochemical sensor. Carbon nanotubes (CNT) as example are a material with remarkable electrochemical characteristics, owing to the activity of edge-plane-like graphite sites at the CNT ends. This character helps in the ability

Table 11 Detection scheme and restrictions (Michalski, 2018; Swartz, 2010; Webster, 2018; Zhang, Kurita, Venkatramani, & Russell, 2019).

Detector technology	Attributes	Mobile phase restriction(s)
Absorption (UV-vis)	Most widely used, nondestructive, sensitive, robust, wide linearity range	Minimize background chromophore, solvent must be transparent and low UV cut-off
Fluorescence	Good selectivity and sensitivity, not all compounds fluorescence	Required derivative formation
Refractive index (RI)	Excellent versatility, nondestructive	No gradient, temperature and flow sensitive. Refractive index difference between mobile phase and compound of interest must be big
Evaporative light scattering (ELSD)	Nebulization of mobile phase and measure the scattering light by the particle	Volatile mobile phases and additives
Mass spectrometry	Electrospray ionization	Volatile mobile phases and additives
Charged aerosol det		
Charged aerosol detection (CAD)	Good sensitivity for non-volatile compounds	Volatile mobile phases, sensitive to contaminant in solvent and buffers
Electrochemical	High selectivity and sensitivity, reliable, easy to use	Charge carrying ion present, must be conductive, applicable for compound that can be oxidize and reduced
Chemical luminescence nitrogen detection (CLND)	Formation of excited NO_2 decays to ground state releasing photon	No ACN or other nitrogen containing additives
Conductivity	Very selective, low cost	Require suppression of mobile phase background conductivity

to detect low analytes concentrations, high sensitivity and fast response due to the signal enhancement provided by high surface area, low overvoltage, and rapid electrode kinetics (Mazloum-Ardakani et al., 2012; Torrinha et al., 2020).

Despite of owning several advantages such as high degree of selectivity and sensitivity, ECDs are not a common detector in chromatographic system. Nonetheless, the application of ECDs has been recognize that offers better chance of miniaturization of chromatography system with attributes like electrode composition and its respective potential, nM detection limits, a linear dynamic range over four magnitudes, and noise levels in the nano to picoamperes (Lynch et al., 2018).

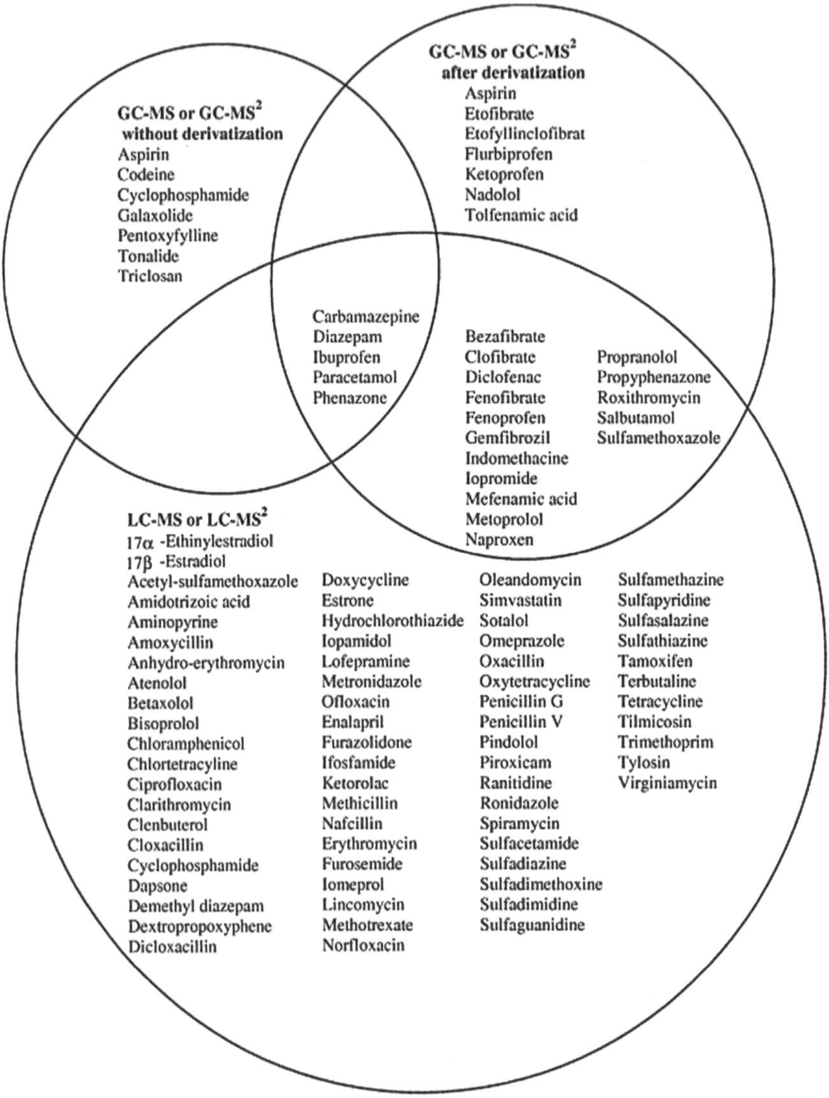

Fig. 9 Analytical methods applied for the most common pharmaceuticals in water and wastewater.
Adapted from Fatta, D., Achilleos, A., Nikolaou, A., & Meriç, S. (2007). Analytical methods for tracing pharmaceutical residues in water and wastewater. *Trends in Analytical Chemistry, 26* (6), 515–533. https://doi.org/10.1016/j.trac.2007.02.001.

4. Conclusions

This chapter has introduced the current employed methods and techniques that are used for the analysis of pharmaceutical compounds in the aquatic environment. The preanalysis or pretreatment protocols, which include sampling, preservation, and clean-up methods, were summarized, and discussed to provide researchers with the adequate information that they might need when conducting their research. Deep insights into the solid-phase extraction method for both liquid and solid analytes were observed in this chapter. The analytical techniques, including the use of chromatography and electrochemical sensors were extensively discussed.

References

Afonso-Olivares, C., Sosa-Ferrera, Z., & Santana-Rodríguez, J. J. (2017). Occurrence and environmental impact of pharmaceutical residues from conventional and natural wastewater treatment plants in Gran Canaria (Spain). *Science of the Total Environment*, *599–600*, 934–943. https://doi.org/10.1016/J.SCITOTENV.2017.05.058.

Baird, R. B., Eaton, A. D., & Rice, E. W. (2017). *Standard methods for the examination of water and wastewater* (23rd ed.). Washington, DC: American Public Health Association.

Basil, K. M. (2017). Gas chromatography—Mass spectrometry, basic concept and instrumentation. In L. M. L. Nollet, & D. A. Lambropoulou (Eds.), *Chromatography analysis of the environment* (4th ed., pp. 3–24). Boca Raton, FL: Taylor & Francis.

Burgess, R. R. (2018). A brief practical review of size exclusion chromatography: Rules of thumb, limitations, and troubleshooting. *Protein Expression and Purification*, *150*(April), 81–85. https://doi.org/10.1016/j.pep.2018.05.007.

Camacho-Muñoz, D., Martín, J., Santos, J. L., Aparicio, I., & Alonso, E. (2009). An affordable method for the simultaneous determination of the most studied pharmaceutical compounds as wastewater and surface water pollutants. *Journal of Separation Science*, *32*(18), 3064–3073. https://doi.org/10.1002/jssc.200900128.

Cantwell, F. F., & Losier, M. (2002). Sampling and sample preparation for field and laboratory: Liquid-liquid extraction. In J. Pawliszyn (Ed.), *Vol. 37. Comprehensive analytical chemistry* (pp. 297–340). Elsevier Science. https://doi.org/10.1016/S0166-526X(02)80048-4.

Daverey, A., Dutta, K., & Sarkar, A. (2019). An overview of analytical methodologies for environmental monitoring. In S. K. Brar, K. Hedge, & V. L. Pachapur (Eds.), *Tools, techniques and protocols for monitoring environmental contaminants* (pp. 3–18). Elsevier.

Fatta, D., Achilleos, A., Nikolaou, A., & Meriç, S. (2007). Analytical methods for tracing pharmaceutical residues in water and wastewater. *Trends in Analytical Chemistry*, *26*(6), 515–533. https://doi.org/10.1016/j.trac.2007.02.001.

Fatta-Kassinos, D., Nikolaou, A., & Ioannou-Ttofa, L. (2019). Advances in analytical methods for the determination of pharmaceutical residues in waters and wastewaters. In *Vol. 1. Encyclopedia of environmental health* (2nd ed.). Elsevier. https://doi.org/10.1016/B978-0-12-409548-9.11247-3.

Gao, P., Ding, Y., Li, H., & Xagoraraki, I. (2012). Occurrence of pharmaceuticals in a municipal wastewater treatment plant: Mass balance and removal processes. *Chemosphere*, *88*(1), 17–24. https://doi.org/10.1016/J.CHEMOSPHERE.2012.02.017.

Gracia-Lor, E., Sancho, J. V., & Hernández, F. (2010). Simultaneous determination of acidic, neutral and basic pharmaceuticals in urban wastewater by ultra high-pressure liquid chromatography-tandem mass spectrometry. *Journal of Chromatography A, 1217*(5), 622–632. https://doi.org/10.1016/j.chroma.2009.11.090.

Gross, J. H. (2017). *Mass spectrometry, a textbook* (3rd ed.). Cham, Switzerland: Springer International Publishing.

Guedes-Alonso, R., Afonso-Olivares, C., Montesdeoca-Esponda, S., Sosa-Ferrera, Z., & Santana-Rodríguez, J. J. (2013). An assessment of the concentrations of pharmaceutical compounds in wastewater treatment plants on the island of Gran Canaria (Spain). *SpringerPlus, 2*(1), 24. https://doi.org/10.1186/2193-1801-2-24.

Hashemi, B., Zohrabi, P., Kim, K. H., Shamsipur, M., Deep, A., & Hong, J. (2017). Recent advances in liquid-phase microextraction techniques for the analysis of environmental pollutants. *Trends in Analytical Chemistry, 97*, 83–95. https://doi.org/10.1016/j.trac.2017.08.014.

Hoh, E., & Mastovska, K. (2008). Large volume injection techniques in capillary gas chromatography. *Journal of Chromatography A, 1186*(1–2), 2–15. https://doi.org/10.1016/j.chroma.2007.12.001.

Hu, W., Ma, L., Guo, C., Sha, J., Zhu, X., & Wang, Y. (2012). Simultaneous extraction and determination of fluoroquinolones, tetracyclines and sulfonamides antibiotics in soils using optimised solid phase extraction chromatography-tandem mass spectrometry. *International Journal of Environmental Analytical Chemistry, 92*(6), 698–713. https://doi.org/10.1080/03067319.2010.520122.

Ismail, B., & Suzanne Nielsen, S. (2010). Basic principles of chromatography. In *Food analysis (Food Science)* Springer Science + Business Media LLC.

Jelić, A., Petrović, M., & Barceló, D. (2009). Multi-residue method for trace level determination of pharmaceuticals in solid samples using pressurized liquid extraction followed by liquid chromatography/quadrupole-linear ion trap mass spectrometry. *Talanta, 80*(1), 363–371. https://doi.org/10.1016/j.talanta.2009.06.077.

K'oreje, K. O., Vergeynst, L., Ombaka, D., De Wispelaere, P., Okoth, M., Van Langenhove, H., et al. (2016). Occurrence patterns of pharmaceutical residues in wastewater, surface water and groundwater of Nairobi and Kisumu city, Kenya. *Chemosphere, 149*, 238–244. https://doi.org/10.1016/j.chemosphere.2016.01.095.

Krakkó, D., Licul-Kucera, V., Záray, G., & Mihucz, V. G. (2019). Single-run ultra-high performance liquid chromatography for quantitative determination of ultra-traces of ten popular active pharmaceutical ingredients by quadrupole time-of-flight mass spectrometry after offline preconcentration by solid phase extraction. *Microchemical Journal, 148*(April), 108–119. https://doi.org/10.1016/j.microc.2019.04.047.

Kruve, A., Rebane, R., Kipper, K., Oldekop, M. L., Evard, H., Herodes, K., et al. (2015). Tutorial review on validation of liquid chromatography-mass spectrometry methods: Part I. *Analytica Chimica Acta, 870*(1), 29–44. https://doi.org/10.1016/j.aca.2015.02.017.

Kumar, N., Bhadwal, A. S., Mizaikoff, B., Singh, S., & Kranz, C. (2019). Electrochemical detection and photocatalytic performance of MoS2/TiO2 nanocomposite against pharmaceutical contaminant: Paracetamol. *Sensing and Bio-Sensing Research, 24*(May), 2–10. https://doi.org/10.1016/j.sbsr.2019.100288.

Lynch, K. B., Chen, A., & Liu, S. (2018). Miniaturized high-performance liquid chromatography instrumentation. *Talanta, 177*(September 2017), 94–103. https://doi.org/10.1016/j.talanta.2017.09.016.

Ma, X., Zhao, M., Zhao, F., Guo, H., Crittenden, J., Zhu, Y., et al. (2016). Application of silica-based monolith as solid-phase extraction sorbent for extracting toxaphene congeners in soil. *Journal of Sol-Gel Science and Technology, 80*(1), 87–95. https://doi.org/10.1007/s10971-016-4054-8.

Marinas, M., Sa, E., Rojas, M. M., Moalem, M., Urbano, F. J., Guillou, C., et al. (2010). Determination of acidic degradation products of nonylphenol polyethoxylates by large-volume injection-port derivatization and gas chromatography/mass spectrometry. *Rapid Communications in Mass Spectrometry*, *24*, 1457–1466. https://doi.org/10.1002/rcm.

Mazloum-Ardakani, M., Sheikh-Mohseni, M. A., Mirjalili, B. F., & Zamani, L. (2012). Simultaneous determination of captopril, acetaminophen and tryptophan at a modified electrode based on carbon nanotubes. *Journal of Electroanalytical Chemistry*, *686*, 12–18. https://doi.org/10.1016/j.jelechem.2012.09.021.

Meyer, V. R. (2013). Chromatography principles. In *Encyclopedia of analytical science* (3rd ed., Vol. 2). Elsevier. https://doi.org/10.1016/B978-0-12-409547-2.00081-0.

Michalski, R. (2018). Ion chromatography applications in wastewater analysis. *Separations*, *5* (16), 1–12. https://doi.org/10.3390/separations5010016.

Mohapatra, S., Huang, C. H., Mukherji, S., & Padhye, L. P. (2016). Occurrence and fate of pharmaceuticals in WWTPs in India and comparison with a similar study in the United States. *Chemosphere*, *159*, 526–535. https://doi.org/10.1016/j.chemosphere.2016.06.047.

Montaseri, H., & Forbes, P. B. C. (2018). Analytical techniques for the determination of acetaminophen: A review. *Trends in Analytical Chemistry*, *108*, 122–134. https://doi.org/10.1016/j.trac.2018.08.023.

Montesdeoca-Esponda, S., Sosa-Ferrera, Z., & Santana-Rodríguez, J. J. (2012). Combination of microwave-assisted micellar extraction with liquid chromatography tandem mass spectrometry for the determination of fluoroquinolone antibiotics in coastal marine sediments and sewage sludges samples. *Biomedical Chromatography*, *26*(1), 33–40. https://doi.org/10.1002/bmc.1621.

Moret, S., Hidalgo, M., & Sanchez, J. M. (2012). Simple and fast methods based on solid-phase extraction coupled to liquid chromatography with UV detection for the monitoring of caffeine in natural, and wastewater as marker of anthropogenic impact. *ISRN Chromatography*, *2012*, 1–7. https://doi.org/10.5402/2012/487138.

Murray, J. A. (2012). Qualitative and quantitative approaches in comprehensive two-dimensional gas chromatography. *Journal of Chromatography A*, *1261*, 58–68. https://doi.org/10.1016/j.chroma.2012.05.012.

Mutavdžić Pavlović, D., Pinušić, T., Periša, M., & Babić, S. (2012). Optimization of matrix solid-phase dispersion for liquid chromatography tandem mass spectrometry analysis of 12 pharmaceuticals in sediments. *Journal of Chromatography A*, *1258*, 1–15. https://doi.org/10.1016/j.chroma.2012.08.025.

Neue, E. D., Alden, B. A., Grover, E. R., Grumbach, E. S., Iraneta, P. C., & Mendez, A. (2007). HPLC column and packings. In S. Ahuja, & H. Rasmussen (Eds.), *Vol. 8. HPLC method development for pharmaceuticals*. Italy: Elsevier Academic Press.

Pérez-Carrera, E., Hansen, M., León, V. M., Björklund, E., Krogh, K. A., Halling-Sørensen, B., et al. (2010). Multiresidue method for the determination of 32 human and veterinary pharmaceuticals in soil and sediment by pressurized-liquid extraction and LC-MS/MS. *Analytical and Bioanalytical Chemistry*, *398*(3), 1173–1184. https://doi.org/10.1007/s00216-010-3862-x.

Płotka-Wasylka, J., Szczepańska, N., de la Guardia, M., & Namieśnik, J. (2016). Modern trends in solid phase extraction: New sorbent media. *Trends in Analytical Chemistry*, *77*, 23–43. https://doi.org/10.1016/j.trac.2015.10.010.

Poole, C. F. (2003). New trends in solid-phase extraction. *Trends in Analytical Chemistry*, *22* (6), 362–373. https://doi.org/10.1016/S0165-9936(03)00605-8.

Romera-Torres, A., Romero-González, R., Martínez Vidal, J. L., & Garrido Frenich, A. (2018). Analytical methods, occurrence and trends of tropane alkaloids and calystegines: An update. *Journal of Chromatography A*, *1564*, 1–15. https://doi.org/10.1016/j.chroma.2018.06.004.

Samaras, V. G., Stasinakis, A. S., Mamais, D., Thomaidis, N. S., & Lekkas, T. D. (2013). Fate of selected pharmaceuticals and synthetic endocrine disrupting compounds during wastewater treatment and sludge anaerobic digestion. *Journal of Hazardous Materials*, *244–245*, 259–267. https://doi.org/10.1016/J.JHAZMAT.2012.11.039.

Schwermer, C. U., Krzeminski, P., Wennberg, A. C., Vogelsang, C., & Uhl, W. (2018). Removal of antibiotic resistant E. coli in two Norwegian wastewater treatment plants and by nano- and ultra-filtration processes. *Water Science and Technology*, *77*(4), 1115–1126. https://doi.org/10.2166/wst.2017.642.

Silvestre, C. I. C., Santos, J. L. M., Lima, J. L. F. C., & Zagatto, E. A. G. (2009). Liquid-liquid extraction in flow analysis: A critical review. *Analytica Chimica Acta*, *652*(1–2), 54–65. https://doi.org/10.1016/j.aca.2009.05.042.

Skoog, D. A., West, D. M., Holler, F. J., & Crouch, S. R. (2014). *Fundamental of analytical chemistry* (9th ed.). Belmont, CA: Mary Finch.

Spataro, F., Ademollo, N., Pescatore, T., Rauseo, J., & Patrolecco, L. (2019). Antibiotic residues and endocrine disrupting compounds in municipal wastewater treatment plants in Rome, Italy. *Microchemical Journal*, *148*, 634–642. https://doi.org/10.1016/J.MICROC.2019.05.053.

Stout, T. H., & Dorsey, J. G. (2002). High-performance liquid chromatography. In L. Ohannesian, & A. J. Streeter (Eds.), *Handbook of pharmaceutical analysis*. Basel, Switzerland: Marcel Dekker, Inc.

Swartz, M. (2010). HPLC detector: A brief review. *Journal of Liquid Chromatography & Related Technology*, *33*(9–12), 1130–1150.

Teske, J., & Engewald, W. (2002). Methods for, and applications of, large-volume injection in capillary gas chromatography. *Trends in Analytical Chemistry*, *21*(9–10), 584–593. https://doi.org/10.1016/S0165-9936(02)00801-4.

Torrinha, Á., Oliveira, T. M. B. F., Ribeiro, F. W. P., Correia, A. N., Lima-Neto, P., & Morais, S. (2020). Application of nanostructured carbon-based electrochemical (bio)sensors for screening of emerging pharmaceutical pollutants in waters and aquatic species: A review. *Nanomaterials*, *10*(7), 1–29. https://doi.org/10.3390/nano10071268.

Tran, N. H., Chen, H., Do, T. V., Reinhard, M., Ngo, H. H., He, Y., & Gin, K. Y.-H. (2016). Simultaneous analysis of multiple classes of antimicrobials in environmental water samples using SPE coupled with UHPLC-ESI-MS/MS and isotope dilution. *Talanta*, *159*, 163–173. https://doi.org/10.1016/j.talanta.2016.06.006.

Vazquez-Roig, P., Blasco, C., & Picó, Y. (2013). Advances in the analysis of legal and illegal drugs in the aquatic environment. *Trends in Analytical Chemistry*, *50*, 65–77. https://doi.org/10.1016/j.trac.2013.04.008.

Vazquez-Roig, P., Segarra, R., Blasco, C., Andreu, V., & Picó, Y. (2010). Determination of pharmaceuticals in soils and sediments by pressurized liquid extraction and liquid chromatography tandem mass spectrometry. *Journal of Chromatography A*, *1217*(16), 2471–2483. https://doi.org/10.1016/j.chroma.2009.11.033.

Veiga-Gómez, M., Nebot, C., Franco, C. M., Miranda, J. M., Vázquez, B., & Cepeda, A. (2017). Identification and quantification of 12 pharmaceuticals in water collected from milking parlors: Food safety implications. *Journal of Dairy Science*, *100*(5), 3373–3383. https://doi.org/10.3168/jds.2016-12227.

Voedisch, B., & Thie, H. (1990). Size exclusion chromatography. In R. Kontermann, & S. Dubel (Eds.), *Vol. 1. Antibody engineering* (pp. 268–303). Berlin: Springer-Verlag Berlin Heidelberg. https://doi.org/10.1021/ac00211a020.

Wang, Y., Li, J., Sun, D., Yang, S., Liu, H., & Chen, L. (2021). Strategies of dispersive liquid-liquid microextraction for coastal zone environmental pollutant determination. *Journal of Chromatography A, 1658*, 462615. https://doi.org/10.1016/j.chroma.2021.462615.

Wang, Q., Ma, L., Yin, C. R., & Xu, L. (2013). Developments in injection port derivatization. *Journal of Chromatography A, 1296*, 25–35. https://doi.org/10.1016/j.chroma.2013.04.036.

Webster, G. K. (2018). Unified approach to reversed-phase and normal-phase chromatographic method development. In G. K. Webster, & L. Kott (Eds.), *Vol. 3. Chromatic method development*. Singapore: Jenny Standford Publishing.

Wu, M., & Janssen, S. (2011). Dosed without prescription: A framework for preventing pharmaceutical contamination of our nations drinking water. *Environmental Science and Technology*, *45*(2), 366–367. https://doi.org/10.1021/es104032a.

Yu, Y., Wu, L., & Chang, A. C. (2013). Seasonal variation of endocrine disrupting compounds, pharmaceuticals and personal care products in wastewater treatment plants. *Science of the Total Environment*, *442*, 310–316. https://doi.org/10.1016/J.SCITOTENV.2012.10.001.

Zhang, K., Kurita, K. L., Venkatramani, C., & Russell, D. (2019). Seeking universal detectors for analytical characterizations. *Journal of Pharmaceutical and Biomedical Analysis*, *162*, 192–204. https://doi.org/10.1016/j.jpba.2018.09.029.

Zwir-Ferenc, A., & Biziuk, M. (2006). Solid phase extraction technique—Trends, opportunities and applications. *Polish Journal of Environmental Studies*, *15*(5), 677–690.

Treatment innovation using solar/UV

6

Waldemar Studziński[a] *and Maciej Przybyłek*[b]
[a]Faculty of Chemical Technology and Engineering, Division of Food Analytics and Environment, Department of Engineering and Chemical and Food Analytics, Bydgoszcz University of Science and Technology, Bydgoszcz, Poland, [b]Pharmacy Faculty, Collegium Medicum of Bydgoszcz, Department of Physical Chemistry, Nicolaus Copernicus University, Bydgoszcz, Poland

1. Introduction

Around 3000 pharmaceuticals are used in the European Union. These compounds can be classified into several types, including analgesics, antibiotics, antidepressants and anticonvulsants, hormone regulators, and β-blockers (Krishnan et al., 2021). Pharmaceuticals and their metabolites are considered emerging pollutants (EP) (Farré, Pérez, Kantiani, & Barceló, 2008) and one of the most environmentally relevant micropollutants (Dhodapkar & Gandhi, 2019). A constantly growing consumption of pharmaceutics, suggests that these compounds are widely distributed in the environment. Noteworthy, the average daily dose in the OECD (Organization for Economic Co-operation and Development) countries increased in case of antihypertensive, anticholesterol, antidiabetic, and antidepressant drugs, in the period 2000–2017, by 68%, 217%, 97%, and 105%, respectively (OECD, 2019). The environmental monitoring of these compounds has been carried out for several decades. The application of the LC-MS method and solid-phase extraction (SPE), which became popular in the 2000s, allowed to qualify and quantify pharmaceutical micropollutants in water samples (Farré et al., 2008; Perret, Gentili, Marchese, Greco, & Curini, 2006; Richardson, 2004; Richardson & Ternes, 2005). Pharmaceutical residues have been detected (usually in the range from ng/L to μg/L) in almost all environmental matrices on every continent including groundwater, urban wastewater, surface water and drinking water (Atugoda et al., 2021; Jodeh, Abdelwahab, Jaradat, Warad, & Jodeh, 2016; Kümmerer, 2009; Patel et al., 2019; Wang, Chu, Wojnárovits, & Takács, 2020; Wang & Wang, 2016). One of the interesting phenomena that have been observed is seasonality. It has been noted that there is a difference in the concentrations of pharmaceutically active compounds in sewage treatment plants in Italy during the winter and summer (Castiglioni et al., 2006). This is probably mainly due to less use of antibiotics and antiinflammatory drugs in the summer. Some popular sources of pharmaceuticals in the environment are animal husbandry, the veterinary industry, industrial production, hospitals and households (Bound & Voulvoulis, 2005; Insani et al., 2020; Kaczala & Blum, 2016;

The Treatment of Pharmaceutical Wastewater. https://doi.org/10.1016/B978-0-323-99160-5.00005-9

Larsson, 2014; Luo, Reimers, Yang, & Lin, 2021; Rogowska, Zimmermann, Muszyńska, Ratajczyk, & Wolska, 2019; Taylor, 2016; Xiang et al., 2018). The latter source is mainly related to the problem of consuming analgesics and antiinflammatory drugs without medical supervision. Recent studies on the migration of pharmaceuticals have led to very interesting and surprising conclusions since apart from the well-known emission sources, there are also quite nonobvious and inconspicuous ones. For instance, pharmaceuticals can enter food products through the agricultural use of treated wastewater and biosolids (Fu et al., 2019). Furthermore, the environmental impact of these compounds was analyzed in the context of their interactions with microplastics, playing the role of transporting vehicles (Atugoda et al., 2021). Noteworthy, new problematic medical and pharmaceutical waste migration pathways have recently emerged due to the COVID-19 pandemic (Das, Islam, Billah, & Sarker, 2021; Ejeromedoghene, Nwosisi, Tesi, Noragbon, & Akinyeye, 2021; Goswami, Goswami, Nautiyal, & Prakash, 2021; Kalantary et al., 2021; Liang et al., 2021; Maalouf & Maalouf, 2021; Mansouri, Chouchene, Roche, & Ksibi, 2021; Patrício Silva et al., 2021).

Pharmaceutical micropollutants can have a significant negative impact on living organisms. Importantly, even small concentrations of pharmaceuticals in treated water affect the functioning of living organisms causing ecotoxic effects (aus Beek et al., 2016; O'Flynn et al., 2021; Taylor, 2016; Tiwari et al., 2017). The negative effects of pharmaceuticals emission on humans and animals are also associated with antibiotic resistance and endocrine diseases (Benotti et al., 2009; Bouki, Venieri, & Diamadopoulos, 2013; Bui, Vo, Ngo, Guo, & Nguyen, 2016; Emara, Lehmann, Siegert, & Finkbeiner, 2019; Jiang, Zhou, & Sharma, 2013; Kotwani, Joshi, & Kaloni, 2021). Therefore, it is necessary to develop more advanced water purification methods. Over the last two decades, there has been a growing interest in finding effective wastewater treatment methods employing sunlight or artificial UV light sources and advanced oxidation methods (Ameta, Chohadia, Jain, & Punjabi, 2018; Baeza & Knappe, 2011; Bartolomeu, Neves, Faustino, & Almeida, 2018; Bermúdez, Pascual, Martínez, & Capilla, 2021; Bui et al., 2016; Deng & Zhao, 2015; Garrido-Cardenas, Esteban-García, Agüera, Sánchez-Pérez, & Manzano-Agugliaro, 2020; Giannakis, 2019; Giannakis et al., 2015; Kokkinos, Venieri, & Mantzavinos, 2021; Rekhate & Srivastava, 2020; Von Gunten, 2018). The main purpose of this chapter is to summarize the current state of knowledge and evaluate the usability of various approaches, taking into account technological, methodological, and economic aspects.

2. Solar radiation

Sunlight regulates many life processes and is essential for most ecosystems on Earth. The natural sunlight irradiation comprises visible light (VIS), ultraviolet A (UVA, 315–400nm), ultraviolet B (UVB, 290–320nm), ultraviolet C (UVC, 100–280nm), whereas the UV contribution to the sunlight spectra approximately consists of 5%–10% of UVB and 90%–95% of UVA (Amaro-Ortiz, Yan, & D'Orazio, 2014).

Sunlight has been widely applied for water purification since the first mentions of this method in antiquity (EPA, 2000; Jadhav, 2014). Nowadays, the use of effective and low-cost solar disinfection (SODIS) procedures is becoming increasingly popular, especially in developing countries and in areas with limited access to drinking water (Burch & Thomas, 1998; Cowie, Porley, & Robertson, 2020; Moreno-SanSegundo et al., 2021). The SODIS process is mainly associated with the protection of water consumers against harmful pathogens. However, sunlight can also play an important role in removing chemicals from wastewater. Indeed, the photodegradation process takes place naturally in the environment being the common mechanism of water self-purification. For instance, it has been found that 17α-Ethinylestradiol (EE2) undergoes photodegradation during sunlight irradiation in Lake Quinsigamond (USA) during summer and its half-life does not exceed two days (Zuo, Zhang, & Zhou, 2013). The studies on the photodegradation of several popular drugs Naproxen, Indomethacin, Acetaminophen, Dexamethasone, Dapsone, Amiodarone, Phenytoin, Sulindac, and Raloxifene, which are likely to be found in wastewater, showed that all these compounds except Acetaminophen and Phenytoin can be completely or partly removed after 6h of exposure to the sunlight (Kawabata, Sugihara, Sanoh, Kitamura, & Ohta, 2013). Relatively high natural photodegradation efficiencies have been demonstrated for several known pharmaceutically active compounds, including beta-blockers, antibiotics, sulfonamides, and nonsteroidal antiinflammatory drugs (NSAIDs) (Yan & Song, 2014). The mechanism of natural photodegradation usually involves the dissolved organic matter (DOM) playing the role of the photosensitizer. However, in some cases (e.g., Tryptophan) DOM can act as both photosensitizer and quencher (Janssen, Erickson, & McNeill, 2014).

One of the obvious benefits of using sunlight in wastewater treatment is its low cost. However, the possibility of its use depends on the season, the level of insolation, and the weather. Although there are many well-documented examples of high degradation rate using natural light, much greater efficiency can be obtained when combining solar radiation with advanced oxidation processes (AOPs). Such techniques will be discussed in detail in the next sections of this chapter.

3. Advanced oxidation processes

According to the Web of Science, there is an upward trend, indicating the growth of the interest in effective methods of eliminating pharmaceuticals from water (Fig. 1). Many of the studied approaches involve photochemical and oxidation processes.

In most cases, anthropogenic wastewater can be effectively purified by conventional methods, such as biological methods, physical processes (reverse osmosis, flocculation) and chemical methods (chlorine treatment or ozonation) (Litter & Quici, 2010; Qi et al., 2020; Saravanan et al., 2021). However, many studies have demonstrated that these methods are not sufficient in the case of pharmaceuticals (Calcio Gaudino, Canova, Liu, Wu, & Cravotto, 2021; Ebele, Abou-Elwafa Abdallah, & Harrad, 2017; Miklos et al., 2018; Montemurro et al., 2018; Ribeiro, Nunes, Pereira, & Silva, 2015). Noteworthy, some of the often detected in wastewater compounds (Sulfamethoxazole, Erythromycin, Naproxen, and Clofibric acid) can be

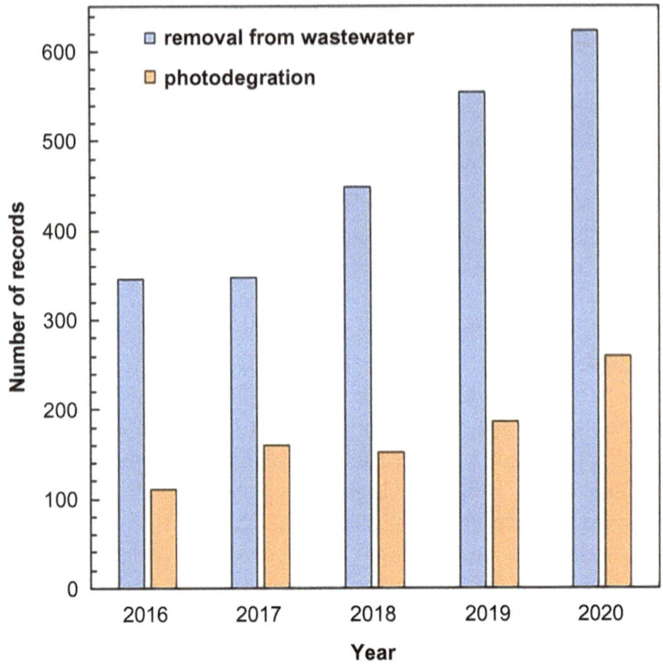

Fig. 1 The number of papers on the removal of pharmaceuticals from wastewater and studies on photodegradation in 2016–2020 according to Web of Sciences.

stable for a period of about one to several years (Patel et al., 2019). For this purpose, AOPs are of great interest.

In general, AOPs utilize oxidants with a high redox potential to oxidize highly stable pharmaceutical compounds (Mansouri et al., 2021; Szabó et al., 2011). These methods involve chemical, electrochemical, photochemical and sonochemical treatment mostly to induce hydroxyl radical ($HO^{\bullet}$) formation (Ameta et al., 2018; Baghirzade, Yetis, & Dilek, 2021; Calcio Gaudino et al., 2021; Kishimoto, 2019; Mansouri et al., 2021; Oturan & Aaron, 2014). Other popular reactive oxygen species formed during AOPs are peroxide radicals, sulfate radicals and singlet oxygen (Wang & Wang, 2020; Wang & Zhuan, 2020; Zhou et al., 2018). Although the singlet oxygen and superoxide radicals are less powerful oxidizers than hydroxyl radicals, they are capable to induce the decomposition of some resistant pharmaceutical compounds (Cai et al., 2019; Mahdi Ahmed, Barbati, Doumenq, & Chiron, 2012; Sirtori, Agüera, Carra, & Sanchéz Pérez, 2014; Wang & Wang, 2018; Wang & Zhuan, 2020). The most promising chemical and photochemical techniques employed in the removal of pharmaceuticals from wastewater are summarized in Fig. 2 (Bartolomeu et al., 2018; Liu, Wang, & Wang, 2020; Mansouri et al., 2021; Wang & Wang, 2020). Probably the most frequently studied methods are based on the combination of popular chlorine-free oxidizers and UV radiation (e.g., UV/H_2O_2 and UV/O_3, heterogeneous photocatalysis).

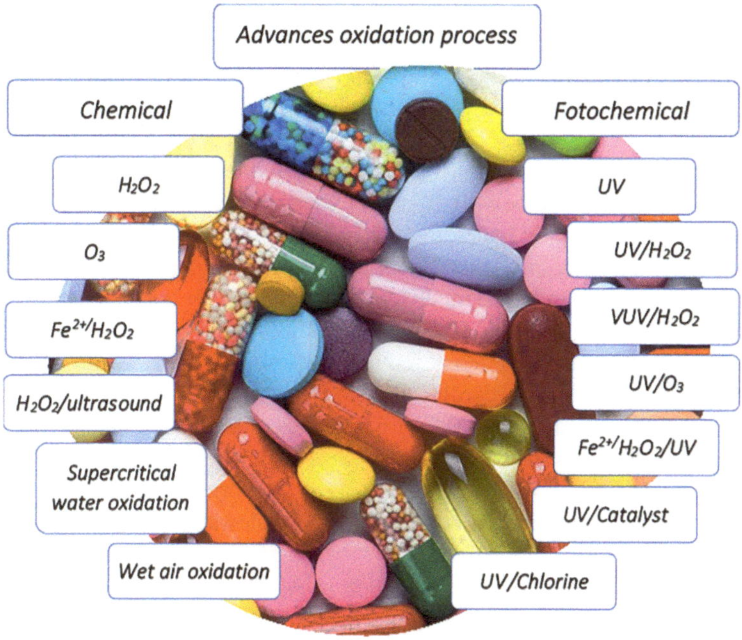

Fig. 2 Classification of advanced oxidation processes.

3.1 UV/H_2O_2

The UV/H_2O_2 system has been one of the most frequently studied AOP used for the treatment of wastewater contaminated with pharmaceuticals (Guo, Wu, & Fang, 2019; Poyatos et al., 2010). The mechanism of UV/H_2O_2 is based on the photolysis of the peroxide bond leading to the hydroxyl radical (HO·) formation. In general, at the low peroxide concentrations level, the number of HO· increases with the increase of H_2O_2 dose. However, in the case of high H_2O_2 concentrations, the hydroxyl radical quenching occurs (Guo et al., 2018; Lester, Avisar, & Mamane, 2010). For instance, Sulfamethoxazole undergoes photodegradation more efficiently, when H_2O_2 was added, however, this effect was not observed for the high concentration of the oxidizer (>50 mg/L) (Lester et al., 2010).

There are many spectacular examples of UV/H_2O_2 application. For instance, in the case of Ibuprofen, about 100% degradation efficiency was achieved by Monteoliva-García, Martín-Pascual, Muñío, and Poyatos (2019). Furthermore, the synergistic effect of H_2O_2 and UV treatment was well-documented (Kim, Yamashita, & Tanaka, 2009; Lopez, Bozzi, Mascolo, & Kiwi, 2003; Alharbi et al., 2017). However, there are also studies showing some disadvantages of this approach or the advantage of other methods. According to some reports, such as the recent Diclofenac photodegradation study by Ledakowicz et al. (2019), there was no significant gain from using UV/H_2O_2 over direct photodegradation. According to earlier research by Lekkerkerker-Teunissen, Benotti, Snyder, and Van Dijk (2012), a large decrease of Diclofenac and Sulfamethoxazole concentration (>85%) was observed when efficient UV lamps (low- and medium-pressure, 300–700 mJ/cm^2 range) were used. Interestingly, H_2O_2 (5 and 10 mg/L) had no significant effect on the removal efficiency. Quite recently, the degradation of a β-blocker drug, Metoprolol was examined by Gao, Zhang, Li, Tian, and Gao (2020). It was found that Metoprolol undergoes rapid degradation. However, according to the same study, the chlorination combined with UV radiation was found to be a significantly cheaper method than UV/H_2O_2. Nevertheless, despite of economic issues, there are some benefits of Metoprolol elimination using H_2O_2 combined with UV treatment as well. According to the cited study UV/H_2O_2 efficiency was found to be much less dependent on pH than UV/chlorine. Interestingly, a relatively high dose of H_2O_2 (27.21 mg/L) did not cause the quenching effect contributing to the efficiency reduction. A similar observation was made in the case of Carbamazepine. According to Alharbi et al. (2017), when the concentration of H_2O_2 was 200 mg/L around 100% of Carbamazepine was removed. For comparison, when the UV radiation was used alone, the efficiency was low (25.2%). Approximately 100% elimination using UV/H_2O_2 was also reported in the latter study, which mimicked realistic conditions (natural pH and wastewater samples) (Monteoliva-García et al., 2019).

In general, one of the most important goals of AOPs development is the removal of persistent pollutants which cannot degrade during biological processes carried out in wastewater treatment plants. Antibiotics certainly belong to this group. In the recent decade, photodegradation of antibiotics in the presence of oxidants and especially H_2O_2 have been particularly frequently studied (Lekkerkerker-Teunissen et al., 2012; Michael et al., 2020; Monteoliva-García et al., 2019; Yang et al., 2017; Yuan et al., 2011). These compounds, due to their high stability, are present in treated water (Rodriguez-Mozaz et al., 2020). Despite this fact, there are some examples showing the high efficiency of UV/H_2O_2 in removing them. An interesting and quite recently studied compound is Ciprofloxacin (Bermúdez et al., 2021; Michael et al., 2020; Monteoliva-García et al., 2019; Yuan et al., 2011). This antibiotic can be removed (>99%) in c.a. 25 min. by using H_2O_2 and UV-C light (Michael et al., 2020). A very similar removal efficiency (c.a. 100%), however, achieved in a shorter time (*c.a.* 10 min.) can be inferred from the data reported in the earlier study (Monteoliva-García et al., 2019). The application of UV-C/H_2O_2 was found to be slightly more effective in case of Sulfamethoxazole (>99% after 20 min) (Michael et al., 2020).

As with all photo-AOPs, the removal efficiency of UV/H_2O_2 depends on many factors including quenching effects. A study on the Metronidazole photodegradation by Johnson and Mehrvar (2008) well illustrates these effects. Noteworthy, Metronidazole contamination is a quite important issue, since the adverse effects of this compound on aquatic organisms have been well described (Lanzky & Halling-Sørensen, 1997). Metronidazole is quite resistant to UV/H_2O_2, as evidenced by the relatively low efficiency (4.9%–41%) (Johnson & Mehrvar, 2008). Furthermore, CO_3^{2-} and HCO_3^- ions were found to play the inhibiting role. The hydroxyl radical scavenging properties of CO_3^{2-} and HCO_3^- and other inorganic ions (Cl^- and Br^-) were observed by various researchers (Beltrán, Rivas, Álvarez, Alonso, & Acedo, 1999; Grebel, Pignatello, & Mitch, 2010; Johnson & Mehrvar, 2008; Liao, Kang, & Wu, 2001; Mehrvar, Anderson, & Moo-Young, 2002). Other important factors affecting the photooxidation efficiency are dissolved organic matter (DOM) and effluent organic matter (EfOM) scavengers (Dong, Mezyk, & Rosario-Ortiz, 2010; Goldstone, Pullin, Bertilsson, & Voelker, 2002; Hwang et al., 2020; Lee, Yoon, & Von Gunten, 2007; Romera-Castillo & Jaffé, 2015; Yang, Pignatello, Ma, & Mitch, 2016) and pH. The role of the former factor is ambiguous, since the reactive oxygen species formation may be induced by the presence of DOM^3 (triplet excited state), which is sometimes a more pronounced effect than quenching (Gmurek, Olak-Kucharczyk, & Ledakowicz, 2015). Another factor, pH is associated with the deactivation of hydroxyl radicals in the strongly alkaline conditions ($pH > 9$), due to the deprotonation reaction (Guo et al., 2018; Li, Jain, Ishida, & Liu, 2017; Yang, Zhou, & Cicek, 2014). The significant pH-sensitivity was observed in the case of Sulfamethoxazole, Sulfamethazine, Sulfadiazine, Trimethoprim, and Diclofenac (Baeza & Knappe, 2011; Li et al., 2017).

3.2 UV/O_3

Another interesting and efficient chlorine-free method is the UV/O_3. Ozone treatment is for sure a very effective method of water disinfection, which has been confirmed by numerous implementations of this technology since the 1930s (Glaze, 1987; Gorito et al., 2021; Lawrence & Cappelli, 1977; Seridou & Kalogerakis, 2021). The ozone oxidizing and disinfecting power is attributed to hydroxyl radical formation according to the following equation (Deng & Zhao, 2015; Gottschalk, Libra, & Saupe, 2010):

$$3O_3 + H_2O \rightarrow 2HO^\bullet + 4O_2$$

The UV light induces the formation of H_2O_2 which then undergoes photodegradation yielding hydroxyl radicals:

$$O_3 + H_2O \xrightarrow{h\nu} H_2O_2 + O_2$$

$$H_2O_2 \xrightarrow{h\nu} 2HO^\bullet$$

When considering the application of UV/O_3 for removing organic compounds, it is worth taking into account that this method may not be the most effective solution in all cases. Recently the stability of 38 pharmaceutical and personal care products (PPCPs) was tested against UV_{65W}/O_3 and five classes of stability were distinguished (Bermúdez et al., 2021). According to this classification system, the group of the most sensitive compounds highly prone to photodegradation comprises Dipyridamole, Diclofenac, Mefenamic acid, Diltiazem Ciprofloxacin, Chloramphenicol, Naproxen, and Isopropylantipyrine. On the other hand, Clarithromycin, Sulpiride, Theophylline,

Azithromycin, Roxithromycin, *N,N*-Diethyl-meta-toluamide (DEET), and Ketoprofen are characterized by the highest resistance to UV_{65W}/O_3 treatment.

Another important issue is to determine whether UV radiation has a significantly beneficial effect on degradation. This question is closely associated with the economic aspects. The main disadvantage of UV/O_3 is the high energy cost compared to the application of O_3 without UV irradiation. According to the study on the PPCPs secondary effluent removal using a bench-scale wastewater treatment system by I. Kim and Tanaka (2011), the UV/O_3 and O_3 treatments required 1.09 kWh and 0.09 kWh of electrical energy per m^3 of effluent respectively. The study on degradation of a set of 21 micropollutants including pharmaceuticals (Carbamazepine, Tramadol, Azithromycin, Diclofenac, Clarithromycin, Diphenhydramine, Venlafaxine, Fluoxetine, Clindamycin) shows that O_3 treatment reduced the concentration of most of the compounds by more than 85% (Gorito et al., 2021). However, according to this work, the combination of O_3 treatment with UV irradiation does not improve significantly the removal rate. Nevertheless, several other reports showed the beneficial effects of combining UV with O_3 (Abrile et al., 2021; Akmehmet Balcioğlu & Ötker, 2004; Kim et al., 2008; Lu, Li, & Feng, 2022; Rao & Chu, 2009).

Due to its efficiency, ozonation is often combined with other AOPs. According to the report focusing on the degradation of 30 pollutants (Kim et al., 2008), among the considered systems, the ones containing O_3 (i.e., UV/O_3, and H_2O_2/O_3) were found to be more efficient, than UV/H_2O_2. This study deserves special attention since the authors examined many different classes of pharmaceuticals, namely analgesic and NSAIDs (Acetaminophen, Fenoprofen, Antipyrine, Ethenzamide, Mefenamic acid, Naproxen, Isopropylantipyrine, Diclofenac, Indomethacin), antiarrhythmia agents (Disopyramide, Metoprolol, Propranolol), anticonvulsant (Ifenprodil, Carbamazepine), antibiotics (Sulfadimethoxine, Sulfadimizine, Sulfamethoxazole, Sulfamonomethoxine, Tetracycline, Chlorotetracycline, Oxytetracycline, Clarithromycin, 2-Quinoxaline carboxylic acid, Ceftiofur), antineoplastic (cyclophosphamide), and bronchodilators (Clenbuterol, Theophylline). Noteworthy, the combination of O_3 with H_2O_2 (peroxone process) has been successfully used in wastewater treatment including works on pharmaceuticals degradation/elimination (S. Patel, Majumder, Mondal, Das, & Ghosh, 2020; Wang et al., 2019; Yao et al., 2016). There are also some other interesting innovations used in ozonation technology. An intriguing example is a study on high-energy photons irradiation using a vacuum ultraviolet (VUV) lamp in the presence of O_3 followed by granular activated carbon (GAC) adsorption (Krakkó et al., 2021). According to this research, the very high removal efficiencies (over 95%) were achieved for a popular set of pharmaceuticals, Carbamazepine, Ciprofloxacin, Clarithromycin, Diclofenac, Metoprolol, Sitagliptin, and Sulfamethoxazole (Krakkó et al., 2021). In general, vacuum UV irradiation is considered a promising wastewater treatment tool providing high-quality disinfection and organic compounds elimination (Zoschke, Börnick, & Worch, 2014). The energy of VUV photons is very high which allows for water photodegradation at 185 nm leading to hydroxyl radicals formation (Huang et al., 2016; Krakkó et al., 2021). This source of UV radiation induces the in situ O_3 formation which finally contributes to hydroxyl and hydroperoxyl radicals (HO_2) formation.

3.3 UV/chlorine

As in the case of reactive oxygen species sources, chlorinating agents (Cl_2, HOCl, ClO_2, NH_2Cl) are often combined with UV irradiation (Kishimoto, 2019). The mechanisms and kinetics of PPCPs transformations during wastewater treatment have been widely studied. According to a comprehensive study taking into account 28 different compounds (Guo et al., 2018), UV/chlorine systems were found to be more efficient than UV/H_2O_2. Furthermore, the study on the chlorination of Clofibric acid (Tang, Shi, Liu, Feng, & Zhang, 2018) showed that UV irradiation is necessary to induce the degradation processes, which involve the hydroxyl radical formation. Noteworthy, according to this report, there are several important factors affecting the photolysis, such as the presence of humic acid and HCO_3^- ions, which are inhibitors and Cl^-, which slightly increase the degradation rate (Tang et al., 2018). A significant removal efficiency advantage from combining radiation with chlorination was noticed for pharmaceutical micropollutants dissolved in biologically treated effluents (Wang, Ying, Ma, Huo, & Yang, 2019). A similar synergism was also observed in the case of Carbamazepine (Wang, Wu, Huang, Wang, & Hu, 2016; Zhou et al., 2016). However, the opposite effect for Carbamazepine and Caffeine was reported by Yang et al. (2016), when the water samples were characterized by a relatively high pH (c.a. 8). In the mentioned study, a quite diverse set of PPCPs comprising not only Caffeine and Carbamazepine but also Ciprofloxacin, Tetracycline, Trimethoprim, Acetaminophen, Triclosan, Benzophenone-3, and 2-Ethylhexyl-4-methoxycinnamate was analyzed in terms of their stability using real water samples and two different AOPs techniques, UV/chlorine and UV/H_2O_2 (Yang, Sun, et al., 2016). In general, the former method was found to be more efficient. However, when comparing the UV/chlorine with UV/H_2O_2 the formation of chloral hydrate, haloketones and trichloronitromethane was intensified, while the haloacetonitriles formation was repressed (Yang, Sun, et al., 2016). The formation of low-molecular weight halogenated pollutants such as trichloromethanes (THMs) is a serious environmental problem closely associated with pharmaceuticals elimination using UV/chlorine treatment. However, the formation of these compounds is usually strongly influenced by the process conditions. According to the study of Wang, Xu, Zhang, Chen, and Gao (2017), the formation potential of THM in the case of Ronidazole chlorination combined with UV irradiation is dependent on the pH and can be very high at pH = 8 (50.6 μg/L).

In general, the chlorination by-products are of particular interest, due to the well-documented negative impact of chlorinated organic compounds on living organisms (Ding et al., 2020). This applies to both experimental and theoretical research. For instance, the theoretical QSPR analysis of Norfloxacin, Sulfamethoxazole, and 17β-Estradiol degradation products formed during the chlorination process indicated the increase of their bioaccumulation, when compared with the precursors (Zhang et al., 2021).

Although chlorination by-products pose various environmental risks, it does not mean that chlorination methods always lead to increased toxicity of treated wastewater effluents, and there are some surprising results. For instance, the studies on the 5,5-Diphenylhydantoin photodegradation (Mansor & Tay, 2017) showed that the application of the UV/chlorine system contributed to the toxicity reduction.

Of course the formation of chlorinated degradation by-products is closely associated with the diversity of chlorinating oxidizing agents formed during the process. The UV/chlorine is a quite complex system affected by temperature, pH and the type of chlorination agent. In the review paper by Kishimoto (2019), 33 primary reactions occurring in chlorine aqueous solutions under the influence of UV irradiation were collected and compiled along with the kinetic data. According to this work, the following compounds, ions and radicals can be distinguished: ClO^-, $ClO^{\cdot}$, H_2O, OH^-, H^+, $HO^{\cdot}$, Cl^-, $Cl_2^{\cdot-}$, $Cl^{\cdot}$, $ClOH^{\cdot-}$, ClO_2^-, $ClO_2^{\cdot}$, ClO_3^-, $ClO_3^{\cdot}$, Cl_2O_2, Cl_2O_4, Cl_2O_6, ClO_3^-, ClO_4^-, O_2, $O^{\cdot-}$, $O(^1D)$ and $O(^3P)$. It might seem that the use of H_2O_2 in the presence of light will be more effective due to the direct hydroxyl radical formation than UV/chlorine systems dominated by chlorine species. However, the effectiveness of chlorination combined with irradiation should not be underestimated. In fact, the most reactive chlorine oxidizers, $ClO^{\cdot}$ (1.5–1.8 V), $Cl_2^{\cdot-}$ (2.0 V), $Cl^{\cdot}$ (2.20–2.60 V) are characterized by a quite similar oxidation potential to hydroxyl radical (1.95–2.80 V) (Alfassi, Huie, Mosseri, & Neta, 1988; Beitz, Bechmann, & Mitzner, 1998; Guo et al., 2018; Kishimoto, 2019; Neta, Huie, & Ross, 1988; Wardman, 1989; Yin & Zhang, 2019). Furthermore, HOCl and OCl^- photodegradation are characterized by a significantly higher quantum yield than H_2O_2 (Feng, Smith, & Bolton, 2007; Guo et al., 2018; Watts & Linden, 2007).

Although $HOCl/OCl^-$ is probably the most popular chlorinating agent, several studies dealing with monochloramine appeared (S. Li et al., 2020; Sun et al., 2019; Tian et al., 2020). According to one of them (Tian et al., 2020), Iopamidol photodegradation with the use of UV/NH_2Cl is more efficient than UV/ClO_2 and UV, whereas lower efficient than UV/Cl_2 and UV/H_2O_2. On the other hand, the studies on the Ibuprofen and Naproxen degradation (Li et al., 2020) showed the comparable performance of UV/NH_2Cl and UV/NaClO systems. According to this study, and other papers (Bu et al., 2018; Sun et al., 2019; Zhang et al., 2018), the NH_2Cl photodegradation leads to the hydroxyl radicals formation playing a crucial role in wastewater treatment, which is in general a common feature of all AOPs. This is a quite complex transformation path, which can be illustrated by the following reactions:

$$NH_2Cl \xrightarrow{h\nu} NH_2^{\bullet} + Cl^{\bullet}$$

$$Cl^{\bullet} + NH_2Cl \rightarrow NHCl^{\bullet} + H^+ + Cl^-$$

$$Cl^{\bullet} + Cl^- \rightarrow Cl_2^{\bullet-}$$

$$OH^{\bullet} + HOCl \rightarrow ClO^{\bullet} + H_2O$$

$$Cl^{\bullet} + HOCl \rightarrow ClO^{\bullet} + Cl^- + H^+$$

$$Cl^{\bullet} + ClO^- \rightarrow ClO^{\bullet} + Cl^-$$

$$Cl^{\bullet} + OH^- \rightarrow ClOH^{\bullet-}$$

$$Cl^{\bullet} + H_2O \rightarrow ClOH^{\bullet-} + H^+$$

$$ClOH^{\bullet-} \rightarrow OH^{\bullet} + Cl^-$$

As it can be inferred from the reactions involving H^+ cations, the NH_2Cl photodegradation process is dependent on the pH (Li et al., 2020). As expected, there are also other factors typical for UV/chlorine systems. For instance, the study on the photochlorination of 5 different micropollutants (Triclosan, Carbamazepine, Sulfamethoxazole and Estradiol) dissolved in tap water matrices (Sun et al., 2019), showed that the process is strongly influenced by the chlorinating agent dose, the presence of Cl^-, HCO_3^- and natural organic matter (NOM).

3.4 Photo-Fenton and related processes

Fenton system and related methods are one of the most popular AOPs. As in other cases, to increase pharmaceuticals decomposition effectiveness, it is often combined with UV irradiation (Giannakis, 2019; Giri & Golder, 2015; Leyva, Moctezuma, Baines, Noriega, & Zarazua, 2018; Raji, Pandiyaraj, Kandavelu, Vasu, & Saravanan, 2020; Rueda-Márquez, Levchuk, Manzano, & Sillanpää, 2020). Another advantage of photo-Fenton is the moderate costs of reactants. Importantly, it can be additionally reduced by the use of wastewater containing the Fe ions and solar radiation (Cuervo Lumbaque et al., 2019). As with other photo-AOPs, the use of artificial UV irradiation is a serious economic disadvantage. Fortunately, solar light source irradiation combined with the Fenton system was found to be a useful tool for pharmaceutical wastewater treatment (Cuervo Lumbaque et al., 2021; Sirtori et al., 2009). Although the use of sunlight in photo-Fenton processes is associated with a slight decrease in the rate of degradation compared to UV radiation, it is a much less expensive way. Studies on the nonsteroidal antiinflammatory drugs Tenoxicam, Meloxicam, and Ketoprofen elimination using simulated sunlight and UV-C-based photo-Fenton processes demonstrated that solar radiation is comparably effective to artificial UV-C light source (de Melo Santos, da Silva, de Lucena, Napoleão, & Duarte, 2020). The solar photo-Fenton (SPF) system was also successfully used to degrade Anastrozole (Sanabria, Scunderlick, Wilde, Lüdtke, & Sirtori, 2021). The process was carried out using different concentrations of iron (Fe^{2+}) and H_2O_2, achieving the best elimination efficiency of 95%. When considering the UV light sources suitable not only for photo-Fenton process, Light Emitting Diode (LED) is of a particular interest. The LED light has become a quite popular tool in wastewater treatment (Bawiec, Pawęska, & Pulikowski, 2020; Nguyen, Suwan, Koottatep, & Beck, 2019). The major advantage of LED is associated with the relatively low operating costs. Furthermore LED is a good alternative for UV lamps using Mercury. In the last few years several well-documented LED applications in photo-Fenton systems were reported. For instance, it was found that the UVA irradiation generated using LED allowed to remove Antipyrine with very high efficiency of 93% in a very short time (2.5 min) (Davididou, Monteagudo, Chatzisymeon, Durán, & Expósito, 2017). The studies on Losartan and Hydrochlorothiazide degradation showed c.a. 75%

removal in the case of both pharmaceuticals after 90 min (Kosman, Monteiro, Lenart, Weinert, & Tiburtius, 2021). Another report showed that Caffeine, Salicylic acid, Diclofenac, Paracetamol, and Ketoprofen can be removed with an efficiency of c.a. 80% after 10 min (Marchetti & Bessa Azevedo, 2020).

As it can be inferred from the above examples, photo-Fenton is in general characterized by a good performance. There are various studies showing the advantage of photo-Fenton over other kinds of AOPs. For instance, in the case of Metronidazole, the application of H_2O_2/Fe^{2+} and $UV/H_2O_2/Fe^{2+}$ systems led to 76% and 94% of removal, respectively (Shemer, Kunukcu, & Linden, 2006), which indicates a quite distinct advantage of the method using UV irradiation. Another interesting comparison was performed for Penicillin G Formulation photo-Fenton and sonolysis treatment (Saghafinia, Emadian, & Vossoughi, 2011). In the case of Dipyrone degradation studies using photo-Fenton, Fenton, photocatalytic method and simple photolysis (Giri & Golder, 2014), the former exhibited the best performance taking into account both the elimination rate and mineralization.

One of the most important factors affecting the photo-Fenton system performance is pH. For instance, higher efficiency of Lincomycin degradation was observed, when acidic conditions were applied (Andreozzi et al., 2006). Noteworthy, this is a general rule important for all Fenton systems since it prevents the formation of insoluble hydroxides, which precipitate out of solution. The highly acidic conditions were used for Diclofenac (initial concentration: 0.62 mM, pH = 2.8), which was completely decomposed in a few minutes using Fe^{3+}/H_2O_2 and 36 W-lamp (366 nm) (Ravina, Campanella, & Kiwi, 2002). The disadvantage of low pH is the need of neutralization of the effluent after the process. This generates additional costs and increases the salinity (Klamerth, Malato, Agüera, & Fernández-Alba, 2013).

As various previously mentioned studies have shown, despite the high degradation efficiency of the photo-Fenton system, the time needed to achieve complete mineralization is often relatively long. This suggests the formation of many stable transformation products. Therefore, more and more attention is paid to analyzing the potential harmfulness of these compounds (Cuervo Lumbaque et al., 2019; de Melo Santos et al., 2020; Della-Flora, Wilde, Pinto, Lima, & Sirtori, 2020; Sanabria et al., 2021). For instance, it was found that the photo-Fenton process used in the case of three pharmaceutics (Ketoprofen, Meloxicam and Tenoxicam) is characterized by relatively high efficiency of degradation, as well as a reduction in ecotoxicity (de Melo Santos et al., 2020). In quite recent work of Cuervo Lumbaque et al. (2019), Dipiron, Diazepam, Fluoxetine, Furosemide, Nimesulide, Paracetamol, Progesterone, and Propranolol were subjected to the photo-Fenton treatment process. Noteworthy, in the mentioned study, most of the transformation products were estimated using the quantitative structure-property relationships (QSPR) approach to be highly toxic and persistent. In case of Anastrozole five transformation products were reported (Sanabria et al., 2021). According to the QSPR analysis performed for these compounds, transformation products are supposed to be nonbiodegradable and may have mutagenic and carcinogenic effects (Sanabria et al., 2021). There is a great deal of research into the application of the Fenton system and its modifications, therefore there is a lot of information on degradation products. Theoretical methods allow assessing whether the actually applied method leads to the formation of less problematic by-products than the precursors. Nowadays, there are many free tools and online applications that enable the prediction of environmentally relevant parameters. The transport properties deserve special attention due to well-defined criteria, which can be used to evaluate the susceptibility of pollutants to migrate in the environment. Two important environmental fate features are long-range transport potential (LRTP) and overall persistence P_{OV}. There are several papers on the environmental fate of pollutants, including AOP by-products in which these characteristics were analyzed (Gackowska, Studziński, Kudlek, Dudziak, & Gaca, 2018; Mackay & Webster, 2006; Mostrąg, Puzyn, & Haranczyk, 2010; Studziński, Gackowska, & Kudlek, 2021; Wegmann, Cavin, MacLeod, Scheringer, & Hungerbühler, 2009). The useful "The OECD POV and LRTP Screening Tool" (Wegmann et al., 2009) and EPI suite (EPA, 2012) freeware software can be used to compute them. Based on the common classification system (Klasmeier et al., 2006) four regions on the LRTP vs. P_{OV} plot can be distinguished, namely region A (high LRTP and high P_{OV}), region B (high LRTP and low P_{OV}), region C (low LRTP and high P_{OV}) and region D (low LRTP and low P_{OV}). This type of analysis may be completed in a short time. Based on the two-dimensional structures (SMILES codes) of Flutamide and its transformation products (Della-Flora et al., 2020) exemplary calculations were performed for this chapter (Fig. 3). As one can see, Flutamide and the majority of degradation products are assigned to area D. On the other hand, three transformation products, denoted as TP4, TP10, and TP11 (Della-Flora et al., 2020), are characterized by a low LRTP and high persistence (area C). Based on this analysis, it can be concluded that the use of the photo-Fenton system favors the possible accumulation of flutamide by-products in regions close to the emission source.

3.5 Heterogeneous photocatalysis

In recent years, there has been a great interest in photocatalysis, especially the photodegradation of pharmaceutical compounds (Valari, Antoniadis, Mantzavinos, & Poulios, 2015; Velempini, Prabakaran, & Pillay, 2021). Noteworthy, photocatalysis is also a natural surface water self-purification process occurring in the presence of sand particles that act as photocatalysts (Yi et al., 2020). The general mechanism of photocatalysis is associated with the electrostatic charges induction on the semiconductor surface by the UV light which mediates in the reactive oxygen species formation (e.g., hydroxyl radical, peroxide anion radical) (Awofiranye, Modise, & Naidoo, 2020).

The benefits of using photocatalysts in the photodegradation of wastewater have been well documented. It has been confirmed that the photocatalysis process proceeded faster and was characterized by a higher degree of removal of pharmaceutical products from wastewater than direct photolysis (He, Sutton, Rijnaarts, & Langenhoff, 2016). There are many spectacular uses of photocatalyst, which are characterized by a very high removal rate. For instance, a very high elimination efficiency (c.a. 100%) was reported for Carbamazepine, Ibuprofen, and Ciprofloxacin by Monteoliva-García et al. (2019). The other study (Calza et al., 2006) showed complete

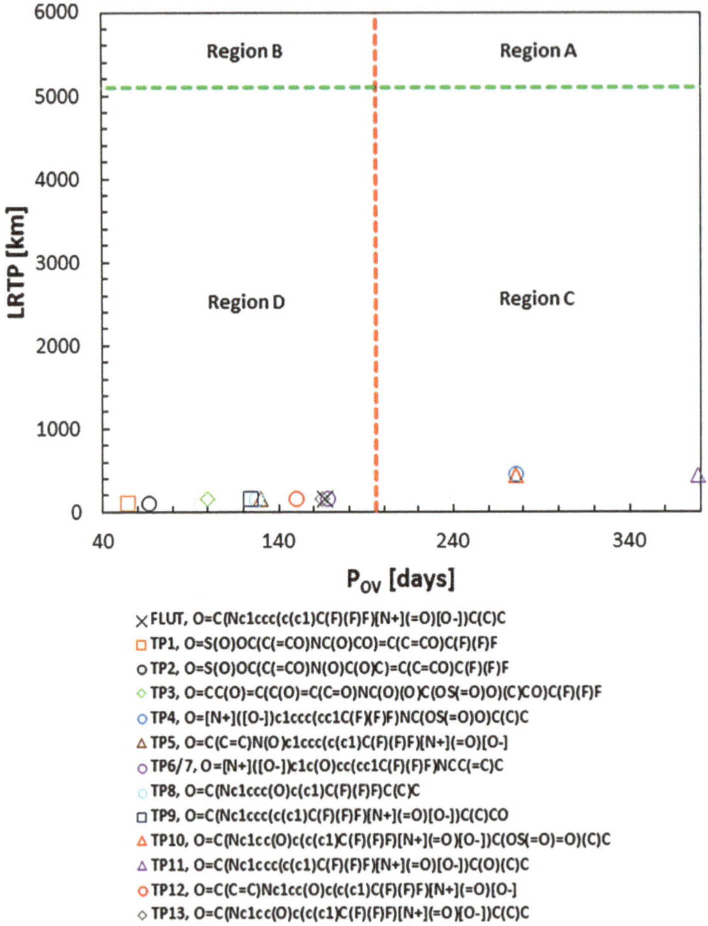

Fig. 3 LRTP vs. P_{OV} plot of Flutamide and the photo-Fenton transformation products detected by Della-Flora et al. (2020). In the legend, the selected SMILES codes assigned to the transformation products were provided. The *green* and *red dashed lines* stand for threshold values (LRTP = 5096.73 km, P_{OV} = 195 days). All calculations were performed for this chapter using "The OECD POV and LRTP Screening Tool" (Wegmann et al., 2009) and the EPI suite (EPA, 2012).

degradation of Diclofenac after 2 h in the presence of TiO_2. The high removal efficiency has also been observed for Furosemide and Ranitidine (Molinari, Pirillo, Loddo, & Palmisano, 2006), Methotrexate and Doxorubicin (Calza et al., 2014), Propranolol (De la Cruz, Dantas, Giménez, & Esplugas, 2013) and Naproxen (Jallouli et al., 2016).

The search for optimal conditions for conducting a photocatalytic process to obtain the maximum efficiency of degradation is not an easy task. It should be taken into account that the efficiency of the photolysis process is highly dependent on the characteristics of the photocatalyst and the reaction conditions (such as pH), which is a general rule in case of all AOPs and photo-AOPs. An interesting example is the photocatalytic removal of Acetaminophen (Borràs-Ferrís et al., 2019; Gómez-Avilés et al., 2019). When the metal-organic Ti-Zr photocatalyst framework was used for Acetaminophen at the initial pH of 6.9 under simulated sunlight, a completely removed was achieved after about 180 min in case of NH_2-MIL-125(Ti) metal-organic framework and after 120 in case of modified Zirconium-containing photocatalyst (Gómez-Avilés et al., 2019). Interestingly, the photocatalytic degradation of the same compound, but with the use of TiO_2 nanotubes was found to be the fastest at pH = 3 (Borràs-Ferrís et al., 2019). However, based on the data in the cited study, it can be inferred that after 240 min, the degradation was not complete. Furthermore, the concentrations of Acetaminophen were relatively high even after a day when the photocatalytic process was carried out at pH = 7 and 9.

An important factors affecting the photodegradation efficiency is the type of light source. As mentioned in the previous section, LED lamp is a very promising tool that is increasingly used in wastewater treatment studies. LED irradiation has also been used in the case of photocatalysis, and there are several interesting reports available in the literature. For example, the combination of the TiO_2 catalyst with the LED contributed to the effective degradation of Ibuprofen (Jallouli et al., 2018). In the mentioned study, the authors found that by using this method, the toxicity of Ibuprofen-containing wastewater (measured on Vibrio fischeri), can be reduced by about 40%. However, the study on the application of UV-LED/TiO_2 system for several pharmaceuticals belonging to the European Union Watch List (Ciprofloxacin, Sulfamethoxazole, Trimethoprim, Venlafaxine, and Desmethylvenlafaxine) (Bertagna Silva, Buttiglieri, Babić, Ašperger, & Babić, 2022) showed that the cost of electricity is still too high to use this method on the large scale. Furthermore, it was found that each compound behaves differently and the elimination efficiency is dependent on several parameters including type of matrix, the irradiation wavelength and pH.

The advancement of photocatalytic methods is inextricably linked to the advancement of photocatalysts. Various effective photocatalysts such as TiO_2, ZnO, ZnS, SiO_2, AgBr, and Al_2O_3 have been used for water purification (Byrne, Subramanian, & Pillai, 2018; Miklos et al., 2018; Wang et al., 2015; Yaqoob, Parveen, Umar, & Ibrahim, 2020). However, the former photocatalyst is much more popular than the others because of its high photoactivity, stability, relatively low cost, and low environmental toxic effects (Hashimoto, Irie, & Fujishima, 2005; Yaqoob et al., 2020).

However, the major limitation of TiO_2 is excited state threshold of 3.2 eV bandgap ($\lambda \leq 388$ nm). Fortunately, this can be overcome by doping (Etacheri, Di Valentin, Schneider, Bahnemann, & Pillai, 2015), using photoactive composites (Bartolomeu et al., 2018; Wang et al., 2020) or by using artificial light sources (Velempini et al., 2021). A combination of TiO_2 with other p-type semiconductors such as Ag_2O, NiO, and CuO (Chen, Wang, Han, & Guo, 2020; Gurung et al., 2019; Kubiak et al., 2020) has been used to improve the conductivity and thus to increase the catalytic activity. Some interesting examples are the photodegradation of Paracetamol using TiO_2/Fe_2O_3 (Abdel-Wahab, Al-Shirbini, Mohamed, & Nasr, 2017), TiO_2/SiO_2 (Zaccariello et al., 2014) and MoS_2/TiO_2 (Kumar, Bhadwal, Mizaikoff, Singh, &

Kranz, 2019), Amoxicillin using, K_2TiO_3 nanotubes/carbon quantum dots (Chen et al., 2019), Diclofenac using TiO_2/WO_3 (Mugunthan, Saidutta, & Jagadeeshbabu, 2018), Ciprofloxacin using TiO_2/ZnO (Silva et al., 2016) and TiO_2/Fe^{2+} (Monteoliva-García et al., 2019) or Oxytetracycline using TiO_2/graphene/ZSM-5 zeolite (Hu, Zhou, Chen, & Chang, 2016).

Despite the various benefits of heterogeneous photocatalysts, there are some disadvantages as well. The main limitation is poor degradation efficiency in turbid water (Lee, Palaniandy, & Dahlan, 2017). However, it can be overcome by using dialysis (Azrague, Aimar, Benoit-Marquié, & Maurette, 2007), which, on the other hand, makes the process more complex and expensive. Another downside is the toxicity of transformation products, which is a common disadvantage of AOPs. Despite the fact that photocatalysis has been found to minimize toxicity in pharmaceutical effluents (Rueda-Marquez, Levchuk, Fernández Ibañez, & Sillanpää, 2020), there are exceptions as well. For instance, it has been demonstrated that Paracetamol transformation products formed during the wastewater treatment process can be more toxic than the precursor (Szabó, Tóth, Homlok, Takács, & Wojnárovits, 2012).

3.6 Combined methods

As it was demonstrated some pharmaceuticals are very stable and resistant to the conventional water treatment methods or standard photo-AOPs. Thus, the combinations of photo-AOPs are often considered as a good alternative. The key elements for determining the selection of wastewater treatment technology are efficiency, energy costs, the types of pollutants occurring at a certain area, and hazardous by-products prevention. Therefore, it is important to find the optimal condition. Some interesting applications of combined AOPs characterized by high removal efficiency are $UV/H_2O_2/O_3$ (Ciprofloxacin, Trimethoprim, Carbamazepine, Fluoxetine, Gemfibrozil, Primidone, Sulfamethoxazole, and Ibuprofen) (Epold, Dulova, Veressinina, & Trapido, 2012; Lester, Avisar, Gozlan, & Mamane, 2011; Sgroi, Anumol, Vagliasindi, Snyder, & Roccaro, 2021), $UV/O_3/(NH_4)_2S_2O_8$ (Diclofenac) (Jabbari, Eslami, & Mahmoudian, 2020), $UV/Cl_2/O_3$ (Carbamazepine, Fluoxetine, Gemfibrozil, Primidone, Sulfamethoxazole and Trimethoprim) (Sgroi et al., 2021), $UV/O_3/Fe^{2+}/Cu^{2+}$ (Paracetamol) (Skoumal et al., 2006), and TiO_2/photo-Fenton (Cloxacillin) (Serna-Galvis, Giraldo-Aguirre, Silva-Agredo, Flórez-Acosta, & Torres-Palma, 2017).

In the previous sections, chemical factors were mainly discussed. However, physical factors such as hydrodynamic effects cannot be underestimated when considering photo-AOPs modifications. Cavitation and sonication are interesting and effective options for enhancing the water treatment process (Abramov et al., 2021; Dular et al., 2016; Wu et al., 2018). The use of ultrasonic cavitation reduces turbidity significantly due to its ability to separate suspensions (Fast, Kokabian, & Gude, 2014; Fast & Gude, 2015; Kowalska, Bien, & Zielewicz-Madej, 1988; Liu, Wang, Zhang, & Ni, 2015; Wu et al., 2018). Another important issue is that cavitation as well as other methods such as thermal treatment, pure UV irradiation or UV treatment in the presence of oxidizer (H_2O_2), alkaline solutions, deionized water and the oxygen treatment in boiling water (Calcio Gaudino et al., 2021; Carp, Huisman, & Reller, 2004; Kabra, Chaudhary, & Sawhney, 2004; Miranda-García, Suárez, Maldonado, Malato, & Sánchez, 2014; Pan, Zhang, Fu, & Xu, 2013; Rueda-Marquez et al., 2020; Wang, Jia, & Wang, 2017; Wang et al., 2015) can be used for the photocatalysts regeneration. Importantly, the photocatalyst may be seriously inactivated by adsorption of contaminants or conversion products (Garcia, Lourinho, Romano, & Brito, 2020; Ollis, 2000; Rueda-Marquez et al., 2020).

It's worth mentioning that there has been well-documented research that show the value of combining ultrasonic (US) or hydrodynamic cavitation (HC) with photochemical oxidation, including removal of hardly degradable antibiotics. For instance, the use of hydrodynamic cavitation with TiO_2 photocatalyst enables the Tetracycline reduction by 78% (Wang, Jia, & Wang, 2017). By using ultrasound and rGO/CdWO4 photocatalyst, complete degradation of Tetracycline was achieved (Ghoreishian et al., 2019). Also, complete degradation was observed in case of Sulfadiazine (Hayati, Isari, Anvaripour, Fattahi, & Kakavandi, 2020). In this study, the heterogeneous hexagonal magnesium oxide (MgO) deposited on carbon nanotubes (CNTs) was used. In all these cases, a synergy of pollutant removal efficiency, resulting from the combination of photocatalysis and hydrodynamic treatment, was observed. Similarly in the case of Amoxicillin, Doxycycline and Sulfadiazine, the combination of ultrasound with UV irradiation and Fenton-type systems was much more successful than using each approach separately (Tao, Cai, Huai, & Liu, 2018).

Interesting results were provided by research exploiting sorption and surface phenomena including photocatalysis, biosorption, and membrane effects. It has been established that biological treatment, along with sorption and photodegradation, is an effective method for removing pharmaceuticals from activated sludge (Salgado et al., 2012). The key advantages of this solution are energy savings and the ability to scale-up the process. Another promising solution is based on the use of an anaerobic-anoxic-oxic membrane bioreactor in conjunction with UV/O_3 (Sui et al., 2014). This method was tested against thirteen pharmaceutical and personal care products dissolved in domestic wastewater (Caffeine, Metoprolol, Gemfibrozil, Sulpiride, *N*,*N*-Diethyl-meta-toluamide, Clofibric acid, Bezafibrate, Carbamazepine, Chloramphenicol, Diclofenac, Mefenamic acid, Propranolol, Trimethoprim) (Sui et al., 2014). The application of adsorption/photocatalysis systems was employed for Ciprofloxacin (Zheng et al., 2018) and Ibuprofen (Lin et al., 2019). In both mentioned cases the TiO_2-based nanocomposites were utilized. Noteworthy, in the case of Ciprofloxacin at the concentration of 15 mg/L, a complete mineralization efficiency was achieved after 1.5 h (Zheng et al., 2018). However, taking into account the toxicity of transformation products, the process can be carried out for a shorter time, namely 120 min.

There has been a variety of combined AOPs characterized by a high efficiency, which has been reported in the last years. A summary and details of the exemplary reaction systems are provided in Table 1.

Table 1 The selected examples of combined AOPs used for pharmaceutical wastewater treatment.

Pharmaceuticals	Combined AOP system	Details and remarks	Reference
Cloxacillin	TiO_2/photo-Fenton	Concentration of pharmaceutical in the matrix: 203 μmol/L. Concentration of H_2O_2: 10 mmol/L. Concentration of Fe(II): 90 μmol/L. Initial pH: 6. Volume: 100 mL. Maximal light power: 150 W. The complete elimination after 120 min. Total organic carbon (TOC) reduction by 45%.	Serna-Galvis et al. (2017)
Diclofenac	$UV/O_3/(NH_4)_2S_2O_8$	Concentration of pharmaceutical in the matrix: 17–51 μmol/L. $(NH_4)_2S_2O_8$ concentration: 88–964 μmol/L. Ozone flow rate: 1 g/h. Initial pH: 3–9. Maximal experiment duration: 40 min. TOC reduction by 19% after 30 min. The humic acid addition reduces the removal rate.	Jabbari et al. (2020)
Carbamazepine, Fluoxetine, Gemfibrozil, Primidone, Sulfamethoxazole, Trimethoprim	Cl_2/O_3/UV	Concentration of pharmaceutical in the matrix: 0.1–7.2 nmol/L. Flow rate: 1200–6000 L/h. Volume: 21 L. O_3 doses: 0.01–0.19 mmol/L. Cl_2 concentration: 0.04 mmol/L. Radiation intensity: 191–981 mJ/cm^2. The presence of transformation products.	Sgroi et al. (2021)
Ciptoflaxacin	Graphitized mesoporous Carbon-TiO_2/UV	Concentration of pharmaceutical in the matrix: 0.045 mmol/L. Volume: 200 mL. The presence of transformation products. Completely mineralization after 1.5 h.	Zheng et al. (2018)
Ibuprofen	TiO_2/boron nitride nanocomposites/ UV	Concentration of pharmaceutical in the matrix: 0.024 mmol/L. Maximal experiment duration: 180 min. Volume: 50 mL. Almost complete elimination of Ibuprofen.	Lin et al. (2019)

Tetracycline	Hydrodynamic Cavitation/UV/ TiO_2	Concentration of pharmaceutical in the matrix: 0.068 mmol/L. UV radiation intensity: 27.8 mW/cm^2. Constant temperature applied (30°C). Volume: 4.0 L. Hydrodynamic cavitation conditions (maximal pressure: 0.34 MPa, maximal flow rate: 105.7 mL/s). Experiment duration: 90 min The drug concentration was reduced by approximately 78%.	Wang, Jia, and Wang (2017)
Tetracycline	Ultrasound/Photocatalyst: Reduced Graphene Oxide/ $CdWO_4$	Concentration of pharmaceutical in the matrix: 0.023–0.068 mmol/L. Maximal experiment duration: 60 min. Volume: 500 mL. Almost complete degradation.	Ghoreishian et al. (2019)
Sulfadiazine	Ultrasound/Photocatalyst: MgO/carbon nanotubes	Concentration of pharmaceutical in the matrix: 0.180 to 0.300 mmol/L. Light irradiation power: 50–150 W. Ultrasound power: 100–250 W/m^2, the optimal sonication: 200 W/m^2. Reactor volume: 200 mL. Complete elimination after 80 min. Complete mineralization after 180 min.	Hayati et al. (2020)
Amoxicillin, Doxycycline, Sulfadiazine	Cavitating jet impingement, UV/Fenton, analogous Fenton, photocatalytic oxidation	Initial chemical oxidation demand (COD) of aqueous solutions containing Amoxicillin, Doxycycline, Sulfadiazine: 236–578 mg/L. pH: 2.1–4.3. Temperature: 22–49°C. Cavitation conditions (jets impingement, nozzle inlet pressure: 6–12 MPa). Volume: 30 L. The reduction of COD by approximately 80%.	Tao et al. (2018)

4. Summary

Pharmaceuticals are often a very problematic component of wastewater due to their relatively high stability and adverse effects on living organisms, such as antibiotic resistance and endocrine disorders. The combination of AOP with UV and solar radiation is a promising solution for eliminating pharmaceuticals from wastewater. This is especially important for highly stable compounds that are resistant to conventional water treatment methods. The most frequently studied methods of this type are UV/H_2O_2, UV/O_3, UV/chlorine, photo-Fenton and heterogeneous photocatalysis. Based on the literature review, the advantages and disadvantages of these approaches were discussed. The pros and cons list of discussed in this chapter methods is briefly presented in Fig. 4. A common advantage of photochemical AOPs is the high efficiency of removing stable pharmaceutical compounds. According to various reports, the removal rate achieved for many pharmaceuticals was close to 100%. Furthermore, in some cases, complete mineralization can be achieved after a relatively short time. However, the wastewater treatment efficiency depends on various agents such as matrix composition including the presence of DOM and pH. Another important issue related to the use is the cost-effectiveness analysis. Oxidizing agents such as hydrogen peroxide, ozone, chlorine compounds, Fenton reagent and TiO_2 are relatively cheap reactants. However, the disadvantage of the photochemical AOP techniques is the high cost of the process associated with electricity consumption. For this reason, they

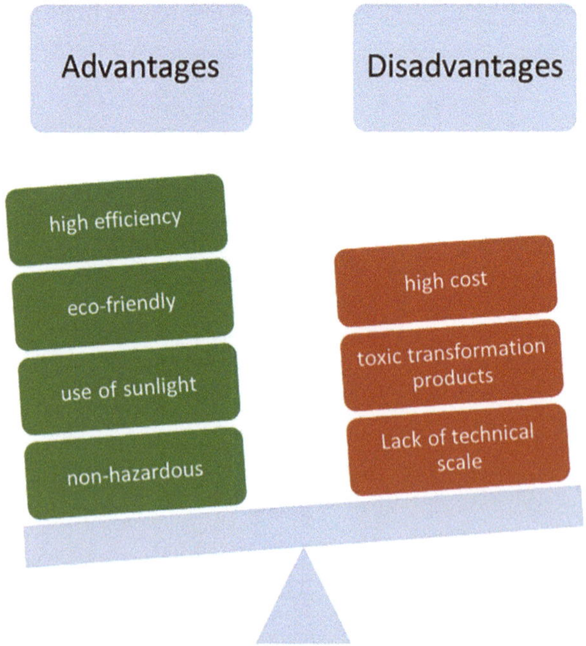

Fig. 4 The advantages and drawbacks of AOP applying UV and solar radiation.

are rarely used on an industrial scale. It is worth mentioning that LED UV irradiation source is a promising tool for photochemical wastewater treatment. Nevertheless, its use is still associated with relatively high costs compared to methods that do not require artificial UV irradiation. On the other hand, the application of natural solar irradiation can help to reduce the cost of wastewater treatment and transfer laboratory-scale testing to wastewater treatment plants. It is worth noting, however, that even in the case of combining AOP with natural UV irradiation, alternative methods should also be available in case of insufficient solar radiation due to weather conditions.

Another disadvantage of the wastewater treatment methods discussed in this chapter is the formation of toxic transformation by-products that may enter the environment and pose a greater risk to the environment than the precursors. Therefore, it is worthwhile to analyze the postreaction mixture in order to assess possible hazards to ecosystems. QSPR approaches are also effective for assessing the toxicity, mutagenicity, carcinogenicity, and transport characteristics of substances.

Combining AOPs with UV irradiation is in general beneficial for removing pharmaceuticals from the wastewater. However, the large-scale application of these methods still requires further research and innovation. Among the methods discussed in this chapter, heterogeneous photocatalysis and photo-Fenton oxidation seem to be the most developed and suitable for implementation in the future. Nevertheless, other AOPs and especially combined ones such as $UV/H_2O_2/O_3$, $UV/O_3/(NH_4)_2S_2O_8$, $UV/Cl_2/O_3$, $UV/O_3/Fe^{2+}/Cu^{2+}$, TiO_2/photo-Fenton are interesting in the context of designing highly effective methods for a specific type of contamination.

References

Abdel-Wahab, A. M., Al-Shirbini, A. S., Mohamed, O., & Nasr, O. (2017). Photocatalytic degradation of paracetamol over magnetic flower-like TiO_2/Fe_2O_3 core-shell nanostructures. *Journal of Photochemistry and Photobiology A: Chemistry*, *347*, 186–198. https://doi.org/10.1016/j.jphotochem.2017.07.030.

Abramov, V. O., Abramova, A. V., Cravotto, G., Nikonov, R. V., Fedulov, I. S., & Ivanov, V. K. (2021). Flow-mode water treatment under simultaneous hydrodynamic cavitation and plasma. *Ultrasonics Sonochemistry*, *70*, 105323. https://doi.org/10.1016/j.ultsonch.2020.105323.

Abrile, M. G., Ciucio, M. M., Demarchi, L. M., Bono, V. M., Fiasconaro, M. L., & Lovato, M. E. (2021). Degradation and mineralization of the emerging pharmaceutical pollutant sildenafil by ozone and UV radiation using response surface methodology. *Environmental Science and Pollution Research*, *28*(19), 23868–23886. https://doi.org/10.1007/s11356-020-11717-9.

Akmehmet Balcioğlu, I., & Ötker, M. (2004). Pre-treatment of antibiotic formulation wastewater by O_3, O_3/H_2O_2, and O_3/UV processes. *Turkish Journal of Engineering and Environmental Sciences*, *28*(5), 325–331.

Alfassi, Z. B., Huie, R. E., Mosseri, S., & Neta, P. (1988). Kinetics of one-electron oxidation by the ClO radical. *International Journal of Radiation Applications and Instrumentation*, *32*(1), 85–88. https://doi.org/10.1016/1359-0197(88)90018-5.

Alharbi, S. K., Kang, J., Nghiem, L. D., van de Merwe, J. P., Leusch, F. D. L., & Price, W. E. (2017). Photolysis and UV/H2O2 of diclofenac, sulfamethoxazole, carbamazepine, and trimethoprim: Identification of their major degradation products by ESI–LC–MS and assessment of the toxicity of reaction mixtures. *Process Safety and Environmental Protection*, *112*, 222–234. https://doi.org/10.1016/j.psep.2017.07.015.

Amaro-Ortiz, A., Yan, B., & D'Orazio, J. A. (2014). Ultraviolet radiation, aging and the skin: Prevention of damage by topical cAMP manipulation. *Molecules*, *19*(5), 6202–6219. https://doi.org/10.3390/molecules19056202.

Ameta, R., Chohadia, A. K., Jain, A., & Punjabi, P. B. (2018). Fenton and photo-Fenton processes. In *Advanced oxidation processes for wastewater treatment: Emerging green chemical technology* (pp. 49–87). Elsevier Inc. https://doi.org/10.1016/B978-0-12-810499-6.00003-6.

Andreozzi, R., Canterino, M., Giudice, R. L., Marotta, R., Pinto, G., & Pollio, A. (2006). Lincomycin solar photodegradation, algal toxicity and removal from wastewaters by means of ozonation. *Water Research*, *40*(3), 630–638. https://doi.org/10.1016/j.watres.2005.11.023.

Atugoda, T., Vithanage, M., Wijesekara, H., Bolan, N., Sarmah, A. K., Bank, M. S., et al. (2021). Interactions between microplastics, pharmaceuticals and personal care products: Implications for vector transport. *Environment International*, *149*, 106367. https://doi.org/10.1016/j.envint.2020.106367.

aus der Beek, T., Weber, F. A., Bergmann, A., Hickmann, S., Ebert, I., Hein, A., et al. (2016). Pharmaceuticals in the environment-global occurrences and perspectives. *Environmental Toxicology and Chemistry*, *35*(4), 823–835. https://doi.org/10.1002/etc.3339.

Awofiranye, O. S., Modise, S. J., & Naidoo, E. B. (2020). Overview of polymer–TiO_2 catalyst for aqueous degradation of pharmaceuticals in heterogeneous photocatalytic process. *Journal of Nanoparticle Research*, *22*(6). https://doi.org/10.1007/s11051-020-04877-9.

Azrague, K., Aimar, P., Benoit-Marquié, F., & Maurette, M. T. (2007). A new combination of a membrane and a photocatalytic reactor for the depollution of turbid water. *Applied Catalysis B: Environmental*, *72*(3–4), 197–204. https://doi.org/10.1016/j.apcatb.2006.10.007.

Baeza, C., & Knappe, D. R. U. (2011). Transformation kinetics of biochemically active compounds in low-pressure UV photolysis and UV/H_2O_2 advanced oxidation processes. *Water Research*, *45*(15), 4531–4543. https://doi.org/10.1016/j.watres.2011.05.039.

Baghirzade, B. S., Yetis, U., & Dilek, F. B. (2021). Imidacloprid elimination by O_3 and O_3/UV: Kinetics study, matrix effect, and mechanism insight. *Environmental Science and Pollution Research*, *28*(19), 24535–24551. https://doi.org/10.1007/s11356-020-09355-2.

Bartolomeu, M., Neves, M. G. P. M. S., Faustino, M. A. F., & Almeida, A. (2018). Wastewater chemical contaminants: Remediation by advanced oxidation processes. *Photochemical and Photobiological Sciences*, *17*(11), 1573–1598. https://doi.org/10.1039/c8pp00249e.

Bawiec, A., Pawęska, K., & Pulikowski, K. (2020). LED light use for the improvement of wastewater treatment in the hydroponic system. *Environmental Technology (United Kingdom)*, *41*(16), 2024–2036. https://doi.org/10.1080/09593330.2018.1554007.

Beitz, T., Bechmann, W., & Mitzner, R. (1998). Investigations of reactions of selected azaarenes with radicals in water. 1. Hydroxyl and sulfate radicals. *Journal of Physical Chemistry A*, *102*(34), 6760–6765. https://doi.org/10.1021/jp980654i.

Beltrán, F. J., Rivas, J., Álvarez, P. M., Alonso, M. A., & Acedo, B. (1999). A kinetic model for advanced oxidation processes of aromatic hydrocarbons in water: Application to phenanthrene and nitrobenzene. *Industrial and Engineering Chemistry Research*, *38*(11), 4189–4199. https://doi.org/10.1021/ie990189r.

Benotti, M. J., Trenholm, R. A., Vanderford, B. J., Holady, J. C., Stanford, B. D., & Snyder, S. A. (2009). Pharmaceuticals and endocrine disrupting compounds in U.S. drinking water.

Environmental Science and Technology, *43*(3), 597–603. https://doi.org/10.1021/es801845a.

Bermúdez, L. A., Pascual, J. M., Martínez, M. D. M. M., & Capilla, J. M. P. (2021). Effectiveness of advanced oxidation processes in wastewater treatment: State of the art. *Water (Switzerland)*, *13*(15). https://doi.org/10.3390/w13152094.

Bertagna Silva, D., Buttiglieri, G., Babić, B., Ašperger, D., & Babić, S. (2022). Performance of TiO2/UV-LED-based processes for degradation of pharmaceuticals: Effect of matrix composition and process variables. *Nanomaterials*, *12*(2), 295. https://doi.org/10.3390/nano12020295.

Borràs-Ferrís, J., Sánchez-Tovar, R., Blasco-Tamarit, E., Muñoz-Portero, M. J., Fernández-Domene, R. M., & García-Antón, J. (2019). TiO_2 nanostructures for photoelectrocatalytic degradation of acetaminophen. *Nanomaterials*, *9*(4). https://doi.org/10.3390/nano9040583.

Bouki, C., Venieri, D., & Diamadopoulos, E. (2013). Detection and fate of antibiotic resistant bacteria in wastewater treatment plants: A review. *Ecotoxicology and Environmental Safety*, *91*, 1–9. https://doi.org/10.1016/j.ecoenv.2013.01.016.

Bound, J. P., & Voulvoulis, N. (2005). Household disposal of pharmaceuticals as a pathway for aquatic contamination in the United Kingdom. *Environmental Health Perspectives*, *113* (12), 1705–1711. https://doi.org/10.1289/ehp.8315.

Bu, L., Zhou, S., Zhu, S., Wu, Y., Duan, X., Shi, Z., et al. (2018). Insight into carbamazepine degradation by UV/monochloramine: Reaction mechanism, oxidation products, and DBPs formation. *Water Research*, *146*, 288–297. https://doi.org/10.1016/j.watres.2018.09.036.

Bui, X. T., Vo, T. P. T., Ngo, H. H., Guo, W. S., & Nguyen, T. T. (2016). Multicriteria assessment of advanced treatment technologies for micropollutants removal at large-scale applications. *Science of the Total Environment*, *563–564*, 1050–1067. https://doi.org/10.1016/j.scitotenv.2016.04.191.

Burch, J. D., & Thomas, K. E. (1998). Water disinfection for developing countries and potential for solar thermal pasteurization. *Solar Energy*, *64*(1–3), 87–97. https://doi.org/10.1016/S0038-092X(98)00036-X.

Byrne, C., Subramanian, G., & Pillai, S. C. (2018). Recent advances in photocatalysis for environmental applications. *Journal of Environmental Chemical Engineering*, *6*(3), 3531–3555. https://doi.org/10.1016/j.jece.2017.07.080.

Cai, M., Li, R., Xie, Z., Huang, J., Zeng, Y., Zhang, Q., et al. (2019). Synthesis of a core-shell heterostructured MoS2/Cd0.9Zn0.1S photocatalyst for the degradation of diclofenac under visible light. *Applied Catalysis B: Environmental*, *259*. https://doi.org/10.1016/j.apcatb.2019.118033.

Calcio Gaudino, E., Canova, E., Liu, P., Wu, Z., & Cravotto, G. (2021). Degradation of antibiotics in wastewater: New advances in cavitational treatments. *Molecules*, *26*(3). https://doi.org/10.3390/molecules26030617.

Calza, P., Medana, C., Sarro, M., Rosato, V., Aigotti, R., Baiocchi, C., et al. (2014). Photocatalytic degradation of selected anticancer drugs and identification of their transformation products in water by liquid chromatography-high resolution mass spectrometry. *Journal of Chromatography A*, *1362*, 135–144. https://doi.org/10.1016/j.chroma.2014.08.035.

Calza, P., Sakkas, V. A., Medana, C., Baiocchi, C., Dimou, A., Pelizzetti, E., et al. (2006). Photocatalytic degradation study of diclofenac over aqueous TiO_2 suspensions. *Applied Catalysis B: Environmental*, *67*(3–4), 197–205. https://doi.org/10.1016/j.apcatb.2006.04.021.

Carp, O., Huisman, C. L., & Reller, A. (2004). Photoinduced reactivity of titanium dioxide. *Progress in Solid State Chemistry*, *32*(1–2), 33–177. https://doi.org/10.1016/j.progsolidstchem.2004.08.001.

Castiglioni, S., Bagnati, R., Fanelli, R., Pomati, F., Calamari, D., & Zuccato, E. (2006). Removal of pharmaceuticals in sewage treatment plants in Italy. *Environmental Science and Technology*, *40*(1), 357–363. https://doi.org/10.1021/es050991m.

Chen, J., Wang, M., Han, J., & Guo, R. (2020). TiO_2 nanosheet/NiO nanorod hierarchical nanostructures: p–n heterojunctions towards efficient photocatalysis. *Journal of Colloid and Interface Science*, *562*, 313–321. https://doi.org/10.1016/j.jcis.2019.12.031.

Chen, Q., Chen, L., Qi, J., Tong, Y., Lv, Y., Xu, C., et al. (2019). Photocatalytic degradation of amoxicillin by carbon quantum dots modified K2Ti6O13 nanotubes: Effect of light wavelength. *Chinese Chemical Letters*, *30*(6), 1214–1218. https://doi.org/10.1016/j.cclet.2019.03.002.

Cowie, B. E., Porley, V., & Robertson, N. (2020). Solar disinfection (SODIS) provides a much underexploited opportunity for researchers in photocatalytic water treatment (PWT). *ACS Catalysis*, *10*(20), 11779–11782. https://doi.org/10.1021/acscatal.0c03325.

Cuervo Lumbaque, E., Cardoso, R. M., de Araújo Gomes, A., Malato, S., Sánchez Pérez, J. A., & Sirtori, C. (2021). Removal of pharmaceuticals in hospital wastewater by solar photo-Fenton with Fe3+-EDDS using a pilot raceway pond reactor: Transformation products and in silico toxicity assessment. *Microchemical Journal*, *164*. https://doi.org/10.1016/j.microc.2021.106014.

Cuervo Lumbaque, E., Wielens Becker, R., Salmoria Araújo, D., Dallegrave, A., Ost Fracari, T., Lavayen, V., et al. (2019). Degradation of pharmaceuticals in different water matrices by a solar homo/heterogeneous photo-Fenton process over modified alginate spheres. *Environmental Science and Pollution Research*, *26*(7), 6532–6544. https://doi.org/10.1007/s11356-018-04092-z.

Das, A. K., Islam, M. N., Billah, M. M., & Sarker, A. (2021). COVID-19 pandemic and healthcare solid waste management strategy – A mini-review. *Science of the Total Environment*, *778*. https://doi.org/10.1016/j.scitotenv.2021.146220.

Davididou, K., Monteagudo, J. M., Chatzisymeon, E., Durán, A., & Expósito, A. J. (2017). Degradation and mineralization of antipyrine by UV-A LED photo-Fenton reaction intensified by ferrioxalate with addition of persulfate. *Separation and Purification Technology*, *172*, 227–235. https://doi.org/10.1016/j.seppur.2016.08.021.

De la Cruz, N., Dantas, R. F., Giménez, J., & Esplugas, S. (2013). Photolysis and TiO_2 photocatalysis of the pharmaceutical propranolol: Solar and artificial light. *Applied Catalysis B: Environmental*, *130–131*, 249–256. https://doi.org/10.1016/j.apcatb.2012.10.003.

de Melo Santos, M. M., da Silva, T. D., de Lucena, A. L. A., Napoleão, D. C., & Duarte, M. M. M. B. (2020). Degradation of ketoprofen, tenoxicam, and meloxicam drugs by photo-assisted peroxidation and photo-Fenton processes: Identification of intermediates and toxicity study. *Water, Air, and Soil Pollution*, *231*(2). https://doi.org/10.1007/s11270-020-4401-9.

Della-Flora, A., Wilde, M. L., Pinto, I. D. F., Lima, É. C., & Sirtori, C. (2020). Degradation of the anticancer drug flutamide by solar photo-Fenton treatment at near-neutral pH: Identification of transformation products and in silico (Q)SAR risk assessment. *Environmental Research*, *183*, 109223. https://doi.org/10.1016/j.envres.2020.109223.

Deng, Y., & Zhao, R. (2015). Advanced oxidation processes (AOPs) in wastewater treatment. *Current Pollution Reports*, *1*(3), 167–176. https://doi.org/10.1007/s40726-015-0015-z.

Dhodapkar, R. S., & Gandhi, K. N. (2019). Pharmaceuticals and personal care products in aquatic environment: Chemicals of emerging concern? In *Pharmaceuticals and personal care products: Waste management and treatment technology emerging contaminants and micro pollutants* (pp. 63–85). Elsevier. https://doi.org/10.1016/B978-0-12-816189-0.00003-2.

Ding, X., Zhu, J., Zhang, J., Dong, T., Xia, Y., Jiao, J., et al. (2020). Developmental toxicity of disinfection by-product monohaloacetamides in embryo-larval stage of zebrafish. *Ecotoxicology and Environmental Safety*, *189*. https://doi.org/10.1016/j.ecoenv.2019.110037.

Dong, M. M., Mezyk, S. P., & Rosario-Ortiz, F. L. (2010). Reactivity of effluent organic matter (EfOM) with hydroxyl radical as a function of molecular weight. *Environmental Science and Technology*, *44*(15), 5714–5720. https://doi.org/10.1021/es1004736.

Dular, M., Griessler-Bulc, T., Gutierrez-Aguirre, I., Heath, E., Kosjek, T., Krivograd Klemenčič, A., et al. (2016). Use of hydrodynamic cavitation in (waste)water treatment. *Ultrasonics Sonochemistry*, *29*, 577–588. https://doi.org/10.1016/j.ultsonch.2015.10.010.

Ebele, A. J., Abou-Elwafa Abdallah, M., & Harrad, S. (2017). Pharmaceuticals and personal care products (PPCPs) in the freshwater aquatic environment. *Emerging Contaminants*, 1–16. https://doi.org/10.1016/j.emcon.2016.12.004.

Ejeromedoghene, O., Nwosisi, M. C., Tesi, G. O., Noragbon, E. J., & Akinyeye, R. O. (2021). Impact of pharmaceutical waste generation and handling on environmental health in developing countries: COVID–19 pandemic in perspective! *Journal of Applied Sciences and Environmental Management*, *25*(3), 385–396. https://doi.org/10.4314/jasem.v25i3.12.

Emara, Y., Lehmann, A., Siegert, M. W., & Finkbeiner, M. (2019). Modeling pharmaceutical emissions and their toxicity-related effects in life cycle assessment (LCA): A review. *Integrated Environmental Assessment and Management*, *15*(1), 6–18. https://doi.org/10.1002/ieam.4100.

EPA, U. (2000). *The history of drinking water treatment* (pp. 1–4). Public Health.

EPA, U. (2012). *Estimation programs interface SuiteTM for Microsoft® Windows, v 4. Vol. 11.*

Epold, I., Dulova, N., Veressinina, Y., & Trapido, M. (2012). Application of ozonation, UV photolysis, Fenton treatment and other related processes for degradation of ibuprofen and sulfamethoxazole in different aqueous matrices. *Journal of Advanced Oxidation Technologies*, *15*(2), 354–364. https://doi.org/10.1515/jaots-2012-0215.

Etacheri, V., Di Valentin, C., Schneider, J., Bahnemann, D., & Pillai, S. C. (2015). Visible-light activation of TiO_2 photocatalysts: Advances in theory and experiments. *Journal of Photochemistry and Photobiology C: Photochemistry Reviews*, *25*, 1–29. https://doi.org/10.1016/j.jphotochemrev.2015.08.003.

Farré, M.l., Pérez, S., Kantiani, L., & Barceló, D. (2008). Fate and toxicity of emerging pollutants, their metabolites and transformation products in the aquatic environment. *TrAC, Trends in Analytical Chemistry*, *27*(11), 991–1007. https://doi.org/10.1016/j.trac.2008.09.010.

Fast, S. A., & Gude, V. G. (2015). Ultrasound-chitosan enhanced flocculation of low algal turbid waters. *Journal of Industrial and Engineering Chemistry*, *24*, 153–160. https://doi.org/10.1016/j.jiec.2014.09.023.

Fast, S. A., Kokabian, B., & Gude, V. G. (2014). Chitosan enhanced coagulation of algal turbid waters—Comparison between rapid mix and ultrasound coagulation methods. *Chemical Engineering Journal*, *244*, 403–410. https://doi.org/10.1016/j.cej.2014.01.081.

Feng, Y., Smith, D. W., & Bolton, J. R. (2007). Photolysis of aqueous free chlorine species (HOCl and OCl-) with 254 nm ultraviolet light. *Journal of Environmental Engineering and Science*, *6*(3), 277–284. https://doi.org/10.1139/S06-052.

Fu, Q., Malchi, T., Carter, L. J., Li, H., Gan, J., & Chefetz, B. (2019). Pharmaceutical an personal care products: From wastewater treatment into agro-food systems. *Environmental Science and Technology*, *53*(24), 14083–14090. https://doi.org/10.1021/acs.est.9b06206.

Gackowska, A., Studziński, W., Kudlek, E., Dudziak, M., & Gaca, J. (2018). Estimation of physicochemical properties of 2-ethylhexyl-4-methoxycinnamate (EHMC) degradation products and their toxicological evaluation. *Environmental Science and Pollution Research*, *25*(16), 16037–16049. https://doi.org/10.1007/s11356-018-1796-6.

Gao, Y.q., Zhang, J., Li, C., Tian, F.x., & Gao, N.y. (2020). Comparative evaluation of metoprolol degradation by UV/chlorine and UV/H_2O_2 processes. *Chemosphere*, *243*. https://doi.org/10.1016/j.chemosphere.2019.125325.

Garcia, B. B., Lourinho, G., Romano, P., & Brito, P. S. D. (2020). Photocatalytic degradation of swine wastewater on aqueous TiO_2 suspensions: Optimization and modeling via Box-Behnken design. *Heliyon*, *6*(1). https://doi.org/10.1016/j.heliyon.2020.e03293.

Garrido-Cardenas, J. A., Esteban-García, B., Agüera, A., Sánchez-Pérez, J. A., & Manzano-Agugliaro, F. (2020). Wastewater treatment by advanced oxidation process and their worldwide research trends. *International Journal of Environmental Research and Public Health*, *17*(1). https://doi.org/10.3390/ijerph17010170.

Ghoreishian, S. M., Raju, G. S. R., Pavitra, E., Kwak, C. H., Han, Y. K., & Huh, Y. S. (2019). Ultrasound-assisted heterogeneous degradation of tetracycline over flower-like rGO/CdWO4 hierarchical structures as robust solar-light-responsive photocatalysts: Optimization, kinetics, and mechanism. *Applied Surface Science*, *489*, 110–122. https://doi.org/10.1016/j.apsusc.2019.05.299.

Giannakis, S. (2019). A review of the concepts, recent advances and niche applications of the (photo) Fenton process, beyond water/wastewater treatment: Surface functionalization, biomass treatment, combatting cancer and other medical uses. *Applied Catalysis B: Environmental*, *248*, 309–319. https://doi.org/10.1016/j.apcatb.2019.02.025.

Giannakis, S., Gamarra Vives, F. A., Grandjean, D., Magnet, A., De Alencastro, L. F., & Pulgarin, C. (2015). Effect of advanced oxidation processes on the micropollutants and the effluent organic matter contained in municipal wastewater previously treated by three different secondary methods. *Water Research*, *84*, 295–306. https://doi.org/10.1016/j.watres.2015.07.030.

Giri, A. S., & Golder, A. K. (2014). Fenton, photo-Fenton, H_2O_2 photolysis, and TiO_2 photocatalysis for dipyrone oxidation: Drug removal, mineralization, biodegradability, and degradation mechanism. *Industrial and Engineering Chemistry Research*, *53*(4), 1351–1358. https://doi.org/10.1021/ie402279q.

Giri, A. S., & Golder, A. K. (2015). Decomposition of drug mixture in Fenton and photo-Fenton processes: Comparison to singly treatment, evolution of inorganic ions and toxicity assay. *Chemosphere*, *127*, 254–261. https://doi.org/10.1016/j.chemosphere.2015.02.010.

Glaze, W. H. (1987). Drinking-water treatment with ozone. *Environmental Science and Technology*, *21*(3), 224–230. https://doi.org/10.1021/es00157a001.

Gmurek, M., Olak-Kucharczyk, M., & Ledakowicz, S. (2015). Influence of dissolved organic matter in natural and simulated water on the photochemical decomposition of butylparaben. *Journal of Environmental Health Science and Engineering*, *13*(1). https://doi.org/10.1186/s40201-015-0185-z.

Goldstone, J. V., Pullin, M. J., Bertilsson, S., & Voelker, B. M. (2002). Reactions of hydroxyl radical with humic substances: Bleaching, mineralization, and production of bioavailable carbon substrates. *Environmental Science and Technology*, *36*(3), 364–372. https://doi.org/10.1021/es0109646.

Gómez-Avilés, A., Peñas-Garzón, M., Bedia, J., Dionysiou, D. D., Rodríguez, J. J., & Belver, C. (2019). Mixed Ti-Zr metal-organic-frameworks for the photodegradation of acetaminophen under solar irradiation. *Applied Catalysis B: Environmental*, *253*, 253–262. https://doi.org/10.1016/j.apcatb.2019.04.040.

Gorito, A. M., Pesqueira, J. F. J. R., Moreira, N. F. F., Ribeiro, A. R., Pereira, M. F. R., Nunes, O. C., et al. (2021). Ozone-based water treatment (O_3, O_3/UV, O_3/H_2O_2) for removal of organic micropollutants, bacteria inactivation and regrowth prevention. *Journal of Environmental Chemical Engineering*, *9*(4). https://doi.org/10.1016/j.jece.2021.105315.

Goswami, M., Goswami, P. J., Nautiyal, S., & Prakash, S. (2021). Challenges and actions to the environmental management of bio-medical waste during COVID-19 pandemic in India. *Heliyon*, *7*(3). https://doi.org/10.1016/j.heliyon.2021.e06313.

Gottschalk, C., Libra, J. A., & Saupe, A. (2010). *Ozonation of water and waste water: A practical guide to understanding ozone and its applications* (2nd edition). Wiley-VCH. https://doi.org/10.1002/9783527628926.

Grebel, J. E., Pignatello, J. J., & Mitch, W. A. (2010). Effect of halide ions and carbonates on organic contaminant degradation by hydroxyl radical-based advanced oxidation processes in saline waters. *Environmental Science and Technology*, *44*(17), 6822–6828. https://doi.org/10.1021/es1010225.

Guo, K., Wu, Z., & Fang, J. (2019). UV-based advanced oxidation process for the treatment of pharmaceuticals and personal care products. In *Contaminants of emerging concern in water and wastewater: Advanced treatment processes* (pp. 367–408). Elsevier. https://doi.org/10.1016/B978-0-12-813561-7.00010-9.

Guo, K., Wu, Z., Yan, S., Yao, B., Song, W., Hua, Z., et al. (2018). Comparison of the UV/chlorine and UV/H2O2 processes in the degradation of PPCPs in simulated drinking water and wastewater: Kinetics, radical mechanism and energy requirements. *Water Research*, *147*, 184–194. https://doi.org/10.1016/j.watres.2018.08.048.

Gurung, K., Ncibi, M. C., Thangaraj, S. K., Jänis, J., Seyedsalehi, M., & Sillanpää, M. (2019). Removal of pharmaceutically active compounds (PhACs) from real membrane bioreactor (MBR) effluents by photocatalytic degradation using composite Ag2O/P-25 photocatalyst. *Separation and Purification Technology*, *215*, 317–328. https://doi.org/10.1016/j.seppur.2018.12.069.

Hashimoto, K., Irie, H., & Fujishima, A. (2005). TiO_2 photocatalysis: A historical overview and future prospects. *Japanese Journal of Applied Physics, Part 1: Regular Papers and Short Notes and Review Papers*, *44*(12), 8269–8285. https://doi.org/10.1143/JJAP.44.8269.

Hayati, F., Isari, A. A., Anvaripour, B., Fattahi, M., & Kakavandi, B. (2020). Ultrasound-assisted photocatalytic degradation of sulfadiazine using MgO@CNT heterojunction composite: Effective factors, pathway and biodegradability studies. *Chemical Engineering Journal*, *381*. https://doi.org/10.1016/j.cej.2019.122636.

He, Y., Sutton, N. B., Rijnaarts, H. H. H., & Langenhoff, A. A. M. (2016). Degradation of pharmaceuticals in wastewater using immobilized TiO_2 photocatalysis under simulated solar irradiation. *Applied Catalysis B: Environmental*, *182*, 132–141. https://doi.org/10.1016/j.apcatb.2015.09.015.

Hu, X. Y., Zhou, K., Chen, B. Y., & Chang, C. T. (2016). Graphene/TiO_2/ZSM-5 composites synthesized by mixture design were used for photocatalytic degradation of oxytetracycline under visible light: Mechanism and biotoxicity. *Applied Surface Science*, *362*, 329–334. https://doi.org/10.1016/j.apsusc.2015.10.192.

Huang, H., Lu, H., Huang, H., Wang, L., Zhang, J., & Leung, D. Y. C. (2016). Recent development of VUV-based processes for air pollutant degradation. *Frontiers in Environmental Science*, *4*. https://doi.org/10.3389/fenvs.2016.00017.

Hwang, T. M., Nam, S. H., Lee, J., Koo, J. W., Kim, E., & Kwon, M. (2020). Hydroxyl radical scavenging factor measurement using a fluorescence excitation-emission matrix and parallel factor analysis in ultraviolet advanced oxidation processes. *Chemosphere*, *259*. https://doi.org/10.1016/j.chemosphere.2020.127396.

Insani, W. N., Qonita, N. A., Jannah, S. S., Nuraliyah, N. M., Supadmi, W., Gatera, V. A., et al. (2020). Improper disposal practice of unused and expired pharmaceutical products in Indonesian households. *Heliyon*, *6*(7). https://doi.org/10.1016/j.heliyon.2020.e04551.

Jabbari, F., Eslami, A., & Mahmoudian, J. (2020). Degradation of diclofenac in water using the O3/UV/S2O8 advanced oxidation process. *Health Scope*. https://doi.org/10.5812/jhealthscope.99436.

Jadhav, A. S. (2014). Advancement in drinking water treatments from ancient times. *International Journal of Science Environmental Technology*, *3*, 1415.

Jallouli, N., Elghniji, K., Hentati, O., Ribeiro, A. R., Silva, A. M. T., & Ksibi, M. (2016). UV and solar photo-degradation of naproxen: TiO_2 catalyst effect, reaction kinetics, products identification and toxicity assessment. *Journal of Hazardous Materials*, *304*, 329–336. https://doi.org/10.1016/j.jhazmat.2015.10.045.

Jallouli, N., Pastrana-Martínez, L. M., Ribeiro, A. R., Moreira, N. F. F., Faria, J. L., Hentati, O., et al. (2018). Heterogeneous photocatalytic degradation of ibuprofen in ultrapure water, municipal and pharmaceutical industry wastewaters using a TiO_2/UV-LED system. *Chemical Engineering Journal*, *334*, 976–984. https://doi.org/10.1016/j.cej.2017.10.045.

Janssen, E. M. L., Erickson, P. R., & McNeill, K. (2014). Dual roles of dissolved organic matter as sensitizer and quencher in the photooxidation of tryptophan. *Environmental Science and Technology*, *48*(9), 4916–4924. https://doi.org/10.1021/es500535a.

Jiang, J. Q., Zhou, Z., & Sharma, V. K. (2013). Occurrence, transportation, monitoring and treatment of emerging micro-pollutants in waste water—A review from global views. *Microchemical Journal*, *110*, 292–300. https://doi.org/10.1016/j.microc.2013.04.014.

Jodeh, S., Abdelwahab, F., Jaradat, N., Warad, I., & Jodeh, W. (2016). Adsorption of diclofenac from aqueous solution using *Cyclamen persicum* tubers based activated carbon (CTAC). *Journal of the Association of Arab Universities for Basic and Applied Sciences*, *20*, 32–38. https://doi.org/10.1016/j.jaubas.2014.11.002.

Johnson, M. B., & Mehrvar, M. (2008). Aqueous metronidazole degradation by UV/H_2O_2 process in single-and multi-lamp tubular photoreactors: Kinetics and reactor design. *Industrial & Engineering Chemistry Research*, *47*(17), 6525–6537. https://doi.org/10.1021/ie071637v.

Kabra, K., Chaudhary, R., & Sawhney, R. L. (2004). Treatment of hazardous organic and inorganic compounds through aqueous-phase photocatalysis: A review. *Industrial and Engineering Chemistry Research*, *43*(24), 7683–7696. https://doi.org/10.1021/ie0498551.

Kaczala, F., & Blum, S. E. (2016). The occurrence of veterinary pharmaceuticals in the environment: A review. *Current Analytical Chemistry*, *12*(3), 169–182. https://doi.org/10.2174/1573411012666151009193108.

Kalantary, R. R., Jamshidi, A., Mofrad, M. M. G., Jafari, A. J., Heidari, N., Fallahizadeh, S., et al. (2021). Effect of COVID-19 pandemic on medical waste management: A case study. *Journal of Environmental Health Science and Engineering*, *19*(1), 831–836. https://doi.org/10.1007/s40201-021-00650-9.

Kawabata, K., Sugihara, K., Sanoh, S., Kitamura, S., & Ohta, S. (2013). Photodegradation of pharmaceuticals in the aquatic environment by sunlight and UV-A,-B and-C irradiation. *Journal of Toxicological Sciences*, *38*(2), 215–223. https://doi.org/10.2131/jts.38.215.

Kim, I., & Tanaka, H. (2011). Energy consumption for ppcps removal by O_3 and O_3/UV. *Ozone: Science & Engineering*, *33*(2), 150–157. https://doi.org/10.1080/01919512.2011.549427.

Kim, I., Yamashita, N., & Tanaka, H. (2009). Photodegradation of pharmaceuticals and personal care products during UV and UV/H_2O_2 treatments. *Chemosphere*, *77*(4), 518–525. https://doi.org/10.1016/j.chemosphere.2009.07.041.

Kim, I. H., Tanaka, H., Iwasaki, T., Takubo, T., Morioka, T., & Kato, Y. (2008). Classification of the degradability of 30 pharmaceuticals in water with ozone, UV and H_2O_2. *Water Science and Technology*, *57*(2), 195–200. https://doi.org/10.2166/wst.2008.808.

Kishimoto, N. (2019). State of the art of UV/chlorine advanced oxidation processes: Their mechanism, byproducts formation, process variation, and applications. *Journal of Water and Environment Technology*, *17*(5), 302–335. https://doi.org/10.2965/jwet.19-021.

Klamerth, N., Malato, S., Agüera, A., & Fernández-Alba, A. (2013). Photo-Fenton and modified photo-Fenton at neutral pH for the treatment of emerging contaminants in wastewater treatment plant effluents: A comparison. *Water Research*, *47*(2), 833–840. https://doi.org/10.1016/j.watres.2012.11.008.

Klasmeier, J., Matthies, M., Macleod, M., Fenner, K., Scheringer, M., Stroebe, M., et al. (2006). Application of multimedia models for screening assessment of long-range transport potential and overall persistence. *Environmental Science and Technology*, *40*(1), 53–60. https://doi.org/10.1021/es0512024.

Kokkinos, P., Venieri, D., & Mantzavinos, D. (2021). Advanced oxidation processes for water and wastewater viral disinfection. A systematic review. *Food and Environmental Virology*, *13*(3), 283–302. https://doi.org/10.1007/s12560-021-09481-1.

Kosman, J., Monteiro, J. F. H. L., Lenart, V. M., Weinert, P. L., & Tiburtius, E. R. L. (2021). UV–vis LED-assisted photo-Fenton process for mineralization of losartan and hydrochlorothiazide: Optimization using desirability function. *Environmental Science and Pollution Research*, *28*(19), 24046–24056. https://doi.org/10.1007/s11356-020-12011-4.

Kotwani, A., Joshi, J., & Kaloni, D. (2021). Pharmaceutical effluent: A critical link in the interconnected ecosystem promoting antimicrobial resistance. *Environmental Science and Pollution Research*, *28*(25), 32111–32124. https://doi.org/10.1007/s11356-021-14178-w.

Kowalska, E., Bien, J., & Zielewicz-Madej, E. (1988). Ultrasound in the suspension separation methods. *Drying Technology*, *6*(3), 447–471. https://doi.org/10.1080/07373938808916393.

Krakkó, D., Illés, Á., Licul-Kucera, V., Dávid, B., Dobosy, P., Pogonyi, A., et al. (2021). Application of (V)UV/O_3 technology for post-treatment of biologically treated wastewater: A pilot-scale study. *Chemosphere*, *275*. https://doi.org/10.1016/j.chemosphere.2021.130080.

Krishnan, R. Y., Manikandan, S., Subbaiya, R., Biruntha, M., Govarthanan, M., & Karmegam, N. (2021). Removal of emerging micropollutants originating from pharmaceuticals and personal care products (PPCPs) in water and wastewater by advanced oxidation processes: A review. *Environmental Technology and Innovation*, *23*. https://doi.org/10.1016/j.eti.2021.101757.

Kubiak, A., Bielan, Z., Kubacka, M., Gabała, E., Zgoła-Grześkowiak, A., Janczarek, M., et al. (2020). Microwave-assisted synthesis of a TiO_2-CuO heterojunction with enhanced photocatalytic activity against tetracycline. *Applied Surface Science*, *520*. https://doi.org/10.1016/j.apsusc.2020.146344.

Kumar, N., Bhadwal, A. S., Mizaikoff, B., Singh, S., & Kranz, C. (2019). Electrochemical detection and photocatalytic performance of MoS2/TiO_2 nanocomposite against pharmaceutical contaminant: Paracetamol. *Sensing and Bio-Sensing Research*, *24*. https://doi.org/10.1016/j.sbsr.2019.100288.

Kümmerer, K. (2009). The presence of pharmaceuticals in the environment due to human use—Present knowledge and future challenges. *Journal of Environmental Management*, *90*(8), 2354–2366. https://doi.org/10.1016/j.jenvman.2009.01.023.

Lanzky, P. F., & Halling-Sørensen, B. (1997). The toxic effect of the antibiotic metronidazole on aquatic organisms. *Chemosphere*, *35*(11), 2553–2561. https://doi.org/10.1016/S0045-6535(97)00324-X.

Larsson, D. G. J. (2014). Pollution from drug manufacturing: Review and perspectives. *Philosophical Transactions of the Royal Society, B: Biological Sciences*, *369*(1656). https://doi.org/10.1098/rstb.2013.0571.

Lawrence, J., & Cappelli, F. P. (1977). Ozone in drinking water treatment: A review. *The Science of the Total Environment*, *7*(2), 99–108. https://doi.org/10.1016/0048-9697(77)90001-8.

Ledakowicz, S., Drozdek, E., Boruta, T., Foszpańczyk, M., Olak-Kucharczyk, M., Zyłła, R., et al. (2019). Impact of hydrogen peroxide on the UVC photolysis of diclofenac and toxicity of the phototransformation products. *International Journal of Photoenergy*, *2019*. https://doi.org/10.1155/2019/1086704.

Lee, C., Yoon, J., & Von Gunten, U. (2007). Oxidative degradation of N-nitrosodimethylamine by conventional ozonation and the advanced oxidation process ozone/hydrogen peroxide. *Water Research*, *41*(3), 581–590. https://doi.org/10.1016/j.watres.2006.10.033.

Lee, C. M., Palaniandy, P., & Dahlan, I. (2017). Pharmaceutical residues in aquatic environment and water remediation by TiO_2 heterogeneous photocatalysis: A review. *Environmental Earth Sciences*, *76*(17). https://doi.org/10.1007/s12665-017-6924-y.

Lekkerkerker-Teunissen, K., Benotti, M. J., Snyder, S. A., & Van Dijk, H. C. (2012). Transformation of atrazine, carbamazepine, diclofenac and sulfamethoxazole by low and medium pressure UV and UV/H_2O_2 treatment. *Separation and Purification Technology*, *96*, 33–43. https://doi.org/10.1016/j.seppur.2012.04.018.

Lester, Y., Avisar, D., Gozlan, I., & Mamane, H. (2011). Removal of pharmaceuticals using combination of UV/H_2O_2/O_3 advanced oxidation process. *Water Science and Technology*, *64*(11), 2230–2238. https://doi.org/10.2166/wst.2011.079.

Lester, Y., Avisar, D., & Mamane, H. (2010). Photodegradation of the antibiotic sulphamethoxazole in water with UV/H_2O_2 advanced oxidation process. *Environmental Technology*, *31*(2), 175–183. https://doi.org/10.1080/09593330903414238.

Leyva, E., Moctezuma, E., Baines, K. M., Noriega, S., & Zarazua, E. (2018). A review on chemical advanced oxidation processes for pharmaceuticals with paracetamol as a model compound. Reaction conditions, intermediates and total mechanism. *Current Organic Chemistry*, *22*(1), 2–17. https://doi.org/10.2174/1385272821666171019145520.

Li, S., Ao, X., Li, C., Lu, Z., Cao, W., Wu, F., et al. (2020). Insight into PPCP degradation by UV/NH2Cl and comparison with UV/NaClO: Kinetics, reaction mechanism, and DBP formation. *Water Research*, *182*. https://doi.org/10.1016/j.watres.2020.115967.

Li, W., Jain, T., Ishida, K., & Liu, H. (2017). A mechanistic understanding of the degradation of trace organic contaminants by UV/hydrogen peroxide, UV/persulfate and UV/free chlorine for water reuse. *Environmental Science: Water Research & Technology*, *3*(1), 128–138. https://doi.org/10.1039/c6ew00242k.

Liang, Y., Song, Q., Wu, N., Li, J., Zhong, Y., & Zeng, W. (2021). Repercussions of COVID-19 pandemic on solid waste generation and management strategies. *Frontiers of Environmental Science and Engineering*, *15*(6). https://doi.org/10.1007/s11783-021-1407-5.

Liao, C. H., Kang, S. F., & Wu, F. A. (2001). Hydroxyl radical scavenging role of chloride and bicarbonate ions in the H_2O_2/UV process. *Chemosphere*, *44*(5), 1193–1200. https://doi.org/10.1016/S0045-6535(00)00278-2.

Lin, L., Jiang, W., Bechelany, M., Nasr, M., Jarvis, J., Schaub, T., et al. (2019). Adsorption and photocatalytic oxidation of ibuprofen using nanocomposites of TiO_2 nanofibers combined with BN nanosheets: Degradation products and mechanisms. *Chemosphere*, *220*, 921–929. https://doi.org/10.1016/j.chemosphere.2018.12.184.

Litter, M. I., & Quici, N. (2010). Photochemical advanced oxidation processes for water and wastewater treatment. *Recent Patents on Engineering*, *4*(3), 217–241. https://doi.org/10.2174/187221210794578574.

Liu, H., Wang, C., & Wang, G. (2020). Photocatalytic advanced oxidation processes for water treatment: Recent advances and perspective. *Chemistry - An Asian Journal*, *15*(20), 3239–3253. https://doi.org/10.1002/asia.202000895.

Liu, W., Wang, S., Zhang, L., & Ni, Z. (2015). Water pollution characteristics of Dianchi Lake and the course of protection and pollution management. *Environmental Earth Sciences, 74*(5), 3767–3780. https://doi.org/10.1007/s12665-015-4152-x.

Lopez, A., Bozzi, A., Mascolo, G., & Kiwi, J. (2003). Kinetic investigation on UV and UV/H_2O_2 degradations of pharmaceutical intermediates in aqueous solution. *Journal of Photochemistry and Photobiology A: Chemistry, 156*(1–3), 121–126. https://doi.org/10.1016/S1010-6030(02)00435-5.

Lu, H., Li, Q., & Feng, W. (2022). Application progress of O_3/UV advanced oxidation technology in the treatment of organic pollutants in water. *Sustainability (Switzerland), 14*(3). https://doi.org/10.3390/su14031556.

Luo, Y., Reimers, K., Yang, L., & Lin, J. (2021). Household drug management practices of residents in a second-tier city in China: Opportunities for reducing drug waste and environmental pollution. *International Journal of Environmental Research and Public Health, 18* (16). https://doi.org/10.3390/ijerph18168544.

Maalouf, A., & Maalouf, H. (2021). Impact of COVID-19 pandemic on medical waste management in Lebanon. *Waste Management and Research, 39*(1), 45–55. https://doi.org/10.1177/0734242X211003970.

Mackay, D., & Webster, E. (2006). Environmental persistence of chemicals. *Environmental Science and Pollution Research, 13*(1), 43–49. https://doi.org/10.1065/espr2006.01.008.

Mahdi Ahmed, M., Barbati, S., Doumenq, P., & Chiron, S. (2012). Sulfate radical anion oxidation of diclofenac and sulfamethoxazole for water decontamination. *Chemical Engineering Journal, 197*, 440–447. https://doi.org/10.1016/j.cej.2012.05.040.

Mansor, N. A., & Tay, K. S. (2017). Degradation of 5,5-diphenylhydantoin by chlorination and UV/chlorination: Kinetics, transformation by-products, and toxicity assessment. *Environmental Science and Pollution Research, 24*(28), 22361–22370. https://doi.org/10.1007/s11356-017-9892-6.

Mansouri, F., Chouchene, K., Roche, N., & Ksibi, M. (2021). Removal of pharmaceuticals from water by adsorption and advanced oxidation processes: State of the art and trends. *Applied Sciences (Switzerland), 11*(14). https://doi.org/10.3390/app11146659.

Marchetti, M. D., & Bessa Azevedo, E. (2020). Degradation of NSAIDs by optimized photo-Fenton process using UV-LEDs at near-neutral pH. *Journal of Water Process Engineering, 35*. https://doi.org/10.1016/j.jwpe.2020.101171.

Mehrvar, M., Anderson, W. A., & Moo-Young, M. (2002). Photocatalytic degradation of aqueous organic solvents in the presence of hydroxyl radical scavengers. *International Journal of Photoenergy, 3*(4), 187–191. https://doi.org/10.1155/S1110662X01000241.

Michael, S. G., Michael-Kordatou, I., Nahim-Granados, S., Polo-López, M. I., Rocha, J., Martínez-Piernas, A. B., et al. (2020). Investigating the impact of UV-C/H_2O_2 and sunlight/H_2O_2 on the removal of antibiotics, antibiotic resistance determinants and toxicity present in urban wastewater. *Chemical Engineering Journal, 388*. https://doi.org/10.1016/j.cej.2020.124383.

Miklos, D. B., Remy, C., Jekel, M., Linden, K. G., Drewes, J. E., & Hübner, U. (2018). Evaluation of advanced oxidation processes for water and wastewater treatment—A critical review. *Water Research, 139*, 118–131. https://doi.org/10.1016/j.watres.2018.03.042.

Miranda-García, N., Suárez, S., Maldonado, M. I., Malato, S., & Sánchez, B. (2014). Regeneration approaches for TiO_2 immobilized photocatalyst used in the elimination of emerging contaminants in water. *Catalysis Today, 230*, 27–34. https://doi.org/10.1016/j.cattod.2013.12.048.

Molinari, R., Pirillo, F., Loddo, V., & Palmisano, L. (2006). Heterogeneous photocatalytic degradation of pharmaceuticals in water by using polycrystalline TiO_2 and a nanofiltration membrane reactor. *Catalysis Today, 118*(1–2), 205–213. https://doi.org/10.1016/j.cattod.2005.11.091.

Montemurro, N., García-Vara, M., Peña-Herrera, J. M., Llada, J., Barcela, D., & Pérez, S. (2018). Conventional and advanced processes for the removal of pharmaceuticals and their human metabolites from wastewater. In *Vol. 1302. ACS symposium series* (pp. 15–67). American Chemical Society. https://doi.org/10.1021/bk-2018-1302.ch002.

Monteoliva-García, A., Martín-Pascual, J., Muñío, M. M., & Poyatos, J. M. (2019). Removal of carbamazepine, ciprofloxacin and ibuprofen in real urban wastewater by using light-driven advanced oxidation processes. *International journal of Environmental Science and Technology, 16*(10), 6005–6018. https://doi.org/10.1007/s13762-019-02365-9.

Moreno-SanSegundo, J., Giannakis, S., Samoili, S., Farinelli, G., McGuigan, K. G., Pulgarín, C., et al. (2021). SODIS potential: A novel parameter to assess the suitability of solar water disinfection worldwide. *Chemical Engineering Journal, 419*. https://doi.org/10.1016/j.cej.2021.129889.

Mostrąg, A., Puzyn, T., & Haranczyk, M. (2010). Modeling the overall persistence and environmental mobility of sulfur-containing polychlorinated organic compounds. *Environmental Science and Pollution Research, 17*(2), 470–477. https://doi.org/10.1007/s11356-009-0257-7.

Mugunthan, E., Saidutta, M. B., & Jagadeeshbabu, P. E. (2018). Visible light assisted photocatalytic degradation of diclofenac using TiO_2-WO_3 mixed oxide catalysts. *Environmental Nanotechnology, Monitoring and Management, 10*, 322–330. https://doi.org/10.1016/j.enmm.2018.07.012.

Neta, P., Huie, R. E., & Ross, A. B. (1988). Rate constants for reactions of inorganic radicals in aqueous solution. *Journal of Physical and Chemical Reference Data, 17*(3), 1027–1284. https://doi.org/10.1063/1.555808.

Nguyen, T. M. H., Suwan, P., Koottatep, T., & Beck, S. E. (2019). Application of a novel, continuous-feeding ultraviolet light emitting diode (UV-LED) system to disinfect domestic wastewater for discharge or agricultural reuse. *Water Research, 153*, 53–62. https://doi.org/10.1016/j.watres.2019.01.006.

O'Flynn, D., Lawler, J., Yusuf, A., Parle-Mcdermott, A., Harold, D., Mc Cloughlin, T., et al. (2021). A review of pharmaceutical occurrence and pathways in the aquatic environment in the context of a changing climate and the COVID-19 pandemic. *Analytical Methods, 13*(5), 575–594. https://doi.org/10.1039/d0ay02098b.

OECD. (2019). Pharmaceutical consumption. In *Health at a glance 2019: OECD indicators*. Paris: OECD Publishing. https://doi.org/10.1787/43146d4b-en.

Ollis, D. F. (2000). Photocatalytic purification and remediation of contaminated air and water. *Comptes Rendus de l'Academie Des Sciences - Series IIc: Chemistry, 3*(6), 405–411. https://doi.org/10.1016/S1387-1609(00)01169-5.

Oturan, M. A., & Aaron, J. J. (2014). Advanced oxidation processes in water/wastewater treatment: Principles and applications. A review. *Critical Reviews in Environmental Science and Technology, 44*(23), 2577–2641. https://doi.org/10.1080/10643389.2013.829765.

Pan, X., Zhang, N., Fu, X., & Xu, Y. J. (2013). Selective oxidation of benzyl alcohol over TiO_2 nanosheets with exposed {001} facets: Catalyst deactivation and regeneration. *Applied Catalysis A: General, 453*, 181–187. https://doi.org/10.1016/j.apcata.2012.12.023.

Patel, M., Kumar, R., Kishor, K., Mlsna, T., Pittman, C. U., & Mohan, D. (2019). Pharmaceuticals of emerging concern in aquatic systems: Chemistry, occurrence, effects, and removal methods. *Chemical Reviews, 119*(6), 3510–3673. https://doi.org/10.1021/acs.chemrev.8b00299.

Patel, S., Majumder, S. K., Mondal, S., Das, P., & Ghosh, P. (2020). Treatment of a pharmaceutical industrial effluent by a hybrid process of advanced oxidation and adsorption. *ACS Omega*, *5*(50), 32305–32317. https://doi.org/10.1021/acsomega.0c04139.

Patrício Silva, A. L., Prata, J. C., Walker, T. R., Duarte, A. C., Ouyang, W., Barcelò, D., et al. (2021). Increased plastic pollution due to COVID-19 pandemic: Challenges and recommendations. *Chemical Engineering Journal*, *405*. https://doi.org/10.1016/j.cej.2020.126683.

Perret, D., Gentili, A., Marchese, S., Greco, A., & Curini, R. (2006). Sulphonamide residues in Italian surface and drinking waters: A small scale reconnaissance. *Chromatographia*, *63* (5–6), 225–232. https://doi.org/10.1365/s10337-006-0737-6.

Poyatos, J. M., Muñio, M. M., Almecija, M. C., Torres, J. C., Hontoria, E., & Osorio, F. (2010). Advanced oxidation processes for wastewater treatment: State of the art. *Water, Air, and Soil Pollution*, *205*(1–4), 187–204. https://doi.org/10.1007/s11270-009-0065-1.

Qi, M., Yang, Y., Zhang, X., Zhang, X., Wang, M., Zhang, W., et al. (2020). Pollution reduction and operating cost analysis of municipal wastewater treatment in China and implication for future wastewater management. *Journal of Cleaner Production*, *253*. https://doi.org/10.1016/j.jclepro.2020.120003.

Raji, A., Pandiyaraj, K. N., Kandavelu, V., Vasu, D., & Saravanan, D. (2020). Efficiency evaluation of the photocatalytic degradation of telmisartan anti-hypertensive drug with Fenton, photo-Fenton and recyclable TiO_2 heterogeneous catalyst. *Reaction Kinetics, Mechanisms and Catalysis*, *130*(2), 1141–1154. https://doi.org/10.1007/s11144-020-01806-8.

Rao, Y. F., & Chu, W. (2009). A new approach to quantify the degradation kinetics of linuron with UV, ozonation and UV/O_3 processes. *Chemosphere*, *74*(11), 1444–1449. https://doi.org/10.1016/j.chemosphere.2008.12.012.

Ravina, M., Campanella, L., & Kiwi, J. (2002). Accelerated mineralization of the drug diclofenac via Fenton reactions in a concentric photo-reactor. *Water Research*, *36*(14), 3553–3560. https://doi.org/10.1016/S0043-1354(02)00075-1.

Rekhate, C. V., & Srivastava, J. K. (2020). Recent advances in ozone-based advanced oxidation processes for treatment of wastewater – A review. *Chemical Engineering Journal Advances*, *3*, 100031. https://doi.org/10.1016/j.ceja.2020.100031.

Ribeiro, A. R., Nunes, O. C., Pereira, M. F. R., & Silva, A. M. T. (2015). An overview on the advanced oxidation processes applied for the treatment of water pollutants defined in the recently launched directive 2013/39/EU. *Environment International*, *75*, 33–51. https://doi.org/10.1016/j.envint.2014.10.027.

Richardson, S. D. (2004). Environmental mass spectrometry: Emerging contaminants and current issues. *Analytical Chemistry*, *76*(12), 3337–3364. https://doi.org/10.1021/ac040060d.

Richardson, S. D., & Ternes, T. A. (2005). Water analysis: Emerging contaminants and current issues. *Analytical Chemistry*, *77*(12), 3807–3838. https://doi.org/10.1021/ac058022x.

Rodriguez-Mozaz, S., Vaz-Moreira, I., Varela Della Giustina, S., Llorca, M., Barceló, D., Schubert, S., et al. (2020). Antibiotic residues in final effluents of European wastewater treatment plants and their impact on the aquatic environment. *Environment International*, *140*. https://doi.org/10.1016/j.envint.2020.105733.

Rogowska, J., Zimmermann, A., Muszyńska, A., Ratajczyk, W., & Wolska, L. (2019). Pharmaceutical household waste practices: Preliminary findings from a case study in Poland. *Environmental Management*, *64*(1), 97–106. https://doi.org/10.1007/s00267-019-01174-7.

Romera-Castillo, C., & Jaffé, R. (2015). Free radical scavenging (antioxidant activity) of natural dissolved organic matter. *Marine Chemistry*, *177*, 668–676. https://doi.org/10.1016/j.marchem.2015.10.008.

Rueda-Marquez, J. J., Levchuk, I., Fernández Ibañez, P., & Sillanpää, M. (2020). A critical review on application of photocatalysis for toxicity reduction of real wastewaters. *Journal of Cleaner Production*, *258*. https://doi.org/10.1016/j.jclepro.2020.120694.

Rueda-Márquez, J. J., Levchuk, I., Manzano, M., & Sillanpää, M. (2020). Toxicity reduction of industrial and municipal wastewater by advanced oxidation processes (photo-Fenton, UVC/H_2O_2, electro-Fenton and galvanic Fenton): A review. *Catalysts*, *10*(6). https://doi.org/10.3390/catal10060612.

Saghafinia, M. S., Emadian, S. M., & Vossoughi, M. (2011). Performances evaluation of photo-Fenton process and Sonolysis for the treatment of penicillin G formulation effluent. In *Vol. 8. Procedia environmental sciences* (pp. 202–208). Elsevier B.V. https://doi.org/10.1016/j.proenv.2011.10.033.

Salgado, R., Marques, R., Noronha, J. P., Carvalho, G., Oehmen, A., & Reis, M. A. M. (2012). Assessing the removal of pharmaceuticals and personal care products in a full-scale activated sludge plant. *Environmental Science and Pollution Research*, *19*(5), 1818–1827. https://doi.org/10.1007/s11356-011-0693-z.

Sanabria, P., Scunderlick, D., Wilde, M. L., Lüdtke, D. S., & Sirtori, C. (2021). Solar photo-Fenton treatment of the anti-cancer drug anastrozole in different aqueous matrices at near-neutral pH: Transformation products identification, pathways proposal, and in silico (Q)SAR risk assessment. *Science of the Total Environment*, *754*. https://doi.org/10.1016/j.scitotenv.2020.142300.

Saravanan, A., Senthil Kumar, P., Jeevanantham, S., Karishma, S., Tajsabreen, B., Yaashikaa, P. R., et al. (2021). Effective water/wastewater treatment methodologies for toxic pollutants removal: Processes and applications towards sustainable development. *Chemosphere*, *280*. https://doi.org/10.1016/j.chemosphere.2021.130595.

Seridou, P., & Kalogerakis, N. (2021). Disinfection applications of ozone micro- and nanobubbles. *Environmental Science: Nano*, *8*(12), 3493–3510. https://doi.org/10.1039/d1en00700a.

Serna-Galvis, E. A., Giraldo-Aguirre, A. L., Silva-Agredo, J., Flórez-Acosta, O. A., & Torres-Palma, R. A. (2017). Removal of antibiotic cloxacillin by means of electrochemical oxidation, TiO_2 photocatalysis, and photo-Fenton processes: Analysis of degradation pathways and effect of the water matrix on the elimination of antimicrobial activity. *Environmental Science and Pollution Research*, *24*(7), 6339–6352. https://doi.org/10.1007/s11356-016-6257-5.

Sgroi, M., Anumol, T., Vagliasindi, F. G. A., Snyder, S. A., & Roccaro, P. (2021). Comparison of the new Cl_2/O_3/UV process with different ozone- and UV-based AOPs for wastewater treatment at pilot scale: Removal of pharmaceuticals and changes in fluorescing organic matter. *Science of the Total Environment*, *765*. https://doi.org/10.1016/j.scitotenv.2020.142720.

Shemer, H., Kunukcu, Y. K., & Linden, K. G. (2006). Degradation of the pharmaceutical metronidazole via UV, Fenton and photo-Fenton processes. *Chemosphere*, *63*(2), 269–276. https://doi.org/10.1016/j.chemosphere.2005.07.029.

Silva, A. R., Martins, P. M., Teixeira, S., Carabineiro, S. A. C., Kuehn, K., Cuniberti, G., et al. (2016). Ciprofloxacin wastewater treated by UVA photocatalysis: Contribution of irradiated TiO_2 and ZnO nanoparticles on the final toxicity as assessed by *Vibrio fischeri*. *RSC Advances*, *6*(98), 95494–95503. https://doi.org/10.1039/c6ra19202e.

Sirtori, C., Agüera, A., Carra, I., & Sanchéz Pérez, J. A. (2014). Identification and monitoring of thiabendazole transformation products in water during Fenton degradation by LC-QTOF-MS. *Analytical and Bioanalytical Chemistry*, *406*(22), 5323–5337. https://doi.org/10.1007/s00216-014-7942-1.

Sirtori, C., Zapata, A., Oller, I., Gernjak, W., Agüera, A., & Malato, S. (2009). Solar photo-Fenton as finishing step for biological treatment of a pharmaceutical wastewater. *Environmental Science and Technology*, *43*(4), 1185–1191. https://doi.org/10.1021/es802550y.

Skoumal, M., Cabot, P. L., Centellas, F., Arias, C., Rodríguez, R. M., Garrido, J. A., et al. (2006). Mineralization of paracetamol by ozonation catalyzed with Fe2+, Cu2+ and UVA light. *Applied Catalysis B: Environmental*, *66*(3–4), 228–240. https://doi.org/10.1016/j.apcatb.2006.03.016.

Studziński, W., Gackowska, A., & Kudlek, E. (2021). Determination of environmental properties and toxicity of octyl-dimethyl-Para-aminobenzoic acid and its degradation products. *Journal of Hazardous Materials*, *403*. https://doi.org/10.1016/j.jhazmat.2020.123856.

Sui, Q., Huang, J., Lu, S., Deng, S., Wang, B., Zhao, W., et al. (2014). Removal of pharmaceutical and personal care products by sequential ultraviolet and ozonation process in a full-scale wastewater treatment plant. *Frontiers of Environmental Science & Engineering*, *8*(1), 62–68. https://doi.org/10.1007/s11783-013-0518-z.

Sun, P., Meng, T., Wang, Z., Zhang, R., Yao, H., Yang, Y., et al. (2019). Degradation of organic micropollutants in UV/NH_2 Cl advanced oxidation process. *Environmental Science & Technology*, *53*(15), 9024–9033. https://doi.org/10.1021/acs.est.9b00749.

Szabó, L., Tóth, T., Homlok, R., Takács, E., & Wojnárovits, L. (2012). Radiolysis of paracetamol in dilute aqueous solution. *Radiation Physics and Chemistry*, *81*(9), 1503–1507. https://doi.org/10.1016/j.radphyschem.2011.11.036.

Szabó, R. K., Megyeri, C. S., Illés, E., Gajda-Schrantz, K., Mazellier, P., & Dombi, A. (2011). Phototransformation of ibuprofen and ketoprofen in aqueous solutions. *Chemosphere*, *84* (11), 1658–1663. https://doi.org/10.1016/j.chemosphere.2011.05.012.

Tang, Y., Shi, X., Liu, Y., Feng, L., & Zhang, L. (2018). Degradation of clofibric acid in UV/chlorine disinfection process: Kinetics, reactive species contribution and pathways. *Royal Society Open Science*, *5*(2), 171372. https://doi.org/10.1098/rsos.171372.

Tao, Y., Cai, J., Huai, X., & Liu, B. (2018). A novel antibiotic wastewater degradation technique combining cavitating jets impingement with multiple synergetic methods. *Ultrasonics Sonochemistry*, *44*, 36–44. https://doi.org/10.1016/j.ultsonch.2018.02.008.

Taylor, D. (2016). The pharmaceutical industry and the future of drug development. *Issues in Environmental Science and Technology*, *2016*(41), 1–33. https://doi.org/10.1039/9781782622345-00001.

Tian, F. X., Ye, W. K., Xu, B., Hu, X. J., Ma, S. X., Lai, F., et al. (2020). Comparison of UV-induced AOPs (UV/Cl_2, UV/NH_2Cl, UV/ClO_2 and UV/H_2O_2) in the degradation of iopamidol: Kinetics, energy requirements and DBPs-related toxicity in sequential disinfection processes. *Chemical Engineering Journal*, *398*. https://doi.org/10.1016/j.cej.2020.125570.

Tiwari, B., Sellamuthu, B., Ouarda, Y., Drogui, P., Tyagi, R. D., & Buelna, G. (2017). Review on fate and mechanism of removal of pharmaceutical pollutants from wastewater using biological approach. *Bioresource Technology*, *224*, 1–12. https://doi.org/10.1016/j.biortech.2016.11.042.

Valari, M., Antoniadis, A., Mantzavinos, D., & Poulios, I. (2015). Photocatalytic reduction of Cr(VI) over titania suspensions. *Catalysis Today*, *252*, 190–194. https://doi.org/10.1016/j.cattod.2014.10.014.

Velempini, T., Prabakaran, E., & Pillay, K. (2021). Recent developments in the use of metal oxides for photocatalytic degradation of pharmaceutical pollutants in water—A review. *Materials Today Chemistry*, *19*. https://doi.org/10.1016/j.mtchem.2020.100380.

Von Gunten, U. (2018). Oxidation processes in water treatment: Are we on track? *Environmental Science and Technology*, *52*(9), 5062–5075. https://doi.org/10.1021/acs.est.8b00586.

Wang, A. Q., Xu, B., Zhang, T. Y., Chen, Y. Y., & Gao, N. Y. (2017). Effect of UV irradiation and UV/chlorine processes on trichloronitromethane formation during chlorination of Ronidazole. *Clean: Soil, Air, Water*, *45*(6). https://doi.org/10.1002/clen.201600163.

Wang, C., Ying, Z., Ma, M., Huo, M., & Yang, W. (2019). Degradation of micropollutants by UV-chlorine treatment in reclaimedwater: pH effects, formation of disinfectant byproducts, and toxicity assay. *Water (Switzerland)*, *11*(12). https://doi.org/10.3390/W11122639.

Wang, H., Su, L., Zhu, S., Zhu, W., Han, X., Cheng, Y., et al. (2019). Optimization of the electro-peroxone process for micropollutant abatement using chemical kinetic approaches. *Molecules*, *24*(14). https://doi.org/10.3390/molecules24142638.

Wang, J., Chu, L., Wojnárovits, L., & Takács, E. (2020). Occurrence and fate of antibiotics, antibiotic resistant genes (ARGs) and antibiotic resistant bacteria (ARB) in municipal wastewater treatment plant: An overview. *Science of the Total Environment*, *744*. https://doi.org/10.1016/j.scitotenv.2020.140997.

Wang, J., & Wang, S. (2016). Removal of pharmaceuticals and personal care products (PPCPs) from wastewater: A review. *Journal of Environmental Management*, *182*, 620–640. https://doi.org/10.1016/j.jenvman.2016.07.049.

Wang, J., & Wang, S. (2018). Activation of persulfate (PS) and peroxymonosulfate (PMS) and application for the degradation of emerging contaminants. *Chemical Engineering Journal*, *334*, 1502–1517. https://doi.org/10.1016/j.cej.2017.11.059.

Wang, J., & Wang, S. (2020). Reactive species in advanced oxidation processes: Formation, identification and reaction mechanism. *Chemical Engineering Journal*, *401*. https://doi.org/10.1016/j.cej.2020.126158.

Wang, J., & Zhuan, R. (2020). Degradation of antibiotics by advanced oxidation processes: An overview. *Science of the Total Environment*, *701*. https://doi.org/10.1016/j.scitotenv.2019.135023.

Wang, W. L., Wu, Q. Y., Huang, N., Wang, T., & Hu, H. Y. (2016). Synergistic effect between UV and chlorine (UV/chlorine) on the degradation of carbamazepine: Influence factors and radical species. *Water Research*, *98*, 190–198. https://doi.org/10.1016/j.watres.2016.04.015.

Wang, X., Jia, J., & Wang, Y. (2017). Combination of photocatalysis with hydrodynamic cavitation for degradation of tetracycline. *Chemical Engineering Journal*, *315*, 274–282. https://doi.org/10.1016/j.cej.2017.01.011.

Wang, Y., Zhu, C., Zuo, G., Guo, Y., Xiao, W., Dai, Y., et al. (2020). 0D/2D Co_3O_4/TiO_2 Z-scheme heterojunction for boosted photocatalytic degradation and mechanism investigation. *Applied Catalysis B: Environmental*, *278*. https://doi.org/10.1016/j.apcatb.2020.119298.

Wang, Y., Zhang, X., Liu, J., Wang, Y., Duan, D., & Fan, C. (2015). Facile regeneration and photocatalytic activity of CuO-modified silver bromide photocatalyst. *Materials Science in Semiconductor Processing*, *40*, 613–620. https://doi.org/10.1016/j.mssp.2015.07.036.

Wardman, P. (1989). Reduction potentials of one electron couples involving free radicals in aqueous solution. *Journal of Physical and Chemical Reference Data*, *18*(4), 1637–1755. https://doi.org/10.1063/1.555843.

Watts, M. J., & Linden, K. G. (2007). Chlorine photolysis and subsequent OH radical production during UV treatment of chlorinated water. *Water Research*, *41*(13), 2871–2878. https://doi.org/10.1016/j.watres.2007.03.032.

Wegmann, F., Cavin, L., MacLeod, M., Scheringer, M., & Hungerbühler, K. (2009). The OECD software tool for screening chemicals for persistence and long-range transport potential.

Environmental Modelling and Software, *24*(2), 228–237. https://doi.org/10.1016/j.envsoft.2008.06.014.

Wu, Z., Yuste-Córdoba, F. J., Cintas, P., Wu, Z., Boffa, L., Mantegna, S., et al. (2018). Effects of ultrasonic and hydrodynamic cavitation on the treatment of cork wastewater by flocculation and Fenton processes. *Ultrasonics Sonochemistry*, *40*, 3–8. https://doi.org/10.1016/j.ultsonch.2017.04.016.

Xiang, J., Wu, M., Lei, J., Fu, C., Gu, J., & Xu, G. (2018). The fate and risk assessment of psychiatric pharmaceuticals from psychiatric hospital effluent. *Ecotoxicology and Environmental Safety*, *150*, 289–296. https://doi.org/10.1016/j.ecoenv.2017.12.049.

Yan, S., & Song, W. (2014). Photo-transformation of pharmaceutically active compounds in the aqueous environment: A review. *Environmental Science: Processes & Impacts*, *16*(4), 697–720. https://doi.org/10.1039/c3em00502j.

Yang, W., Zhou, H., & Cicek, N. (2014). Treatment of organic micropollutants in water and wastewater by UV-based processes: A literature review. *Critical Reviews in Environmental Science and Technology*, *44*(13), 1443–1476. https://doi.org/10.1080/10643389.2013.790745.

Yang, X., Sun, J., Fu, W., Shang, C., Li, Y., Chen, Y., et al. (2016). PPCP degradation by UV/chlorine treatment and its impact on DBP formation potential in real waters. *Water Research*, *98*, 309–318. https://doi.org/10.1016/j.watres.2016.04.011.

Yang, Y., Lu, X., Jiang, J., Ma, J., Liu, G., Cao, Y., et al. (2017). Degradation of sulfamethoxazole by UV, UV/H_2O_2 and UV/persulfate (PDS): Formation of oxidation products and effect of bicarbonate. *Water Research*, *118*, 196–207. https://doi.org/10.1016/j.watres.2017.03.054.

Yang, Y., Pignatello, J. J., Ma, J., & Mitch, W. A. (2016). Effect of matrix components on UV/H_2O_2 and UV/S2O82-advanced oxidation processes for trace organic degradation in reverse osmosis brines from municipal wastewater reuse facilities. *Water Research*, *89*, 192–200. https://doi.org/10.1016/j.watres.2015.11.049.

Yao, W., Wang, X., Yang, H., Yu, G., Deng, S., Huang, J., et al. (2016). Removal of pharmaceuticals from secondary effluents by an electro-peroxone process. *Water Research*, *88*, 826–835. https://doi.org/10.1016/j.watres.2015.11.024.

Yaqoob, A. A., Parveen, T., Umar, K., & Ibrahim, M. N. M. (2020). Role of nanomaterials in the treatment of wastewater: A review. *Water (Switzerland)*, *12*(2). https://doi.org/10.3390/w12020495.

Yi, C., Song, L., Wu, Q., Li, Z., Zhang, W., & Yin, K. (2020). Enhanced photodegradation of diphenhydramine in aqueous solution containing natural sand particles. *RSC Advances*, *10*(29), 17228–17234. https://doi.org/10.1039/d0ra02019b.

Yin, J., & Zhang, X. (2019). Technologies for bHRPs and risk control. In *High-risk pollutants in wastewater* (pp. 237–258). Elsevier. https://doi.org/10.1016/B978-0-12-816448-8.00010-1.

Yuan, F., Hu, C., Hu, X., Wei, D., Chen, Y., & Qu, J. (2011). Photodegradation and toxicity changes of antibiotics in UV and UV/H_2O_2 process. *Journal of Hazardous Materials*, *185*(2–3), 1256–1263. https://doi.org/10.1016/j.jhazmat.2010.10.040.

Zaccariello, G., Moretti, E., Storaro, L., Riello, P., Canton, P., Gombac, V., et al. (2014). TiO_2–mesoporous silica nanocomposites: Cooperative effect in the photocatalytic degradation of dyes and drugs. *RSC Advances*, *4*(71), 37826–37837. https://doi.org/10.1039/C4RA06767C.

Zhang, R., Meng, T., Huang, C. H., Ben, W., Yao, H., Liu, R., et al. (2018). PPCP degradation by chlorine-UV processes in ammoniacal water: New reaction insights, kinetic modeling, and DBP formation. *Environmental Science and Technology*, *52*(14), 7833–7841. https://doi.org/10.1021/acs.est.8b00094.

Zhang, Y., Zhu, Y., Shao, Y., Rong, C., Pan, Z., & Deng, J. (2021). Toxicity of disinfection byproducts formed during the chlorination of sulfamethoxazole, norfloxacin, and 17β-estradiol in the presence of bromide. *Environmental Science and Pollution Research*, *28*(36), 50718–50730. https://doi.org/10.1007/s11356-021-14161-5.

Zheng, X., Xu, S., Wang, Y., Sun, X., Gao, Y., & Gao, B. (2018). Enhanced degradation of ciprofloxacin by graphitized mesoporous carbon (GMC)-TiO_2 nanocomposite: Strong synergy of adsorption-photocatalysis and antibiotics degradation mechanism. *Journal of Colloid and Interface Science*, *527*, 202–213. https://doi.org/10.1016/j.jcis.2018.05.054.

Zhou, S., Xia, Y., Li, T., Yao, T., Shi, Z., Zhu, S., et al. (2016). Degradation of carbamazepine by UV/chlorine advanced oxidation process and formation of disinfection by-products. *Environmental Science and Pollution Research*, *23*(16), 16448–16455. https://doi.org/10.1007/s11356-016-6823-x.

Zhou, Y., Wang, X., Zhu, C., Dionysiou, D. D., Zhao, G., Fang, G., et al. (2018). New insight into the mechanism of peroxymonosulfate activation by sulfur-containing minerals: Role of sulfur conversion in sulfate radical generation. *Water Research*, *142*, 208–216. https://doi.org/10.1016/j.watres.2018.06.002.

Zoschke, K., Börnick, H., & Worch, E. (2014). Vacuum-UV radiation at 185nm in water treatment—A review. *Water Research*, *52*, 131–145. https://doi.org/10.1016/j.watres.2013.12.034.

Zuo, Y., Zhang, K., & Zhou, S. (2013). Determination of estrogenic steroids and microbial and photochemical degradation of 17α-ethinylestradiol (EE2) in lake surface water, a case study. *Environmental Science: Processes & Impacts*, *15*(8), 1529–1535. https://doi.org/10.1039/c3em00239j.

7 Treatment innovation using biological methods in combination with physical treatment methods

Azmatullah Noor[a,b], *Shamsul Rahman Mohamed Kutty*[a,b], *Mohamed Hasnain Isa*[c], *Izharul Haq Farooqi*[d], *Augustine Chioma Affam*[e], *Abdullahi Haruna Birniwa*[f], and *Ahmad Hussaini Jagaba*[a,b,g]
[a]Department of Civil and Environmental Engineering, Universiti Teknologi PETRONAS, Bandar Seri Iskandar, Perak Darul Ridzuan, Malaysia, [b]Centre of Urban Resource Sustainability, Institute of Self-Sustainable Building, Universiti Teknologi PETRONAS, Bandar Seri Iskandar, Perak Darul Ridzuan, Malaysia, [c]Civil Engineering Programme, Faculty of Engineering, Universiti Teknologi Brunei, Gadong, Brunei Darussalam, [d]Department of Civil Engineering, Zakir Husain College of Engineering and Technology, Aligarh Muslim University, Aligarh, Uttar Pradesh, India, [e]Civil Engineering Department, School of Engineering and Technology, University of Technology Sarawak, Sibu, Sarawak, Malaysia, [f]Department of Chemistry, Sule Lamido University, Kafin-Hausa, Nigeria, [g]Department of Civil Engineering, Abubakar Tafawa Balewa University, Bauchi, Nigeria

1. Introduction

The pharmaceutical chemical industry's rapid growth and the widespread use of healthcare items in human medical treatment, veterinary and agriculture have increased pharmaceutical production leading to the generation of a huge amount of toxic wastewater labeled "pharmaceutical wastewater (PWW)" (Khan, Khan, Tiwari, Zubair, & Naushad, 2021). A huge amount of water is utilized to sterilize and wash the relevant equipment throughout the pharmaceutics production chain, which is then dumped as wastewater (Mohammed et al., 2021; Rashid et al., 2021). Different type of water used for manufacturing various pharmaceutical products is summarized in Table 1. PWW exhibits complex characteristics that cause serious pollutions to the water environment (Khan, Khan, Zubair, Azfar Shaida, et al., 2021). Chemical oxygen demand (COD), ammonia-nitrogen (NH_4^+-N), phosphate, salinity, color, and heavy metals are frequently present in significant concentrations in this type of industrial wastewater (Zhang et al., 2013). Thus, the organizations responsible for environmental regulation such as European Union (EU) Water Framework Directive, US Environmental Protection Agency (US EPA), and World Health Organization (WHO) studied the issues occurring due to PWW discharge into

The Treatment of Pharmaceutical Wastewater. https://doi.org/10.1016/B978-0-323-99160-5.00010-2

Table 1 Application of various type of water for manufacturing of pharmaceutical products.

Type of product	Category	Subcategory	Type of water used
Active pharmaceutical ingredients	Intermediates	NA	Portable
	Final product		Purified
Final dosage form	Sterile medicine product	Ophthalmic product	Portable
		Nasal/ear preparations	
		Cutaneous preparations	
		Biologics (incl. vaccines)	Water for injections
		Parenteral products	
		Hemodiafiltration solutions	
		Hemodiafiltration solutions	
		Peritoneal dialysis solution	
		Irrigation solutions	
	Nonsterile medicinal product	Nonparental vaccines	Purified
		Oral presentations	
		Nebulizer solution	
		Cutaneous preparation	
		Rectal/vaginal preparations	
		Nasal/ear preparations	
		Herbal drugs extracts	Extract preparation water

water bodies. US EPA and EU Water Framework Directive enlisted approximately 130 and 35 compounds, respectively, as compounds of emerging concern (Khan, Aziz, Khan, Dhingra, et al., 2021).

Pharmaceutical metabolites can enter water bodies through a variety of routes, both direct and indirect (Fig. 1). Following the use of pharmaceutical items, pharmaceutical remnants can be discharged into soil, surface, and groundwater by living organisms defecation (Khan, Abutaleb, Khan, et al., 2021). Furthermore, because traditional water treatment was not really designed especially to remove medications, pharmaceutical metabolites might still be found in drinking water following treatment (Khan, Tirth, Fawzy, et al., 2021). X-ray contrast media, antibiotics, beta-blockers, anticonvulsants, and cytostatic medicines are among the pharmaceuticals available (Rosman et al., 2018). The studies conducted for the treatment of PWW indicate that the current wastewater treatment processes are incapable of removing pharmaceutical contaminants efficiently (Ji et al., 2013). The conventional treatment process can only eliminate the organics, nutrients, and suspended solids to meet the permissible discharge limits (Zhao et al., 2019). Consequently, the effluent from municipal

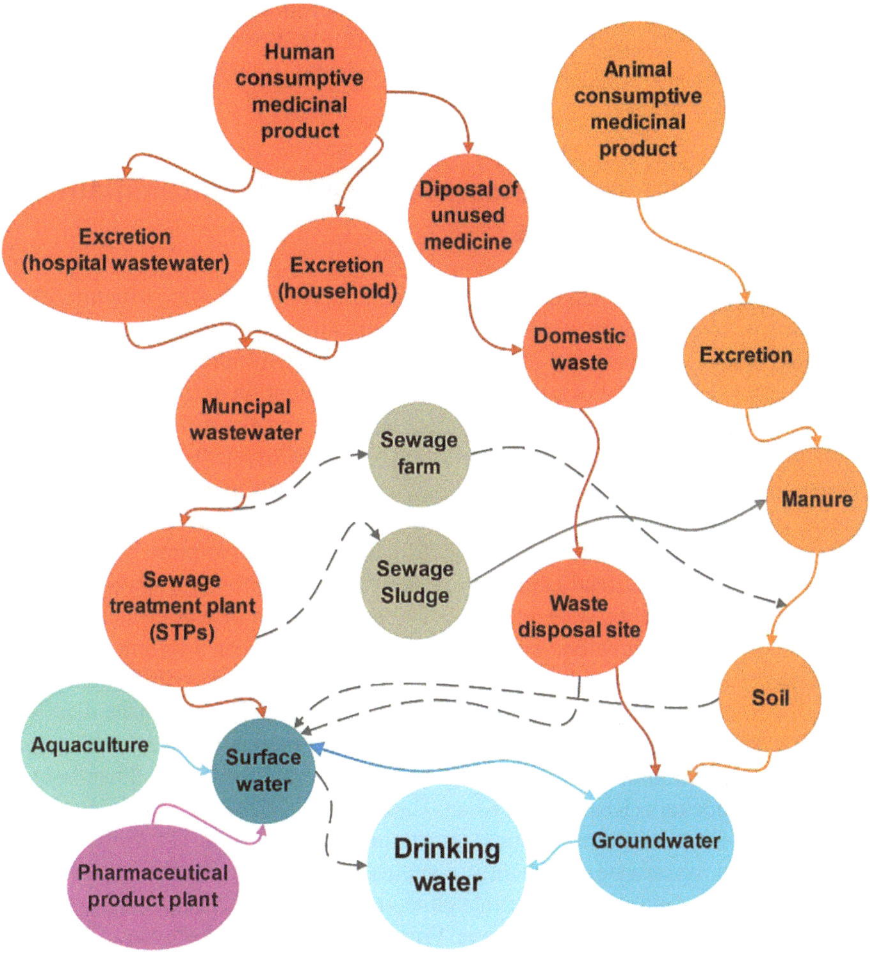

Fig. 1 Schematic representation of likely sources and paths for pharmaceutical residues presence in aquatic environment.

wastewater treatment plants is discharged into the water bodies with high loads of pharmaceutical contaminants (Khan et al., 2020). The residue of veterinary medicines can leach into the ground, through the application of fertilizers from animal excreta (Chen et al., 2020), contaminating the groundwater which is potentially used for drinking (Zhao et al., 2007).

The other reason for the ineffective treatment is due to antibiotic resistant bacteria (ARB) and genes growth because of the persistent release of antibiotics into water bodies. Presently, the effective treatment of PWW is the key concern. Much research has been conducted on hybrid (biological treatment coupled with physical treatment) treatment technologies. The status of research on the biological treatment of PWW coupled with physical treatment technologies is a primary concern of discussion. Thus, this chapter will discuss the performance evaluation of hybrid technologies for the treatment of PWW. Moreover, the fate and proliferation of Antibiotic resistance genes (ARGs) during biological treatment of PWW is also highlighted. Additionally, the isolation of bacterial strains for the biodegradation of certain antibiotics is also stated.

2. Pharmaceutical wastewater: Existing contaminants, toxicity, and effects

Antibiotics, in particular, have an impact on microbial community structure in the environment. They've been widely utilized as a treatment to manage or lessen infectious infections in humans and animals, as well as growth hormones for cattle and poultry on farms. Antibiotics have gained increased attention as a class of emergent environmental hazards as a result of their excessive use and abuse (Husain Khan et al., 2020). They may have a negative influence on aquatic ecosystems and human health due to endocrine disruption, as well as a negative effect on the preservation or development of ARB (Khan, 2021). Degradation was found to be the best process for removing antibiotics in mass balance investigations, while sorption was found to be less important. Additional laboratory research is needed to investigate the important influencing elements, such as the design and operation of wastewater treatment plants' hydrologic parameters, on the fate and elimination of different antibiotics. Antibiotics have been found in a variety of urban, livestock, and agricultural effluent, as well as surface water (Hou, Wang, Mao, & Luo, 2016). The pharmaceutical contaminants are of four categories namely: antibiotics, hormones and steroids/estrogens, nonsteroidal antiinflammatory drugs (NSAIDs) (Amari & Boshrouyeh Ghandashtani, 2020), and β-blockers (Khan, Aziz, Khan, Hasan, et al., 2021). The categorical demonstration of different pharmaceutical contaminants is shown in Fig. 2. In addition, pharmaceutical wastewater frequently contains a variety of hazardous chemicals, including benzene, nitrobenzene, tetracycline, oxytetracycline, fluoxetine, carbamazepine, oxcarbazepine (Dwivedi, Morone, Pratape, Chakrabarti, & Pandey, 2017), fluoroquinolone (Guo et al., 2017), naphthalene, colchicine, progesterone, raloxifene, berberine, p-xylene (Wang, Wang, Li, & Li, 2015), lincosamides, aminoglycosides, macrolides, spectinomycin, norfloxacin, lincomycin, gemifloxacin, amoxicillin, streptomycin, levofloxacin, vancomycin, aureomycin, ampicillin, ciprofloxacin, chloromycetin, nonribosomal peptide antibiotic zinc bacitracin (zn-bc), tetrahydrofuran, nicotinamide, quinolones, penicillin g-na, procaine benzylpenicillin, butyl acetate, *n*-butyl alcohol, phenylacetic acid (Wang, Zhang, Tian, Sun, & Xie, 2007), ketoprofen (Ismail, Essam, Ragab, & Mourad, 2016), ibuprofen, omethoate, organonitriles (Li, Liu, Bai, Ohandja, & Wong, 2007), sulfamethoxazole/sulfamethazine (Azimi, Hassani, Darzi, & Borghei, 2017), phenoxyacetic acid, metronidazole, endotoxins, cephalosporins, diclofenac, β-lactam, persulfate, meloxicam, acetaminophen, PAHs, phenolic and heterocyclic compounds, estradiol valerate, ethinylestradiol, atenolol, chloramphenicol, polymyxin b sulfate, and kanamycin sulfate (Husain Khan et al., 2020). Due to their relative persistency in the environment, these chemical elements are inadequately biodegradable and cannot be entirely eliminated

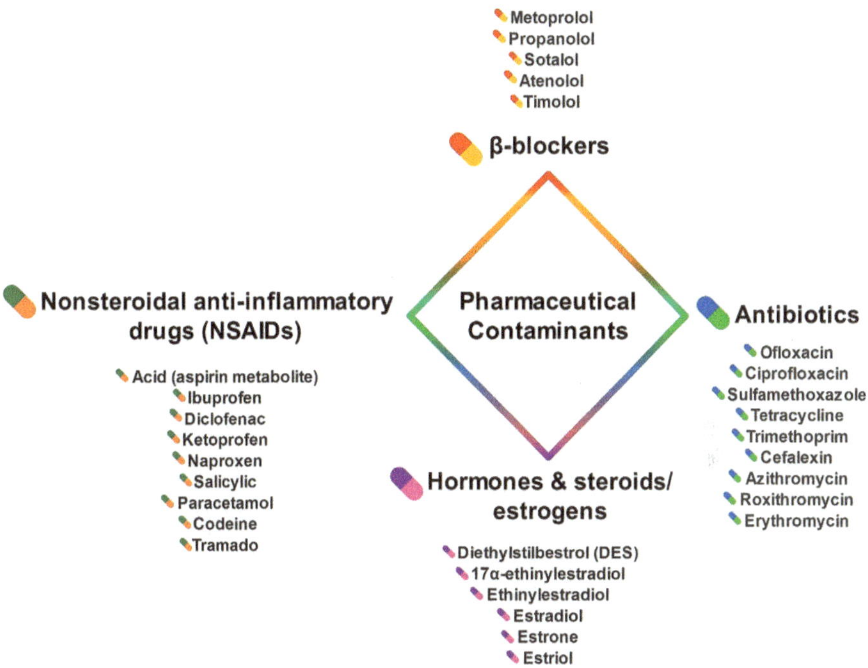

Fig. 2 Classification of different types of pharmaceutical contaminants.

from wastewater. Thus, amounting to secondary contaminants and ongoing toxicant release into aquatic habitats (Ji et al., 2013). The majority of these pollutants can enter the body through the respiratory tract, intestines, or skin. In animals, they can cause irritation of the central nervous system, narcosis, edema, and colonic muscular spasms (Chen, Zhou, Chen, & Jin, 2010).

Antibiotics present in PWW have indeed been identified as one of the major causes of the issues, as they can impede or even destroy anaerobic digesting bacteria (ADMs). As antibiotics are found in PWW and are not generated during the anaerobic digestion process, they are referred to as *exo*-toxicants, which are among the other major causes of anaerobic digestion worsening or even breakdown (Chen et al., 2020). Antibiotics that can coexist with two substrates in PWW but aren't used in the ANAMMOX process will be referred to as exogenous inhibitory chemicals from now on (Tang et al., 2011). Endogenous and exogenous inhibitory chemicals are significant variables that degrade or perhaps even interrupt the ANAMMOX process undergoing PWW treatment (Jagaba, Kutty, Lawal, Abubakar, et al., 2021).

2.1 Toxicity

The public's attention has recently been drawn to the toxicity of pharmaceutical wastewaters around the world. COD and NH_3-N were shown to be strongly and positively linked with toxicity (Noor, Kutty, Baloo, et al., 2021). Since pharmaceutical wastewater comprises undiscovered chemicals and compounds available at low concentrations, harmful compound analysis is costly, time-consuming, and needs high-tech technologies (Jagaba, Kutty, Isa, Affam, et al., 2021).

The standard tests with luminous bacteria, daphnids, and green algae, reveal acute and chronic toxicity (Zhao et al., 2007), are undoubtedly the most famous ecotoxicological procedures for wastewater testing. The Luminescent Bacteria Test is a simple and reliable tool for predicting the viability of the ANAMMOX procedure for the treatment of pharmaceutical wastewater. It has been discovered to be a rapid and sensitive tool for monitoring the toxicity of pharmaceutical wastewater discharge (Ismail & Habeeb, 2017; Jagaba, Kutty, Hayder, Baloo, Noor, et al., 2021b). Unidentified compounds that elicit additive, synergistic, or antagonistic effects can be detected using bioassays like the unicellular algal assay, which can provide comprehensive information about the harmful effects induced by all of the components in the wastewater. The necessity of evaluating the negative impacts of wastewaters on water bodies has been highlighted by the use of bioassays (Ji et al., 2013).

Distinct chemical tests for acute toxicity could be performed using a standard toxicity measuring technique stated in Relative Luminosity Units (RLU). The samples' mean luminescence unit (LU) and the controls' mean luminescence unit (LU0) (without inhibitory compounds) can determine the RLU value by using the following expression.

$$\mathrm{RLU} = \frac{\mathrm{LU}}{\mathrm{LU_0}} \times 100\% \tag{1}$$

By utilizing linear regression analysis of chemical concentrations and RLU, the doses that cause 50% inhibition of bioluminescence (IC50) for individual and joint acute toxicity experiments could be determined (Ding et al., 2015; Zhao et al., 2019). Pharmaceutical wastewater has been shown to cause reproductive damage in male mice. The number and size of Leydig cells were reduced in the testicles of treated mice, indicating an expansion of interstitial space (Zhao et al., 2007). The differences between different forms of pharmaceutical wastewater are numerous. To safeguard aquatic organisms as well as human health, specific and continuous monitoring actions should be taken to detect and forecast the consequences of wastewater toxicity during PWW treatment and to reduce the impact of effluent discharge on receiving waterways (Jagaba, Kutty, Hayder, Baloo, Ghaleb, et al., 2021a).

2.2 Effect

PWW is characterized by huge concentration of refractory chemicals, a complex water quality, and high salinity (Kutty et al., 2019). There are polar, bioactive, and persistent chemicals in the wastewater. It contains pathogenic germs, partially metabolized pharmaceuticals, radioactive elements, and other harmful chemical compounds. This effluent, which contains nonbiodegradable organic chemicals as well as a variety of antimicrobial agents, invariably results in plasmolysis and a loss of cell function, making treatment difficult (Ghaleb et al., 2020b). They may impede the growth of microorganisms in bioreactors, resulting in the deterioration of bacterial functions for the removal of organic matter and nutrients, unstable operation, and low system efficiency (Zhao et al., 2019). Because of the stubborn and poisonous chemicals, they may also disrupt biological treatment processes in sewage treatment facilities, resulting in the discharge of inadequately degraded contaminants into the atmosphere Jagaba, Kutty, Isa, et al. (2022). Pharmaceutical pollution has been linked to aquatic toxicity, pathogenic microorganisms developing resistance, genotoxicity, and endocrine disruption (Jagaba et al., 2020). As a result of the significant contamination capability of pharmaceutical wastewaters, pharmaceutics and their active components must be entirely removed from aquatic systems (Chen et al., 2020).

3. Pharmaceutical wastewater treatment

Technologies for pharmaceutical wastewater treatment ranges from the physical, chemical, biological, and integration of more than one method. Several existing physicochemical treatment methods such as coagulation-sedimentation, lime precipitation, flocculation, membrane filtration, adsorption, incineration, electrolysis, and oxidation are not very good in removing the large concentration of organic matter from the antibiotic wastewater (Almahbashi et al., 2021; Christofilopoulos, Syranidou, Gkavrou, Manousaki, & Kalogerakis, 2016). Because of their high cost and risk of creating new contaminants, they have a limited application. Furthermore, hazardous antibiotic residues in wastewater substantially obstruct treatment operations. As a result, scientists are looking at improved oxidation methods such as semiconductor photo-catalysis oxidation, supercritical water oxidation, Fenton reagent oxidation, ozone, and mixed oxidation, ferrate oxidation, and photochemical oxidation (Luo et al., 2019).

In response to increasingly strict effluent quality criteria, biological treatment processes that include aerobic membrane bioreactors (MBR), pressurized biochemical processes, up-flow anaerobic sludge blanket reactors, up-flow anaerobic packed bed reactor (Khan, Khan, et al., 2022), modified extended aeration, activated sludge processes, biological enhanced activated sludge (Wang, Wang, Xu, Li, & Huang, 2020), anaerobic-aerobic, anaerobic digestion processes, sequencing batch reactors (SBR), oxygen-ditch, hydrolysis-acidification, and combined biological technologies have been extensively studied (Jagaba, Kutty, Noor, Birniwa, et al., 2021). Biological treatment approaches have played a vital role in pharmaceutical wastewater treatment due to their high efficiencies and cheap rate (Guo, Yin, & Zhou, 2013). However, due to the wastewater's microbiological toxicity and weak biodegradability, it's hard to treat it directly with a biological method (Omar, Toemen, Muda, & Sulaiman, 2019). As a result, it must be pre-treated with a physicochemical approach (Noor, Kutty, Almahbashi, et al., 2021).

3.1 Biological treatment: Technologies and crisis

Biological treatment is most recommended due to its reduced cost and possibility for total mineralization. The main biological approach for treating pharmaceutical wastewater is the activated sludge process that utilizes bacterial metabolic functions for eliminating different kinds of organic pollutants and carbon (Alsubih, Khan, Khan, Khan, & Sharma, 2021; Srinivasan & Nambi, 2020). The predominant fate of organics is biodegradation. Thus, to boost biodegradability in biological treatment process, pharmaceutical effluent can be combined with food, industrial, and municipal wastes. Organics including organosilicon compounds, esters, alcohols, heterocycles, and acids have been found to break down well after biological treatment, as evidenced by a rise in the quantity of alkanes in the effluent (Khan, Bokhari, et al., 2022; Li, Wang, & Li, 2021). Organic substances, on the other hand, can be found in both the influent and effluent. This is due to the fact that some pharmaceutical compounds are organic molecules with complex structures that are resistant to biological breakdown (Jagaba, Kutty, Hayder, Elsadig, et al., 2021). As a result, prior to being discharged into aquatic ecosystems, the study of their transformation products is advised (Luo et al., 2019). The vulnerability of many connected growth systems to shear pressures in the system limits their performance, resulting in inconsistent findings.

3.2 Treatment technologies

Microorganisms use organic substances in wastewater to extract nutrients and organic matter under aerobic conditions. SBR and deep well aeration (DWA) are the most used aerobic biological treatment techniques for pharmaceutical wastewater. Compared to traditional aeration, DWA has a greater oxygen utilization rate and activated sludge loading (Jagaba, Kutty, Fauzi, Razali, et al., 2021). As a result, aerobic processes consume heaps of oxygen and generate too much sludge. Antibiotic residues also have an inhibiting effect on aerobic microorganisms. Therefore, when using aerobic treatments, wastewater must always be pretreated (Guo et al., 2013). Anaerobic systems can handle high volumetric loads, minimize air pollution, need fewer nutrients and energy, produce minimal sludge, and gather methane when compared to aerobic methods (Ghaleb et al., 2020a). Up-flow anaerobic sludge-bed (UASB) and up-flow blanket filter are the two main anaerobic techniques utilized in PWW treatment (UBF) (Jagaba, Kutty, Baloo, Noor, et al., 2021). UBF combines the advantages of membrane bioreactors and activated sludge. Huge amounts of biomass, rapid velocity, good separation impacts and high loading, are all advantages. As a result, they've been widely used in the treatment of nonbiodegradable industrial effluents (Saeed et al., 2021). UBF proved to be more efficient than UASB at treating large volumetric wastewater with high SS concentrations. Anaerobic reactors, on the other hand, take considerably more time to start-up and are more sensitive to low temperatures and harmful compounds. Simultaneously, the COD of anaerobic process effluents continues to rise, exceeding drainage standards. As a result, aerobic processing must be added afterward (Jagaba, Kutty, Salih, Noor, et al., 2021). Researchers discovered that neither anaerobic nor aerobic approaches are adequate for handling pharmaceutical waste. Anaerobic procedures can break down macromolecular refractory chemicals into tiny molecules, which aerobic processes can easily remove from wastewater (Ghaleb et al., 2021). As a result, mixed anaerobic-aerobic procedures are increasingly being used to treat antibiotic effluent. UASB-SBR, combined hydrolysis and acidification-aerobic process, and MBR are the most common anaerobic-aerobic technologies (Guo et al., 2013).

3.3 Treatment by strains isolated from PWW

Biological strains could be identified and isolated from activated sludge in a pharmaceutical wastewater treatment plant (PWWTP) to be utilized for the degradation of certain antibiotics. Visual observation, as well as biochemical and phylogenetic analyses, could be used to identify strains. Several strains have been tried by researchers and outcomes are discussed in the literature. A pure bacterial strain B16 determined as a *Sphingopyxis* sp., exhibited considerable capability for Berberine (BBR) degradation. 40 mg/L BBR initial concentration was completely degraded within 48 h (Liu, Zhang, et al., 2019). Similarly, a bacterial strain known as *Pseudomonas oleovorans* (DT4) was obtained from a PWWTP with a quicker generation time and the ability to thrive aerobically on tetrahydrofuran (THF) as the only carbon and energy source. At a generation time of 2.7 h and a maximum degradation rate of 203.9 mg THF/h, DT4 could completely breakdown 5 mM THF even without lag phase. The concentration of 100 mM THF was eliminated in a continuous flow reactor at a maximum specific growth rate of 0.113/h and a half-saturation constant of 1.224 mg/L (Chen et al., 2010).

The bacteria that degrade nitrobenzene (NBA-1) were extracted from a PWWTP and then subjected to sludge. The microbial community became more diverse, and NBA-1 remained one of the most prevalent strains. The rate of nitrobenzene decomposition under aerobic settings was only 20% of that under anaerobic conditions. This indicates that the bacterial community remained constant in anaerobic circumstances and that the microorganisms were more robust to the harsh environment (Ren, Yang, & Chen, 2015). A benzoic acid capable of degrading bacterial strain (Bz) was isolated from a PWWTP and found to have similarities to *Pseudomonas putida* in terms of cytochrome oxidase positivity and substrate assimilation. It was also identified as a *P. putida* member. *P. putida* members shared 98%–99% homologies with the bacterium. The Bz grew only on benzoic acid. The benzoic acid breakdown rate in the basal medium reached 94% after 96 h at 30°C and pH 7.5 (Zhang, Li, Zhang, & Shao, 2012).

3.4 Crisis: Regeneration of ARGs during biological treatment

Bacterial growth and genetic interchange are aided by biological treatment procedures that could aid the spread of ARB and ARGs. The aeration and anoxic tanks of conventional activated sludge systems, cyclic activated sludge systems (CASS), biocontact oxidation systems, membrane bioreactors (MBR), and SBRs are where ARGs thrive. This suggests that microbial proliferation may be responsible for considerable replication of some ARG subtypes. In diverse treatment units, there is no obvious trend of ARG proliferation or attenuation. ARGs have been found in soil, groundwater, biofilm, sediment, the deep-ocean, wastewater, and activated sludge, among other places dynamic transport of antibiotics and antibiotic resistance genes under different treatment processes in a typical pharmaceutical (Li et al., 2018).

PWWTP is regarded to be a source, seedbed, and reservoir for multiantibiotic resistant pathogenic bacteria that can be spread through aeration into the air (Tao et al., 2016). The potential for ARB and pathogens emitted from PWWTP to spread through bioaerosols poses a health danger as well. Because of their hygienic implications, airborne pathogenic microorganisms have recently received a lot of attention. Infectious disorders, cancer, acute poisoning, stress, and allergies can all be caused by these germs. Gram-negative bacteria can also produce endotoxins into the air, which can persist in bioaerosols for long periods of time and induce inflammation, emphysema, and lung fibrosis when inhaled (Zhang et al., 2018).

Antibiotics and bacterial populations at high levels create selective pressure on ARG maintenance and propagation. It facilitates the acquisition of resistance by bacteria and the spread of ARGs (Zhang et al., 2013). Therefore, antibiotic manufacturing techniques should be taken into consideration before PWW discharge into a treatment facility. That's because the quantity of ARGs in fermentation wastewater is substantially greater than in chemical synthesis and preparation wastewater (Guo et al., 2018). Among these five activities, fermentation and synthesis usually yield the largest organic burden. ARGs found in PWWTP effluent or sludges may hasten the spread of antibiotic resistance among indigenous bacteria. To completely understand the antibiotic distribution, partition, and elimination, it is necessary to evaluate antibiotics in both the sludge and water phases at each stage of PWWTPs. Identification of typical ARG proliferation/attenuation processes at various treatment stages would aid in the development of strategies and treatment technologies for preventing ARG spread (Guo et al., 2018).

ARGs may have an impact on microbial structures in the environment, posing a threat to public health. ARGs were found in abundance in the biological treatment unit in a study done by Li et al. (2018). Antibiotics were partially eliminated in the anaerobic/aerobic phases, where ARGs multiplied, according to the findings. Horizontal transfer worked as a significant factor in the formation of the ARGs, while physicochemical variables impacted the ARG level. The results demonstrate that the ARG subtypes levels in the final effluents ranged from 103 to 106 copies/mL. Only 0.6%–59.8% of the influents of the PWWTPs studied were from these sources. In raw influents, the total amounts of ARGs emitted by dewatered sludge were 7–308 times greater, and in final effluents, they were 16–638 times higher. Antibiotic residues in pharmaceutical WWTPs changed the proliferation and fate of the associated ARG subtypes, as well as bacterial growth (Wang, Mao, Mu, & Luo, 2015). A related study found that the ARGs abundance in effluents accounted for just 0.03%–78.1% of PWWTP influents. ARG loads emitted by dewatered sludge plus effluent were 1.01–14.09 times higher than those in raw influents, implying that ARGs proliferated during wastewater treatment. Despite a 1–3 order of ARGs magnitude decrease in the final effluent, large quantities of ARGs settle and transfer into sludge, increasing the danger of ARG propagation to endogenous soil bacteria via the sludge disposal procedure (Zhai, Yang, Mao, & Luo, 2016).

A comparison of ARG abundance in PWWTP and STP sludge revealed that ARG abundance in PWWTP sludge varied from 55 to 585 ppm, which was greater than that in STP sludge (27.2–86.4 ppm). Furthermore, ARG diversity was higher in PWWTP aerobic sludge than in STP aerobic sludge. It was hypothesized that the microbial community and mobile genetic elements (MGEs) in PWWTP sludge, both of which could be influenced by DO, could be the key determinants influencing ARG profiles. High levels of antibiotic resistance genes and their correlations with the bacterial community and mobile genetic elements (Tao et al., 2016). In all studies found from literature, most ARGs were transported to the dewatered sludge. More so, the antibiotic concentrations and the corresponding ARG subtypes present in the effluent had substantial relationships, according to partial correlation analysis.

4. Hybrid technologies

Conventional PWW treatment technologies are unable to comply with the regulatory limits for effluent discharge that safeguard the environment successfully (Luo et al., 2019). Pharmaceutical wastewater treatment techniques based on biotechnology are now unable to adequately deal with pharmaceutical pollutants (Ismail & Habeeb, 2015). First, improving biodegradability via pretreatment, and then inventing novel combination treatment methods, is the most efficient way of treating pharmaceutical wastewater in an energy-efficient manner. In spite of the fact that several well-known anaerobic, aerobic, and effective operation units exist in PWW treatment, hybrid methods with high-performance efficiency and cheap energy have to be explored (Guo et al., 2013). The membrane bioreactor (MBR) technology was developed as a result of a hybrid process that combined physical and biological wastewater treatment approaches. It has a lot of promise for treating many types of wastewater. The biological and hybrid treatment system efficiencies are summarized in Table 2.

4.1 Membrane technology

Membrane technologies have been well advanced in world practice. These technologies include microfiltration, nanofiltration, ultrafiltration, and reverse osmosis membranes (Liu, Liu, et al., 2019). Membrane bioreactor (MBR) systems are commonly utilized in biological wastewater treatment as a combination of membrane filtration and biological treatment using CAS, with the membrane replacing the clarifier in the wastewater treatment system, as shown in Fig. 3. Tertiary procedures such as sand filtering are not necessary in MBRs. The activated sludge microorganisms in the bioreactor eat up the biodegradable organic matter in the wastewater, while the membrane set apart colloids, biomass flocs, and suspended particulates from the activated sludge using low-pressure membrane filtration, such as microfiltration or ultrafiltration (Alizad Oghyanous, Etemadi, & Yegani, 2020). Membranes operate as a selective barrier in MBRs, allowing treated water to pass through while particles bigger than the membrane pore size are filtered. Membrane properties have a significant impact on the separation process quality.

MBRs have gotten a lot of attention and are the most widely utilized water and wastewater treatment methods because of their benefits, which include a compact structure, higher effluent quality, less area requirement, higher contaminant removal efficiency, better solid-liquid separation efficiency, inorganic salt repulsion, more compacted volume, longer sludge retention time (SRT), lower sludge production, high water flow, hydraulic rate (HRT), and higher organic loading rate (OLR). MBR separates HRT from SRT, allowing for more versatile operation parameter management (Shariati et al., 2010; Yusuf et al., 2021). A membrane can also operate as an

Table 2 Efficiency of biological and hybrid treatment technology for PWW treatment.

Wastewater	Wastewater/ contaminant	Influent concentration	Treatment system	Reactor operating capacity (*L*)	Reactor operating condition	Removal efficiency (%)	References
Synthetic	Tetrahydrofuran (THF)	5–100 Mm	Continuous flow system in jacketed tubular reactor	4 L	30°C, 160 rpm, pH 7.2, 2.3 mg/L DO, 2.7 h generation time	68	Chen et al. (2010)
	Butyl acetate (BA), *n*-butyl alcohol (n-BA), and phenylacetic acid (PA)	2200 mg/m^3 BA, 2400 mg/m^3 *n*-BA and 370 mg/m^3 PA	Biotrickling filter packed with fibrous balls, ZX02	2800 mm × 300 mm ×1650 mm	27–35°C, 9.5–10.5 L/h optimal flow rate	95 BA, 92 *n*-BA and 100 PA	Wang et al. (2007)
	Ketoprofen	0.25–16 mM	Combined isolated bacterial and microalgal strains	NA	30 ± 2°C, 150 rpm, 10 h lag phase, diurnal cycle of 12 h	41 COD, 82 toxicity	Ismail et al. (2016)
	Omethoate	1000 mg/L		6 L	30°C, 7 pH, 150 rpm	77.24	Li et al. (2021)
	Organonitriles	0.5–10 g/L	Activated sludge consortium	2 L	25°C, 7 pH, optimal time 8 h	97.3	Li et al. (2007)
	Sulfamethoxazole	5–120 mg/L	Net-like rotating biological contactor	78.75 L	0–21.3 g COD/L d OLR, 12–72 h HRT	89.84	Azimi et al. (2017)
	Carbamazepine	2.21 mg/L	Activated sludge process	NA	NA	50.17	Dwivedi et al. (2017)
	p-Xylene	200 mg/L		NA	NA	91.8	

			Biotrickling filter				Wang, Wang, et al. (2015)
Pharmaceutical wastewater	Pharmaceutical wastewater	NA	Graphite based microbial fuel cell	NA	22 ± 2 °C, 7.4–7.7 pH, 4 days HRT, 5460 mg/L COD	80.55 COD, 35.62 TDS	Rashid et al. (2021)
Pharmaceutical wastewater	Pharmaceutical wastewater	1.5 L	Sequential batch reactor system	3 L	6–7 mg/L DO, 6 h cycle time	>90 COD	Omar et al. (2019)
Pharmaceutical wastewater	Pharmaceutical wastewater	5 L	Upflow anaerobic sludge blanket (UASB) reactors	10 L	1–1.3 mg/L DO, 35 ± 1°C, 6–7 mg/L, 12 h cycle time	77.8 ammonium, 84.3 nitrite	Tang et al. (2011)
Pharmaceutical wastewater	Pharmaceutical wastewater	NA	Dual chamber microbial fuel cell with granular activated carbon	NA	32 h HRT, BOD/COD = 0.05, 800 mg/L COD, 7–7.2 pH	83 COD	Ismail and Habeeb (2017)
Pharmaceutical wastewater	Pharmaceutical wastewater	NA	UASB	27.44 m^3	6.26–10.33 kg COD/m^3/day OLR, 33–43 HRT	86.2–91.6 COD, 90–95.2 BOD, 62.6–68 SS	Satyanarayan, Karambe, and Vanerkar (2009)
Synthetic	Benzoic acid	1 g/L	NA	NA	30°C, 7.5 pH	94	Zhang et al. (2012)
Pharmaceutical wastewater	Pharmaceutical wastewater	NA	Two-chambered microbial fuel cell	100 mL	BOD/COD = 0.53, 7440 mg/L COD, 7–7.2 pH	93	Amari and Boshrouyeh Ghandashtani (2020)

Continued

Table 2 Continued

Wastewater	Wastewater/ contaminant	Influent concentration	Treatment system	Reactor operating capacity (*L*)	Reactor operating condition	Removal efficiency (%)	References
Pharmaceutical wastewater	Pharmaceutical wastewater	NA	Up-flow anaerobic packed bed reactor	22.5 L	1.5–4.6 kg COD/ m^3/day OLR, 37°C, 464 mg/L TKN, 5.2–6.8 pH, 2–5.6 HRT	60–70 COD	Chelliapan and Sallis (2010)
Pharmaceutical wastewater	Pharmaceutical wastewater	NA	Microbial fuel cell	NA	800 mg/L COD, 38 HRT	81 COD	Ismail and Habeeb (2015)
Synthetic	Fluoroquinolone	0.32 μg/L to 5.7 mg/L	MBBR	NA	NA	94.5–99.9	Guo et al. (2017)
Synthetic	Naphthalene	0–60 mg/L	MBBR	4 L	8 h HRT, aeration was 0.12 m^3/h, 25 ± 1°C, 6.8–7.5 pH, 240–410 mg/L COD, 28 h cycle time	96	Wang et al. (2020)
Synthetic	Tetracyclines	2 L	Expanded granular sludge bed (EGSB) and a membrane bioreactor (MBR)	NA	24 h HRT, 40 days and 6 months sludge age for MBR and EGSB, respectively	62–78.3 tetracycline, 80 COD, 90 TSS	Hou et al. (2016)
Synthetic	Ciprofloxacin (CIP), sulfamethoxazole (SMX)	10 mg/L CIP, 50 mg/L SMX	Constructed wetland	1.5 L	NA	92 CIP, 61 SMX	Christofilopoulos et al. (2016)

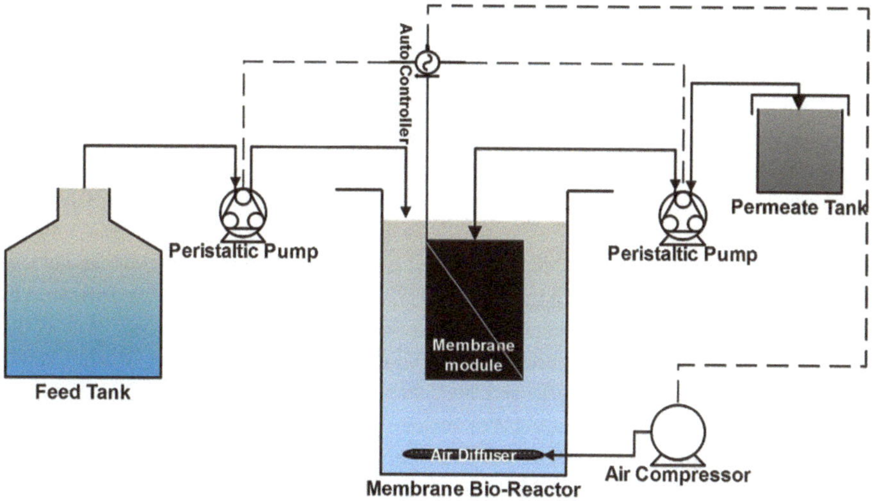

Fig. 3 Schematic diagram of the MBR system.

impediment for molecules that are decomposed in their solution in the bioreactor system, not only in separating the particle from the deteriorated medium but also in separating the particle from the degraded medium (Zakeritabar, Jahanshahi, & Peyravi, 2018).

The usage of MBRs has been limited, however, due to membrane fouling, which leads to an increase in membrane hydraulic resistance and a decrease in membrane permeability. Membrane materials deteriorate as a result of fouling, resulting in a decrease in membrane lifetime. Owing to the cleaning and replacement of membranes, it also raises operational costs (Hosseinpour, Azimian-Kivi, Jafarzadeh, & Yegani, 2021). Biofilm formation on membrane surface due to adsorption of microbes has been proven to be the main source of fouling in MBR systems. Membrane parameters have been shown to play a role in membrane fouling in studies. Hydrophilic membranes are known to be more fouling resistant. As a result, any effort to increase the membrane's hydrophilicity can be considered a successful strategy for reducing membrane fouling (Javadi, Jafarzadeh, Yegani, & Kazemi, 2018). Another drawback of the MBR system is membrane foaming which results in the considerable reduction of the biological reactor performance (Alizad Oghyanous et al., 2020). Other shortcomings are very less gas holdup, low COD removal rate, dead bioreactor angle, weak fluid flow, expensive cost of membranes module and aeration, etc.

The system OLR and SRT were modified to reduce the EPS generation used for pharmaceutical wastewater treatment in a submerged MBR (SMBR) that had been operating for more than 8 months. The lowest foamy amount was found at an OLR of 0.32 gCOD/L day and a SRT of 0.5 month, which was linked to low EPS protein concentrations. In addition, the findings revealed at an OLR of 0.03 gCOD/L day, the foaming quality was better than at higher OLRs in all SRTs. The study found that an OLR of 0.1 gCOD/L day was the best operating condition for activated sludge in SMBRs. A 92.8% COD elimination efficiency and 3.708/day max (highest specific growth rate) were attained under these ideal conditions (Alizad Oghyanous et al., 2020). Likewise, PWW effluent treatment system was treated with a submerged membrane bioreactor. COD and NH_4^+-N were removed at a rate of 73% and 63%, respectively, with the final effluent remaining steady at less than 50mg/L COD and 40mg/L ammonia-nitrogen. High quantities of organic chemicals, ammonia nitrogen, and biomass caused nitrifying bacteria activity suppression and a rapid decline in filtration/permeate flux (Ding, Wang, & Ren, 2013; Noor, Kutty, Jagaba, et al., 2021).

The high-loading mixed PWW was treated over 260days with a membrane-aerated biofilm reactor (MABR) system that included acidification pre-treatment, followed by postprocessing using activated carbon adsorption and the MABR process. The effluent of the integrated treatment MABR system remained steady during post-processing, with COD 200mg/L and NH_4^+-N 3mg/L, equating to >90% COD and 98% NH_4^+-N removal efficiency (Wei et al., 2012). To remove organic matter, a 10m^3/day MBR plant was erected, and its performance was examined for a period of 4.67months. COD and BOD_5 levels were measured at 95% and 99%, respectively. Notwithstanding membrane fouling the system has been demonstrated to be a viable option for treating high-strength wastewater (Jagaba et al., 2022). During 180days of operation under 0.5–6mg/L DOs, MBR was designed and operated to treat PWW containing 6-APA and ceftriaxone sodium. The results revealed that the best DO range was 1.5–2.1mg/L, with COD and BOD_5 removal efficiencies of 87% and 95%, respectively (Hu et al., 2020).

Anaerobic MBR technology can be used to treat PWW. Anaerobic membrane bioreactors (AnMBRs), like aerobic MBRs, offer similar benefits for wastewater treatment, notwithstanding the fact that oil and grease, fat, and/or inhibiting chemicals are present (Almahbashi et al., 2021). However, as compared to aerobic MBRs, the AnMBR has the disadvantage of poorer biomass filterability, resulting in lower filtration fluxes. In an anaerobic MBR that run for 580days, anaerobic treatment of PWW as the only carbon source was studied. COD elimination effectiveness was found to be 78%. Adding waste organic solvents higher than 2.5g/L as COD to the influent, on the other hand, resulted in low COD removal efficiency or even anaerobic digestion breakdown. Changes in the population of anaerobic bacteria were detected as a result of the influent composition (Svojitka et al., 2017). An AnMBR and an anaerobic bio-entrapped membrane reactor (AnBEMR) were used to treat PWW. Under various OLR circumstances, the AnBEMR result to a 5%–10% higher total COD removal efficiency and a 15% higher methane production than the AnMBR. The AnBEMR's increased abundance of the leading archaeal genus *Methanimicrococcus* should aid in the breakdown of the principal organic contaminant found in pharmaceutical wastewater (Ng, Shi, Ong, & Ng, 2016).

The treatment performance of PWW was investigated using a bioentrapped membrane reactor (BEMR) and a salt-marsh sediment MBR (SMSMBR). The SMSMBR revealed marine microorganisms' ability to degrade complex and resistant chemicals and use hydrocarbon compounds, resulting in high organic removal efficiency. However, because to the saline effect, specific nitrifying activity declined and was hindered, resulting in poor ammonia nitrogen removal. The SMSMBR outperformed the BEMR in terms of TCOD and TOC removal efficiency, obtaining 23.6 1.8% greater TCOD and 26.5 3.1% greater TOC removal efficiency at an OLR of 3.1–8.2 kg COD/m^3 day. This is because traditional biomass in the BEMR could not sustain hypersaline circumstances, whereas microorganisms present in the SMSMBR, which were seeded from the coast, worked hard and were able to effectively decompose complex organics (Ng, Shi, Yao, & Ng, 2014). On a full-scale of 800 t/day, a coal cinder filter was integrated with a biological treatment process for pharmaceutical wastewater treatment, and the system's performance was tested. The coal cinder's large specific surface area and great adsorption capabilities served to lower the pollutant load that had to be transported to the biological treatment process afterward. As a result, the combined process' typical removal efficiencies for COD, BOD_5, SS, and color were around 99.7%, 98.2%, 98.5%, and 96.3%, respectively. The findings suggested that the combined process could provide a cost-effective alternative for pharmaceutical wastewater treatment (Jagaba, Kutty, Noor, et al., 2022; Zheng et al., 2010).

The efficiency of an upflow anaerobic sludge blanket (UASB) and a built wetlands system for PWW treatment were examined separately and in combination. In UASB and built wetlands, COD removal efficiencies were 75% and 94%, respectively. The hybrid (UASB and CW) system, on the other hand, showed promising and steady as a cheap and efficient method, with COD elimination efficiency ranging from 90% to 96% (Vistanty & Crisnaningtyas, 2021). The mixed PWW was treated using a combined anaerobic/oxic (A/O) and biological filtration oxygenated reactor (BIOFOR) technique dubbed (A/O + BIOFOR). The BIOFOR process has a high biofilm thickness and retention, whereas the anaerobic/oxic (A/O) process for biological phosphorus removal has sequential anaerobic and aerobic stages. The advantages of both technologies are incorporated in the combined A/O + BIOFOR process. The COD, NH_4^+-N, and SS concentrations in the effluent met standard discharge limits, according to the findings (Ouyang, Lu, Ouyang, Liu, & Wang, 2017). The hybrid hydrolysis acidification method (HHAT) processed pharmaceutical wastewater in an anaerobic-low DO environment, and the hybrid aerobic biotechnology included an alternate-flow biological reactor (ABR) and a two-way-flow biological aerated filter (TBAF). The results show that using the micro aerobic HHAT-ABR-TBAF method on raw PWW treatment removed a significant number of refractory chemicals while also lowering toxicity and enhancing biodegradability (Chen, Huang, Liu, & Sun, 2012).

The impact of embedding pristine and modified nanodiamond (ND) nanoparticles into polyvinylidene fluoride (PVDF) membranes at varying percentages on structure and performance were investigated. The performance of 1.5 wt% PVDF/ND and 1.5 wt% PVDF/P-ND membranes in an MBR system with PWW feed was investigated, and it was discovered that the PVDF/P-ND membrane had better antifouling capabilities than the PVDF/ND membrane. This could be because the PVDF/P-ND membrane's pore size distribution curve has been tampered toward tiny pores. PVDF/P-ND membranes had a higher pure water flux than PVDF/ND membranes at the same nanoparticle content (Javadi et al., 2018). In a related investigation, the effect of neat and modified ND particles implanted in high-density polyethylene (HDPE) immersed MBRs for pharmaceutical wastewater treatment under subcritical conditions was investigated. Thermal carboxylation (ND-COOH) and grafting with polyethylene glycol were used to modify the ND (ND-PEG). The best combination of ND-COOH and ND-PEG nanoparticles was found to be 0.5 wt% ND-COOH and 0.75 wt% ND-PEG nanoparticles. The 0.75 wt% HDPE/ND-PEG membrane exhibited the largest water flux and the best antifouling capabilities, according to the findings. The 0.75 wt% ND-PEG showed the highest hydrophilicity, lowest filtration flux decline, lowest total fouling ratio, and also highest flux recovery ratio (Kivi, Alinia, Jafarzadeh, & Yegani, 2019). In MBR treating pharmaceutical wastewater, the performance of polypropylene (PP), PP/ND-carboxylated (–COOH) (0.75 wt%), and PP/ND-polyethylene glycol (–PEG) (0.75 wt%) membranes was investigated. In all cases, COD elimination was greater than 91%, while membranes containing 0.75 wt% nanoparticles had the maximum water flux, porosity, and mechanical strength. The antifouling properties, membrane resistance (R_m), and specific cake layer resistance of nanocomposite membranes were shown to be superior to those of a plain PP membrane. The PP/ND-PEG (0.75 wt%) membrane had the highest hydrophilicity, critical flux, and antifouling properties. The antifouling properties of the COOH and PEG functionalized ND membrane were superior to those of the clean PP membrane (Hosseinpour et al., 2021). Based on the findings, it can be inferred that adding PVP-immobilized ND nanoparticles to PVDF membranes improves their hydrophilicity, water flow, and antifouling capabilities (Javadi et al., 2018). Silver nanoparticles were incorporated into polyvinyl chloride/polycarbonate hollow fiber membranes (PVC/PC/Mag). The size of macrovoids in the inner layer shrank, while the thickness of the sponge-like outer layer expanded, changing the shape of hollow fiber membranes. For 28 days, the modified membranes were utilized in an MBR to handle pharmaceutical wastewater. Hollow fiber membranes' mechanical qualities, thermal resistance, and antifouling properties were all improved as a result of this. As the nanoparticle content increased, so did the number of pores on the membrane surface, resulting in improved hydrophilicity, higher COD removal and better pure water flux in the MBR system (Behboudi, Jafarzadeh, & Yegani, 2018).

Under different temperatures, the performance and removal methods of a multiple draught tubes airlift loop membrane bioreactor (Mt-ALMBR) for processing ampicillin PWW were studied. The merits of Mt-ALMBR were identified to aid in enhancing microorganism activity in the reactor, resulting in good COD and ampicillin removal performance. The ideal average COD and ampicillin removal rates for 150 days of operation were 98.2% (36°C) and 63.2% (7°C), respectively. In this reactor, there were primarily four removal mechanisms: adsorption and membrane rejection were the primary methods at low temperatures, but as temperatures rose, hydrolysis and biodegradation eventually took over (Chen et al., 2020). A bioreactor system using an external loop airlift membrane bioreactor (ELAMBR) was used to treat acetaminophen-containing pharmaceutical effluent. The ELAMBR system has a very high removal efficiency, with 100% acetaminophen eliminated after 2 days. This research proves the utility of the ELAMBR system, highlighting its advantages of ease of use and maintenance, effective COD and acetaminophen removal, and low energy usage (Shariati et al., 2010).

4.2 Advanced oxidation process coupled to biological process

In recent years, combining AOP with biological processes has gotten a lot of attention as a possible alternative treatment for resistant wastewater. The ozonation method can efficiently lower the concentration of medicinal components that are difficult to degrade. The goal of this investigation was to see if ozonation could be used with medicinal substances on MBR. When ozonation was used as a supplementary unit following MBR, the treatment efficiency improved. The findings also revealed that high amounts of pharmaceutical pollutants caused cell wall damage, lowering MLSS and MLVSS. The presence of non-biodegradable pharmaceutical compounds in MBR also resulted in large levels of biopolymers. As a result of the greater SMP concentrations in the solution, biological clearance was reduced and membrane fouling occurred. Acetaminophen, fluoxetine, and COD elimination could be as high as 98.4%, 57.83%, and 99.9%, respectively, using the MBR-ozonation technique (Shahbeig, Mehrnia, Mohammadi, Moghaddam, & Rouini, 2017). PWW treatment was investigated using a combination of internal microelectrolysis and coagulation as pretreatment in order to lower the pollutants' toxicity to microbes and increase the biodegradability of the wastewater in order to ensure the biological process operated smoothly. This was followed by biological treatment and activated carbon adsorption. The residual nonbiodegradable particles were removed using activated carbon adsorption as a posttreatment technique. When the HRT was 4–6 h and the temperature was 38°C, the coupled hydrolysis acidification and SBR performed a crucial role in COD decomposition. In effluent wastewater, the COD concentration was <300 mg/L, with an overall removal rate of up to 96% (Khan, Khan, Zubair, Azfar Shaida, et al., 2021).

The practicality of treating genuine oxytetracycline (OTC) wastewater from a pharmaceutical wastewater treatment facility with a combination method of ClO_2 oxidation and biological activated carbon (BAC) was also explored. The combination has been shown to be a long-term solution for the treatment of pharmaceutical wastewater. As the BOD5/COD ratio rises from 0.04 to 0.23, experimental data demonstrate that the ClO_2 preoxidation method can improve the biodegradability of OTC wastewater. At pH 2 and a ClO_2/OTC ratio of 1.5:1, the OTC reacts stoichiometrically with ClO_2, with the maximum OTC removal rate of 98.3%. The COD elimination efficiency was 68.1% with 0.2 mM ClO_2 oxidation, 8 h of HRT, and an OLR of 0.8 kg COD/m^3 day (Liu, Li, Zhang, Liu, & Li, 2017). The efficacy of a continuous coupled wet oxidation/biological process for treating phenolic chemicals in pharmaceutical wastewater was assessed. The connected industrial operation is depicted in simplified form in Fig. 4. Wet oxidation pretreatment efficiencies for each chemical increased to 95%. For chemicals with kinetic constants higher than 8.4/h during the wet oxidation process, the wet oxidation process also demonstrated low sensitivity. In the case of the biotreatment, phenolic compounds were almost completely removed, while COD levels were reduced by 95%. For residence times in the bioreactor of less than 25 days, the system was stable to fluctuations in the flow rate. The elimination of phenolic compounds was unaffected by changes in the oxygen flow rate. For the elimination of phenolic compounds, dampening factors near 1 were achieved, confirming the

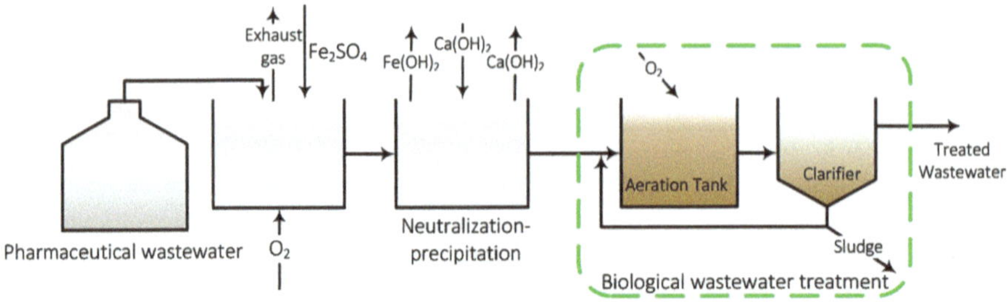

Fig. 4 Schematic diagrams for the coupled wet oxidation and biological process.

outstanding behavior of the combined wet oxidation/biological system when quick and continuous variations in the contaminant load were assessed (Collado, Quero, Laca, & Díaz, 2013).

A bioelectrochemical treatment train system was used to treat a complicated pharmaceutical effluent. Findings revealed that the bioelectroperoxone system inactivated 99.99% of the microorganisms, and removed 92.2%, 84.7%, 89% of color, TSS, and TOC, respectively. As a result, bioelectroperoxone creates path for a more efficient and long-term method of removing biodegradable and refractory, persistent organic pollutants from pharmaceutical effluent (Srinivasan & Nambi, 2020). A study that used Integrated Moving bed bioreactor attempted to solve the fouling problem in membrane bioreactors (MBRs) by combining adsorptive cake, ozonation, bio-film carriers, electrooxidation and (IMBBR). At 300 kPa, the developed IMBBR removes 95%, 95%, and 70% TOC, COD, and TDS, respectively, with a 35% increase in permeate flux. The research found that less energy was consumed without the issue of filamentous bulking (Palani, Ramasamy, Palaniappan, Huh, & Natesan, 2019). To extract antibiotics and ARGs from genuine PWW, a reactor with an UASB, A/O, and four distinct AOPs was built, as shown in Fig. 5.

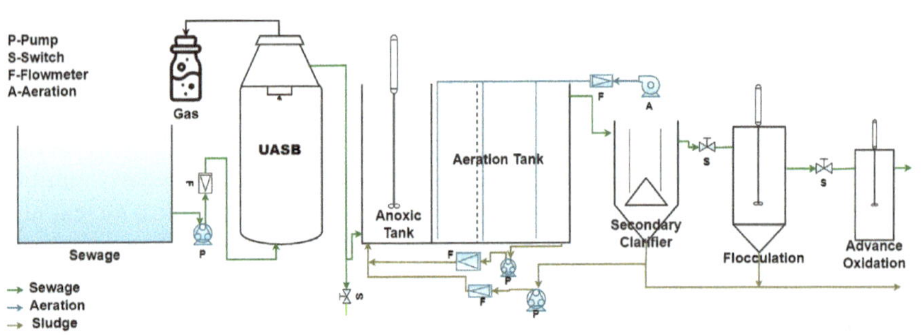

Fig. 5 Schematic representation of the designed simulation reactor for simultaneous removal of antibiotics and ARGs from PWW.
Adapted from Hou, J., Chen, Z., Gao, J., Xie, Y., Li, L., Qin, S., Wang, Q., Mao, D., & Luo, Y. (2019). Simultaneous removal of antibiotics and antibiotic resistance genes from pharmaceutical wastewater using the combinations of up-flow anaerobic sludge bed, anoxic-oxic tank, and advanced oxidation technologies. *Water Research, 159*, 511–520. https://doi.org/10.1016/j.watres.2019.05.034.

It was discovered that within 180 days of operation, all antibiotics were totally removed from the reactor, and the biological treatment units contributed more to the removal of examined antibiotics, as ARGs were primarily removed by AOPs. Fenton/UV followed by Fenton oxidation were the most effective techniques for simultaneous elimination of total bacteria and ARGs among the studied AOPs. Sulfamethoxazole, ampicillin, and tetracyclines were removed primarily through degradation (62.5%–81%), while enrofloxacin was removed primarily through sorption to sludge (74%) in UASB (Hou et al., 2019).

5. Conclusions

Generally, the existence of contaminants in PWW is a surging issue that can cause water quality problems. Due to its potential toxicity, PWW appears to pose a significant risk to the quality of water bodies and aquatic life, as well as to human health. Along with their acute and recalcitrant characteristics throughout the treatment stages, these constituents may be detected in WWTP effluents at amounts equivalent to those reported at the influent. The variety of PWW sources has made it even more difficult to restrict their existence, thus there is an urgent need to explore and develop a treatment to ease the stress caused by pharmaceutical pollutants on water resources. Various treatment methods have been employed. However, the performance of biological processes for PWW appears to be inconsistent; removal efficiencies varied greatly. The literature indicates the following scenario precisely, (1) no removals at all due to the existence of refractory compounds, (2) incomplete removals, and (3) considerable removals of PWW relying on the ability of degradation of pharmaceutical compounds and its tendency to be adsorbed to the biomass in the biological reactors. After biological treatment such as attached growth, suspended growth, as well as natural systems, a group of compounds still remain unchanged in the treated PWW, which exhibits recalcitrant behavior of those compounds while treatment. Carbamazepine and diclofenac are found to be refractory compounds in PWW as stated in most studies. These recalcitrant compounds are not degraded after biological treatment instead transform into more hazardous than parent compounds. Although research into biological methods for PWW treatment will continue, the potential of these systems to remove chemicals that are more complex will remain uncertain. However, integrating physical treatment methods such as bio-electrochemical, adsorption, advanced oxidation, or membrane technologies with biological processes is found to be more effective in degrading recalcitrant compounds. In most of the studies conducted on hybrid treatment systems, antibiotics and ARGs were fully eliminated. The future prospect of studies related to hybrid treatment is to find a suitable combination of biological and physical treatment methods to achieve sustainable and cost-effective treatment, which can be employed practically for PWW treatment.

References

Alizad Oghyanous, F., Etemadi, H., & Yegani, R. (2020). Foaming control and determination of biokinetic coefficients in membrane bioreactor system under various organic loading rate and sludge retention time. *Biochemical Engineering Journal*, 107491. https://doi.org/10.1016/j.bej.2020.107491.

Almahbashi, N. M. Y., Kutty, S. R. M., Ayoub, M., Noor, A., Salihi, I. U., Al-Nini, A., et al. (2021). Optimization of preparation conditions of sewage sludge based activated carbon. *Ain Shams Engineering Journal*, *12*(2), 1175–1182. https://doi.org/10.1016/j.asej.2020.07.026.

Alsubih, M., Khan, R. A., Khan, N. A., Khan, A. R., & Sharma, G. (2021). Performance evaluation of aerobic fluidized bed bioreactor coupled with tube-settler for hospital wastewater treatment. *Journal of Environmental Chemical Engineering*, *9*(5), 105896.

Amari, S., & Boshrouyeh Ghandashtani, M. (2020). Non-steroidal anti-inflammatory pharmaceutical wastewater treatment using a two-chambered microbial fuel cell. *Water Environment Journal*, *34*(3), 413–419. https://doi.org/10.1111/wej.12476.

Azimi, N., Hassani, A. H., Darzi, G. N., & Borghei, S. M. (2017). Biodegradation of wastewater containing high concentration of sulfamethoxazole by antibiotic adopted biofilm in attached growth bioreactor. *Polish Journal of Environmental Studies*, *26*(6), 2463–2469. https://doi.org/10.15244/pjoes/67686.

Behboudi, A., Jafarzadeh, Y., & Yegani, R. (2018). Incorporation of silica grafted silver nanoparticles into polyvinyl chloride/polycarbonate hollow fiber membranes for pharmaceutical wastewater treatment. *Chemical Engineering Research and Design*, *135*, 153–165. https://doi.org/10.1016/j.cherd.2018.03.019.

Chen, Z., Huang, X., Liu, Z., & Sun, X. (2012). A novel application of combined biotechnologies on chemical synthesis-based pharmaceutical wastewater. *Advanced Materials Research*, *455–456*, 1261–1266. https://doi.org/10.4028/www.scientific.net/AMR.455-456.1261.

Chen, Z., Min, H., Hu, D., Wang, H., Zhao, Y., Cui, Y., et al. (2020). Performance of a novel multiple draft tubes airlift loop membrane bioreactor to treat ampicillin pharmaceutical wastewater under different temperatures. *Chemical Engineering Journal*, *380*, 122521. https://doi.org/10.1016/j.cej.2019.122521.

Chelliapan, S., & Sallis, P. J. (2010). Performance of an up-flow anaerobic packed bed reactor system treating pharmaceutical wastewater. In *vol.1. International conference on biology, environment and chemistry* IPCBEE.

Chen, J. M., Zhou, Y. Y., Chen, D. Z., & Jin, X. J. (2010). A newly isolated strain capable of effectively degrading tetrahydrofuran and its performance in a continuous flow system. *Bioresource Technology*, *101*(16), 6461–6467. https://doi.org/10.1016/j.biortech.2010.03.064.

Christofilopoulos, S., Syranidou, E., Gkavrou, G., Manousaki, E., & Kalogerakis, N. (2016). The role of halophyte Juncus acutus L. in the remediation of mixed contamination in a hydroponic greenhouse experiment. *Journal of Chemical Technology and Biotechnology*, *91*(6), 1665–1674. https://doi.org/10.1002/jctb.4939.

Collado, S., Quero, D., Laca, A., & Díaz, M. (2013). Efficiency and sensitivity of the wet oxidation/biological steps in coupled pharmaceutical wastewater treatment. *Chemical Engineering Journal*, *234*, 484–490. https://doi.org/10.1016/j.cej.2013.07.075.

Ding, Y. H., Wang, Q., & Ren, H. Q. (2013). Activities of activated sludge affected by pharmaceutical wastewater in a membrane bioreactor. *Advanced Materials Research*, *610*, 1426–1431. Trans Tech Publ https://doi.org/10.4028/www.scientific.net/AMR.610-613.1426.

Ding, S., Wu, J., Zhang, M., Lu, H., Mahmood, Q., & Zheng, P. (2015). Acute toxicity assessment of ANAMMOX substrates and antibiotics by luminescent bacteria test. *Chemosphere*, *140*, 174–183. https://doi.org/10.1016/j.chemosphere.2015.03.057.

Dwivedi, K., Morone, A., Pratape, V., Chakrabarti, T., & Pandey, R. A. (2017). Carbamazepine and oxcarbazepine removal in pharmaceutical wastewater treatment plant using a mass balance approach: A case study. *Korean Journal of Chemical Engineering*, *34*(10), 2662–2671. https://doi.org/10.1007/s11814-017-0190-2.

Ghaleb, A. A. S., Kutty, S. R. M., Ho, Y.-C., Jagaba, A. H., Noor, A., Al-Sabaeei, A. M., et al. (2020a). Response surface methodology to optimize methane production from mesophilic anaerobic co-digestion of oily-biological sludge and sugarcane bagasse. *Sustainability*, *12* (5), 2116. https://doi.org/10.3390/su12052116.

Ghaleb, A. A. S., Kutty, S. R. M., Ho, Y. C., Jagaba, A. H., Noor, A., Al-Sabaeei, A. M., et al. (2020b). Anaerobic co-digestion for oily-biological sludge with sugarcane bagasse for biogas production under mesophilic condition. *IOP Conference Series: Materials Science and Engineering*, *991*(1), 012084. https://doi.org/10.1088/1757-899x/991/1/012084.

Ghaleb, A. A. S., Kutty, S. R. M., Salih, G. H. A., Jagaba, A. H., Noor, A., Kumar, V., et al. (2021). Sugarcane bagasse as a co-substrate with oil-refinery biological sludge for biogas production using batch mesophilic anaerobic co-digestion technology: Effect of carbon/nitrogen ratio. *Water*, *13*(5), 590. https://doi.org/10.3390/w13050590.

Guo, X., Yan, Z., Zhang, Y., Kong, X., Kong, D., Shan, Z., et al. (2017). Removal mechanisms for extremely high-level fluoroquinolone antibiotics in pharmaceutical wastewater treatment plants. *Environmental Science and Pollution Research*, *24*(9), 8769–8777. https://doi.org/10.1007/s11356-017-8587-3.

Guo, X., Yan, Z., Zhang, Y., Xu, W., Kong, D., Shan, Z., et al. (2018). Behavior of antibiotic resistance genes under extremely high-level antibiotic selection pressures in pharmaceutical wastewater treatment plants. *Science of the Total Environment*, *612*, 119–128. https://doi.org/10.1016/j.scitotenv.2017.08.229.

Guo, W. Q., Yin, R. L., & Zhou, X. J. (2013). Current trends for biological antibiotic pharmaceutical wastewater treatment. *Advanced Materials Research*, *726–731*, 2140–2145. https://doi.org/10.4028/www.scientific.net/AMR.726-731.2140.

Hosseinpour, S., Azimian-Kivi, M., Jafarzadeh, Y., & Yegani, R. (2021). Pharmaceutical wastewater treatment using polypropylene membranes incorporated with carboxylated and PEG-grafted nanodiamond in membrane bioreactor (MBR). *Water Environment Journal*, *35*(4), 1249–1259. https://doi.org/10.1111/wej.12715.

Hou, J., Chen, Z., Gao, J., Xie, Y., Li, L., Qin, S., et al. (2019). Simultaneous removal of antibiotics and antibiotic resistance genes from pharmaceutical wastewater using the combinations of up-flow anaerobic sludge bed, anoxic-oxic tank, and advanced oxidation technologies. *Water Research*, *159*, 511–520. https://doi.org/10.1016/j.watres.2019.05.034.

Hou, J., Wang, C., Mao, D., & Luo, Y. (2016). The occurrence and fate of tetracyclines in two pharmaceutical wastewater treatment plants of Northern China. *Environmental Science and Pollution Research*, *23*(2), 1722–1731. https://doi.org/10.1007/s11356-015-5431-5.

Hu, D., Luo, K., Ma, H., Min, H., Zhao, Y., Cui, Y., et al. (2020). A sustainability anti-infective pharmaceutical wastewater treatment technology: Multi-stage vertical variable diameter membrane bioreactor with DO online controlling. *Bioresource Technology*, *311*. https://doi.org/10.1016/j.biortech.2020.123507.

Husain Khan, A., Abdul Aziz, H., Khan, N. A., Ahmed, S., Mehtab, M. S., Vambol, S., et al. (2020). Pharmaceuticals of emerging concern in hospital wastewater: Removal of ibuprofen and Ofloxacin drugs using MBBR method. *International Journal of Environmental Analytical Chemistry*, 1–15.

Ismail, M. M., Essam, T. M., Ragab, Y. M., & Mourad, F. E. (2016). Biodegradation of ketoprofen using a microalgal–bacterial consortium. *Biotechnology Letters*, *38*(9), 1493–1502. https://doi.org/10.1007/s10529-016-2145-9.

Ismail, Z. Z., & Habeeb, A. A. (2015). Pharmaceutical wastewater treatment associated with renewable energy generation in microbial fuel cell based on mobilized electroactive biofilm on zeolite bearer. *Journal of Engineering*, *21*, 35–44.

Ismail, Z. Z., & Habeeb, A. A. (2017). Experimental and modeling study of simultaneous power generation and pharmaceutical wastewater treatment in microbial fuel cell based on mobilized biofilm bearers. *Renewable Energy, 101*, 1256–1265. https://doi.org/10.1016/j.renene.2016.10.008.

Jagaba, A. H., Kutty, S. R. M., Baloo, L., Noor, A., Abubakar, S., Lawal, I. M., et al. (2021). Effect of hydraulic retention time on the treatment of pulp and paper industry wastewater by extended aeration activated sludge system. In *2021 Third international sustainability and resilience conference: Climate change* (pp. 221–224). IEEE.

Jagaba, A. H., Kutty, S. R. M., Fauzi, M. A. H. M., Razali, M. A., Hafiz, M. F. U. M., & Noor, A. (2021). Organic and nutrient removal from pulp and paper industry wastewater by extended aeration activated sludge system. *IOP Conference Series: Earth and Environmental Science, 842*(1), 012021. https://doi.org/10.1088/1755-1315/842/1/012021.

Jagaba, A. H., Kutty, S. R. M., Hayder, G., Baloo, L., Abubakar, S., Ghaleb, A. A. S., et al. (2020). Water quality hazard assessment for hand dug wells in Rafin Zurfi, Bauchi State, Nigeria. *Ain Shams Engineering Journal, 11*(4), 983–999. https://doi.org/10.1016/j.asej.2020.02.004.

Jagaba, A. H., Kutty, S. R. M., Hayder, G., Baloo, L., Ghaleb, A. A. S., Lawal, I. M., et al. (2021a). Degradation of Cd, Cu, Fe, Mn, Pb and Zn by Moringa-oleifera, zeolite, ferric-chloride, chitosan and alum in an industrial effluent. *Ain Shams Engineering Journal, 12*(1), 57–64. https://doi.org/10.1016/j.asej.2020.06.016.

Jagaba, A. H., Kutty, S. R. M., Hayder, G., Baloo, L., Noor, A., Yaro, N. S. A., et al. (2021b). A systematic literature review on waste-to-resource potential of palm oil clinker for sustainable engineering and environmental applications. *Materials, 14*(16), 4456.

Jagaba, A., Kutty, S., Hayder, G., Elsadig, E. O., Lawal, I., Sayed, K., et al. (2021). Evaluation of the physical chemical bacteriological and trace metals concentrations in different brands of packaged drinking water. *Engineering Letters, 29*(4), 1552–1560.

Jagaba, A. H., Kutty, S. R. M., Isa, M. H., Affam, A. C., Aminu, N., Abubakar, S., et al. (2021). *Effect of environmental and operational parameters on sequential batch reactor Systems in dye degradation* (pp. 193–225). Springer Science and Business Media LLC. https://doi.org/10.1007/978-981-16-5932-4_8.

Jagaba, A. H., Kutty, S. R. M., Isa, M. H., Ghaleb, A. A. S., Lawal, I. M. U., Kilaco, A., et al. (2022). Toxic effects of xenobiotic compounds on the microbial community of activated sludge. *ChemBioEng Reviews, 9*(5). https://doi.org/10.1002/cben.202100055.

Jagaba, A. H., Kutty, S. R. M., Lawal, I. M., Abubakar, S., Hassan, I., Zubairu, I., et al. (2021). Sequencing batch reactor technology for landfill leachate treatment: A state-of-the-art review. *Journal of Environmental Management, 282*, 111946. https://doi.org/10.1016/j.jenvman.2021.111946.

Jagaba, A. H., Kutty, S. R. M., Noor, A., Affam, A. C., Ghfar, A. A., Usman, A. K., et al. (2022). Parametric optimization and kinetic modelling for organic matter removal from agro-waste derived paper packaging biorefinery wastewater. *Biomass Conversion and Biorefinery*, 1–18.

Jagaba, A. H., Kutty, S. R. M., Noor, A., Birniwa, A. H., Affam, A. C., Lawal, I. M., et al. (2021). A systematic literature review of biocarriers: Central elements for biofilm formation, organic and nutrients removal in sequencing batch biofilm reactor. *Journal of Water Process Engineering, 42*, 102178. https://doi.org/10.1016/j.jwpe.2021.102178.

Jagaba, A. H., Kutty, S. R. M., Salih, G. H. A., Noor, A., Bin Md Hafiz, M. F. U., Yaro, N. S. A., et al. (2021). Palm oil clinker as a waste by-product: Utilization and circular economy potential. *Elaeis guineensis*. IntechOpen.

Jagaba, A. H., Kutty, S. R. M., Naushad, M., Lawal, I. M., Noor, A., Affam, A. C., et al. (2022). Removal of nutrients from pulp and paper biorefinery effluent: Operation, kinetic modelling and optimization by response surface methodology. *Environmental Research, 214*(Pt 4), 114091. https://doi.org/10.1016/j.envres.2022.114091.

Javadi, M., Jafarzadeh, Y., Yegani, R., & Kazemi, S. (2018). PVDF membranes embedded with PVP functionalized nanodiamond for pharmaceutical wastewater treatment. *Chemical Engineering Research and Design, 140*, 241–250. https://doi.org/10.1016/j.cherd.2018.10.029.

Ji, J. Y., Xing, Y. J., Ma, Z. T., Cai, J., Zheng, P., & Lu, H. F. (2013). Toxicity assessment of anaerobic digestion intermediates and antibiotics in pharmaceutical wastewater by luminescent bacterium. *Journal of Hazardous Materials, 246–247*, 319–323. https://doi.org/10.1016/j.jhazmat.2012.12.025.

Khan, A. H. (2021). *Comparison among pharmaceutical wastewater treatment schemes for the removal of particular type of drugs and their residues* (pp. 271–303). IWA Publishing. https://doi.org/10.2166/9781789061338_0271.

Khan, A. H., Abutaleb, A., Khan, N. A., El Din Mahmoud, A., Khursheed, A., & Kumar, M. (2021). Co-occurring indicator pathogens for SARS-CoV-2: A review with emphasis on exposure rates and treatment technologies. *Case Studies in Chemical and Environmental Engineering, 4*, 100113. https://doi.org/10.1016/j.cscee.2021.100113.

Khan, A. H., Aziz, H. A., Khan, N. A., Dhingra, A., Ahmed, S., & Naushad, M. (2021). Effect of seasonal variation on the occurrences of high-risk pharmaceutical in drain-laden surface water: A risk analysis of Yamuna River. *Science of the Total Environment, 794*. https://doi.org/10.1016/j.scitotenv.2021.148484.

Khan, A., Aziz, H., Khan, N., Hasan, M., Ahmed, S., Farooqi, I., et al. (2021). Impact, disease outbreak and the eco-hazards associated with pharmaceutical residues: A critical review. *International Journal of Environmental Science and Technology*, 1–12.

Khan, N. A., Bokhari, A., Mubashir, M., Klemeš, J. J., El Morabet, R., Khan, R. A., et al. (2022). Treatment of hospital wastewater with submerged aerobic fixed film reactor coupled with tube-settler. *Chemosphere, 286*, 131838.

Khan, A. H., Khan, N. A., Ahmed, S., Dhingra, A., Singh, C. P., Khan, S. U., et al. (2020). Application of advanced oxidation processes followed by different treatment technologies for hospital wastewater treatment. *Journal of Cleaner Production, 269*. https://doi.org/10.1016/j.jclepro.2020.122411.

Khan, N. A., Khan, A. H., Ahmed, S., Farooqi, I.u. H., Alam, S. S., Ali, I., et al. (2022). Efficient removal of ibuprofen and ofloxacin pharmaceuticals using biofilm reactors for hospital wastewater treatment. *Chemosphere*, 134243. https://doi.org/10.1016/j.chemosphere.2022.134243.

Khan, N. A., Khan, A. H., Tiwari, P., Zubair, M., & Naushad, M. (2021). New insights into the integrated application of Fenton-based oxidation processes for the treatment of pharmaceutical wastewater. *Journal of Water Process Engineering, 44*. https://doi.org/10.1016/j.jwpe.2021.102440.

Khan, A. H., Khan, N. A., Zubair, M., Azfar Shaida, M., Manzar, M. S., Abutaleb, A., et al. (2021). Sustainable green nanoadsorbents for remediation of pharmaceuticals from water and wastewater: A critical review. *Environmental Research, 112243*. https://doi.org/10.1016/j.envres.2021.112243.

Khan, A. H., Tirth, V., Fawzy, M., Mahmoud, A. E. D., Khan, N. A., Ahmed, S., et al. (2021). COVID-19 transmission, vulnerability, persistence and nanotherapy: A review. *Environmental Chemistry Letters*, 1–15.

Kivi, M. A., Alinia, H., Jafarzadeh, Y., & Yegani, R. (2019). High-density polyethylene membranes embedded with carboxylated and polyethylene glycol-grafted nanodiamond to be used in membrane bioreactors. *Journal of Applied Polymer Science, 136*(35), 47914.

Kutty, S. R. M., Almahbashi, N. M. Y., Nazrin, A. A. M., Malek, M. A., Noor, A., Baloo, L., et al. (2019). Adsorption kinetics of colour removal from palm oil mill effluent using wastewater sludge carbon in column studies. *Heliyon*, *5*(10), e02439. https://doi.org/10.1016/j.heliyon.2019.e02439.

Li, L., Guo, C., Fan, S., Lv, J., Zhang, Y., Xu, Y., et al. (2018). Dynamic transport of antibiotics and antibiotic resistance genes under different treatment processes in a typical pharmaceutical wastewater treatment plant. *Environmental Science and Pollution Research*, *25*(30), 30191–30198. https://doi.org/10.1007/s11356-018-2913-2.

Li, T., Liu, J., Bai, R., Ohandja, D. G., & Wong, F. S. (2007). Biodegradation of organonitriles by adapted activated sludge consortium with acetonitrile-degrading microorganisms. *Water Research*, *41*(15), 3465–3473. https://doi.org/10.1016/j.watres.2007.04.033.

Li, R., Wang, J., & Li, S. (2021). Biodegradation of omethoate by Bacillus sp. YB-10: Optimization of culture conditions and degradation characteristics. *Environmental Engineering Research*, *26*(4). https://doi.org/10.4491/eer.2020.235.

Liu, Y., Li, G., Zhang, Z., Liu, H., & Li, Z. (2017). Combined chlorine dioxide oxidation and biological activated carbon processes for treatment of oxytetracycline wastewater. *Desalination and Water Treatment*, *89*, 111–117. https://doi.org/10.5004/dwt.2017.21382.

Liu, C., Liu, Y., Guo, Y., Wang, C., Hu, Z., & Zhang, C. (2019). High-hydrophilic and salt rejecting PA-g/co-PVP RO membrane via bionic sand-fixing grass for pharmaceutical wastewater treatment. *Chemical Engineering Journal*, *357*, 269–279. https://doi.org/10.1016/j.cej.2018.09.151.

Liu, S., Zhang, Y., Zeng, P., Wang, H., Song, Y., & Li, J. (2019). Isolation and characterization of a bacterial strain capable of efficient berberine degradation. *International Journal of Environmental Research and Public Health*, *16*(4). https://doi.org/10.3390/ijerph16040646.

Luo, Q., Wang, J., Wang, J. H., Shen, Y., Yan, P., Chen, Y. P., et al. (2019). Fate and occurrence of pharmaceutically active organic compounds during typical pharmaceutical wastewater treatment. *Journal of Chemistry*, *2019*. https://doi.org/10.1155/2019/2674852.

Mohammed, S., Gill, A. R., Alsafadi, K., Hijazi, O., Yadav, K. K., Hasan, M. A., et al. (2021). An overview of greenhouse gases emissions in Hungary. *Journal of Cleaner Production*, *314*, 127865. https://doi.org/10.1016/j.jclepro.2021.127865.

Ng, K. K., Shi, X., Ong, S. L., & Ng, H. Y. (2016). Pyrosequencing reveals microbial community profile in anaerobic bio-entrapped membrane reactor for pharmaceutical wastewater treatment. *Bioresource Technology*, *200*, 1076–1079. https://doi.org/10.1016/j.biortech.2015.10.100.

Ng, K. K., Shi, X., Yao, Y., & Ng, H. Y. (2014). Bio-entrapped membrane reactor and salt marsh sediment membrane bioreactor for the treatment of pharmaceutical wastewater: Treatment performance and microbial communities. *Bioresource Technology*, *171*(1), 265–273. https://doi.org/10.1016/j.biortech.2014.08.078.

Noor, A., Kutty, S. R., Almahbashi, N. M., Kumar, V., Ghaleb, A. A., & Al-Dhawi, B. N. (2021). Integrated submerged media extended aeration activated sludge (ISmEAAS) reactor start-up and biomass acclimatization. *Journal of Hunan University Natural Sciences*, *48*.

Noor, A., Kutty, S. R. M., Baloo, L., Almahbashi, N. M. Y., Kumar, V., & Ghaleb, A. A. S. (2021). Bio-kinetics of organic removal in EAAS reactor for co-treatment of refinery wastewater with municipal wastewater. *IOP Conference Series: Materials Science and Engineering*, *1092*(1), 012068. https://doi.org/10.1088/1757-899x/1092/1/012068.

Noor, A., Kutty, S. R. M., Jagaba, A. H., Yusuf, M., Akram, M. W., Adil, M. R., et al. (2021). Kinetic modelling of nutrient removal of petroleum industry wastewater remediation. In

2021 Third international sustainability and resilience conference: Climate change (pp. 216–220). IEEE.

Omar, A. H., Toemen, S., Muda, K., & Sulaiman, S. F. (2019). Effect of static magnetic field on the efficiency of granular sludge development for pharmaceutical wastewater treatment. *International Journal of Integrated Engineering*, *11*(1), 151–159. https://doi.org/10.30880/ijie.2019.11.01.018.

Ouyang, E., Lu, Y., Ouyang, J., Liu, X., & Wang, X. (2017). Bacterial communities in a full-scale combined A/O+BIOFOR system treating pharmaceutical wastewater. *Polish Journal of Environmental Studies*, *26*(6), 2661–2666. https://doi.org/10.15244/pjoes/70480.

Palani, K. N., Ramasamy, N., Palaniappan, K. V., Huh, Y. S., & Natesan, B. (2019). Development of integrated membrane bioreactor and numerical modeling to mitigate fouling and reduced energy consumption in pharmaceutical wastewater treatment. *Journal of Industrial and Engineering Chemistry*, *76*, 150–159. https://doi.org/10.1016/j.jiec.2019.03.028.

Rashid, T., Sher, F., Hazafa, A., Hashmi, R. Q., Zafar, A., Rasheed, T., et al. (2021). Design and feasibility study of novel paraboloid graphite based microbial fuel cell for bioelectrogenesis and pharmaceutical wastewater treatment. *Journal of Environmental Chemical Engineering*, *9*(1), 104502. https://doi.org/10.1016/j.jece.2020.104502.

Ren, Y., Yang, J., & Chen, S. (2015). The fate of a nitrobenzene-degrading bacterium in pharmaceutical wastewater treatment sludge. *Chemosphere*, *141*, 13–18. https://doi.org/10.1016/j.chemosphere.2015.05.098.

Rosman, N., Salleh, W. N. W., Mohamed, M. A., Jaafar, J., Ismail, A. F., & Harun, Z. (2018). Hybrid membrane filtration-advanced oxidation processes for removal of pharmaceutical residue. *Journal of Colloid and Interface Science*, *532*, 236–260. https://doi.org/10.1016/j.jcis.2018.07.118.

Saeed, A. A. H., Harun, N. Y., Sufian, S., Bilad, M. R., Nufida, B. A., Ismail, N. M., et al. (2021). Modeling and optimization of biochar based adsorbent derived from Kenaf using response surface methodology on adsorption of Cd2+. *Water*, *13*(7), 999. https://doi.org/10.3390/w13070999.

Satyanarayan, S., Karambe, A., & Vanerkar, A. P. (2009). Herbal pharmaceutical wastewater treatment by a pilot scale upflow anaerobic sludge blanket (UASB) reactor. *Water Science and Technology*, *59*(11), 2265–2272. https://doi.org/10.2166/wst.2009.273.

Shahbeig, H., Mehrnia, M. R., Mohammadi, A. R., Moghaddam, P. E., & Rouini, M. R. (2017). Pharmaceutical wastewater treatment using membrane bioreactor-ozonation system. *Water Environment Journal*, *31*(1), 57–63. https://doi.org/10.1111/wej.12222.

Shariati, F. P., Mehrnia, M. R., Salmasi, B. M., Heran, M., Wisniewski, C., & Sarrafzadeh, M. H. (2010). Membrane bioreactor for treatment of pharmaceutical wastewater containing acetaminophen. *Desalination*, *250*(2), 798–800. https://doi.org/10.1016/j.desal.2008.11.044.

Srinivasan, R., & Nambi, I. M. (2020). Liquid crystal display electrode-assisted bio-electroperoxone treatment train for the abatement of organic contaminants in a pharmaceutical wastewater. *Environmental Science and Pollution Research*, *27*(24), 29737–29748. https://doi.org/10.1007/s11356-019-06898-x.

Svojitka, J., Dvořák, L., Studer, M., Straub, J. O., Frömelt, H., & Wintgens, T. (2017). Performance of an anaerobic membrane bioreactor for pharmaceutical wastewater treatment. *Bioresource Technology*, *229*, 180–189. https://doi.org/10.1016/j.biortech.2017.01.022.

Tang, C. J., Zheng, P., Chen, T. T., Zhang, J. Q., Mahmood, Q., Ding, S., et al. (2011). Enhanced nitrogen removal from pharmaceutical wastewater using SBA-ANAMMOX process. *Water Research*, *45*(1), 201–210. https://doi.org/10.1016/j.watres.2010.08.036.

Tao, W., Zhang, X.-X., Zhao, F., Huang, K., Ma, H., Wang, Z., et al. (2016). High levels of antibiotic resistance genes and their correlations with bacterial community and mobile genetic elements in pharmaceutical wastewater treatment bioreactors. *PLoS One*, *11*(6), e0156854. https://doi.org/10.1371/journal.pone.0156854.

Vistanty, H., & Crisnaningtyas, F. (2021). Integration of upflow anaerobic sludge blanket and constructed wetlands for pharmaceutical wastewater treatment. *IOP Conference Series: Earth and Environmental Science*, *623*(1), 012082. https://doi.org/10.1088/1755-1315/623/1/012082.

Wang, J., Mao, D., Mu, Q., & Luo, Y. (2015). Fate and proliferation of typical antibiotic resistance genes in five full-scale pharmaceutical wastewater treatment plants. *Science of the Total Environment*, *526*, 366–373. https://doi.org/10.1016/j.scitotenv.2015.05.046.

Wang, X., Wang, Q., Li, S., & Li, W. (2015). Degradation pathway and kinetic analysis for p-xylene removal by a novel Pandoraea sp. strain WL1 and its application in a biotrickling filter. *Journal of Hazardous Materials*, *288*, 17–24. https://doi.org/10.1016/j.jhazmat.2015.02.019.

Wang, G., Wang, D., Xu, Y., Li, Z., & Huang, L. (2020). Study on optimization and performance of biological enhanced activated sludge process for pharmaceutical wastewater treatment. *Science of the Total Environment*, *739*, 140166. https://doi.org/10.1016/j.scitotenv.2020.140166.

Wang, Q. H., Zhang, L., Tian, S., Sun, P. T. C., & Xie, W. (2007). A pilot-study on treatment of a waste gas containing butyl acetate, n-butyl alcohol and phenylacetic acid from pharmaceutical factory by bio-trickling filter. *Biochemical Engineering Journal*, *37*(1), 42–48. https://doi.org/10.1016/j.bej.2007.03.005.

Wei, X., Li, B., Zhao, S., Wang, L., Zhang, H., Li, C., et al. (2012). Mixed pharmaceutical wastewater treatment by integrated membrane-aerated biofilm reactor (MABR) system—A pilot-scale study. *Bioresource Technology*, *122*, 189–195. https://doi.org/10.1016/j.biortech.2012.06.041.

Yusuf, M., Bazli, L., Alam, M. A., Masood, F., Keong, L. K., Noor, A., et al. (2021). Hydrogen production via natural gas reforming: A comparative study between DRM, SRM and BRM techniques. In *2021 Third international sustainability and resilience conference: Climate change* (pp. 155–158). IEEE.

Zakeritabar, S. F., Jahanshahi, M., & Peyravi, M. (2018). Photocatalytic behavior of induced membrane by ZrO2–SnO2 nanocomposite for pharmaceutical wastewater treatment. *Catalysis Letters*, *148*(3), 882–893. https://doi.org/10.1007/s10562-018-2303-x.

Zhai, W., Yang, F., Mao, D., & Luo, Y. (2016). Fate and removal of various antibiotic resistance genes in typical pharmaceutical wastewater treatment systems. *Environmental Science and Pollution Research*, *23*(12), 12030–12038. https://doi.org/10.1007/s11356-016-6350-9.

Zhang, C., Li, C., Zhang, & Shao, X. (2012). Isolation and characterization of a benzoic acid degrading pseudomonas putida from a pharmaceutical wastewater treatment plant. *African Journal of Microbiology Research*, *6*, 4608–4613.

Zhang, Y., Xie, J., Liu, M., Tian, Z., He, Z., van Nostrand, J. D., et al. (2013). Microbial community functional structure inresponse to antibiotics in pharmaceutical wastewater treatment systems. *Water Research*, *47*(16), 6298–6308. https://doi.org/10.1016/j.watres.2013.08.003.

Zhang, M., Zuo, J., Yu, X., Shi, X., Chen, L., & Li, Z. (2018). Quantification of multi-antibiotic resistant opportunistic pathogenic bacteria in bioaerosols in and around a pharmaceutical wastewater treatment plant. *Journal of Environmental Sciences*, *72*, 53–63. https://doi.org/10.1016/j.jes.2017.12.011.

Zhao, F., Ju, F., Huang, K., Mao, Y., Zhang, X. X., Ren, H., et al. (2019). Comprehensive insights into the key components of bacterial assemblages in pharmaceutical wastewater treatment plants. *Science of the Total Environment*, *651*, 2148–2157. https://doi.org/10.1016/j.scitotenv.2018.10.101.

Zhao, D., Zhu, C., Sun, S., Yu, H., Zhang, L., Pan, W., et al. (2007). Toxicity of pharmaceutical wastewater on male reproductive system of Mus musculus. *Toxicology and Industrial Health*, *23*(1), 47–54. https://doi.org/10.1177/0748233707077446.

Zheng, W., Li, X. M., Hao, Z. M., Wang, D. B., Yang, Q., & Zeng, G. M. (2010). Coal cinder filtration as pretreatment with biological processes to treat pharmaceutical wastewater. *Water Science and Technology*, *62*(1), 15–20. https://doi.org/10.2166/wst.2010.244.

Application of hybrid advanced oxidation and adsorption processes for pharmaceutical wastewater treatment

8

Ivanna Lapchuk[a], *Alexander Shyichuk*[b,c], *and Tetiana Tatarchuk*[d]
[a]Educational and Scientific Center of Material Science and Nanotechnology, Vasyl Stefanyk Precarpathian National University, Ivano-Frankivsk, Ukraine, [b]Department of Chemistry, Vasyl Stefanyk Precarpathian National University, Ivano-Frankivsk, Ukraine, [c]Faculty of Chemical Technology and Engineering, Bydgoszcz University of Science and Technology, Bydgoszcz, Poland, [d]Faculty of Chemistry, Jagiellonian University, Kraków, Poland

1. Introduction

Water pollution is an important environmental problem. The reason is the continuous development of industry, which has caused huge amounts of wastewater. Purification of surface and wastewater, as well as drinking water, is becoming an increasingly important task. Chemical and biological pollution of water resources leads to negative consequences (Adegoke & Stenström, 2019). The ingress of chemicals into water bodies has a negative effect on flora, fauna, and humans.

Among chemical pollutants, pharmaceuticals have a high resistance to destruction and harmful effects on living organisms in water bodies (Kosma, Nannou, Boti, & Albanis, 2019; Luo et al., 2014). Pharmaceuticals enter water bodies with insufficiently treated wastewater. Usually, the sources of pharmaceutically active compounds are the waters of the pharmaceutical industry and medical institutions (Antonopoulou, Kosma, Albanis, & Konstantinou, 2021). Another equally important source of contamination by pharmaceuticals is sewage system (Adegoke & Stenström, 2019; Kosma et al., 2019). The significant emissions of drugs are due to their widespread use. The coronavirus pandemic and other diseases require use of large amounts of pharmaceuticals. The pharmaceutical industry produces a large number of pharmaceutical compounds used for the treatment and prevention of diseases (Rosman et al., 2018). The development of medicine and pharmacy promotes good health, prevents disease and mortality in severe cases (Chavoshani, Hashemi, Mehdi Amin, & Ameta, 2020). However, excessive use of pharmaceuticals has a negative impact on water resources. As a result, polluted water bodies cause new diseases and increase health care costs. Therefore, there is a growing interest in removing pharmaceuticals from waters (Chavoshani et al., 2020).

The Treatment of Pharmaceutical Wastewater. https://doi.org/10.1016/B978-0-323-99160-5.00008-4

The most common pharmaceuticals contained in water sources are antibiotics, beta-blockers, antimicrobials, analgesics, anesthetics, disinfectants, lipid regulators, and antiinflammatory drugs (Antonopoulou et al., 2021; Dhangar & Kumar, 2020). Fig. 1 shows the main polluting drugs and ways of their release into the environment. Pharmaceuticals differ from other pollutants such as heavy metals, dyes, fertilizers, radionuclides (Faisal, Al-Wakel, Assi, Naji, & Naushad, 2020; Naushad, 2014; Naushad et al., 2019; Naushad, ALOthman, Awual, Alam, & Eldesoky, 2015; Naushad, Mittal, Rathore, & Gupta, 2015; Sharma & Naushad, 2020). The main difference is large diversity in structure, molecular weight and properties. For example, pharmaceuticals may be sparingly soluble in water or be lipophilic (Chavoshani et al., 2020). Lipophilicity plays an important role in the penetration of drugs across the cell membrane (Fatta-Kassinos, Meric, & Nikolaou, 2011). In turn, the ionic charge determines the role of the drug in ion exchange processes occurring in the environment (Fatta-Kassinos et al., 2011). A negative factor is the high resistance of drugs which leads to their accumulation in the environment and living organisms (Adegoke & Stenström, 2019; Chavoshani et al., 2020). An example is clofibric acid, which retains its structure and properties for several years (Chavoshani et al., 2020). This agent was found in rivers and groundwater (Boyd, Reemtsma, Grimm, & Mitra, 2003). In contrast, paracetamol and ibuprofen have low resistance and decompose within a few days (Löffler, Römbke, Meller, & Ternes, 2005).

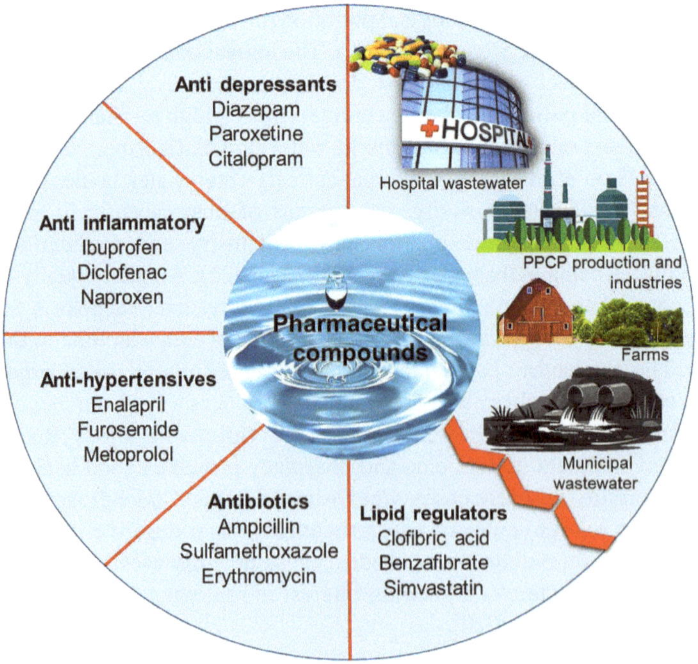

Fig. 1 The main polluting drugs and ways of their release into the environment.

The largest amount of pharmaceutical emissions comes from hospital wastewater. Hospital wastewater contains analgesics, antibiotics, hormonal drugs, and radionuclides. These drugs have a significant impact on human body. The greatest attention is paid to endocrine active drugs. For example, the World Health Organization has published a list of endocrine disruptors (ED). These compounds affect the endocrine system and can even lead to cancer and infertility. The list includes substances of natural origin (phytoestrogens and mycotoxins), as well as synthetic compounds (diethylstilbestrol and bisphenol A), which are components of antibacterial agents and detergents (Ebele, Abou-Elwafa Abdallah, & Harrad, 2017; Wielogórska, Elliott, Danaher, Chevallier, & Connolly, 2015). Antibiotics are another group of dangerous pharmaceutical contaminants. High concentrations of common antibiotics, such as ampicillin, ciprofloxacin, erythromycin, and tetracycline, were reported in hospital wastewater, livestock effluents, and sewage systems (Zainab, Junaid, Xu, & Malik, 2020). Unused antibiotics often end up in water bodies and cause the formation of antibiotic-resistant bacteria (Jamil, 2019; Yakubu, 2017). There are many examples of bacteria that have become resistant: *Staphylococcus aureus* is resistant to methicillin, enterococci and multiresistant pseudomonads are resistant to vancomycin. Infections with the resistant bacteria are difficult to treat (Fatta-Kassinos et al., 2011). The number of resistant bacteria may be reduced through the proper use of antibiotics and effective wastewater treatment.

The content of pharmaceutical contaminants in water bodies usually varies from a few ng/L to a few mg/L. Table 1 shows the exemplary content of pharmaceutical contaminants in water bodies. Dangerous concentrations of pharmaceuticals in water are the result of imperfect water purification by standard methods (Luo et al., 2014). Pharmaceuticals in water reservoirs may or may not cause adverse effects depending on their concentrations. Water contaminated with various types of pharmaceutical compounds can become toxic. The reason is that mixtures of drugs in water can have synergistic effects even at low concentrations. As a result, the impact on humans becomes much stronger (Ebele et al., 2017). An example is the combination of carbamazepine and clofibric acid, which in the mixture had a stronger effect on *Daphnia magna* (Thorpe et al., 2001). Therefore, it is important to use effective purification methods to minimize water contamination by pharmaceuticals.

2. Purification of pharmaceutical wastewaters by advanced oxidation processes

Like other contaminants, pharmaceuticals require appropriate removal methods. The optimal choice of water treatment method is dependent on the volume of effluents, the concentration of pollutants, as well as operating costs (Morone, Mulay, & Kamble, 2019). In real conditions, there are various difficulties. Sometimes it is difficult to achieve perfect wastewaters treatment. Therefore, improving the methods of removing pharmaceuticals remains still a relevant task (Rostam & Taghizadeh, 2020).

The process of removing contaminants consists of several stages. At the initial stage, the wastewater is purified by filtration, precipitation, coagulation, and

Table 1 Concentrations of pharmaceuticals in water reservoirs.

Pollutant	Reservoir	Concentration (ng/L)	References
Ibuprofen	Atlantic Ocean	326.1–2094.4	Pereira et al. (2016)
Ibuprofen	Surface waters	2796	Wu, Witter, Spongberg, and Czajkowski (2009)
Ibuprofen	Waste water treatment plant influents	3.877–19.118	Paíga et al. (2016)
Acetaminophen	Ocean surface and bottom water samples	17.4–34.6	Pereira et al. (2016)
Carbamazepine	Surface waters	1238	Wu et al. (2009)
Mefenamic acid	Sewage effluents	15–225	McEneff, Barron, Kelleher, Paull, and Quinn (2014)
	Sea surface waters	3–38	McEneff et al. (2014)
	Surface waters	366	Roberts and Thomas (2006)
Erythromycin	Surface waters	438	Wu et al. (2009)
Nonylphenol	Surface waters	372	Duong et al. (2010)
Sulfamethazine	Surface waters	10	Wu et al. (2009)
Acetaminophen	Hospital effluents	211–58,857	Santos et al. (2013)
Diazepam	Drinking waters	1.9	Wu, Xiang, Que, Chen, and Xu (2015)
	Safe fluid waste	1180	Van Der Ven et al. (2006)
	Surface waters	53	Ternes, Bonerz, and Schmidt (2001)
	Hospital effluents	29.6–49.1	Santos et al. (2013)
Oxazepam	Surface waters	40	Hass, Duennbier, and Massmann (2012)
Dextropropoxyphene	Surface waters	682	Roberts and Thomas (2006)
Naproxen	Hospital effluents	36.5–6042	Santos et al. (2013)
Pravastatin	Hospital effluents	332–2086	Santos et al. (2013)
Salicylic acid	Hospital effluents	6.85–4681	Santos et al. (2013)
	Waste water treatment plant influents	33.535	Paíga et al. (2016)

flocculation. The second stage involves the removal of organic compounds by biological treatment. Usually, activated sludge is used at this stage. The final stage is the use of advanced oxidation and adsorption as the most effective methods for removing contaminants (Morone et al., 2019).

In addition to advantages, standard cleaning methods have also a number of disadvantages. Micro-, ultra- and nanofiltration are effective for purifying water from pathogens, heavy metals, and dyes. However, frequent contamination of membranes, their utilization and high-energy consumption make this method too expensive. The use of activated sludge requires a long time, which is unacceptable for the removal of pharmaceuticals from wastewater. The reason is the sludge bacteria become resistance to antibiotics. In addition, activated sludge is ineffective in removing beta-blockers. Activated carbon is less effective in removing polar organic matter through its hydrophobic surface (Dhangar & Kumar, 2020).

As we can see, usual treatment methods are often ineffective in removing persistent pharmaceutical contaminants from wastewater (Chavoshani et al., 2020). Improved methods provide higher cleaning efficiency. In the study (Watkinson, Murby, & Costanzo, 2007), the wastewater was purified using activated sludge and an improved microfiltration/reverse osmosis method. Both the methods reduced the initial concentrations of antibiotics such as cephalexin, sulfamethoxazole, ciprofloxacin, cefaclor, and trimethoprim. However, the second method was 25% more efficient.

In contrast to the above methods, advanced oxidation processes (AOPs) are highly efficient and provide almost 100% destruction of pharmaceuticals. The active species in advanced oxidation are hydroxyl radicals, superoxide radicals, and singlet oxygen, which cause the decomposition of pharmaceuticals (Morone et al., 2019). Advanced oxidation processes include ozonation, Fenton and photo-Fenton processes, photocatalysis, and electrochemical oxidation (Fig. 2). These methods are common for the purification of wastewaters; however, their use is becoming more diverse

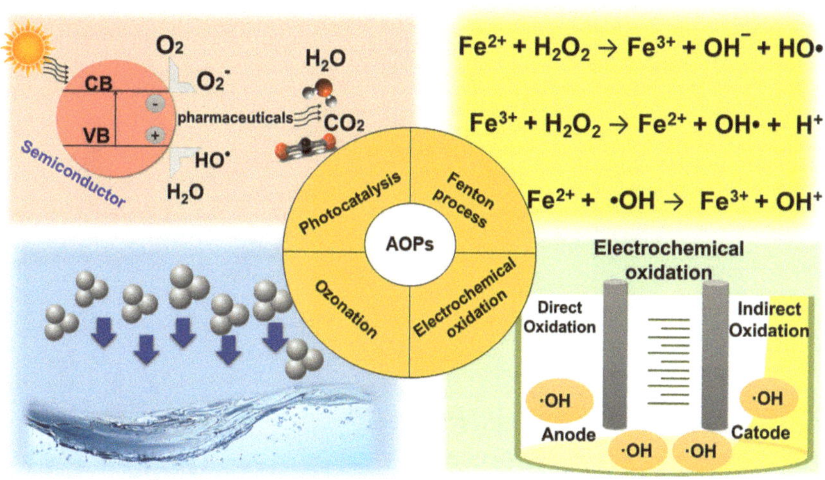

Fig. 2 Types of advanced oxidation processes.

(Ahmed, Naji, Faisal, Al-Ansari, & Naushad, 2020; Kumar et al., 2020; Muthusaravanan et al., 2018; Naushad & ALOthman, 2015; Naushad, Sharma, & Alothman, 2019; Sharma et al., 2019, 2020). Advanced oxidation processes are also used to treat microbial skin infections (Zeng & Lu, 2018), obtain energy (Gellé & Moores, 2019), synthesize organic compounds (Michelin & Hoffmann, 2018). The following are some examples of the application of advanced oxidation processes to wastewater treatment.

2.1 Photocatalytic oxidation

The photocatalytic reaction is the chemical reaction that occurs under the action of light and substances that absorb light quanta (Danyliuk, Tatarchuk, & Shyichuk, 2020). The result is the formation of reactive oxygen species. The type and amount of reactive oxygen species depend on the irradiation used and the photocatalyst used. A photocatalytic reaction includes two half-reactions: reduction and oxidation. The oxidation reaction occurs when the photocatalyst absorbs quanta having energy equal to or larger than the energy of band gap. Electrons pass from the conduction band to the valence band. As a result, positively charged holes are formed (Darwish & Mohammadi, 2018). The holes initiate oxidation reactions that lead to the deep destruction of pharmaceuticals to CO_2 and H_2O. Fig. 3 shows the basic mechanisms of the oxidative destruction.

Among the many possible photocatalysts, the most common is titanium dioxide. The rate of photocatalytic reaction depends on the surface area of the photocatalyst, light intensity and wavelength (Morone et al., 2019). Light intensity has a key effect on the degradation of pharmaceuticals. Both low and high intensity modes are used with a power density of <20 and >25 mW/cm^2, respectively (Nguyen et al., 2020).

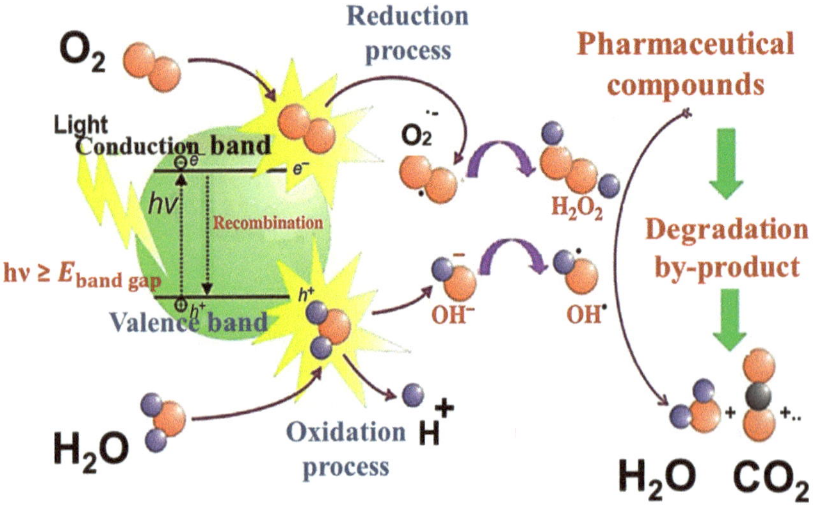

Fig. 3 Probable mechanism of the photocatalytic destruction reaction (Nguyen, Bui, & Nguyen, 2020).

The antibiotic moxifloxacin was successfully destroyed using a commercial titania photocatalyst Degussa P25 and near ultraviolet irradiation (UV-A, 300–440 nm) with a density of $4\,mW/cm^2$ (Van Doorslaer et al., 2015). The formed by-products had lower toxicity than moxifloxacin. The irradiation time is also an important factor. The antiinflammatory drug diclofenac has reached full mineralization after 2 h of sun exposure over TiO_2 photocatalyst (Calza et al., 2006). The method of material synthesis directly affects the performance of the photocatalyst. Titanium dioxide nanotubes were obtained by electrochemical anodizing of titanium metal and proved to be a very active photocatalyst. The decomposition of caffeine and salicylic acid in acidic and alkaline media was about 50% (Arfanis et al., 2017).

Zinc oxide is also well known for its photocatalytic properties. ZnO was obtained by precipitation using the pullulan biomaterial and showed a high percentage of degradation of amoxicillin and paracetamol (Mohamed Isa, Shameli, Ch'ng, Che-Jusoh, & Hazan, 2021). ZnO concentrations during degradation experiments were 0.17, 0.67, 0.84, and 1.01 M. The fastest decomposition of drugs was observed at a ZnO concentration equal to 0.67 M. The degradation of amoxicillin and paracetamol was 85.7% and 96.8%, respectively. Details are presented in Fig. 4. Good photocatalytic activity of ZnO led to the destruction of 63% of ciprofloxacin in an aqueous solution (Silva et al., 2016). Photocatalytic zinc oxide also showed high toxicity against gram-negative bacteria *Vibrio fischeri*. After 45 min of contact, the toxicity was 70%.

In addition to ZnO and TiO_2, a number of other compounds have photocatalytic properties. Examples are the following: $Bi_4Ti_3O_{12}$, carbon nanotubes, polyesters, reduced graphene oxide (rGO), graphene, polyaniline, WO_3, Fe_2O_3, Fe_3O_4, and Bi_2O_3. These substances can be doped with Pt, Bi, Sr, Ag, Bi, Sn, N, Fe, N, and Ru (Velempini, Prabakaran, & Pillay, 2021).

2.2 Fenton oxidation

One more effective AOP is Fenton oxidation. This reaction has a high rate and does not require expensive chemicals. The Fenton reaction is the decomposition of hydrogen peroxide using ferrous ions as a catalyst. The optimal pH value is in the range of 2 to 4. The hydroxyl radicals formed are key active agents in the decomposition of pollutants in wastewater. During Fenton oxidation, the following reactions occur (Morone et al., 2019; Rosman et al., 2018):

$$Fe^{2+} + H_2O_2 = Fe^{3+} + OH^- + HO^\bullet$$

$$Fe^{3+} + H_2O_2 = Fe^{2+} + HOO^\bullet + H^+$$

$$Fe^{2+} + OH^\bullet = Fe^{3+} + OH^-$$

Because the Fenton reaction is effective in an acidic environment, wastewater needs to be acidified. Further neutralization of effluents requires additional costs (Morone et al., 2019; Rosman et al., 2018). The efficiency of the Fenton reaction depends on the concentration of hydrogen peroxide used. The higher the concentration of H_2O_2, the greater the number of active radicals $OH^\bullet$ (Kumar et al., 2019).

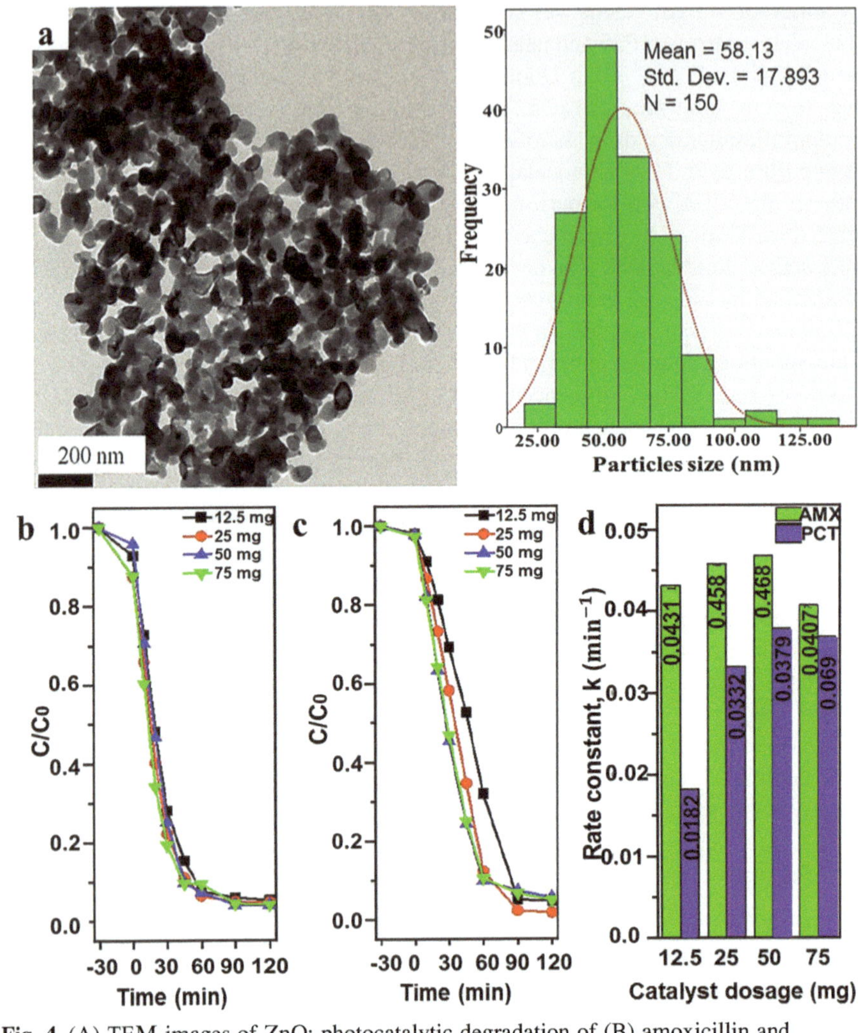

Fig. 4 (A) TEM images of ZnO; photocatalytic degradation of (B) amoxicillin and (C) paracetamol depending on dose; and (D) the rate constant of degradation reaction (Mohamed Isa et al., 2021).

The disadvantages of the homogeneous Fenton reaction are the limited pH range and the formation of insoluble iron oxy-hydroxides. The combination of Fenton oxidation with other methods of pollutant removal can minimize these disadvantages. Fenton oxidation in combination with ultraviolet radiation is more effective for pharmaceutical contaminants removing. Such methods are called photo-Fenton processes (Bello & Raman, 2019). Removal of pharmaceuticals from wastewater was tested by UV radiation at 254 nm (UV_{254}), Fenton oxidation in dark (Fe^{2+}, Fe^{3+}/H_2O_2), and photo-Fenton (Fe^{2+}, Fe^{3+}/H_2O_2/light) (De la Cruz et al., 2012). Table 2 shows the obtained results. The highest purification efficiency of effluent coming from a municipal wastewater treatment plant was observed when irradiated with ultraviolet radiation in combination with H_2O_2 (50 mg/L) and Fe^{2+} (5 mg/L).

Table 2 Removal of pharmaceuticals by treatment with UV and UV/H_2O_2 (De la Cruz et al., 2012).

Pharmaceutical		Removal (%)		
		UV_{254}, 10 min	**UV + H_2O_2 (mg/L), 10 min**	**UV + H_2O_2 (mg/L), 30 min**
Lipid regulators	Bezafibrate	24	75	100
	Gemfibrozil	18	75	100
Antidiabetic drug	Metformin	6	34	11
Antiepileptic drugs	Carbamazepine	23	75	100
Analgesics	Diclofenac	100	100	100
	Ibuprofen	34	100	100
	Ketoprofen	100	100	100
	Mefenamic acid	100	100	100
	Primidone	5	65	100
Beta Blockers	Atenolol	20	58	100
	Metoprolol	14	73	100
Antibiotics	Azithromycin	0	50	100
	Ciprofloxacin	48	69	100
	Metronidazole	24	52	100
	Sulfamethoxazole	51	98	100

The Fenton oxidation can also use heterogeneous catalysts. Usually, the heterogeneous Fenton catalysts contain iron compounds. An example of an improved Fenton catalyst is reported by Tang and Wang (2018). Magnetic nanoparticles were embedded in a mesoporous carbon hybrid (Fe@MesoC) by pyrolysis. The catalytic activity in the Fenton process was tested by the decomposition of sulfamethoxazole. Fe@MesoC showed 100% catalytic degradation of the drug. The probable mechanism of catalytic degradation is shown in Fig. 5A. Such a high decomposition efficiency is due to the large number of Fe-based nanoparticles and the synergistic effect of the two compounds. The study (Lai et al., 2019) also showed almost complete degradation of pharmaceuticals. The composite $MnFe_2O_4$/biochar provided 95% degradation of tetracycline when exposed to visible light for 2 h at the concentration of H_2O_2 equal to 100 mmol L^{-1}. The general scheme of redox reactions is shown in Fig. 5B.

Another type of Fenton reaction is electro-Fenton oxidation. Here, H_2O_2 is formed due to the electrochemical reduction of oxygen at the cathode. The electrochemical decomposition of carbamazepine (CBZ) by the electro-Fenton reaction was studied (Komtchou, Dirany, Drogui, & Bermond, 2015). The optimal working conditions for the degradation of carbamazepine were pH 3.0 and current density 0.2 A. The decomposition degree of 73% was reached for 120 min. Instead of iron catalysts, alternative catalysts may contain copper, chromium, and cobalt. For example,

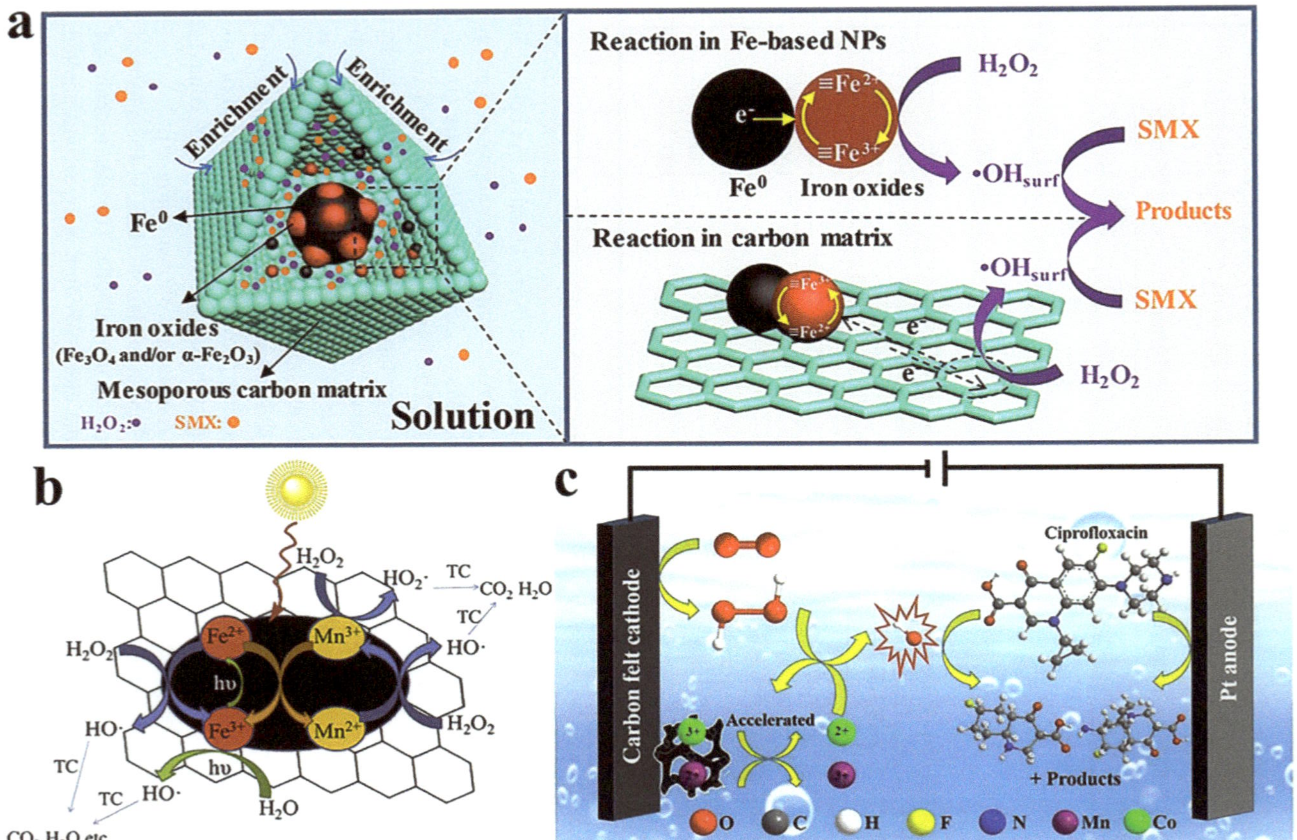

Fig. 5 (A) The main stages of H_2O_2 action and mechanism of sulfamethoxazole removal with Fe@MesoC/H_2O_2 (Tang & Wang, 2018); (B) schematic mechanism of tetracycline degradation with H_2O_2 and $MnFe_2O_4$/biochar composite (Lai et al., 2019); and (C) the mechanism of electro-Fenton oxidation on the $MnCo_2O_4$-CF cathode (Mi et al., 2019).

$Mn_xCo_{3-x}O_4$ was tested as potential stable electro-Fenton catalyst for the degradation of the antibiotic ciprofloxacin (CIP) (Mi et al., 2019). This material provided 100% removal of ciprofloxacin within 5 h. The high percentage of pollutant removal is associated with an increase in the electron transfer rate due to the synergistic effect of manganese and cobalt. Fig. 5C shows the mechanism of ciprofloxacin degradation.

2.3 Ozonation

Ozone is well known as a strong oxidant (Klavarioti, Mantzavinos, & Kassinos, 2009). Ozone can destroy organic matter in wastewater in two ways: directly in the form of molecular ozone O_3 (direct oxidation) or by the formation of hydroxyl radicals (indirect oxidation). The two possible reactions have different kinetics and different products formed (Malik, Ghosh, Vaidya, & Mudliar, 2020). Ozone is used in a variety of industries, including textile and paper processing, as well as for air disinfection. The combination of ozonation with light irradiation or hydrogen peroxide can increase the efficiency of degradation (Klavarioti et al., 2009). The dose of H_2O_2 is important because an excessive amount of H_2O_2 will reduce the percentage of degradation. As a rule, the effectiveness of ozonation in the decomposition of pharmaceuticals increases with increasing pH. For that reason, constant monitoring of ozonized wastewater is required (Klavarioti et al., 2009; Malik et al., 2020). The most significant disadvantage of ozonation in water treatment is the high operating costs. First, this process requires a lot of energy. About 85% of energy is released in the form of heat, which leads to overheating of the reactor. Second, O_3 has a short half-life—15–20 min in water at a temperature of 20–25 °C. These shortcomings result in high operating costs (Morone et al., 2019). The amount of electricity used to break down the antibiotics ciprofloxacin and trimethoprim, as well as the drug cyclophosphamide is decreased in the following order: $UV > UV/O_3 > UV/H_2O_2/O_3 > O_3 > H_2O_2/O_3$ (Lester, Avisar, Gozlan, & Mamane, 2011).

Table 3 shows the efficiency of the removal of various pharmaceutical contaminants. The efficiency of ozonation is dependent on the concentration of O_3 and on the pH value. Typical ozone concentrations range from 5 to 15 mg/L (Patel et al., 2019). However, higher concentrations of O_3 are also used. For example, 256 mg of ozone per liter was used to degrade 97% of cyclophosphamide (Ferre-Aracil et al., 2016).

Many studies have been conducted on wastewater treatment by ozonation. Snyder et al. showed that drugs such as acetaminophen, carbamazepine, diclofenac, naproxen, sulfamethoxazole were decomposed by ozonation to 80% while diazepam and ibuprofen were removed by only 50%. The addition of H_2O_2 did not change the efficiency of degradation (Snyder et al., 2006). The methods of ozonation and ozonation with the addition of H_2O_2 were tested for the pre-treatment of hospital wastewater before discharge to the wastewater treatment plant (Khan et al., 2020). It was found that the combination of O_3 and H_2O_2 is less effective in the degradation of drugs than conventional ozonation. Common contaminants diclofenac and carbamazepine were removed from wastewater by ozonation by 95% (Sui, Huang, Deng, Yu, & Fan, 2010). The concentrations of pharmaceuticals in wastewater ranged from 4.4 ng/L

Table 3 The efficiency of pharmaceuticals removal by ozonation.

Method	Pollutant	Removal (%)	References
O_3/H_2O_2	Carbamazepine	55	Wert, Rosario-Ortiz, and Snyder (2009)
Ozonation	Diazepam	50	Snyder, Wert, Rexing, Zegers, and Drury (2006)
Ozonation	Ibuprofen	50	Snyder et al. (2006)
Ozonation	Ibuprofen	36.1	Nakada et al. (2007)
Ozonation	Acetaminophen	80	Snyder et al. (2006)
Ozonation	Carbamazepine	>81	Nakada et al. (2007)
Ozonation	Diclofenac	80	Snyder et al. (2006)
Ozonation	Naproxen	80	Snyder et al. (2006)
Ozonation	Naproxen	99.7	Nakada et al. (2007)
Ozonation	Etodolac	99	Kaya et al. (2017)
Ozonation	Sulfamethoxazole	80	Snyder et al. (2006)
Ozonation	Sulfamethoxazole	100	Martins, Dantas, Sans, Esplugas, and Quinta-Ferreira (2015)
Ozonation	Mefenamic acid	54.3	Nakada et al. (2007)
Ozonation	Ketoprofen	93.2	Nakada et al. (2007)

to 6.6mg/L and from 2.2 to 320ng/L. Primary water treatment provided a fairly low (12%) rate of degradation, while secondary treatment provided almost complete degradation. Removal of Procaine Penicillin G (PPG) by ozonation was tested in Arslan-Alaton and Caglayan (2005). The study was performed in a semibatch ozone reactor with pH values of 3, 7, 12. PPG was removed by 37% and 82% of COD within 60 and 120min, respectively. Higher dosing of ozone improved the efficiency of wastewater treatment.

2.4 Electrochemical oxidation

Electrochemical oxidation is a very efficient advanced oxidation method used to degrade pharmaceuticals in wastewater (Anglada, Urtiaga, & Ortiz, 2009). This method involves two mechanisms. The first mechanism is direct anodic oxidation by removing electrons at the anode. The second possible mechanism is indirect anodic oxidation by aggressive intermediates. The electrode material must have high electrical conductivity, high chemical resistance, and low cost (Anglada et al., 2009).

Electrochemical oxidation was used for the degradation of the antibiotics ofloxacin and lincomycin (Carlesi Jara, Fino, Specchia, Saracco, & Spinelli, 2007). The anode materials tested were titanium-platinum and graphite. Dimensionally Stable Anodes (DSA) were also tested. Ofloxacin was destroyed at all the mentioned anodes. The electrochemical oxidation reaction took place according to the first-order kinetic equation. Boron doped diamond anode was used to destroy ciprofloxacin, sulfamethoxazole та salbutamol (Lan, Coetsier, Causserand, & Groenen Serrano, 2017). The role of sulfates and chlorides in electrochemical oxidation has been studied. The addition of sulfates increased the rates of degradation of both ciprofloxacin and sulfamethoxazole. Chloride ions showed good results in the degradation of ciprofloxacin. However, the increase in current density has led to the formation of a large amount of halogenated organic compounds. Almost 100% degradation of ciprofloxacin was reported in the study (Mu et al., 2019). Electrochemical oxidation was performed with Sb-doped SnO_2 electrodes at a current density of 20 mA/cm^2. Under these conditions, 100% degradation of ciprofloxacin was achieved within 60 min.

The antiinflammatory drug aspirin is often a component of hospital wastewater. Three anodes were applied for the anodic oxidation of aspirin: PbO_2, boron-doped diamond (BDD) and porous Ti/BDD (He et al., 2015). Aspirin degradation is a pseudo-first order reaction. With PbO_2 anode, aspirin was removed only by 70%. The best removal efficiency was observed with the Ti/BDD anode. The mechanism of degradation probably includes both direct and indirect anodic oxidation on the BDD surface. In turn, the PbO_2 electrode promotes indirect electrochemical oxidation.

3. Adsorption as a method of pharmaceutical substances removing

Adsorption processes are known for their wide application in various fields (Balasubramani, Sivarajasekar, & Naushad, 2020; Sharma et al., 2018, 2020; Sharma, Pathania, Naushad, & Kothiyal, 2014). Molecules of sorbate are attached to the surface mainly by Van der Waals forces, electrostatic interactions and hydrogen bonds (Patel et al., 2019). These interactions provide so-called physical adsorption. In addition, sorbates can also be attached by strong chemical bonds. In this case, chemisorption takes place (Tatarchuk et al., 2021). Adsorption is described by the Langmuir, Freundlich, and Dubinin-Radushkevich models. These models characterize various aspects of an adsorption process.

An important property of adsorbents is regeneration. Reuse of adsorbents reduces operating costs (Bello & Raman, 2019). Adsorption is used to purify water from heavy metals (Latif et al., 2020; Tatarchuk et al., 2021), dyes (Mironyuk, Myslin, Lapchuk, Tatarchuk, & Olkhovyy, 2021), and pharmaceuticals (Bello & Raman, 2019). The purification result depends on surface area and polarity of the adsorbent, the nature and concentration of the sorbate as well as the pH value (Patel et al., 2019). The main advantages of adsorption over other methods are simple equipment, low energy consumption and high efficiency of water treatment.

Various materials are tested as adsorbents for pharmaceuticals. The metal-organic framework modified with ethylenediamine was used as an adsorbent to remove clofibric acid and naproxen (Hasan, Choi, & Jhung, 2013). The adsorbent was well regenerated by washing with C_2H_5OH. Materials derived from household waste can also be adsorbents for the removal of pharmaceuticals. Affordable carbon sorbents sometimes have higher adsorption capacity than commercial ones (Mestre et al., 2009). Nanometer-sized materials also have high adsorption efficiency.

Magnetization may be an additional advantage that simplifies the separation of spent adsorbents. Magnetic nanoparticles of γ-Fe_2O_3 have the adsorption capacity toward diclofenac as high as 261 mg/g (Leone et al., 2018). The temperature did not affect the sorption efficiency. Increasing the pH decreased the adsorption capacity. This fact suggests that OH^- anions compete with diclofenac molecules for adsorption centers. Magnetic particles with graphene oxide showed a moderate adsorption capacity toward tetracyclines ($q_{max} = 45$ mg/g). The pH of the solution had little effect on the adsorption of antibiotics (Lin, Xu, & Li, 2013). Modification of materials allows improving the adsorption properties. The combination of activated carbon with $CuFe_2O_4$ ferrite nanoparticles led to good removal of tetracycline from wastewater (Akkaya Sayğılı & Sayğılı, 2020). The obtained composite material sorbed 384.6 mg/g of antibiotics. As we see, materials with sorption properties are constantly improved.

4. Hybrid oxidation and adsorption processes

Methods of pharmaceutical wastewater treatment alone do not provide complete removal of contaminants. Each method of water purification has its advantages and disadvantages. Thus, a combination of the two methods may minimize the disadvantages and improve wastewater treatment. Advanced oxidation effectively destroys organic pollutants. However, complete mineralization of pharmaceuticals is not always possible. Adsorption effectively removes the pharmaceuticals. However, adsorbent materials gradually lose their functionality and require regeneration. This leads to additional costs and incomplete wastewater treatment. Thus, the combination of AOPs with adsorption can provide highly effective purification of reservoirs from pharmaceuticals. Both methods together provide a synergistic effect that results in improved water purification. Adsorption is often combined with the following advanced oxidation methods: photocatalysis, Fenton oxidation, and ozonation (Bello & Raman, 2019; Moravvej, Soroush, & Rahimpour, 2020).

4.1 Photocatalysis and adsorption

The integration of adsorption and photocatalysis is a new topic in wastewater treatment. It is known that photocatalytic materials, due to their heterogeneous surface, can adsorb a certain amount of contaminants. Therefore, it will be economically advantageous to use materials that have both high photocatalytic and adsorption properties. Materials with high adsorption and photocatalytic characteristics have been reported to be titanium dioxide, zinc compounds, carbon, nanometer-sized silica, and certain polymers. Table 4 shows materials used for the hybrid process and their effectiveness in water treatment. The disadvantage of TiO_2 as a photocatalyst and adsorbent is the insufficiently studied technology for large-scale applications (Rostam & Taghizadeh, 2020).

The use of TiO_2 and ZnO in hybrid methods has been reported in the works (Di et al., 2017; El Mouchtari et al., 2020; Lin, Wang, & Xu, 2017). The degree of

Table 4 Removal of pharmaceuticals by hybrid photocatalysis and adsorption processes.

Material	Pollutant	Removal (%)	References
p(HEA/NMMA)-CuS	Sulfamethoxazole	95.91	Yang, Li, and Zhu (2017)
g-C_3N_4	Berberine hydrochloride	84.68	Yu et al. (2021)
AC/TiO_2–9%	Sulfamethoxazole	67	El Mouchtari et al. (2020)
AC/TiO_2–9%	Carbamazepine	85	El Mouchtari et al. (2020)
AC/TiO_2–9%	Diclofenac	100	El Mouchtari et al. (2020)
Zn-Fe mixed metal oxides	Diclofenac	100	Di et al. (2017)
Reed-based composite beads	Ciprofloxacin	92.95	Karoui, Ben Arfi, Ghorbal, Amrane, and Assadi (2021)
Zn-Fe mixed metal oxides	Ibuprofen	100	Di et al. (2017)
Zn-Fe mixed metal oxides	Acetaminophen	100	Di et al. (2017)

degradation of sulfamethoxazole, diclofenac, carbamazepine was 50%–100% (El Mouchtari et al., 2020). The hybrid material of activated carbon and titanium dioxide showed good adsorption and photocatalytic results. Activated carbon with 9% of TiO_2 showed the maximum adsorption capacity equal to 153.8, 125.0, and 105.3 mg/g to diclofenac, sulfamethoxazole and carbamazepine, respectively. The photocatalyst concentration of 0.1 g/L provided the degradation of the pharmaceuticals by 67%–100% within 6 h of sunlight irradiation. The removal efficiency of carbamazepine by photocatalytic degradation was significantly lower in the study (Lin et al., 2017). However, the combination of TiO_2 with reduced graphene oxide improved the degradation degree up to 54% (after 180 min of ultraviolet radiation). It was concluded that the combination of adsorption and advanced oxidation is very effective for the removal of pharmaceuticals. A new material that has a synergistic effect in the purification of water from tetracycline hydrochloride was presented in the work (Hao, Xiao, Zuo, Nan, & Zhang, 2012). Mixed Zn-Fe metal oxides provided 100% removal of ibuprofen, diclofenac, and acetaminophen from wastewater (Di et al., 2017). Simultaneously with the pharmaceuticals, arsenic was also removed, mainly by adsorption. The domination of photocatalytic degradation of drugs indicates that zinc oxide plays important role in the hybrid material.

A hybrid material with a hydrogel catalyst was used to remove pharmaceuticals in the work (Yang et al., 2017). A novel three-dimensional polymer (*N*-methyl maleic acid/2-hydroxyethyl acrylate) hydrogel loaded with copper sulfide nanoparticles p(HEA/NMMA)-CuS was obtained by precipitation combined with irradiation polymerization. The model contaminant was sulfamethoxazole. The adsorption process was well described by the Langmuir model. The photocatalytic reaction occurred under visible radiation. A day later, the drug sulfamethoxazole was removed up to 95.91%. Hydroxyl radicals were detected during photodegradation. The final reaction mixture contained the decomposition products.

A very similar method was used to remove the antibiotic berberine hydrochloride (Yu et al., 2021). The g-C_3N_4 material was both catalyst and sorbent. A synergistic effect was registered. Adsorption has accelerated the photocatalytic decomposition of berberine hydrochloride, thereby increasing the removal efficiency. Photocatalytic oxidation of the organic sorbate led to the regeneration of active adsorption centers, resulting in the increased productivity of the photocatalyst. The efficiency of berberine hydrochloride removal after 120min of irradiation was 84.68%. Fig. 6 shows the results in details.

BiOI microspheres proved to have good adsorption properties. The adsorption isotherms agree well with both the Langmuir model and the Freundlich model. The high ratio of surface area to volume resulted in a high adsorption capacity (29.31 mg/g). The material has also high photocatalytic activity in the decomposition of tetracycline. Reed-based lignocellulose composite beads were effective in the removal of ciprofloxacin by hybrid method (Karoui et al., 2021). Given the hospital wastewater

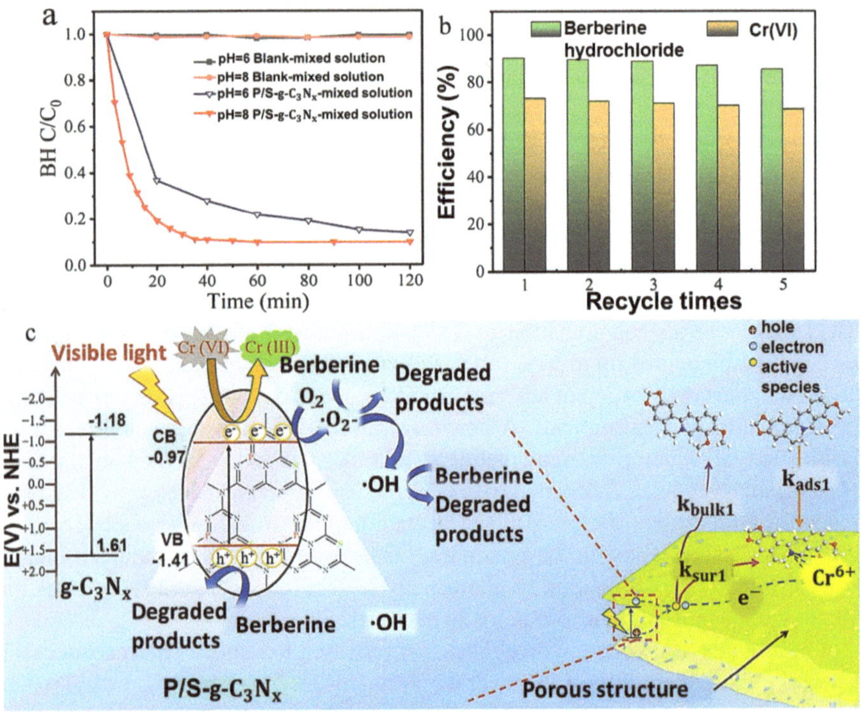

Fig. 6 (A) Synergistic effect of photocatalytic oxidation of berberine hydrochloride with P/S-g-C_3N_4. (B) The efficiency of berberine hydrochloride and Cr(V) removal after several cycles of degradation. (C) Probable decomposition mechanism of berberine hydrochloride (Yu et al., 2021).

contains pharmaceuticals in very low concentrations, the authors used the concentration of ciprofloxacin close to real ones—4 mg/L. Adsorption was performed using a flow column system. The sorption capacity was recorded to be 59.92 mg/g. The photocatalytic oxidation for 6 h of UV radiation provided the removal of ciprofloxacin up to 93%.

4.2 Fenton oxidation and adsorption

The Fenton oxidation combined with adsorption is also a very effective method of removing organic contaminants. The oxidation reaction can occur both when using a heterogeneous catalyst having adsorption properties, and when using an adsorbent combined with a homogeneous Fenton catalyst (Bello & Raman, 2019). Therefore, specific materials are required to ensure the synergistic effect of adsorption and Fenton oxidation.

In the study (Dwivedi, Morone, Chakrabarti, & Pandey, 2018), Fenton oxidation alone removed 49.39% of carbamazepine (an anticonvulsant), whereas Fenton oxidation along with adsorption resulted in the removal of up to 99.51%. The combined catalyst-adsorbent was granular activated carbon (GAC). The optimal concentration of H_2O_2 was 8.5 g/L. Removal of nine drugs from hospital wastewater was carried out using activated avocado seed charcoal and a working concentration of H_2O_2 of 100 mg/L (Della-Flora, Wilde, Lima, Lima, & Sirtori, 2021). Most pharmaceuticals were removed by 99% and fluconazole by 80%. Mesoporous combined catalyst CuO/MnO_2 was more effective than individual oxides of MnO_2 and CuO (Zhang, Liu, Xu, Qi, & Chu, 2016). One hour of oxidation in deionized water at neutral pH resulted in the removal of 89% benzotriazole (BZA). The use of the Fenton-like catalyst in real wastewater conditions reduced the efficiency to 56%. Fig. 7A shows the mechanism proposed by the authors (Zhang et al., 2016).

Usually, the Fenton oxidation occurs with the iron-based catalysts. Magnetic material $FeNi_3/SiO_2$ was used for Fenton oxidation and adsorption of metronidazole (Nasseh et al., 2019). The amount of composite catalyst/sorbent ranged from 0.005 to 0.1 g/L. The concentrations of hydrogen peroxide and metronidazole ranged from 50 to 200 mg/L and from 10 to 30 mg/L, respectively. Increasing the dose of the $FeNi_3/SiO_2$ composite led to an increase in the degradation degree from 40.96% to 80.29%. The catalyst performance was high, as after six test cycles the efficiency decreased by only 8%. Fig. 7B shows the detailed mechanism of the Fenton oxidation combined with the adsorption.

Tetracycline is removed from wastewater by adsorption on $AC@Fe_3O_4$ by 50% (Jonidi Jafari et al., 2017). The combination of adsorption with Fenton oxidation increased the removal up to >80%. The active intermediates of oxidative decomposition were oxygen-containing radicals. Adsorption and photo-Fenton oxidation were used with nanoparticles of zeolite NaX (61 нм) and $CoFe_2O_4$ (16.7 нм) to purify water from paracetamol (Irani, Rad, Pourahmad, & Haririan, 2015). The concentration of paracetamol was in the range of 20–100 mg/L, and pH values were 3 and 4. The combination of adsorption and photo-Fenton oxidation resulted in almost complete (99.80%) removal of the drug.

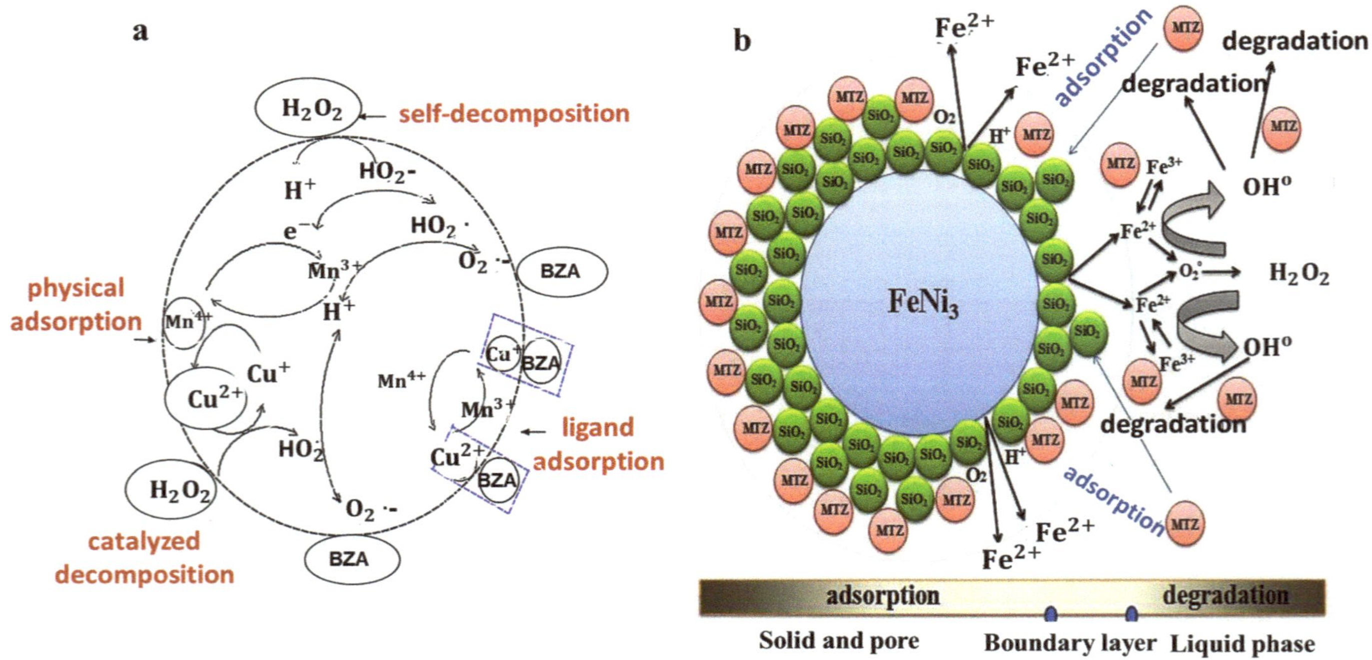

Fig. 7 (A) Reaction mechanism of Fenton-like reaction with mesoporous Cu/MnO_2 (Zhang et al., 2016). (B) The mechanism of metronidazole degradation by $FeNi_3/SiO_2$ (Nasseh, Taghavi, Barikbin, Nasseri, & Allahresani, 2019).

Photo-Fenton oxidation in combination with adsorption showed high efficacy in the work (Michael et al., 2019). The initial concentration of hydrogen peroxide was 100 mg/L. Adsorption on granular activated carbon (500 mg/L) was used to remove the residuals remaining after photo-Fenton oxidation. The following pharmaceuticals were removed: sulfamethoxazole, ampicillin, tetracycline, ofloxacin, clarithromycin, erythromycin within 30 min of photo-Fenton oxidation and 15 min of adsorption. Unfortunately, photo-Fenton degradation stage took place in an acidic environment (pH 2.8–2.9). It can be concluded that new hybrid methods are promising, but additional research is needed to overcome technological problems.

4.3 Ozonation and adsorption

Ozone has strong oxidizing properties, which allows it to be used to degrade pollutants. However, its effectiveness is not always satisfactory. In addition, the formation of by-products can cause toxic effects. In turn, the adsorbents can remove these by-products, thereby providing a synergistic effect in the wastewater remediation. Typically, this approach is a combination of sequential processes of ozonation and adsorption or catalytic ozonation using composite catalyst combined with adsorbent (Bello & Raman, 2019). Table 5 shows examples of combined ozonation and adsorption to remove the pharmaceuticals.

Combined adsorption and catalytic ozonation were used to remove substances that adversely affect the endocrine system (Huang et al., 2017). Magnetic particles of $MWCNT/Fe_3O_4$ were obtained by hydrothermal synthesis. The high surface area and the presence of a large number of surface groups have led to the good adsorption capacity of the composites. As the number of surface groups increased, more hydroxyl

Table 5 Effectiveness of water decontamination by ozonation and adsorption.

Material	Pollutant	Removal (%)	References
Powdered activated carbon/O_3	Carbamazepine	90	Altmann, Ruhl, Zietzschmann, and Jekel (2014)
	Diclofenac	90	Altmann et al. (2014)
Activated carbon/O_3	Carbamazepine	100	Nebout, Cagnon, Delpeux, Di Giusto, and Chedeville (2016)
	Fluoxetine	100	
	Terbutaline	90	
	Sulfamethoxazole	100	
	Metoprolol	100	
γ-Al_2O_3/O_3	Ciprofloxacin	93	Nemati Sani, Fezabady, Yazdani, and Taghavi (2019)
TiO_2/O_3	Atenolol	100	Márquez, Rodríguez, Maldonado, and Álvarez (2014)
	Hydrochlorothiazide	100	
	Ofloxacin	100	
	Trimethoprim	100	

radicals were formed. Increasing the dose of MWCNT/Fe_3O_4 and the pH of solution improved the drug removal. The adsorption itself provided 50% removal, while combined adsorption/catalytic ozonation provided 91% removal. Sulfamethoxazole is removed from the wastewater within 50 min (Akhtar, Amin, & Aris, 2011). Activated carbon and Fe_2O_3/CeO_2 composite with activated carbon showed high efficiency of the adsorption-catalytic ozonation method. Nitroimidazole was removed from the water with activated carbon and ozone (Sánchez-Polo, Rivera-Utrilla, Prados-Joya, Ferro-García, & Bautista-Toledo, 2008). Activated carbon increases the rate of oxidation and adsorbs the formed by-products. As a result, total organic carbon (TOC) was reduced by about 20%.

Wastewater treatment of 90% of carbamazepine and diclofenac was carried out using O_3 and powdered activated carbon (Altmann et al., 2014). Doses of 20 mg/L of carbon and 7 mg/L of ozone were optimal for 100% drug decomposition. Ozonation led to the conversion of drugs into low molecular weight compounds. Six pharmaceuticals were removed by adsorption-ozonation with activated carbon (Nebout et al., 2016). Removal of terbutaline was 17%, 22%, and 90% by adsorption, by ozonation and by hybrid adsorption-ozonation methods, respectively. Terbutaline, carbamazepine, fluoxetine, sulfamethoxazole, and metoprolol were quickly removed from wastewater. However, the formed by-products remained toxic. It can be concluded that hybrid methods are more effective in purifying contaminated water.

5. Conclusions

Water pollution is a global problem that needs to be addressed quickly and effectively. Pharmaceutical substances in wastewater can be dangerous to human health as well as flora and fauna. Conventional methods of pharmaceutical wastewater treatment have disadvantages. Advanced oxidation alone does not have significant removal efficiency. In addition, operating costs are quite high. The combination of advanced oxidation and adsorption compensates the disadvantages of both methods. The formed active radicals interact with the pollutant sorbed on the adsorbent surface, and destroy them to CO_2 and H_2O. The combined method of adsorption and photo-Fenton oxidation requires special materials that have both high adsorption and photocatalytic properties. Ozone has strong oxidizing properties, but the formed by-products are often toxic. Materials with good adsorption capacity can reduce residual toxicity. Due to the synergistic effects, the new hybrid methods are cost-effective and provide complete removal of potentially dangerous pharmaceuticals.

Acknowledgments

IL and AS thank the Ministry of Education and Science of Ukraine for financial support in the framework of project number 0121U109476 ("Engineering of metal oxide catalysts with the regulating activity function for water disinfection with hydroxyl radicals," 2021–23). TT would like to acknowledge the support given by the National Science Centre (Poland) (project UMO-2022/01/3/ST5/00136).

References

Adegoke, A. A., & Stenström, T. A. (2019). Metal oxide nanoparticles in removing residual pharmaceutical products and pathogens from water and wastewater. In *Nanoparticles in pharmacotherapy* (pp. 561–589). Elsevier. https://doi.org/10.1016/B978-0-12-816504-1.00016-8.

Ahmed, D. N., Naji, L. A., Faisal, A. A. H., Al-Ansari, N., & Naushad, M. (2020). Waste foundry sand/MgFe-layered double hydroxides composite material for efficient removal of Congo red dye from aqueous solution. *Scientific Reports*, *10*, 1–12. https://doi.org/10.1038/s41598-020-58866-y.

Akhtar, J., Amin, N. A. S., & Aris, A. (2011). Combined adsorption and catalytic ozonation for removal of sulfamethoxazole using Fe_2O_3/CeO_2 loaded activated carbon. *Chemical Engineering Journal*, *170*, 136–144. https://doi.org/10.1016/j.cej.2011.03.043.

Akkaya Sayğılı, G., & Sayğılı, H. (2020). Pharmaceutical analysis by a novel spinel ferrite nanocomposite derived from a biomaterial-based activated carbon. *Journal of Pharmaceutical and Biomedical Analysis*, *179*. https://doi.org/10.1016/j.jpba.2019.112957.

Altmann, J., Ruhl, A. S., Zietzschmann, F., & Jekel, M. (2014). Direct comparison of ozonation and adsorption onto powdered activated carbon for micropollutant removal in advanced wastewater treatment. *Water Research*, *55*, 185–193. https://doi.org/10.1016/j.watres.2014.02.025.

Anglada, Á., Urtiaga, A., & Ortiz, I. (2009). Contributions of electrochemical oxidation to waste-water treatment: Fundamentals and review of applications. *Journal of Chemical Technology and Biotechnology*, *84*, 1747–1755. https://doi.org/10.1002/jctb.2214.

Antonopoulou, M., Kosma, C., Albanis, T., & Konstantinou, I. (2021). An overview of homogeneous and heterogeneous photocatalysis applications for the removal of pharmaceutical compounds from real or synthetic hospital wastewaters under lab or pilot scale. *Science of the Total Environment*, *765*, 144163. https://doi.org/10.1016/j.scitotenv.2020.144163.

Arfanis, M. K., Adamou, P., Moustakas, N. G., Triantis, T. M., Kontos, A. G., & Falaras, P. (2017). Photocatalytic degradation of salicylic acid and caffeine emerging contaminants using titania nanotubes. *Chemical Engineering Journal*. https://doi.org/10.1016/j.cej.2016.06.098.

Arslan-Alaton, I., & Caglayan, A. E. (2005). Ozonation of procaine penicillin G formulation effluent part I: Process optimization and kinetics. *Chemosphere*, *59*, 31–39. https://doi.org/10.1016/j.chemosphere.2004.10.014.

Balasubramani, K., Sivarajasekar, N., & Naushad, M. (2020). Effective adsorption of antidiabetic pharmaceutical (metformin) from aqueous medium using graphene oxide nanoparticles: Equilibrium and statistical modelling. *Journal of Molecular Liquids*, *301*, 112426. https://doi.org/10.1016/j.molliq.2019.112426.

Bello, M. M., & Raman, A. A. A. (2019). Synergy of adsorption and advanced oxidation processes in recalcitrant wastewater treatment. *Environmental Chemistry Letters*, *17*, 1125–1142. https://doi.org/10.1007/s10311-018-00842-0.

Boyd, G. R., Reemtsma, H., Grimm, D. A., & Mitra, S. (2003). Pharmaceuticals and personal care products (PPCPs) in surface and treated waters of Louisiana, USA and Ontario, Canada. *Science of the Total Environment*, *311*, 135–149. https://doi.org/10.1016/S0048-9697(03)00138-4.

Calza, P., Sakkas, V. A., Medana, C., Baiocchi, C., Dimou, A., Pelizzetti, E., et al. (2006). Photocatalytic degradation study of diclofenac over aqueous TiO2 suspensions. *Applied Catalysis B: Environmental*, *67*, 197–205. https://doi.org/10.1016/j.apcatb.2006.04.021.

Carlesi Jara, C., Fino, D., Specchia, V., Saracco, G., & Spinelli, P. (2007). Electrochemical removal of antibiotics from wastewaters. *Applied Catalysis B: Environmental*, *70*, 479–487. https://doi.org/10.1016/j.apcatb.2005.11.035.

Chavoshani, A., Hashemi, M., Mehdi Amin, M., & Ameta, S. C. (2020). Pharmaceuticals as emerging micropollutants in aquatic environments. In *Micropollutants and challenges* Elsevier. https://doi.org/10.1016/b978-0-12-818612-1.00002-7.

Danyliuk, N. V., Tatarchuk, T. R., & Shyichuk, A. V. (2020). Batch microreactor for photocatalytic reactions monitoring. *Physics and Chemistry of Solid State*, *21*, 338–346. https://doi.org/10.15330/PCSS.21.2.338-346.

Darwish, M., & Mohammadi, A. (2018). Functionalized nanomaterial for environmental techniques. In (Vols. 1–2). *Nanotechnology in environmental science* (pp. 315–350). Elsevier. https://doi.org/10.1002/9783527808854.ch10.

De la Cruz, N., Giménez, J., Esplugas, S., Grandjean, D., De Alencastro, L. F., & Pulgarín, C. (2012). Degradation of 32 emergent contaminants by UV and neutral photo-Fenton in domestic wastewater effluent previously treated by activated sludge. *Water Research*, *46*, 1947–1957. https://doi.org/10.1016/j.watres.2012.01.014.

Della-Flora, A., Wilde, M. L., Lima, D., Lima, E. C., & Sirtori, C. (2021). Combination of tertiary solar photo-Fenton and adsorption processes in the treatment of hospital wastewater: The removal of pharmaceuticals and their transformation products. *Journal of Environmental Chemical Engineering*, *9*, 105666. https://doi.org/10.1016/j.jece.2021.105666.

Dhangar, K., & Kumar, M. (2020). Tricks and tracks in removal of emerging contaminants from the wastewater through hybrid treatment systems: A review. *Science of the Total Environment*, *738*, 140320. https://doi.org/10.1016/j.scitotenv.2020.140320.

Di, G., Zhu, Z., Zhang, H., Zhu, J., Lu, H., Zhang, W., et al. (2017). Simultaneous removal of several pharmaceuticals and arsenic on Zn-Fe mixed metal oxides: Combination of photocatalysis and adsorption. *Chemical Engineering Journal*, *328*, 141–151. https://doi.org/10.1016/j.cej.2017.06.112.

Duong, C. N., Ra, J. S., Cho, J., Kim, S. D., Choi, H. K., Park, J. H., et al. (2010). Estrogenic chemicals and estrogenicity in river waters of South Korea and seven Asian countries. *Chemosphere*, *78*, 286–293. https://doi.org/10.1016/j.chemosphere.2009.10.048.

Dwivedi, K., Morone, A., Chakrabarti, T., & Pandey, R. A. (2018). Evaluation and optimization of Fenton pretreatment integrated with granulated activated carbon (GAC) filtration for carbamazepine removal from complex wastewater of pharmaceutical industry. *Journal of Environmental Chemical Engineering*, *6*, 3681–3689. https://doi.org/10.1016/j.jece.2016.12.054.

Ebele, A. J., Abou-Elwafa Abdallah, M., & Harrad, S. (2017). Pharmaceuticals and personal care products (PPCPs) in the freshwater aquatic environment. *Emerging Contaminants*, *3*, 1–16. https://doi.org/10.1016/j.emcon.2016.12.004.

El Mouchtari, E. M., Daou, C., Rafqah, S., Najjar, F., Anane, H., Piram, A., et al. (2020). TiO2 and activated carbon of Argania Spinosa tree nutshells composites for the adsorption photocatalysis removal of pharmaceuticals from aqueous solution. *Journal of Photochemistry and Photobiology A: Chemistry*, *388*, 112183. https://doi.org/10.1016/j.jphotochem.2019.112183.

Faisal, A. A. H., Al-Wakel, S. F. A., Assi, H. A., Naji, L. A., & Naushad, M. (2020). Waterworks sludge-filter sand permeable reactive barrier for removal of toxic lead ions from contaminated groundwater. *Journal of Water Process Engineering*, *33*, 101112. https://doi.org/10.1016/j.jwpe.2019.101112.

Fatta-Kassinos, D., Meric, S., & Nikolaou, A. (2011). Pharmaceutical residues in environmental waters and wastewater: Current state of knowledge and future research. *Analytical and Bioanalytical Chemistry*, *399*, 251–275. https://doi.org/10.1007/s00216-010-4300-9.

Ferre-Aracil, J., Valcárcel, Y., Negreira, N., de Alda, M. L., Barceló, D., Cardona, S. C., et al. (2016). Ozonation of hospital raw wastewaters for cytostatic compounds removal. Kinetic modelling and economic assessment of the process. *Science of the Total Environment*, *556*, 70–79. https://doi.org/10.1016/j.scitotenv.2016.02.202.

Gellé, A., & Moores, A. (2019). Plasmonic nanoparticles: Photocatalysts with a bright future. *Current Research in Green and Sustainable Chemistry*, *15*, 60–66. https://doi.org/10.1016/j.cogsc.2018.10.002.

Hao, R., Xiao, X., Zuo, X., Nan, J., & Zhang, W. (2012). Efficient adsorption and visible-light photocatalytic degradation of tetracycline hydrochloride using mesoporous BiOI microspheres. *Journal of Hazardous Materials*, *209–210*, 137–145. https://doi.org/10.1016/j.jhazmat.2012.01.006.

Hasan, Z., Choi, E. J., & Jhung, S. H. (2013). Adsorption of naproxen and clofibric acid over a metal–organic framework MIL-101 functionalized with acidic and basic groups. *Chemical Engineering Journal*, *219*, 537–544. https://doi.org/10.1016/j.cej.2013.01.002.

Hass, U., Duennbier, U., & Massmann, G. (2012). Occurrence and distribution of psychoactive compounds and their metabolites in the urban water cycle of Berlin (Germany). *Water Research*, *46*, 6013–6022. https://doi.org/10.1016/j.watres.2012.08.025.

He, Y., Huang, W., Chen, R., Zhang, W., Lin, H., & Li, H. (2015). Anodic oxidation of aspirin on PbO2, BDD and porous Ti/BDD electrodes: Mechanism, kinetics and utilization rate. *Separation and Purification Technology*, *156*, 124–131. https://doi.org/10.1016/j.seppur.2015.09.036.

Huang, Y., Xu, W., Hu, L., Zeng, J., He, C., Tan, X., et al. (2017). Combined adsorption and catalytic ozonation for removal of endocrine disrupting compounds over MWCNTs/Fe3O4 composites. *Catalysis Today*, *297*, 143–150. https://doi.org/10.1016/j.cattod.2017.05.097.

Irani, M., Rad, L. R., Pourahmad, H., & Haririan, I. (2015). Optimization of the combined adsorption/photo-Fenton method for the simultaneous removal of phenol and paracetamol in a binary system. *Microporous and Mesoporous Materials*, *206*, 1–7. https://doi.org/10.1016/j.micromeso.2014.12.009.

Jamil, K. (2019). *Health effects of pharmaceuticals and personal care products*. Elsevier Inc. https://doi.org/10.1016/B978-0-12-816189-0.00005-6.

Jonidi Jafari, A., Kakavandi, B., Jaafarzadeh, N., Rezaei Kalantary, R., Ahmadi, M., & Akbar Babaei, A. (2017). Fenton-like catalytic oxidation of tetracycline by AC@Fe3O4 as a heterogeneous persulfate activator: Adsorption and degradation studies. *Journal of Industrial and Engineering Chemistry*, *45*, 323–333. https://doi.org/10.1016/j.jiec.2016.09.044.

Karoui, S., Ben Arfi, R., Ghorbal, A., Amrane, A., & Assadi, A. A. (2021). Innovative sequential combination of fixed bed adsorption/desorption and photocatalysis cost-effective process to remove antibiotics in solution. *Progress in Organic Coatings*, *151*, 106014. https://doi.org/10.1016/j.porgcoat.2020.106014.

Kaya, Y., Bacaksiz, A. M., Bayrak, H., Gönder, Z. B., Vergili, I., Hasar, H., et al. (2017). Treatment of chemical synthesis-based pharmaceutical wastewater in an ozonation-anaerobic membrane bioreactor (AnMBR) system. *Chemical Engineering Journal*, *322*, 293–301. https://doi.org/10.1016/j.cej.2017.03.154.

Khan, A. H., Khan, N. A., Ahmed, S., Dhingra, A., Singh, C. P., Khan, S. U., et al. (2020). Application of advanced oxidation processes followed by different treatment technologies for hospital wastewater treatment. *Journal of Cleaner Production*, *269*. https://doi.org/10.1016/j.jclepro.2020.122411.

Klavarioti, M., Mantzavinos, D., & Kassinos, D. (2009). Removal of residual pharmaceuticals from aqueous systems by advanced oxidation processes. *Environment International*, *35*, 402–417. https://doi.org/10.1016/j.envint.2008.07.009.

Komtchou, S., Dirany, A., Drogui, P., & Bermond, A. (2015). Removal of carbamazepine from spiked municipal wastewater using electro-Fenton process. *Environmental Science and Pollution Research*, *22*, 11513–11525. https://doi.org/10.1007/s11356-015-4345-6.

Kosma, C. I., Nannou, C. I., Boti, V. I., & Albanis, T. A. (2019). Psychiatrics and selected metabolites in hospital and urban wastewaters: Occurrence, removal, mass loading, seasonal influence and risk assessment. *Science of the Total Environment*, *659*, 1473–1483. https://doi.org/10.1016/j.scitotenv.2018.12.421.

Kumar, A., Kumari, A., Sharma, G., Du, B., Naushad, M., & Stadler, F. J. (2020). Carbon quantum dots and reduced graphene oxide modified self-assembled S@C3N4/B@C3N4 metal-free nano-photocatalyst for high performance degradation of chloramphenicol. *Journal of Molecular Liquids*, *300*, 112356. https://doi.org/10.1016/j.molliq.2019.112356.

Kumar, A., Rana, A., Sharma, G., Naushad, M., Dhiman, P., Kumari, A., et al. (2019). Recent advances in nano-Fenton catalytic degradation of emerging pharmaceutical contaminants. *Journal of Molecular Liquids*, *290*, 111177. https://doi.org/10.1016/j.molliq.2019.111177.

Lai, C., Huang, F., Zeng, G., Huang, D., Qin, L., Cheng, M., et al. (2019). Fabrication of novel magnetic MnFe2O4/bio-char composite and heterogeneous photo-Fenton degradation of tetracycline in near neutral pH. *Chemosphere*, *224*, 910–921. https://doi.org/10.1016/j.chemosphere.2019.02.193.

Lan, Y., Coetsier, C., Causserand, C., & Groenen Serrano, K. (2017). On the role of salts for the treatment of wastewaters containing pharmaceuticals by electrochemical oxidation using a boron doped diamond anode. *Electrochimica Acta*, *231*, 309–318. https://doi.org/10.1016/j.electacta.2017.01.160.

Latif, A., Sheng, D., Sun, K., Si, Y., Azeem, M., Abbas, A., et al. (2020). Remediation of heavy metals polluted environment using Fe-based nanoparticles: Mechanisms, influencing factors, and environmental implications. *Environmental Pollution*, *264*, 114728. https://doi.org/10.1016/j.envpol.2020.114728.

Leone, V. O., Pereira, M. C., Aquino, S. F., Oliveira, L. C. A., Correa, S., Ramalho, T. C., et al. (2018). Adsorption of diclofenac on a magnetic adsorbent based on maghemite: Experimental and theoretical studies. *New Journal of Chemistry*, *42*, 437–449. https://doi.org/10.1039/c7nj03214e.

Lester, Y., Avisar, D., Gozlan, I., & Mamane, H. (2011). Removal of pharmaceuticals using combination of $UV/H_2O_2/O_3$ advanced oxidation process. *Water Science and Technology*, *64*, 2230–2238. https://doi.org/10.2166/wst.2011.079.

Lin, L., Wang, H., & Xu, P. (2017). Immobilized TiO2-reduced graphene oxide nanocomposites on optical fibers as high performance photocatalysts for degradation of pharmaceuticals. *Chemical Engineering Journal*, *310*, 389–398. https://doi.org/10.1016/j.cej.2016.04.024.

Lin, Y., Xu, S., & Li, J. (2013). Fast and highly efficient tetracyclines removal from environmental waters by graphene oxide functionalized magnetic particles. *Chemical Engineering Journal*, *225*, 679–685. https://doi.org/10.1016/j.cej.2013.03.104.

Löffler, D., Römbke, J., Meller, M., & Ternes, T. A. (2005). Environmental fate of pharmaceuticals in water/sediment systems. *Environmental Science & Technology*, *39*, 5209–5218. https://doi.org/10.1021/es0484146.

Luo, Y., Guo, W., Ngo, H. H., Nghiem, L. D., Hai, F. I., Zhang, J., et al. (2014). A review on the occurrence of micropollutants in the aquatic environment and their fate and removal during wastewater treatment. *Science of the Total Environment*, *473–474*, 619–641. https://doi.org/10.1016/j.scitotenv.2013.12.065.

Malik, S. N., Ghosh, P. C., Vaidya, A. N., & Mudliar, S. N. (2020). Hybrid ozonation process for industrial wastewater treatment: Principles and applications: A review. *Journal of Water Process Engineering*, *35*. https://doi.org/10.1016/j.jwpe.2020.101193.

Márquez, G., Rodríguez, E. M., Maldonado, M. I., & Álvarez, P. M. (2014). Integration of ozone and solar TiO2-photocatalytic oxidation for the degradation of selected pharmaceutical compounds in water and wastewater. *Separation and Purification Technology*, *136*, 18–26. https://doi.org/10.1016/j.seppur.2014.08.024.

Martins, R. C., Dantas, R. F., Sans, C., Esplugas, S., & Quinta-Ferreira, R. M. (2015). Ozone/ H_2O_2 performance on the degradation of sulfamethoxazole. *Ozone Science and Engineering*, *37*, 509–517. https://doi.org/10.1080/01919512.2015.1053427.

McEneff, G., Barron, L., Kelleher, B., Paull, B., & Quinn, B. (2014). A year-long study of the spatial occurrence and relative distribution of pharmaceutical residues in sewage effluent, receiving marine waters and marine bivalves. *Science of the Total Environment*, *476–477*, 317–326. https://doi.org/10.1016/j.scitotenv.2013.12.123.

Mestre, A. S., Pires, J., Nogueira, J. M. F., Parra, J. B., Carvalho, A. P., & Ania, C. O. (2009). Waste-derived activated carbons for removal of ibuprofen from solution: Role of surface chemistry and pore structure. *Bioresource Technology*, *100*, 1720–1726. https://doi.org/10.1016/j.biortech.2008.09.039.

Mi, X., Li, Y., Ning, X., Jia, J., Wang, H., Xia, Y., et al. (2019). Electro-Fenton degradation of ciprofloxacin with highly ordered mesoporous MnCo2O4-CF cathode: Enhanced redox capacity and accelerated electron transfer. *Chemical Engineering Journal*, *358*, 299–309. https://doi.org/10.1016/j.cej.2018.10.047.

Michael, S. G., Michael-Kordatou, I., Beretsou, V. G., Jäger, T., Michael, C., Schwartz, T., et al. (2019). Solar photo-Fenton oxidation followed by adsorption on activated carbon for the minimisation of antibiotic resistance determinants and toxicity present in urban wastewater. *Applied Catalysis B: Environmental*, *244*, 871–880. https://doi.org/10.1016/j.apcatb.2018.12.030.

Michelin, C., & Hoffmann, N. (2018). Photocatalysis applied to organic synthesis—A green chemistry approach. *Current Research in Green and Sustainable Chemistry*, *10*, 40–45. https://doi.org/10.1016/j.cogsc.2018.02.009.

Mironyuk, I., Myslin, M., Lapchuk, I., Tatarchuk, T., & Olkhovyy, O. (2021). Adsorption of azo dye Congo red on the Sn-doped TiO2 surface. *Physics and Chemistry of Solid State*, *22*, 561–567. https://doi.org/10.15330/pcss.22.3.561-567.

Mohamed Isa, E. D., Shameli, K., Ch'ng, H. J., Che-Jusoh, N. W., & Hazan, R. (2021). Photocatalytic degradation of selected pharmaceuticals using green fabricated zinc oxide nanoparticles. *Advanced Powder Technology*, *32*, 2398–2409. https://doi.org/10.1016/j.apt.2021.05.021.

Moravvej, Z., Soroush, E., & Rahimpour, M. R. (2020). *Achievements in hybrid processes for wastewater and water treatment*. Elsevier Inc. https://doi.org/10.1016/b978-0-12-817378-7.00009-4.

Morone, A., Mulay, P., & Kamble, S. P. (2019). *Removal of pharmaceutical and personal care products from wastewater using advanced materials*. Elsevier Inc. https://doi.org/10.1016/B978-0-12-816189-0.00008-1.

Mu, Y., Huang, C., Li, H., Chen, L., Zhang, D., & Yang, Z. (2019). Electrochemical degradation of ciprofloxacin with a Sb-doped SnO2 electrode: Performance, influencing factors and degradation pathways. *RSC Advances*, *9*, 29796–29804. https://doi.org/10.1039/c9ra04860j.

Muthusaravanan, S., Sivarajasekar, N., Vivek, J. S., Paramasivan, T., Naushad, M., Prakashmaran, J., et al. (2018). Phytoremediation of heavy metals: Mechanisms, methods and enhancements. *Environmental Chemistry Letters*, *16*, 1339–1359. https://doi.org/10.1007/s10311-018-0762-3.

Nakada, N., Shinohara, H., Murata, A., Kiri, K., Managaki, S., Sato, N., et al. (2007). Removal of selected pharmaceuticals and personal care products (PPCPs) and endocrine-disrupting chemicals (EDCs) during sand filtration and ozonation at a municipal sewage treatment plant. *Water Research, 41*, 4373–4382. https://doi.org/10.1016/j.watres.2007.06.038.

Nasseh, N., Taghavi, L., Barikbin, B., Nasseri, M. A., & Allahresani, A. (2019). $FeNi_3$/SiO_2 magnetic nanocomposite as an efficient and recyclable heterogeneous Fenton-like catalyst for the oxidation of metronidazole in neutral environments: Adsorption and degradation studies. *Composites. Part B, Engineering, 166*, 328–340. https://doi.org/10.1016/j.compositesb.2018.11.112.

Naushad, M. (2014). Surfactant assisted nano-composite cation exchanger: Development, characterization and applications for the removal of toxic Pb2+ from aqueous medium. *Chemical Engineering Journal, 235*, 100–108. https://doi.org/10.1016/j.cej.2013.09.013.

Naushad, M., & ALOthman, Z. A. (2015). Separation of toxic Pb2+ metal from aqueous solution using strongly acidic cation-exchange resin: Analytical applications for the removal of metal ions from pharmaceutical formulation. *Desalination and Water Treatment, 53*, 2158–2166. https://doi.org/10.1080/19443994.2013.862744.

Naushad, M., ALOthman, Z. A., Awual, M. R., Alam, M. M., & Eldesoky, G. E. (2015). Adsorption kinetics, isotherms, and thermodynamic studies for the adsorption of Pb2+ and Hg2+ metal ions from aqueous medium using Ti(IV) iodovanadate cation exchanger. *Ionics, 21*, 2237–2245. https://doi.org/10.1007/s11581-015-1401-7.

Naushad, M., Alqadami, A. A., AlOthman, Z. A., Alsohaimi, I. H., Algamdi, M. S., & Aldawsari, A. M. (2019). Adsorption kinetics, isotherm and reusability studies for the removal of cationic dye from aqueous medium using arginine modified activated carbon. *Journal of Molecular Liquids, 293*, 111442. https://doi.org/10.1016/j.molliq.2019.111442.

Naushad, M., Mittal, A., Rathore, M., & Gupta, V. (2015). Ion-exchange kinetic studies for Cd(II), Co(II), Cu(II), and Pb(II) metal ions over a composite cation exchanger. *Desalination and Water Treatment, 54*, 2883–2890. https://doi.org/10.1080/19443994.2014.904823.

Naushad, M., Sharma, G., & Alothman, Z. A. (2019). Photodegradation of toxic dye using gum Arabic-crosslinked-poly(acrylamide)/Ni(OH)2/FeOOH nanocomposites hydrogel. *Journal of Cleaner Production, 241*, 118263. https://doi.org/10.1016/j.jclepro.2019.118263.

Nebout, P., Cagnon, B., Delpeux, S., Di Giusto, A., & Chedeville, O. (2016). Comparison of the efficiency of adsorption, ozonation, and ozone/activated carbon coupling for the removal of pharmaceuticals from water. *Journal of Environmental Engineering, 142*, 04015074. https://doi.org/10.1061/(asce)ee.1943-7870.0001042.

Nemati Sani, O., Fezabady, A. A. N., Yazdani, M., & Taghavi, M. (2019). Catalytic ozonation of ciprofloxacin using γ-Al_2O_3 nanoparticles in synthetic and real wastewaters. *Journal of Water Process Engineering, 32*, 100894. https://doi.org/10.1016/j.jwpe.2019.100894.

Nguyen, D. N., Bui, H. M., & Nguyen, H. Q. (2020). Heterogeneous photocatalysis for the removal of pharmaceutical compounds. In *Current developments in biotechnology and bioengineering emerging organic micro-pollutants* Elsevier B.V. https://doi.org/10.1016/b978-0-12-819594-9.00007-3.

Paíga, P., Santos, L. H. M. L. M., Ramos, S., Jorge, S., Silva, J. G., & Delerue-Matos, C. (2016). Presence of pharmaceuticals in the Lis river (Portugal): Sources, fate and seasonal variation. *Science of the Total Environment, 573*, 164–177. https://doi.org/10.1016/j.scitotenv.2016.08.089.

Patel, M., Kumar, R., Kishor, K., Mlsna, T., Pittman, C. U., & Mohan, D. (2019). Pharmaceuticals of emerging concern in aquatic systems: Chemistry, occurrence, effects, and removal methods. *Chemical Reviews, 119*, 3510–3673. https://doi.org/10.1021/acs.chemrev.8b00299.

Pereira, C. D. S., Maranho, L. A., Cortez, F. S., Pusceddu, F. H., Santos, A. R., Ribeiro, D. A., et al. (2016). Occurrence of pharmaceuticals and cocaine in a Brazilian coastal zone. *Science of the Total Environment, 548–549*, 148–154. https://doi.org/10.1016/j.scitotenv.2016.01.051.

Roberts, P. H., & Thomas, K. V. (2006). The occurrence of selected pharmaceuticals in wastewater effluent and surface waters of the lower Tyne catchment. *Science of the Total Environment, 356*, 143–153. https://doi.org/10.1016/j.scitotenv.2005.04.031.

Rosman, N., Salleh, W. N. W., Mohamed, M. A., Jaafar, J., Ismail, A. F., & Harun, Z. (2018). Hybrid membrane filtration-advanced oxidation processes for removal of pharmaceutical residue. *Journal of Colloid and Interface Science, 532*, 236–260. https://doi.org/10.1016/j.jcis.2018.07.118.

Rostam, A. B., & Taghizadeh, M. (2020). Advanced oxidation processes integrated by membrane reactors and bioreactors for various wastewater treatments: A critical review. *Journal of Environmental Chemical Engineering, 8*, 104566. https://doi.org/10.1016/j.jece.2020.104566.

Sánchez-Polo, M., Rivera-Utrilla, J., Prados-Joya, G., Ferro-García, M. A., & Bautista-Toledo, I. (2008). Removal of pharmaceutical compounds, nitroimidazoles, from waters by using the ozone/carbon system. *Water Research, 42*, 4163–4171. https://doi.org/10.1016/j.watres.2008.05.034.

Santos, L. H. M. L. M., Gros, M., Rodriguez-Mozaz, S., Delerue-Matos, C., Pena, A., Barceló, D., et al. (2013). Contribution of hospital effluents to the load of pharmaceuticals in urban wastewaters: Identification of ecologically relevant pharmaceuticals. *Science of the Total Environment, 461–462*, 302–316. https://doi.org/10.1016/j.scitotenv.2013.04.077.

Sharma, G., Dionysiou, D. D., Sharma, S., Kumar, A., Al-Muhtaseb, A. H., Naushad, M., et al. (2019). Highly efficient Sr/Ce/activated carbon bimetallic nanocomposite for photoinduced degradation of rhodamine B. *Catalysis Today, 335*, 437–451. https://doi.org/10.1016/j.cattod.2019.03.063.

Sharma, G., Kumar, A., Sharma, S., Naushad, M., Dhiman, P., Vo, D. V. N., et al. (2020). Fe_3O_4/ZnO/Si_3N_4 nanocomposite based photocatalyst for the degradation of dyes from aqueous solution. *Materials Letters, 278*, 128359. https://doi.org/10.1016/j.matlet.2020.128359.

Sharma, G., & Naushad, M. (2020). Adsorptive removal of noxious cadmium ions from aqueous medium using activated carbon/zirconium oxide composite: Isotherm and kinetic modelling. *Journal of Molecular Liquids, 310*, 113025. https://doi.org/10.1016/j.molliq.2020.113025.

Sharma, G., Pathania, D., Naushad, M., & Kothiyal, N. C. (2014). Fabrication, characterization and antimicrobial activity of polyaniline Th(IV) tungstomolybdophosphate nanocomposite material: Efficient removal of toxic metal ions from water. *Chemical Engineering Journal, 251*, 413–421. https://doi.org/10.1016/j.cej.2014.04.074.

Sharma, G., Sharma, S., Kumar, A., Al-Muhtaseb, A. H., Naushad, M., Ghfar, A. A., et al. (2018). Guar gum and its composites as potential materials for diverse applications: A review. *Carbohydrate Polymers, 199*, 534–545. https://doi.org/10.1016/j.carbpol.2018.07.053.

Silva, A. R., Martins, P. M., Teixeira, S., Carabineiro, S. A. C., Kuehn, K., Cuniberti, G., et al. (2016). Ciprofloxacin wastewater treated by UVA photocatalysis: Contribution of irradiated TiO2 and ZnO nanoparticles on the final toxicity as assessed by Vibrio fischeri. *RSC Advances, 6*, 95494–95503. https://doi.org/10.1039/c6ra19202e.

Snyder, S. A., Wert, E. C., Rexing, D. J., Zegers, R. E., & Drury, D. D. (2006). Ozone oxidation of endocrine disruptors and pharmaceuticals in surface water and wastewater. *Ozone Science and Engineering, 28*, 445–460. https://doi.org/10.1080/01919510601039726.

Sui, Q., Huang, J., Deng, S., Yu, G., & Fan, Q. (2010). Occurrence and removal of pharmaceuticals, caffeine and DEET in wastewater treatment plants of Beijing, China. *Water Research*, *44*, 417–426. https://doi.org/10.1016/j.watres.2009.07.010.

Tang, J., & Wang, J. (2018). Fenton-like degradation of sulfamethoxazole using Fe-based magnetic nanoparticles embedded into mesoporous carbon hybrid as an efficient catalyst. *Chemical Engineering Journal*, *351*, 1085–1094. https://doi.org/10.1016/j.cej.2018.06.169.

Tatarchuk, T., Myslin, M., Lapchuk, I., Shyichuk, A., Murthy, A. P., Gargula, R., et al. (2021). Magnesium-zinc ferrites as magnetic adsorbents for Cr(VI) and Ni(II) ions removal: Cation distribution and antistructure modeling. *Chemosphere*, *270*. https://doi.org/10.1016/j.chemosphere.2020.129414.

Ternes, T., Bonerz, M., & Schmidt, T. (2001). Determination of neutral pharmaceuticals in wastewater and rivers by liquid chromatography-electrospray tandem mass spectrometry. *Journal of Chromatography A*, *938*, 175–185. https://doi.org/10.1016/S0021-9673(01)01205-5.

Thorpe, K. L., Hutchinson, T. H., Hetheridge, M. J., Scholze, M., Sumpter, J. P., & Tyler, C. R. (2001). Assessing the biological potency of binary mixtures of environmental estrogens using vitellogenin induction in juvenile rainbow trout (*Oncorhynchus mykiss*). *Environmental Science & Technology*, *35*, 2476–2481. https://doi.org/10.1021/es001767u.

Van Der Ven, K., Keil, D., Moens, L. N., Van Leemput, K., Van Remortel, P., & De Coen, W. M. (2006). Neuropharmaceuticals in the environment: Mianserin-induced neuroendocrine disruption in zebrafish (*Danio rerio*) using cDNA microarrays. *Environmental Toxicology and Chemistry*, *25*, 2645–2652. https://doi.org/10.1897/05-495R.1.

Van Doorslaer, X., Haylamicheal, I. D., Dewulf, J., Van Langenhove, H., Janssen, C. R., & Demeestere, K. (2015). Heterogeneous photocatalysis of moxifloxacin in water: Chemical transformation and ecotoxicity. *Chemosphere*, *119*, S75–S80. https://doi.org/10.1016/j.chemosphere.2014.03.048.

Velempini, T., Prabakaran, E., & Pillay, K. (2021). Recent developments in the use of metal oxides for photocatalytic degradation of pharmaceutical pollutants in water—A review. *Materials Today Chemistry*, *19*, 100380. https://doi.org/10.1016/j.mtchem.2020.100380.

Watkinson, A. J., Murby, E. J., & Costanzo, S. D. (2007). Removal of antibiotics in conventional and advanced wastewater treatment: Implications for environmental discharge and wastewater recycling. *Water Research*, *41*, 4164–4176. https://doi.org/10.1016/j.watres.2007.04.005.

Wert, E. C., Rosario-Ortiz, F. L., & Snyder, S. A. (2009). Using ultraviolet absorbance and color to assess pharmaceutical oxidation during ozonation of wastewater. *Environmental Science & Technology*, *43*, 4858–4863. https://doi.org/10.1021/es803524a.

Wielogórska, E., Elliott, C. T., Danaher, M., Chevallier, O., & Connolly, L. (2015). Validation of an ultra high performance liquid chromatography-tandem mass spectrometry method for detection and quantitation of 19 endocrine disruptors in milk. *Food Control*, *48*, 48–55. https://doi.org/10.1016/j.foodcont.2014.06.001.

Wu, C., Witter, J. D., Spongberg, A. L., & Czajkowski, K. P. (2009). Occurrence of selected pharmaceuticals in an agricultural landscape, western Lake Erie basin. *Water Research*, *43*, 3407–3416. https://doi.org/10.1016/j.watres.2009.05.014.

Wu, M., Xiang, J., Que, C., Chen, F., & Xu, G. (2015). Occurrence and fate of psychiatric pharmaceuticals in the urban water system of Shanghai, China. *Chemosphere*, *138*, 486–493. https://doi.org/10.1016/j.chemosphere.2015.07.002.

Yakubu, O. H. (2017). Pharmaceutical wastewater effluent-source of contaminants of emerging concern: Phytotoxicity of metronidazole to soybean (*Glycine max*). *Toxics*, *5*. https://doi.org/10.3390/toxics5020010.

Yang, J., Li, Z., & Zhu, H. (2017). Adsorption and photocatalytic degradation of sulfamethoxazole by a novel composite hydrogel with visible light irradiation. *Applied Catalysis B: Environmental*, *217*, 603–614. https://doi.org/10.1016/j.apcatb.2017.06.029.

Yu, Y., Wu, K., Xu, W., Chen, D., Fang, J., Zhu, X., et al. (2021). Adsorption-photocatalysis synergistic removal of contaminants under antibiotic and Cr(VI) coexistence environment using non-metal g-C3N4 based nanomaterial obtained by supramolecular self-assembly method. *Journal of Hazardous Materials*, *404*, 124171. https://doi.org/10.1016/j.jhazmat.2020.124171.

Zainab, S. M., Junaid, M., Xu, N., & Malik, R. N. (2020). Antibiotics and antibiotic resistant genes (ARGs) in groundwater: A global review on dissemination, sources, interactions, environmental and human health risks. *Water Research*, *187*, 116455. https://doi.org/10.1016/j.watres.2020.116455.

Zeng, J., & Lu, J. (2018). Mechanisms of action involved in ozone-therapy in skin diseases. *International Immunopharmacology*, *56*, 235–241. https://doi.org/10.1016/j.intimp.2018.01.040.

Zhang, Y., Liu, C., Xu, B., Qi, F., & Chu, W. (2016). Degradation of benzotriazole by a novel Fenton-like reaction with mesoporous Cu/MnO_2: Combination of adsorption and catalysis oxidation. *Applied Catalysis B: Environmental*, *199*, 447–457. https://doi.org/10.1016/j.apcatb.2016.06.003.

Microalgae-based bioremediation of pharmaceuticals wastewater

9

Sasmitha Aulia Zahra[a], Imran Ahmad[a], Norhayati Abdullah[b], Koji Iwamoto[b], and Ali Yuzir[b]
[a]Algae and Biomass Research Laboratory, Malaysia-Japan International Institute of Technology (MJIIT), Universiti Teknologi Malaysia, Kuala Lumpur, Malaysia, [b]Department of Chemical and Environmental Engineering, Malaysia-Japan International Institute of Technology (MJIIT), Universiti Teknologi Malaysia, Kuala Lumpur, Malaysia

1. Introduction

Bioremediation technologies have been growing in interest exponentially as it is a sustainable method for the wastewater treatment process. Bioremediation is defined as a metabolic process utilizing microorganisms such as microalgae, bacteria, and fungi to remove harmful substances found in the wastewater (Kaushik & Malik, 2015; Maqbool, Bhatti, Sadaf, Zahid, & Shahid, 2019; Zhang, Xue, Zhang, & Hu, 2020). The microorganisms used can be indigenous or nonindigenous, whereby specific biological organisms were added to the contaminated water (Verma & Jaiswal, 2016). Compared to both chemical and physical remediation approaches, bioremediation is viewed as a more practical and cost-effective remediation method (Hussain et al., 2021).

Microalgae can be eukaryotic or prokaryotic photosynthetic, usually several micrometers to hundreds of micrometers in size (Mustafa, Bhatti, Maqbool, & Iqbal, 2021). They can exist as individual cells or in the form of chains and flocs. In addition, microalgae have a very high growth rate due to their simple unicellular or multicellular structure (Tyoker Kukwa & Chetty, 2021). According to various literature, microalgae are a promising candidate for bioremediation techniques due to their ability to change their nutritional type based on the carbon source and nutrients available in the surrounding environment (Xiong, Kurade, & Jeon, 2018). In this context, microalgae growth mode can be phototrophic, heterotrophic, or mixotrophic. Autotrophic, also called phototrophic microalgae, uses light as the energy source and carbon dioxide (CO_2) as the inorganic carbon source to produce chemical energy through photosynthesis (Khan, Shin, & Kim, 2018). Heterotrophic microalgae can grow in the absence of light, but they use organic carbon, such as glucose and acetate, as the source of energy (Chojnacka & Noworyta, 2004). Mixotrophic microalgae can shift between autotrophic and heterotrophic based on the carbon sources and nutrients available in the environment. These intrinsic revolutionary advantages give them the flexibility to survive and prosper in harsh conditions (Xiong et al., 2018).

In addition, microalgae are excellent for bioremediation due to their ability to remove a wide range of pollutants or toxic substances from industrial, agricultural,

The Treatment of Pharmaceutical Wastewater. https://doi.org/10.1016/B978-0-323-99160-5.00014-X

and domestic sectors (Mustafa et al., 2021). Microalgae can remediate toxic chemicals such as phosphorus (P), Nitrogen (N), ammonium, and recalcitrant contaminants such as pharmaceuticals (PCs), endocrine-disrupting chemicals (EDC), heavy metals due to the presence of carbohydrates in microalgae's cell wall (Olguin, 2003; Schwarzenbach et al., 2006). In addition, the microalgae will use those toxic substances or pollutants as nutrients, converted by bioaccumulation, bioadsorption, and biodegradation process. Microalgae's photosynthetic and CO_2 biofixation ability can also help decrease the CO_2 amount in the atmosphere (Kurano et al., 1995). Moreover, microalgae photosynthesis can produce oxygen which helps in reducing the BOD level in the wastewater (Sivasubramanian, Subramanian, Raghavan, & Ranjithkumar, 2009).

From an economic viewpoint, microalgae-based bioremediation is cost-effective as it is driven by light energy, such as solar energy, which is chargeless and sustainable. Furthermore, microalgae-based bioremediation also yields economic benefits in terms of cost reduction due to their relatively small operational inputs and potential revenue from valuable by-products from microalgae biomass like biofuel, nutraceuticals, cosmetics, low-value food products for aquaculture (Ahmad, Yuzir, Mohamad, Iwamoto, & Abdullah, 2021). Recently, microalgae-mediated bioremediation in treating effluents containing PCs has been gaining scientific attention (Ahmad, Abdullah, Koji, Yuzir, & Mohamad, 2021). This chapter provides previous studies and a summary regarding removal mechanisms of PCs by microalgae, factors impacting PCs removal, conventional and novel approaches of phycoremediation of PCs, integrated processes for removing PCs, and future recommendations.

2. Removal mechanisms of pharmaceutical contaminants (PCs) by microalgae

Microalgae remove PCs by various methods, including bioadsorption, bioaccumulation, biodegradation, photodegradation, and volatilization. This section discusses the intricate ways via which microalgae remove PCs. Table 2 summarizes current studies on the removal of PCs by microalgae-based bioremediation.

2.1 Bioadsorption

Bioadsorption is a mechanism in which the PCs are adsorbed to the cell surface of the microalgae or adsorbed onto organic compounds excreted by the cells to the surrounding environment, such as extracellular polymeric substances or EPS (Kaplan, n.d.; Naja & Volesky, 2011). EPS consists mainly of polysaccharides, lipids, proteins, and humic and nucleic acids. EPS attached to the microalgae cells is called bonded EPS, while EPS excreted into the surrounding environment is called free EPS. Both bonded and free EPS provide sites for PCs to adsorb (Fomina & Gadd, 2014; Hena, Gutierrez, & Croué, 2021). Bioadsorption is a rapid and passive extracellular process due to the ionic or hydrophobic interaction between the PCs and the microalgae surface or EPS.

The effectiveness of a PC to adsorb onto microalgae cell surfaces varies depending on its hydrophobicity, structure, the existence of a functional group, and the microalgae species. The microalgae cell surfaces are predominantly negatively charged due to functional groups such as carboxyl, hydroxyl, sulfate, and phosphoryl; therefore, facilitating electrostatic interactions with positively charged (hydrophobic) PCs molecules. On the other hand, PCs that are hydrophilic will repel (Leng et al., 2020; Xiong et al., 2018). Environmental conditions, including seasons, temperature, and pH, also determine the rate of the adsorption process.

Bioadsorption is also a nonmetabolic process. Therefore, dry (nonliving) microalgae biomass can also be used as a bioadsorbent due to its surface properties and retention capacities. In addition, most cell receptors are still available even after the cell has died (Bilal et al., 2018; Choi & Lee, 2015). Therefore, bioadsorption using dry microalgae biomass possess many advantages over living microalgae as it is not limited by contaminant toxicity. Moreover, the bioadsorbed PCs can be desorbed using an appropriate desorbing agent, and the dry biomass can be reused. This way, the operational cost can be significantly reduced as the is no need to maintain a living microalgae culture (Mane, Kadam, & Chaudhari, 2013).

Studies showed that the adsorption process alone could effectively remove PCs. For example, the bioadsorption rate of tramadol is at 91% using *Scenedesmus obliquus* (Ali, Abd El-Aty, Badawy, & Ali, 2018). Cefalexin can be bioadsorbed by *Chlorella* sp. biomass and lipid-extracted *Chlorella* sp. biomass at a rate of 82.7% and 71.2%, respectively (Angulo, Bula, Mercado, Montano, & Cubillan, 2018). Moreover, a high adsorption capacity of lipid-extracted *Scenedesmus quadricauda* and *Tetraselmis suecica* biomass is reported to remove tetracycline (Daneshvar et al., 2018). Similarly, it was reported that the biomass of *Scenedesmus* sp. and *Synechocystis* sp. can remove 28 and 20mg/g of diclofenac, respectively (Coimbra et al., 2018). However, several studies also reported that bioadsorption was not a dominant PCs removal mechanism. Yu et al. (2017) reported that bioadsorption was the first process in treating ceftazidime and 7-aminocephalosporanic acid, followed by accumulation and biodegradation (Yu et al., 2017). According to de Wilt et al. (2016), the adsorption rates of six PCs on *C. sorokiniana* were less than 20% (de Wilt et al., 2016). Similarly, bioadsorption had a negligible impact on removing progesterone and norgestrel by *S. obliquus* and *Chlorella pyrenoidosa*, accounting for roughly 10% of the removal (Peng et al., 2014). Therefore, further research needs to be conducted to optimize bioadsorption process and to identify suitable PCs for bioadsorption treatment (Sutherland & Ralph, 2019).

2.2 Bioaccumulation

According to various studies, bioaccumulation by microalgae is intracellular (occurs inside cells) and an active process that has been identified as a critical pathway for PCs removal process (Bai & Acharya, 2016, 2017; Prata et al., 2018). Unlike bioadsorption, bioaccumulation is metabolic. Therefore, living microalgae and biomass is used. Certain PCs may get through the membranes of microalgae and be absorbed into the cells. PCs entering the microalgae cells through bioaccumulation can influence reactive oxygen species (ROS) production. ROS governs the metabolism of microalgae cells at normal concentrations, including cell growth, cell death, and defense against harmful organisms. However, when excessive, their highly oxidative potential can cause severe cell damage, thereby obstructing the cell composition. Moreover, protein and lipid levels will be altered, resulting in cell death (Leng et al., 2020; Mustafa et al., 2021; Xiong et al., 2018).

Bioaccumulation may also serve as a prestep to biodegradation, and the combined process of accumulation and biodegradation in microalgae may considerably contribute to the elimination of certain PCs. Sun et al. (2017) observed the accumulation of sulfamethazine followed by biodegradation in *C. pyrenoidosa* (Sun et al., 2017). Similarly, microalgae *C. vulgaris* could aid in removing levofloxacin through bioaccumulation and followed by biodegradation (Xiong, Kurade, Kim, et al., 2017).

2.3 Biodegradation (biotransformation)

Biodegradation, also known as biotransformation, is one of the most successful and promising bioremediation strategies for PCs. Microalgae can catalytically break down complex parent compounds into simpler compounds either inside or outside the cells. The microalgae will then use the simpler compounds as nutrients for growth (Naghdi et al., 2018). For example, Prosenc et al. (2021) reported the removal of metoprolol and ibuprofen in 7 days using *C. vulgaris* culture (Prosenc et al., 2021). At first, metoprolol acid and ibuprofen carboxylic acid were detected at the highest concentration on day 2 and 4. The concentration of both compounds then decreased significantly after 7 days, indicating that metoprolol and ibuprofen were mainly biodegraded to metoprolol acid and ibuprofen carboxylic acid and then underwent further biodegradation. Other transformation products (TP) of PCs by microalgae are described in Table 1.

Microalgae may be biodegraded intracellularly, extracellularly, or a mix of the two. Degradation may begin extracellularly and then move intracellularly to further degrade the breakdown products (Naghdi et al., 2018; Tiwari et al., 2017). Yu et al. (2017) reported intracellular biodegradation of ceftazidime and its parent compound 7-aminocephalosporanic acid by *C. pyrenoidosa* (Yu et al., 2017). Microalgae *C. vulgaris* and *S. obliquus* were also reported to remove levofloxacin and enrofloxacin by intracellular degradation (Xiong, Kurade, & Jeon, 2017a, 2017b). There are two principal metabolisms in which the biodegradation of PCs can occur. First is by cometabolism, in which nonspecific enzymes are used to degrade PCs with the help of additional carbon input or microbes present in the surrounding that secretes enzymes that degrade the PCs. The second is by metabolic degradation, in which targeted PCs serve as carbon or energy source for the microalgae. Adding electric donors, such as sodium acetate, was reported to promote the biodegradation of ciprofloxacin due to cometabolism (Xiong et al., 2018). Cometabolism will be explained further in Section 4.2.2.

Based on enzyme functionalities, there are three distinct phases of the enzymatic catalysis process for PC biodegradation (Wang, Liu, et al., 2017). Phase I is identified by the initial attacks on the PCs, which transform lipophilic PCs into hydrophilic molecules that are easily excreted. An enzyme called cytochrome P450 is normally

Table 1 Transformation products (TP) of pharmaceuticals (PCs) by microalgae.

Pharmaceuticals (PCs)	Transformation products (TP)	Microalgae species	References
Metoprolol	Metoprolol acid	*C. vulgaris*	Prosenc et al. (2021)
Sulfamethoxazole	Desamino sulfamethoxazole *N*-acetyl sulfamethoxazole		
Ibuprofen	2-Hydroxy ibuprofen		
Carbamazepine	Ibuprofen carboxylic acid Epoxy carbamazepine, 2-hydroxy-carbamazepine		
Estradiol	Estrone Estriol Dihydroxyestrone	*Selenastrum capricornutum* and *Chlamydomonas reinhardtii*	Hom-Diaz et al. (2015), Wang, Liu, Kang, Wu, and Wu (2017), and Zhang et al. (2014)
Triclosan	Methyl-triclosan	*M. aeruginosa*	Hwang, Church, Lee, Park, and Lee (2016)
Progesterone	b-Hydroxy-5a-pregnan-20-one, 3,20-allopregnanedione and 1,4-pregnadiene-3,20-dione	*Scenedesmus obliquus* and *Chlorella pyrenoidosa*	Peng et al. (2014)
Norgestrel	4,5-Dihydronorgestrel and 6,7-dehydronorgestrel		
Sulfamethazine and Sulfamethoxazole	4-Nitro SMX, desmino-SMX, benzenesulfonamide, and toluenesulfonamide	*S. obliquus*	Xiong et al. (2019)
Ibuprofen	Carboxy-ibuprofen and hydroxy-ibuprofen	*Chlorella* sp. and *Scenedesmus* sp.	Matamoros et al. (2016)

responsible during this phase which involves hydrolysis, reduction reactions, or oxidation (Zangar, Davydov, & Verma, 2004). In phase II, the hydrophilic PC molecule may be catalyzed by enzymes such as glutathione-*S*-transferases, opening the epoxide ring to protect the microalgae cell from oxidative damage (Nakajima et al., 2007). Lastly, in phase III, PCs are compartmentalized in the microalgae's vacuoles or cell wall fractions (Petroutsos, Katapodis, Samiotaki, Panayotou, & Kekos, 2008). Microalgae generally have a complex enzyme system; therefore, identifying enzymes' exact role in biodegradation has not yet been thoroughly studied (Du et al., 2017). Recently, omic analyses such as metagenomics, metatranscriptomics, proteomics, and metabolomics started gaining attention to discover enzymatic mechanisms in the biodegradation of PCs (Du et al., 2017; Khona et al., 2016; Ufarté, Laville, Duquesne, & Potocki-Veronese, 2015).

2.4 Photobiodegradation

Illumination is essential in the microalgae-based bioremediation of PCs, as light is vital for microalgae photosynthesis. Sunlight is a desirable light source as it is renewable, sustainable and inexhaustible. Several PCs have been observed to photodegrade when exposed to light (Hanamoto, Kawakami, Nakada, Yamashita, & Tanaka, 2014; Matamoros, Duhec, Albaigés, & Bayona, 2009). PCs may be photodegraded directly or indirectly through photooxidative degradation (Abo, Kummer, & Merkel, 2016). Photolysis happens when PCs absorb light directly, resulting in chemical change and subsequent degradation of the PCs. Due to the high absorption of light irradiation, PCs containing conjugated systems, heteroatoms, aromatic rings, and other functional groups are responsive to direct photolysis (Challis, Hanson, Friesen, & Wong, 2014). Photooxidative degradation, or indirect photolysis, occurs when free radicals (peroxyl radicals, hydroxyl radicals, singlet oxygen, and other oxidants) are produced by photosensitizers under light irradiation (Jiménez & Meent, n.d.; Abo et al., 2016). Photooxidative degradation depicts a more significant role in the photodegradation of PCs due to the omnipresence of photosensitizers in natural water and wastewater (Challis et al., 2014). Several factors affect the photodegradation of PCs, such as the physicochemical properties of the PCs and the water, light intensity, and light wavelength. Dissolved organic molecules (DOM) refer to organic substances such as humic acids, hydrophilic organic acids, fulvic acids, and hemicellulose. Photodegradation is a microalgae-based treatment system that can be amplified by intensifying the presence of DOM, as microalgae release DOM into the surrounding environment. DOM may accelerate photodegradation through various pathways, including catabolic or redox cycling (Norvill et al., 2017). Table 2 summarizes current studies on the removal of PCs by microalgae-based bioremediation.

2.5 Volatilization

Sutherland and Ralph (2019) defined the volatilization of PCs as the loss of volatile organic compounds from a liquid phase to the atmosphere (Sutherland & Ralph, 2019). Volatilization depends on the microalgae treatment system operating

Table 2 Removal of pharmaceutical compounds using different microalgal species.

Pharmaceuticals (PCs)	Microalgae species	Experimental conditions	Mechanisms involved	Removal (%)	References
Ibuprofen, Metoprolol, Sulfamethoxazole, and Carbamazepine	C. *vulgaris*	Bench-scale Medium: BBM Light intensity: $59 \pm 3\,\mu mol\,m^{-2}\,s^{-1}$, 16:8 h light:dark cycle Duration: 12 days Temperature PCs were spiked into reaction mixtures at 20 μg/L	Biodegradation	Ibuprofen, Metoprolol and Sulfamethoxazole, >66% Sulfamethoxazole, >66% Carbamazepine, 34%	Prosenc et al. (2021)
2,6-Xylidine, Caffeine, Paracetamol, Fosfomycin, and Duloxetine	Mixed culture (green algae, blue-green algae, and diatoms) dominated by *Chlorella* sp.	Domestic wastewater Light intensity: $200 \pm 12\,\mu mol\,m^{-2}\,s^{-1}$ 12:12 h light:dark cycle Duration: 6 h Temperature: 25–30°C	Bioadsorption	2,6-Xylidine, Caffeine, Paracetamol and Fosfomycin, 100% Duloxetine, 83%	Katam, Bhattacharyya, Soda, and Shimizu (2021)
Estradiol, Diclofenac, and Triclosan	Microalgae consortium dominated by *Chlorella* sp., *Merismopedia* sp., *Closteriopsis* sp., and *Scenedesmus* sp.	Bench-scale Medium: BG-11 Light intensity: 5 ± 0.5 Klux Florescent white light, 12:12 h light:dark cycle Temperature: 25 ± 1°C	Biodegradation	Estradiol, 91.73% Diclofenac, 74.68% Triclosan, 78.4%	Bano, Malik, and Ahammad (2021)

Continued

Table 2 Continued

Pharmaceuticals (PCs)	Microalgae species	Experimental conditions	Mechanisms involved	Removal (%)	References
Dipyridamole, Metoprolol, Atorvastatin, Codeine, Atenolol, Rosuvastatin, Ranitadine, Diltiazem, Mirtazapine, Diclofenac, Bisoprolol, and Amitriptyline	Microalgae consortia in high rate algal pond (HRAP)	Outdoor HRAP	Photodegradation, passive bioadsorption, bioaccumulation, and biodegradation	Dipyridamole, metoprolol, atorvastatin, codeine, and atenolol, 90%–96% Rosuvastatin, Ranitidine, Diltiazem, Mirtazapine, and Diclofenac, 64%–80% Bisoprolol and Amitriptyline, 55%–58%	Lindberg et al. (2021)
Ibuprofen, Oxybenzone, Triclosan, Bisphenol A, and *N*,*N*-Diethyl-3-methylbenzamide (DEET)	Mixed culture (green algae, blue-green algae, and diatoms)	Revolving algal biofilm (RAB) reactor Medium: BBM Light intensity: 130 μmol m^{-2} s^{-1}, 12:12 h light: dark cycle Duration: 20 days Temperature: 25 °C	Bioaccumulation and biodegradation	Ibuprofen and oxybenzone, 91.4%–94.2% Triclosan and bisphenol A, 72.5%–75.7% DEET, 99.5%	Chen Xie, and Wen (2021)

Tetracycline	*C. vulgaris*	HRAP Synthetic wastewater Tetracycline was spiked into the wastewater mixtures at 2 mg/L	Bioadsorption and photodegradation	69%	de Godos, Munoz, and Guieysse (2012)
Tramadol	*S. obliquus*	Medium: BG-11 Light intensity: 33.8 μE/m^2/s. Temperature: 24 ± 2°C	Bioadsorption	91%	Ali et al. (2018)
7-Amino cephalosporanic acid	*Chlorella* sp. *Chlamydomonas* sp. *Mychonastes* sp.	Medium: BG-11 Light intensity: 200 μmol m^{-2} s^{-1} Temperature: 26 ± 1°C	Bioadsorption, hydrolysis, and photobiodegradation	100%	Guo et al. (2016)
Cephalexin	*Chlorella* sp.	Medium: Deionized water enriched with Nutrifoliar Light intensity: 86.2 ± 5 μmol m-^{2} s^{-}1, Duration: 20 days Temperature: 27 ± 2 °C	Bioadsorption	71%–87%	Angulo et al. (2018)
Tetracycline	*Scenedesmus quadricauda Tetraselmis suecica*	Dairy wastewater Medium: BBM Light intensity: 110 μmol photon/m^2 s Duration: 12 days Temperature: 25°C	Bioadsorption	50%–60%	Daneshvar, Zarrinmehr, Hashtjin, Farhadian, and Bhatnagar (2018)

Continued

Table 2 Continued

Pharmaceuticals (PCs)	Microalgae species	Experimental conditions	Mechanisms involved	Removal (%)	References
Ciprofloxacin	Microalgae consortia in high rate algal pond (HRAP)	Photobioreactor and HRAP Domestic wastewater Temperature: 20 ± 2°C	Photobiodegradation (daytime) and bioadsorption (night)	–	Hom-Diaz et al. (2017)
Diclofenac	*Synechocystis* sp. and *Scenedesmus* sp.	Medium: Mann and Myers Light intensity: 175 μmol photon/m^2 s 12:12 h light:dark cycle Duration: 20 days Temperature: 25 ± 1°C	Adsorption	–	Coimbra, Escapa, Vázquez, Noriega-Hevia, and Otero (2018)
Ceftazidime and 7-Aminocephalosporanic acid	*Chlorella pyrenoidosa*	Medium: BG-11 Light intensity: 12:12 h light:dark cycle Duration: 20 days Temperature: 25 ± 1°C	Bioadsorption, bioaccumulation, and biodegradation	Ceftazidime, 92.70% 7-Aminoce phalosporanic acid, 96.07%	Yu et al. (2017)
Diclofenac, Ibuprofen, Paracetamol, Metoprolol, Carbamazepine, Trimethoprim, Estrone, 17-Estradiol, and Ethinylestradiol	*Chlorella sorokiniana*	Medium: Urine, synthetic urine, anaerobically treated black water, and M8a HRT: 25 days Temperature: 25°C PCs were spiked at 100–350 μg/L	Biodegradation, photolysis, and bioadsorption	–	de Wilt et al. (2016)

Progesterone and Norgestrel	*Scenedesmus obliquus* and *C. pyrenoidosa*	Medium: BG-11 Light intensity: 3000 lux, 12:12 h light:dark cycle Duration: 5 days Shaking incubator at 150 rpm Temperature: 25°C	Biodegradation and bioadsorption	Progesterone, 95% Norgestrel, 100% by *S. obliquus and* 60% by *C. pyrenoidosa*	Peng et al. (2014)
Procainamide and Doxycycline (44–128 μg/L)	*Tetraselmis chuii*	Microplastics (1–5 μm diameter) were added Medium: F/2 Guillard medium intensity: 90 $\mu E/m^2/s$ 16:8 h light:dark cycle Temperature: 20 ± 1°C	Bioaccumulation	–	Prata, Lavorante, Montenegro, and Guilhermino (2018)
17b-Estradiol, 17a-Ethinylestradiol, Salicylic acid, and Triclosan	*Nannochloris* sp.	Autoclaved wastewater 24 h light cycle Duration: 7 days	Photodegradation and biodegradation (17b-estradiol, 17a-ethinylestradiol, salicylic acid) Photodegradation, bioadsorption, and bioaccumulation (Triclosan)	17b-Estradiol, 17a-ethinylestradiol, and Salicylic acid, 60% Triclosan, 63%–100%	Bai and Acharya (2019)
Trimethoprim, Sulfamethoxazole, Carbamazepine, Ciprofloxacin, and Triclosan	*Nannochloris* sp.	Lake mead water Temperature: 20 °C	Photolysis (Ciprofloxacin) Bioadsorption (Sulfamethoxazole, Carbamazepine, and Triclosan)	Ciprofloxacin and Triclosan, 100% Trimethoprim, Sulfamethoxazole, and Carbamazepine are persistent	Bai and Acharya (2017)

Continued

Table 2 Continued

Pharmaceuticals (PCs)	Microalgae species	Experimental conditions	Mechanisms involved	Removal (%)	References
Trimethoprim, Sulfamethoxazole, and Triclosan	*Nannochloris* sp.	Medium: F/2 24 h light cycle	Photolysis (Sulfamethoxazole) Bioaccumulation (Triclosan)	–	Bai and Acharya (2016)
Florfenicol	*Chlorella* sp.	Medium: BG-11 Light intensity: 5000 ± 100 lux temperature: 30 ± 1°C	Biodegradation, bioaccumulation, and bioadsorption	97%	Song et al. (2019)
Sulfamethazine and Sulfamethoxazole	*S. obliquus*	Medium: Bold's basal medium (BBM) Light intensity: 45–50 μmol photon/m^2 s 16:8 h light:dark cycle Shaking incubator at 150 rpm Duration: 14 days	Hydrolysis and biodegradation	15%–29%	Xiong et al. (2019)
Enrofloxacin	*S. obliquus*, *Chlamydomonas mexicana*, *Chlorella vulgaris*, *Ourococcus multisporus*, and *Micractinium resseri*	Medium: BBM Light intensity: 45–50 μmol photon/m^2 s 16:8 h light:dark cycle Shaking incubator at 150 rpm Duration: 14 days	Biodegradation	18%–26%	Xiong, Kurade, and Jeon (2017b)
Levofloxacin	*C. vulgaris*	Medium: BBM Light intensity: 45–50 μmol photon/m^2 s 16:8 h light:dark cycle Shaking incubator at 150 rpm Duration: 14 days	Bioaccumulation and biodegradation	12%–16%	Xiong, Kurade, and Jeon (2017a)

Sulfamethazine	*C. pyrenoidosa*	Medium: f/2 Light intensity: 3000 lux 13:12 h light: dark cycle Duration: 13 days Temperature: 27 °C	Bioaccumulation and biodegradation	48.45%–69.93%	Sun, Lin, Guo, Zhao, and Li (2017)
Ciprofloxacin	*Chlamydomonas mexicana*	Medium: BBM Light intensity: 45–50 μmol photon/m^2 s 16:8 h light:dark cycle Shaking incubator at 150 rpm Duration: 11 days	Biodegradation	1%	Xiong, Kurade, Kim, Roh, and Jeon (2017)
Estradiol and Ethinylestradiol Caffeine, Acetaminophen, Ibuprofen, Hydrocinnamic acid, Ketoprofen, Naproxen, Diclofenac, and Carbamazepine	*Green microalgae* (*C. vulgaris* and *S. capricornutum*, *S. quadricauda*, *Chlamydomonas* sp., and *Chlorella* sp.)	(BM) or SE medium Light intensity: 40 μmol photon/m^2 s 16:8 h light:dark cycle Medium: Bristol Temperature: 23 ± 1°C	Biodegradation	75.3%–93.9% by *Selenastrum capricornutum*	Wang, Wong, and Tam (2017)
Caffeine, Acetaminophen, Ibuprofen, Hydrocinnamic acid, Ketoprofen, Naproxen, Diclofenac, and Carbamazepine	Dominant microalgae in warm-season: *Stigeoclonium* sp., diatoms, *Chlorella* sp., and *Monoraphidium* sp. Dominant algae in the cold season: *Chlorella* sp., diatoms, and *Stigeoclonium* sp.	HRAP (volume: 0.47 m^3, surface area: 1.54 m^2, water depth: 0.3 m) HRT: 4 and 8 days Aeration at 5 rpm, velocity of recirculation of 11 cm s^{-1} Cold and warm season Duration: 10 days	Biodegradation and photodegradation	Caffeine, Acetaminophen, Ibuprofen, and Hydrocinnamic acid, >90% Oxybenzone and Ketoprofen, 60%–90% Diclofenac, 40%–60% Carbamazepine, >40%	Matamoros, Gutiérrez, Ferrer, García, and Bayona, (2015)

Continued

Table 2 Continued

Pharmaceuticals (PCs)	Microalgae species	Experimental conditions	Mechanisms involved	Removal (%)	References
Caffeine, Ibuprofen, and Carbamazepine	*Chlorella* sp. and *Scenedesmus* sp.	HRAP Condition: 1. aerated and nonaerated 2. light vs darkness 3. wastewater vs synthetic wastewater	Biodegradation	–	Matamoros, Uggetti, García, and Bayona (2016)

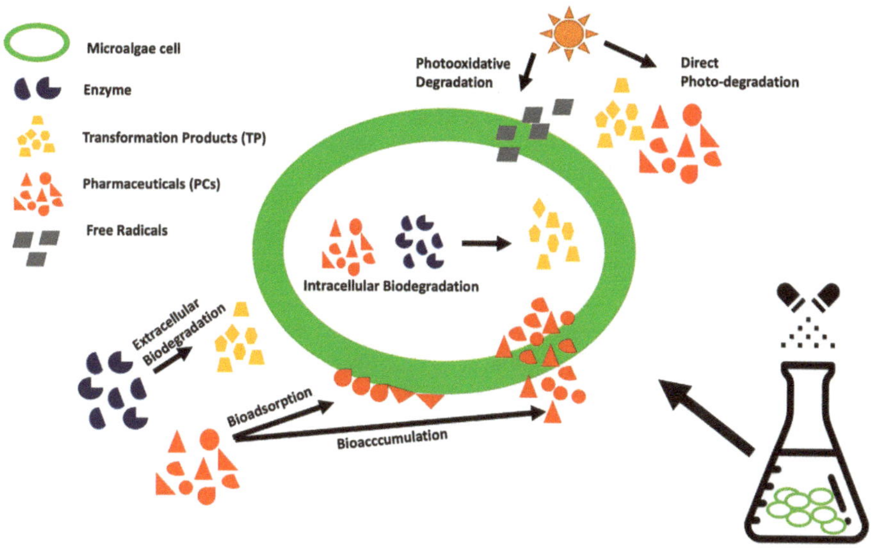

Fig. 1 Microalgae bioremediation mechanisms for the removal of PCs.

conditions and the Henry's law constant (H) values of the PCs. The high aeration rate, combined with sunlight and temperature in microalgae treatment systems, might help to escalate the volatilization of PCs. A study by Matamoros et al. (2016) reported 4-octylphenol, HHCB, and tributyl phosphate removal by volatilization with a high H value in an aerated microalgae batch reactor (Matamoros et al., 2016). The overall mechanism of pharmaceutical contaminants removal from wastewater using microalgae is shown in Fig. 1.

3. Factors influencing the removal of PCs

3.1 Microalgae growth condition

Microalgae growth parameters, such as temperature, seasonal changes and light intensity, are crucial for antibiotic elimination effectiveness (Norvill, Shilton, & Guieysse, 2016). Seasonal and temperature variations have an impact on the biodegradability of PCs as they are tied to enzymatic and microbial activity. In general, a higher temperature is preferred as it will improve microalgae's ability to remove nutrients as well as their physiological metabolism (Gonçalves, Pires, & Simões, 2017). During summer, the rate of bezafibrate, ibuprofen, atenolol, and sulfamethoxazole removal is greater than in winter, owing to increased microbial activity at a warmer temperatures (Castiglioni et al., 2006). The degradation of norfloxacin by photolysis was higher at 35°C than at 25°C (Zhang, Fu, & Wu, 2012). Similarly, warmer weather resulted in a 30% increase in carbamazepine biodegradation when microalgae were present, compared to cold weather (Matamoros et al., 2015).

Moreover, light is a crucial environmental component for microalgae survival. Light intensity influenced the development of microalgae, thus affecting nutrient and pollutant removal from wastewater. Increased light intensity has been shown to increase biomass production, and it has been found that the exponential phase of cells removes the most nutrients and organic substances (Hena, Znad, Heong, & Judd, 2018). Moreover, strong light irradiation may also raise the pH and dissolved oxygen level attributed to microalgae photosynthesis. This results in the production of reactive oxygen species and a rise in tetracycline removal (Norvill et al., 2017). However, the extreme light intensity can lower the metabolic activities of the microalgae as well as their efficacy at removing PCs. Guo et al. (2016) and Zhang et al. (2012) reported a link between *Chlorella vulgaris* biomass and norfloxacin photolysis. Within 60 min, an increase in photodegradation from 35.9% to 40.9% was detected, accompanied by an increase in cell count. However, a decreased photodegradation was detected when the cell count increased further due to the self-shading effects in microalgae cells (Guo et al., 2016; Zhang et al., 2012). Escapa et al. found a substantial drop in salicylic acid and paracetamol photodegradation during the exponential phase of *Chlorella sorokiniana*. Increased biomass reduces the available irradiance in the culture. Therefore, a high number of microalgae cells can reduce photodegradation due to the targeted compounds' adsorption, hence reducing the number of compounds accessible for photooxidation (Escapa, Coimbra, Paniagua, García, & Otero, 2017).

3.2 Microalgae species

Microalgae have been successfully grown in various wastewaters and have demonstrated tolerance to a variety of PCs. However, one microalgae species is not capable sufficient of completely eliminating PCs from wastewater. Depending on the kind of PCs present in the wastewater, microalgae strain selection is a critical stage. Certain PCs are removed by microalgae are compound-specific, whereas others are on a species-specific basis. Some PCs are compound-specific, indicating that they are removed consistently independent of the microalgae employed for phycoremediation. For example, Escapa, Coimbra, Paniagua, García, and Otero (2016) reported two distinct *Chlorella* species, *C. sorokiniana*, and *C. vulgaris* removed almost identical amounts of diclofenac (Escapa et al., 2016). Similarly, various microalgae are able to completely remove 7-amino cephalosporanic acid (Guo et al., 2016). On the other hand, Yang et al. (2017) reported a cefradine removal of 60% using *S. obliquus* and less than 30% of removal by *Chlorella pyrenoidosa* (Yang et al., 2017). As for species-specific removal of PCs, different microalgae have varying removal efficiencies. For example, *Nannochloris* sp. can remove triclosan completely from an aqueous medium, but *Microcystis aeruginosa* only demonstrated 46% of removal (Bai & Acharya, 2016; Huang et al., 2016). Therefore, it is very difficult to remove PCs from actual wastewater utilizing an axenic (mono) growth system of microalgae. Polyculture of various algal strains is capable of comprehensively removing PCs and may also improve the efficacy and durability of the system (Godwin et al., 2017).

Additionally, genetic alteration and metabolic development of the algal strain may increase the efficacy of PC removal (Shen, Jarboe, Brown, & Wen, 2015).

4. Different approaches for the removal of PCs by microalgae phytoremediation

4.1 Conventional approaches

Phycoremediation conventional approaches in the removal of PCs are mainly classified into two groups based on the control over abiotic factors: closed and open culture techniques. Closed and open culture techniques are further categorized into two subcategories based on the motility of the cells: suspended or immobilized culture techniques.

4.1.1 Open culture techniques

Generally, an open culture system does not include an air supply or a CO_2 source. For example, high-rate algal ponds (HRAPs) are shallow raceway ponds equipped with a paddlewheel to keep the microalgae suspended and aid in the mixing of CO_2 with the medium. Advanced Pond Systems (APS) are usually found in WWTPs, consisting of HRAP, an anaerobic digester, and microalgae settling pond. In APS, anaerobic bacteria are used to remove biological oxygen demand (BOD) from the wastewater by anaerobic digestion. The treated wastewater is then sent to HRAP for microalgae treatment, which involves inoculating the wastewater with a single or consortium of microalgae to remove other nutrients such as COD, nitrogen, phosphorus, and other contaminants such as PCs. Finally, the microalgal settling pond receives treated wastewater from HRAP, and a variety of techniques are used to remove biomass from the wastewater.

Lindberg et al. (2021) reported more than 90% removal of dipyridamole, metoprolol, atorvastatin, codeine, and atenolol using HRAP. Norvill et al. reported a 93% removal of tetracycline using the same technique (Lindberg et al., 2021; Norvill et al., 2017). However, the open culture technique has some significant drawbacks, high water evaporation rate, a large area requirement, and limited light consumption by microalgae cells. In addition, the efficiency of this technique is highly dependent on environmental factors (Ugwu, Aoyagi, & Uchiyama, 2008; Wen et al., 2016). To address the issues associated with open culture methods, current research efforts are concentrated on the development of appropriate closed cultivation systems.

4.1.2 Close culture techniques

Photobioreactors are closed microalgae cultivation systems in which microalgae grow under close monitoring. The system is circulated by pumps and equipped with light sources and a gas exchange unit (Ahmad, Abdullah, Koji, Yuzir, & Eva Muhammad, 2021). The closed culture system is more efficient than the open culture system in terms of lipid and biomass production and wastewater removal efficacy (Gupta, Pawar, & Pandey, 2019; Posten, 2009). The currently enclosed photobioreactors are tubular, flat-plate, and vertical-column bioreactors (Ugwu et al., 2008; Xu, Weathers, Xiong, & Liu, 2009). Tubular photobioreactors are the most prevalent, efficient, and promising closed photobioreactor but overheating, construction costs and problems in upscaling are some of the evident disadvantages of using a closed system for large-scale wastewater treatment. Furthermore, closed culture systems are often used only on a small scale in order to determine the suitability of specific microalgae strains to treat real or synthetic wastewater (Escapa et al., 2017; Hena, Gutierrez, & Croué, 2020; Zhou et al., 2011). Moreover, closed bioreactors might impede the volatilization mechanism of PC removal.

4.1.3 Suspended culture techniques

Suspended culture is prevalent in WWTPs compared to immobilized culture because it is more cost-efficient. Suspended culture methods may be used in different modes, such as in semibatch, batch, or continuous mode in both open and closed systems. Studies reported a high removal of ciprofloxacin and tetracycline in HRAP using suspended microalgae cells techniques. Suspended cells of *C. sorokiniana* were used to remove salicylic acid and paracetamol in semicontinuous and batch modes, in which the latter showed better removal efficiency (Escapa et al., 2017). Although the cultivation of suspended microalgae is well-known in wastewater treatment, microalgae cell collection is an additional step required to complete the treatment process. To circumvent the energy and time constraints associated with suspended culture, the immobilized culture approach for microalgae has shown potential as a viable alternative (He & Xue, 2010; Pires, Alvim-Ferraz, Martins, & Simões, 2013).

4.1.4 Immobilized culture techniques

Immobilized cultures are advantageous due to their capability to carry out two processes concurrently: bioprocessing and harvesting. According to Xie et al. (2020), immobilized culture methods are more suited for treating PCs than suspended cell culture techniques, as immobilized culture can prevent antibiotics' damaging impact on microalgae (Xie et al., 2020). Wu et al. (2020) reported an 89.6% removal of carbamazepine using immobilized monoculture of *C. vulgaris* on sodium alginate gel, while only 48.8% removal of carbamazepine was reported using suspended culture of *C. vulgaris* (Wu et al., 2020). Ethinylestradiol was reported to have a considerably greater rate of removal in immobilized microalgae culture than in suspended culture. The light-blocking effect of free microalgae in the suspended culture technique may prevent PCs from photodegradation (Solé & Matamoros, 2016).

4.2 Novel approaches

4.2.1 Acclimation of microalgae

Microalgae's response to harsh environments is critical for wastewater treatment, particularly when it contains PCs. Numerous microalgae may alter their metabolic pathways in response to environmental changes in order to survive and thrive. Acclimated microalgae strains are tolerant to a given condition, and the process is referred to as acclimatization. When PCs are gradually exposed to microalgae, their tolerance limit rises, and their capacity to extract PCs from wastewater improves. *C. pyrenoidosa* that has been acclimated with 60 mg/L cefradine, showed increased removal efficiency (Chen et al., 2021). Acclimatization of wild *C. vulgaris* strains to levofloxacin concentrations ranging from low to high concentration improved the levofloxacin removal effectiveness (Xiong, Kurade, Kim, et al., 2017). *C. pyrenoidosa* was acclimated for 2 weeks in 100 mg/L of cefradine before being added to activated sludge. This increased the system's removal efficacy (Guo & Chen, 2015). Yang et al. (2017) studied *C. pyrenoidosa*'s response to ceftazidime removal under harsh and limiting conditions (limited light, nitrogen, and phosphorus). The removal efficiencies of ceftazidime were improved by up to 99% (Yang et al., 2017). Controlling the salinity of the medium can also acclimatize microalgae cells via genetic and physiological adaptation. Matamoros et al. (2016) acclimated *Nitzschia acicularis* and *Chlorella* sp. to secondary treated wastewater prior to coimmobilizing the microalgae culture to remove estradiol and octylphenol from the wastewater.

4.2.2 Cometabolism of microalgae

Tran, Urase, Ngo, Hu, and Ong (2013) defined cometabolism as the concurrent breakdown of two compounds, in which the second compound's degradation (nongrowth substrate) is conditional on the existence of the first component (growth substrate) (Tran et al., 2013). Supplemental organic substrates can serve as electron donors for the cometabolism of nongrowth substrates, as well as sustaining biomass production (Xiong et al., 2018). When sodium acetate was introduced, the removal of ciprofloxacin by *Chlamydomonas mexicana* increased from 13% to 56% (Xiong, Kurade, Kim, et al., 2017). Similar findings were also observed by Xiong et al. (2020) in which the addition of sodium acetate increased sulfamethoxazole removal from 6.05% to 99.3% by *C. pyrenoidosa* (Xiong et al., 2020). Cometabolism may be regulated by various parameters, including target compound's concentration, response time, microalgal species, as well as carbon source's type and concentration.

5. Integrated processes for removal of PCs

5.1 Integration of microalgae-based biotechnologies and advanced oxidation processes

The advanced oxidation process (AOP) features the production of free radicals generated by the use of ozone (O_3), peroxides, ultraviolet light (UV), and iron salts (Verma, Kundu, & Pandey, 2021). This process can effectively degrade PC but has several downsides, including the production of toxic by-products and high energy consumption. Therefore, AOP has been integrated with a biological method to eliminate the aforesaid difficulties and provide a cost-effective, ecologically sustainable, and efficient way of removing PCs. When coupled with a biological method, the by-products created by AOP can biodegrade more effectively. In this context, AOP coupled with microalgae's capacity to biodegrade PPCPs, and their derivatives are widely reported in the literature. For instance, the removal of amoxicillin by microalgae rose from 70% to over 90% when H_2O_2/Fe(II)-based Fenton processes were added. PCs can also be eliminated by AOP-induced indirect photodegradation, such as the addition of photochemically reactive Fe(III), which results in reactive oxygen species formation (Batista, Cottrell, & Nogueira, 2014). The inclusion of Fe(III) was reported to improve norfloxacin photodegradation by microalgae (Zhang et al., 2012). Ge and Deng (2015) demonstrated that enrofloxacin and ciprofloxacin hydrochloride could be successfully removed using a Fe(III)-microalgae system (Ge & Deng, 2015). However, caution should be taken when utilizing Fenton reagents since high quantities of H_2O_2 or Fe(II) may be fatal to microalgae (Li et al., 2015). Several literatures have also reported microalgae treatment of PCs coupled with AOP using UV-irradiation. Du, Zhang, Guo, and Chen (2015) described using a hybrid UV-microalgae treatment system to reduce the toxicity of cefradine (Du et al., 2015). Yang et al. (2017) showed almost a complete removal at 99.84% of cefradine and amoxicillin within 24 h utilizing *S. obliquus*, along with UV-irradiation (Yang et al., 2017).

5.2 Integration of microalgae-based biotechnologies and constructed wetlands

Constructed wetlands (CWs) are a practical approach for treating PCs and have garnered considerable interest because of their inexpensive construction, operating, and maintenance costs (Wang, Liu, et al., 2017). Constructed wetlands are classified according to their flow regimes, such as horizontal subsurface flow, horizontal surface flow, and vertical flow systems. These systems may be combined to maximize the benefits of each (Vymazal, 2011). The removal of PPCPs from artificial wetlands is highly reliant on the physical-chemical characteristics of the PCs, CW's configuration and operation, and environmental conditions (Garcia-Rodríguez, Matamoros, Fontàs, & Salvadó, 2014; Zhang et al., 2014). In addition, nitrification and denitrification are critical in the formation of CWs. However, because of the scarcity of oxygen in CWs, nitrification becomes the rate-limiting process since it needs an appropriate supply of oxygen and carbon sources. While artificial aeration may increase oxygen levels, it is expensive and prevents denitrification when anaerobic conditions are absent. Incorporating a microalgae pretreatment unit into a CW water treatment system is an appealing alternative as microalgae produce oxygen through photosynthesis. Zhang et al. (2013) demonstrated that a CW system containing *Scirpus validus* removed 74% of carbamazepine and nearly eradicated naproxen (Zhang et al., 2013). Moreover, pretreatment of wastewater containing up to 150 mg/L amoxicillin was reported to be removed at >99% by *Chlorella* sp.-bacteria consortium in CW (Shi, Yeap, Huang, Chen, & Ng, 2018). Extensive study is necessary to fully comprehend the properties of microalgae treatment and CW for PCs containing wastewater. The configuration design, hydraulic mode, temperature and seasonality, pH, oxygen, and redox potential are critical parameters that significantly affect the combined system's removal capacity.

5.3 Integration of microalgae-based biotechnologies and microbial fuel cells

Microbial fuel cells (MFCs) have been proposed as a viable method of treating wastewater as they can produce renewable energy and valuable by-products (He, Kan, Mansfeld, Angenent, & Nealson, 2009; Li, Yu, & He, 2014; Zhang et al., 2014). In MFCs, the conversion of carbon to energy is a simple process. However, the membrane used in this process is expensive, and a high amount of energy is needed for nitrification during mechanical aeration. In addition, MFCs have been reported to have low phosphorus removal (Zhang, Noori, & Angelidaki, 2011). Therefore, the development of microalgae technology may serve as an excellent complement to MFCs. Microalgae and electrochemically active bacteria form a symbiotic relationship that enables the simultaneous removal of nutrients, nitrogen, and phosphorus while also producing energy. While these combined procedures demonstrated advantages and increased flexibility, few studies have been done regarding MFCs and microalgae-based technologies for PC removal. A novel microalgae and MFCs system capable of concurrently degrading florfenicol, removing nitrogen, and generating bioelectricity was also designed (Sun et al., 2020). In addition, studies reported that the MFCs system could aid in the elimination of antibiotic-resistant bacteria and genes (ARB/ARG) (Yan et al., 2019). Thus, microalgae-based processes in conjunction with MFC technology may increase wastewater treatment efficiency while enabling sustainable wastewater treatment. Additionally, this approach may produce biofuels or other valuable products from microalgae biomass (Nagendranatha Reddy, Nguyen, Noori, & Min, 2019).

5.4 Integration of microalgae-based biotechnologies and conventional activated sludge

For more than a century, municipal wastewater has been treated using the standard activated sludge approach. However, the technology has received much criticism for its excessive energy usage and poor efficacy in removing micropollutants Consortiums of microalgae and bacteria are more successful in eliminating heavy metals, micropollutants, and nutrients. The combination of microalgae and activated sludge systems are economically feasible for wastewater treatment due to its cost-effectiveness, low energy needs, and resource recovery potential (Gonçalves et al., 2017; Quijano, Arcila, & Buitrón, 2017). However, the combination of microalgae and activated sludge systems is a relatively new concept, and not much study has been conducted regarding PC removals. Guo and Chen (2015) studied the removal of cephalosporins using a microalgae-activated sludge system, reaching a removal efficiency of up to 97.91% (Guo & Chen, 2015). Cephalosporins are one of the most often used antibiotics in humans and animals such as cattle or aquaculture animals. It was also found that antibiotics pretreated by microalgae do not come into direct contact with bacteria, thus significantly aiding in preventing the establishment of ARBs.

5.5 Integration of microalgae-based biotechnologies and membrane photobioreactor

Numerous research has concentrated on using membrane bioreactors (MBRs) in conjunction with activated sludge to remove PCs from wastewater (Alvarino, Suárez, Garrido, Lema, & Omil, 2016; Kovalova, Siegrist, Singer, Wittmer, & McArdell, 2012). MBR is reported to be more effective than conventional treatment in eliminating PCs (Spring, Bagley, Andrews, Lemanik, & Yang, 2007). Just like MBR, membrane photobioreactor technology (MPBR) utilizes light and membrane units to help the growth of microalgae. However, current MPBR research has focused on removing phosphorus and nitrogen (Gao et al., 2016; Lee, Baek, & Kim, 2018). Derakhshan et al. (2018) showed atrazine removal of up to 95%, confirming the MPBR reactor's capacity to treat wastewater while cultivating and developing microalgae. The decrease in atrazine concentrations in this system may have resulted from microalgae-bacterial symbiosis and cometabolism (Derakhshan et al., 2018). The summary of the different hybrid technologies in the microalgae-based remediation of wastewater containing pharmaceutical contaminants is incorporated in Table 3.

6. Concluding remarks and future perspectives

The removal performance of PCs in wastewater indicates that employing microalgae-based remediation has enormous potential. Bioadsorption, bioaccumulation, biodegradation, photodegradation, and volatilization are all viable mechanisms for PC removal. Biodegradation is the major PC removal pathway and has been intensively studied. However, they remain poorly understood, particularly at the molecular level. Therefore, omics technologies concerning biodegradation of PCs should be intensively studied as they might help to explain the functional genes linked with the PC removal pathways. There are many factors affecting PC treatment using microalgae. Therefore, process optimization and its combination with additional treatment methods such as AOPs, CWs, MBPRs, and activated sludge, may result in an effective antibiotic removal procedure. However, most combined systems are currently being studied at the laboratory-scale size. There are still significant technological, environmental, and economic barriers to expanding their use in full-scale WWTPs. In addition, the prospect of PC removal using microalgae-based remediation needs to be explored further using actual wastewater. Long-term operating performance is also required to assess the compounding impacts on the environment.

Table 3 Summary of different hybrid approaches in microalgae-based bioremediation of PCs.

Techniques	Type of PCs and microalgae (initial concentration and removal efficiency)	Details	References
Microalgae and AOPs	**Fenton** Amoxicillin and Chlorella pyrenoidosa(100–300 mg/L, 90%) Norfloxacin and Chlorella vulgaris (1.0 mg/L, 80.3%) **UV irradiation** Cefradine, amoxicillin, and Scenedesmus obliquus (100 mg/L, 99.84%)	Microalgae demonstrated a high tolerance for PCs and have been shown to reduce the toxicity of the parent compounds and their by-products while also being a prospective source of biodiesel	Li et al. (2015,)Zhang et al. (2012), and Yang et al. (2017)
Microalgae and CWs	Amoxicillin and Chlorella sp. (150 mg/L, 99%) Carbamazepine and Scirpus validus (0.5–2.0 mg/L, 74%) Naproxen and Scirpus validus (0.5–2.0 mg/L, 98%)	Photodegradation and biodegradation played a major role in the removal of naproxen but played a minor color the removal of carbamazepine	Shi et al. (2018) and Zhang et al. (2013)

Continued

Table 3 Continued

Techniques	Type of PCs and microalgae (initial concentration and removal efficiency)	Details	References
Microalgae and MFCs	Florfenicol (0.2–1.0 mg/L, 80%–90%) Sulfamethoxazole (20 mg/L, 85%) Chloramphenicol (32 mg/L, 96%)	Florfenicol degradation was mostly due to the bioelectrochemical increased cometabolic reductive dehalogenation 0.5 mg/L florfenicol to the anode increases Pseudomonas growth, resulting in a 3.2-fold increase in energy output A prolonged acclimation time significantly increases the rate of sulfamethoxazole biodegradation	Sun et al. (2020), Wang and Wang (2016), and Liang et al. (2013)
Microalgae and conventional activated sludge	Cephalosporin (100 mg/L, 97.9%) **HRAP** Norfloxacin (1332 μg/L, 82%) Clarithromycin (1304 μg/L, 95%) Sulfamethoxazole (164 μg/L, 38%) Trimethoprim (159 μg/L, 100%)	The removal efficiency was achieved in the combined system using unacclimated *C. pyrenoidosa* and unacclimated activated sludge	Guo and Chen (2015) and Villar-Navarro, Baena-Nogueras, Paniw, Perales, and Lara-Martin (2018)
Microalgae and MBRs	Atrazine (0.1 mg/L, 95%) Sulfamethoxazole (20 mg/L, 85%) Chloramphenicol (32 mg/L, 96%)	Harmful algal biomass may also be transformed into bioelectricity	Derakhshan et al. (2018), Wang and Wang (2016), and Liang et al. (2013)

Acknowledgments

The author would like to acknowledge and thank the financial support from AUN/SEED-Net Grant no: 4B403 under Collaborative Education Program for Sustainable Environmental Engineering Network (CEP-SEEN).

References

Abo, R., Kummer, N. A., & Merkel, B. J. (2016). Optimized photodegradation of bisphenol A in water using ZnO, TiO2 and SnO2 photocatalysts under UV radiation as a decontamination procedure. *Drinking Water Engineering and Science*, *9*(2), 27–35. https://doi.org/10.5194/dwes-9-27-2016.

Ahmad, I., Abdullah, N., Koji, I., Yuzir, A., & Eva Muhammad, S. (2021). Evolution of Photobioreactors: A review based on microalgal perspective. *IOP Conference Series: Materials Science and Engineering*, *1142*(1), 012004. https://doi.org/10.1088/1757-899x/1142/1/012004.

Ahmad, I., Abdullah, N., Koji, I., Yuzir, A., & Mohamad, S. E. (2021). Potential of microalgae in bioremediation of wastewater. *Bulletin of Chemical Reaction Engineering & Catalysis*, *16*(2), 413–429. https://doi.org/10.9767/bcrec.16.2.10616.413-429.

Ahmad, I., Yuzir, A., Mohamad, S. E., Iwamoto, K., & Abdullah, N. (2021). Role of microalgae in sustainable energy and environment. *IOP Conference Series: Materials Science and Engineering*, *1051*(1), 012059. https://doi.org/10.1088/1757-899x/1051/1/012059.

Ali, M. E. M., Abd El-Aty, A. M., Badawy, M. I., & Ali, R. K. (2018). Removal of pharmaceutical pollutants from synthetic wastewater using chemically modified biomass of green alga *Scenedesmus obliquus*. *Ecotoxicology and Environmental Safety*, *151*, 144–152. https://doi.org/10.1016/j.ecoenv.2018.01.012.

Alvarino, T., Suárez, S., Garrido, M., Lema, J. M., & Omil, F. (2016). A UASB reactor coupled to a hybrid aerobic MBR as innovative plant configuration to enhance the removal of organic micropollutants. *Chemosphere*, *144*, 452–458. https://doi.org/10.1016/j.chemosphere.2015.09.016.

Angulo, E., Bula, L., Mercado, I., Montano, A., & Cubillan, N. (2018). Bioremediation of Cephalexin with non-living *Chlorella* sp., biomass after lipid extraction. *Bioresource Technology*, *257*, 17–22. https://doi.org/10.1016/j.biortech.2018.02.079.

Bai, X., & Acharya, K. (2016). Removal of trimethoprim, sulfamethoxazole, and triclosan by the green alga Nannochloris sp. *Journal of Hazardous Materials*, *315*, 70–75. https://doi.org/10.1016/j.jhazmat.2016.04.067.

Bai, X., & Acharya, K. (2017). Algae-mediated removal of selected pharmaceutical and personal care products (PPCPs) from Lake Mead water. *Science of the Total Environment*, *581–582*, 734–740. https://doi.org/10.1016/j.scitotenv.2016.12.192.

Bai, X., & Acharya, K. (2019). Removal of seven endocrine disrupting chemicals (EDCs) from municipal wastewater effluents by a freshwater green alga. *Environmental Pollution*, *247*, 534–540. https://doi.org/10.1016/j.envpol.2019.01.075.

Bano, F., Malik, A., & Ahammad, S. Z. (2021). Removal of estradiol, diclofenac, and triclosan by naturally occurring microalgal consortium obtained from wastewater. *Sustainability (Switzerland)*, *13*. https://doi.org/10.3390/su13147690.

Batista, A. P. S., Cottrell, B. A., & Nogueira, R. F. P. (2014). Photochemical transformation of antibiotics by excitation of Fe(III)-complexes in aqueous medium. *Journal of*

Photochemistry and Photobiology A: Chemistry, 274, 50–56. https://doi.org/10.1016/j.jphotochem.2013.09.017.

Bilal, M., Rasheed, T., Sosa-Hernández, J. E., Raza, A., Nabeel, F., & Iqbal, H. M. N. (2018). Biosorption: An interplay between marine algae and potentially toxic elements—A review. *Marine Drugs, 16*(2). https://doi.org/10.3390/md16020065.

Castiglioni, S., Bagnati, R., Fanelli, R., Pomati, F., Calamari, D., & Zuccato, E. (2006). Removal of pharmaceuticals in sewage treatment plants in Italy. *Environmental Science and Technology, 40*(1), 357–363. https://doi.org/10.1021/es050991m.

Challis, J. K., Hanson, M. L., Friesen, K. J., & Wong, C. S. (2014). A critical assessment of the photodegradation of pharmaceuticals in aquatic environments: Defining our current understanding and identifying knowledge gaps. *Environmental Science: Processes and Impacts, 16*(4), 672–696. https://doi.org/10.1039/c3em00615h.

Chen, S., Xie, J., & Wen, Z. (2021). Removal of pharmaceutical and personal care products (PPCPs) from waterbody using a revolving algal biofilm (RAB) reactor. *Journal of Hazardous Materials, 406*. https://doi.org/10.1016/j.jhazmat.2020.124284.

Choi, H. J., & Lee, S. M. (2015). Heavy metal removal from acid mine drainage by calcined eggshell and microalgae hybrid system. *Environmental Science and Pollution Research, 22*(17), 13404–13411. https://doi.org/10.1007/s11356-015-4623-3.

Chojnacka, K., & Noworyta, A. (2004). Evaluation of Spirulina sp. growth in photoautotrophic, heterotrophic and mixotrophic cultures. *Enzyme and Microbial Technology, 34*(5), 461–465. https://doi.org/10.1016/j.enzmictec.2003.12.002.

Coimbra, R. N., Escapa, C., Vázquez, N. C., Noriega-Hevia, G., & Otero, M. (2018). Utilization of non-living microalgae biomass from two different strains for the adsorptive removal of diclofenac from water. *Water, 10*(10). https://doi.org/10.3390/w10101401.

Daneshvar, E., Zarrinmehr, M. J., Hashtjin, A. M., Farhadian, O., & Bhatnagar, A. (2018). Versatile applications of freshwater and marine water microalgae in dairy wastewater treatment, lipid extraction and tetracycline biosorption. *Bioresource Technology, 268*, 523–530. https://doi.org/10.1016/j.biortech.2018.08.032.

de Godos, I., Munoz, R., & Guieysse, B. (2012). Tetracycline removal during wastewater treatment in high-rate algal ponds. *Journal of Hazardous Materials, 229*, 446–449. https://doi.org/10.1016/j.jhazmat.2012.05.106.

de Wilt, A., Butkovskyi, A., Tuantet, K., Leal, L. H., Fernandes, T. V., Langenhoff, A., et al. (2016). Micropollutant removal in an algal treatment system fed with source separated wastewater streams. *Journal of Hazardous Materials, 304*, 84–92. https://doi.org/10.1016/j.jhazmat.2015.10.033.

Derakhshan, Z., Mahvi, A. H., Ehrampoush, M. H., Mazloomi, S. M., Faramarzian, M., Dehghani, M., et al. (2018). Studies on influence of process parameters on simultaneous biodegradation of atrazine and nutrients in aquatic environments by a membrane photobioreactor. *Environmental Research, 161*, 599–608. https://doi.org/10.1016/j.envres.2017.11.045.

Du, Y., Zhang, S., Guo, R., & Chen, J. (2015). Understanding the algal contribution in combined UV-algae treatment to remove antibiotic cefradine. *RSC Advances, 5*(74), 59953–59959. https://doi.org/10.1039/c5ra10806c.

Du, C., Zhang, B., He, Y., Hu, C., Ng, Q. X., Zhang, H., et al. (2017). Biological effect of aqueous C60 aggregates on Scenedesmus obliquus revealed by transcriptomics and non-targeted metabolomics. *Journal of Hazardous Materials, 324*, 221–229. https://doi.org/10.1016/j.jhazmat.2016.10.052.

Escapa, C., Coimbra, R. N., Paniagua, S., García, A. I., & Otero, M. (2016). Comparative assessment of diclofenac removal from water by different microalgae strains. *Algal Research, 18*, 127–134. https://doi.org/10.1016/j.algal.2016.06.008.

Escapa, C., Coimbra, R. N., Paniagua, S., García, A. I., & Otero, M. (2017). Paracetamol and salicylic acid removal from contaminated water by microalgae. *Journal of Environmental Management*, *203*, 799–806. https://doi.org/10.1016/j.jenvman.2016.06.051.

Fomina, M., & Gadd, G. M. (2014). Biosorption: Current perspectives on concept, definition and application. *Bioresource Technology*, *160*, 3–14. https://doi.org/10.1016/j.biortech.2013.12.102.

Gao, F., Li, C., Yang, Z. H., Zeng, G. M., Feng, L. J., Liu, J. Z., et al. (2016). Continuous microalgae cultivation in aquaculture wastewater by a membrane photobioreactor for biomass production and nutrients removal. *Ecological Engineering*, *92*, 55–61. https://doi.org/10.1016/j.ecoleng.2016.03.046.

Garcia-Rodríguez, A., Matamoros, V., Fontàs, C., & Salvadó, V. (2014). The ability of biologically based wastewater treatment systems to remove emerging organic contaminants—A review. *Environmental Science and Pollution Research*, *21*(20), 11708–11728. https://doi.org/10.1007/s11356-013-2448-5.

Ge, L., & Deng, H. (2015). Degradation of two fluoroquinolone antibiotics photoinduced by Fe (III)-microalgae suspension in an aqueous solution. *Photochemical and Photobiological Sciences*, *14*(4), 693–699. https://doi.org/10.1039/c3pp50149c.

Godwin, C. M., Hietala, D. C., Lashaway, A. R., Narwani, A., Savage, P. E., & Cardinale, B. J. (2017). Algal polycultures enhance coproduct recycling from hydrothermal liquefaction. *Bioresource Technology*, *224*, 630–638. https://doi.org/10.1016/j.biortech.2016.11.105.

Gonçalves, A. L., Pires, J. C. M., & Simões, M. (2017). A review on the use of microalgal consortia for wastewater treatment. *Algal Research*, *24*, 403–415. https://doi.org/10.1016/j.algal.2016.11.008.

Guo, R., & Chen, J. (2015). Application of alga-activated sludge combined system (AASCS) as a novel treatment to remove cephalosporins. *Chemical Engineering Journal*, *260*, 550–556. https://doi.org/10.1016/j.cej.2014.09.053.

Guo, W. Q., Zheng, H. S., Li, S., Du, J. S., Feng, X. C., Yin, R. L., et al. (2016). Removal of cephalosporin antibiotics 7-ACA from wastewater during the cultivation of lipid-accumulating microalgae. *Bioresource Technology*, *221*, 284–290. https://doi.org/10.1016/j.biortech.2016.09.036.

Gupta, S., Pawar, S. B., & Pandey, R. A. (2019). Current practices and challenges in using microalgae for treatment of nutrient rich wastewater from agro-based industries. *Science of the Total Environment*, *687*, 1107–1126. https://doi.org/10.1016/j.scitotenv.2019.06.115.

Hanamoto, S., Kawakami, T., Nakada, N., Yamashita, N., & Tanaka, H. (2014). Evaluation of the photolysis of pharmaceuticals within a river by 2 year field observations and toxicity changes by sunlight. *Environmental Science: Processes and Impacts*, *16*(12), 2796–2803. https://doi.org/10.1039/c4em00448e.

He, Z., Kan, J., Mansfeld, F., Angenent, L. T., & Nealson, K. H. (2009). Self-sustained phototrophic microbial fuel cells based on the synergistic cooperation between photosynthetic microorganisms and heterotrophic bacteria. *Environmental Science and Technology*, *43*(5), 1648–1654. https://doi.org/10.1021/es803084a.

He, S., & Xue, G. (2010). Algal-based immobilization process to treat the effluent from a secondary wastewater treatment plant (WWTP). *Journal of Hazardous Materials*, *178*(1–3), 895–899. https://doi.org/10.1016/j.jhazmat.2010.02.022.

Hena, S., Gutierrez, L., & Croué, J. P. (2020). Removal of metronidazole from aqueous media by *C. vulgaris*. *Journal of Hazardous Materials*, *384*. https://doi.org/10.1016/j.jhazmat.2019.121400.

Hena, S., Gutierrez, L., & Croué, J. P. (2021). Removal of pharmaceutical and personal care products (PPCPs) from wastewater using microalgae: A review. *Journal of Hazardous Materials*, *403*. https://doi.org/10.1016/j.jhazmat.2020.124041.

Hena, S., Znad, H., Heong, K. T., & Judd, S. (2018). Dairy farm wastewater treatment and lipid accumulation by Arthrospira platensis. *Water Research*, *128*, 267–277. https://doi.org/10.1016/j.watres.2017.10.057.

Hom-Diaz, A., Llorca, M., Rodríguez-Mozaz, S., Vicent, T., Barceló, D., & Blánquez, P. (2015). Microalgae cultivation on wastewater digestate: β-estradiol and 17α-ethynylestradiol degradation and transformation products identification. *Journal of Environmental Management*, *155*, 106–113. https://doi.org/10.1016/j.jenvman.2015.03.003.

Huang, X., Tu, Y., Song, C., Li, T., Lin, J., Wu, Y., et al. (2016). Interactions between the antimicrobial agent triclosan and the bloom-forming cyanobacteria Microcystis aeruginosa. *Aquatic Toxicology*, *172*, 103–110. https://doi.org/10.1016/j.aquatox.2016.01.002.

Hussain, F., Shah, S. Z., Ahmad, H., Abubshait, S. A., Abubshait, H. A., Laref, A., et al. (2021). Microalgae an ecofriendly and sustainable wastewater treatment option: Biomass application in biofuel and bio-fertilizer production. A review. *Renewable and Sustainable Energy Reviews*, *137*. https://doi.org/10.1016/j.rser.2020.110603.

Jiménez, Meent, For a photo of IP-A & chemicals. (n.d.).

Kaplan, D. (n.d.). 32 Absorption and adsorption of heavy metals by microalgae.

Hwang, J. H., Church, J., Lee, S. J., Park, J., & Lee, W. H. (2016). Use of microalgae for advanced wastewater treatment and sustainable bioenergy generation. *Environmental Engineering Science*, *33*, 882–897. https://doi.org/10.1089/ees.2016.0132.

Katam, K., Bhattacharyya, D., Soda, S., & Shimizu, T. (2021). Effect of hydraulic retention time on the performance of trickling photo-bioreactor treating domestic wastewater: Removal of carbon, nutrients, and micropollutants. *Journal of Industrial and Engineering Chemistry*, *102*, 351–362. https://doi.org/10.1016/j.jiec.2021.07.022.

Kaushik, P., & Malik, A. (2015). Mycoremediation of synthetic dyes: An insight into the mechanism, process optimization and reactor design. *Environmental Science and Engineering*, *9783319109411*, 1–25. https://doi.org/10.1007/978-3-319-10942-8_1.

Khan, M. I., Shin, J. H., & Kim, J. D. (2018). The promising future of microalgae: Current status, challenges, and optimization of a sustainable and renewable industry for biofuels, feed, and other products. *Microbial Cell Factories*, *17*(1). https://doi.org/10.1186/s12934-018-0879-x.

Khona, D. K., Shirolikar, S. M., Gawde, K. K., Hom, E., Deodhar, M. A., & D'Souza, J. S. (2016). Characterization of salt stress-induced palmelloids in the green alga, Chlamydomonas reinhardtii. *Algal Research*, *16*, 434–448. https://doi.org/10.1016/j.algal.2016.03.035.

Kovalova, L., Siegrist, H., Singer, H., Wittmer, A., & McArdell, C. S. (2012). Hospital wastewater treatment by membrane bioreactor: Performance and efficiency for organic micropollutant elimination. *Environmental Science and Technology*, *46*(3), 1536–1545. https://doi.org/10.1021/es203495d.

Kurano, N., Ikemoto, H., Miyashita, H., Hasegawa, T., Hata, H., & Miyachi, S. (1995). Fixation and utilization of carbon dioxide by microalgal photosynthesis. *Energy Conversion and Management*, *36*(6–9), 689–692. https://doi.org/10.1016/0196-8904(95)00099-Y.

Lee, J. C., Baek, K., & Kim, H. W. (2018). Semi-continuous operation and fouling characteristics of submerged membrane photobioreactor (SMPBR) for tertiary treatment of livestock wastewater. *Journal of Cleaner Production*, *180*, 244–251. https://doi.org/10.1016/j.jclepro.2018.01.159.

Leng, L., Wei, L., Xiong, Q., Xu, S., Li, W., Lv, S., et al. (2020). Use of microalgae based technology for the removal of antibiotics from wastewater: A review. *Chemosphere*, *238*. https://doi.org/10.1016/j.chemosphere.2019.124680.

Li, H., Pan, Y., Wang, Z., Chen, S., Guo, R., & Chen, J. (2015). An algal process treatment combined with the Fenton reaction for high concentrations of amoxicillin and cefradine. *RSC Advances*, *5*(122), 100775–100782. https://doi.org/10.1039/c5ra21508k.

Li, W. W., Yu, H. Q., & He, Z. (2014). Towards sustainable wastewater treatment by using microbial fuel cells-centered technologies. *Energy and Environmental Science*, *7*(3), 911–924. https://doi.org/10.1039/c3ee43106a.

Liang, B., Cheng, H. Y., Kong, D. Y., Gao, S. H., Sun, F., Cui, D., et al. (2013). Accelerated reduction of chlorinated nitroaromatic antibiotic chloramphenicol by biocathode. *Environmental Science & Technology*, *47*, 5353–5361. https://doi.org/10.1021/es400933h.

Lindberg, R. H., Namazkar, S., Lage, S., Östman, M., Gojkovic, Z., Funk, C., et al. (2021). Fate of active pharmaceutical ingredients in a northern high-rate algal pond fed with municipal wastewater. *Chemosphere*, *271*. https://doi.org/10.1016/j.chemosphere.2021.129763.

Mane, P., Kadam, D., & Chaudhari. (2013). Biochemical responses of some\nfreshwater algal species to selenium: A laboratory study. *Central European Journal of Experimental Biology*, *2*(4), 27–33.

Maqbool, M., Bhatti, H. N., Sadaf, S., Zahid, M., & Shahid, M. (2019). A robust approach towards green synthesis of polyaniline-Scenedesmus biocomposite for wastewater treatment applications. *Materials Research Express*, *6*(5). https://doi.org/10.1088/2053-1591/ab025d.

Matamoros, V., Duhec, A., Albaigés, J., & Bayona, J. M. (2009). Photodegradation of carbamazepine, ibuprofen, ketoprofen and 17α-ethinylestradiol in fresh and seawater. *Water, Air, and Soil Pollution*, *196*(1–4), 161–168. https://doi.org/10.1007/s11270-008-9765-1.

Matamoros, V., Gutiérrez, R., Ferrer, I., García, J., & Bayona, J. M. (2015). Capability of microalgae-based wastewater treatment systems to remove emerging organic contaminants: A pilot-scale study. *Journal of Hazardous Materials*, *288*, 34–42. https://doi.org/10.1016/j.jhazmat.2015.02.002.

Matamoros, V., Uggetti, E., García, J., & Bayona, J. M. (2016). Assessment of the mechanisms involved in the removal of emerging contaminants by microalgae from wastewater: A laboratory scale study. *Journal of Hazardous Materials*, *301*, 197–205. https://doi.org/10.1016/j.jhazmat.2015.08.050.

Mustafa, S., Bhatti, H. N., Maqbool, M., & Iqbal, M. (2021). Microalgae biosorption, bioaccumulation and biodegradation efficiency for the remediation of wastewater and carbon dioxide mitigation: Prospects, challenges and opportunities. *Journal of Water Process Engineering*, *41*, 102009. https://doi.org/10.1016/j.jwpe.2021.102009.

Nagendranatha Reddy, C., Nguyen, H. T. H., Noori, M. T., & Min, B. (2019). Potential applications of algae in the cathode of microbial fuel cells for enhanced electricity generation with simultaneous nutrient removal and algae biorefinery: Current status and future perspectives. *Bioresource Technology*, *292*. https://doi.org/10.1016/j.biortech.2019.122010.

Naghdi, M., Taheran, M., Brar, S. K., Kermanshahi-pour, A., Verma, M., & Surampalli, R. Y. (2018). Removal of pharmaceutical compounds in water and wastewater using fungal oxidoreductase enzymes. *Environmental Pollution*, *234*, 190–213. https://doi.org/10.1016/j.envpol.2017.11.060.

Naja, G., & Volesky, B. (2011). The mechanism of metal cation and anion biosorption. In *Microbial biosorption of metals* (pp. 19–58). Netherlands: Springer. https://doi.org/10.1007/978-94-007-0443-5_3.

Nakajima, N., Teramoto, T., Kasai, F., Sano, T., Tamaoki, M., Aono, M., et al. (2007). Glycosylation of bisphenol A by freshwater microalgae. *Chemosphere*, *69*(6), 934–941. https://doi.org/10.1016/j.chemosphere.2007.05.088.

Norvill, Z. N., Shilton, A., & Guieysse, B. (2016). Emerging contaminant degradation and removal in algal wastewater treatment ponds: Identifying the research gaps. *Journal of Hazardous Materials*, *313*, 291–309. https://doi.org/10.1016/j.jhazmat.2016.03.085.

Norvill, Z. N., Toledo-Cervantes, A., Blanco, S., Shilton, A., Guieysse, B., & Muñoz, R. (2017). Photodegradation and sorption govern tetracycline removal during wastewater treatment in algal ponds. *Bioresource Technology*, *232*, 35–43. https://doi.org/10.1016/j.biortech.2017.02.011.

Olguin, E. J. (2003). Phycoremediation: Key issues for cost-effective nutrient removal processes. *Biotechnology Advances*, *22*(03), 130–137. https://doi.org/10.1016/S0734-9750.

Peng, F. Q., Ying, G. G., Yang, B., Liu, S., Lai, H. J., Liu, Y. S., et al. (2014). Biotransformation of progesterone and norgestrel by two freshwater microalgae (Scenedesmus obliquus and Chlorella pyrenoidosa): Transformation kinetics and products identification. *Chemosphere*, *95*, 581–588. https://doi.org/10.1016/j.chemosphere.2013.10.013.

Petroutsos, D., Katapodis, P., Samiotaki, M., Panayotou, G., & Kekos, D. (2008). Detoxification of 2,4-dichlorophenol by the marine microalga Tetraselmis marina. *Phytochemistry*, *69*(3), 707–714. https://doi.org/10.1016/j.phytochem.2007.09.002.

Pires, J. C. M., Alvim-Ferraz, M. C. M., Martins, F. G., & Simões, M. (2013). Wastewater treatment to enhance the economic viability of microalgae culture. *Environmental Science and Pollution Research*, *20*(8), 5096–5105. https://doi.org/10.1007/s11356-013-1791-x.

Posten, C. (2009). Design principles of photo-bioreactors for cultivation of microalgae. *Engineering in Life Sciences*, *9*(3), 165–177. https://doi.org/10.1002/elsc.200900003.

Prata, J. C., Lavorante, B. R. B. O., Montenegro, M. C. B. S. M., & Guilhermino, L. (2018). Influence of microplastics on the toxicity of the pharmaceuticals procainamide and doxycycline on the marine microalgae Tetraselmis chuii. *Aquatic Toxicology*, *197*, 143–152. https://doi.org/10.1016/j.aquatox.2018.02.015.

Prosenc, F., Piechocka, J., Škufca, D., Heath, E., Griessler Bulc, T., Istenič, D., et al. (2021). Microalgae-based removal of contaminants of emerging concern: Mechanisms in Chlorella vulgaris and mixed algal-bacterial cultures. *Journal of Hazardous Materials*, *418*. https://doi.org/10.1016/j.jhazmat.2021.126284.

Quijano, G., Arcila, J. S., & Buitrón, G. (2017). Microalgal-bacterial aggregates: Applications and perspectives for wastewater treatment. *Biotechnology Advances*, *35*(6), 772–781. https://doi.org/10.1016/j.biotechadv.2017.07.003.

Schwarzenbach, R. P., Escher, B. I., Fenner, K., Hofstetter, T. B., Johnson, C. A., Von Gunten, U., et al. (2006). The challenge of micropollutants in aquatic systems. *Science*, *313*(5790), 1072–1077. https://doi.org/10.1126/science.1127291.

Shen, Y., Jarboe, L., Brown, R., & Wen, Z. (2015). A thermochemical-biochemical hybrid processing of lignocellulosic biomass for producing fuels and chemicals. *Biotechnology Advances*, *33*(8), 1799–1813. https://doi.org/10.1016/j.biotechadv.2015.10.006.

Shi, X., Yeap, T. S., Huang, S., Chen, J., & Ng, H. Y. (2018). Pretreatment of saline antibiotic wastewater using marine microalga. *Bioresource Technology*, *258*, 240–246. https://doi.org/10.1016/j.biortech.2018.02.110.

Sivasubramanian, V., Subramanian, V. V., Raghavan, B. G., & Ranjithkumar, R. (2009). Large scale phycoremediation of acidic effluent from an alginate industry. *Science Asia*, *35*(3), 220–226. https://doi.org/10.2306/scienceasia1513-1874.2009.35.220.

Solé, A., & Matamoros, V. (2016). Removal of endocrine disrupting compounds from wastewater by microalgae co-immobilized in alginate beads. *Chemosphere*, *164*, 516–523. https://doi.org/10.1016/j.chemosphere.2016.08.047.

Song, C., Wei, Y., Qiu, Y., Qi, Y., Li, Y., & Kitamura, Y. (2019). Biodegradability and mechanism of florfenicol via *Chlorella* sp. UTEX1602 and L38: Experimental study. *Bioresource Technology*, *272*, 529–534. https://doi.org/10.1016/j.biortech.2018.10.080.

Spring, A. J., Bagley, D. M., Andrews, R. C., Lemanik, S., & Yang, P. (2007). Removal of endocrine disrupting compounds using a membrane bioreactor and disinfection. *Journal of Environmental Engineering and Science*, *6*(2), 131–137. https://doi.org/10.1139/S06-049.

Sun, J., Li, N., Yang, P., Zhang, Y., Yuan, Y., Lu, X., et al. (2020). Simultaneous antibiotic degradation, nitrogen removal and power generation in a microalgae-bacteria powered biofuel cell designed for aquaculture wastewater treatment and energy recovery. *International Journal of Hydrogen Energy*, *45*(18), 10871–10881. https://doi.org/10.1016/j.ijhydene.2020.02.029.

Sun, M., Lin, H., Guo, W., Zhao, F., & Li, J. (2017). Bioaccumulation and biodegradation of sulfamethazine in Chlorella pyrenoidosa. *Journal of Ocean University of China*, *16*(6), 1167–1174. https://doi.org/10.1007/s11802-017-3367-8.

Sutherland, D. L., & Ralph, P. J. (2019). Microalgal bioremediation of emerging contaminants—Opportunities and challenges. *Water Research*, *164*. https://doi.org/10.1016/j.watres.2019.114921.

Tiwari, B., Sellamuthu, B., Ouarda, Y., Drogui, P., Tyagi, R. D., & Buelna, G. (2017). Review on fate and mechanism of removal of pharmaceutical pollutants from wastewater using biological approach. *Bioresource Technology*, *224*, 1–12. https://doi.org/10.1016/j.biortech.2016.11.042.

Tran, N. H., Urase, T., Ngo, H. H., Hu, J., & Ong, S. L. (2013). Insight into metabolic and cometabolic activities of autotrophic and heterotrophic microorganisms in the biodegradation of emerging trace organic contaminants. *Bioresource Technology*, *146*, 721–731. https://doi.org/10.1016/j.biortech.2013.07.083.

Tyoker Kukwa, D., & Chetty, M. (2021). *Microalgae: The multifaceted biomass of the 21st century*. IntechOpen. https://doi.org/10.5772/intechopen.94090.

Ufarté, L., Laville, É., Duquesne, S., & Potocki-Veronese, G. (2015). Metagenomics for the discovery of pollutant degrading enzymes. *Biotechnology Advances*, *33*(8), 1845–1854. https://doi.org/10.1016/j.biotechadv.2015.10.009.

Ugwu, C. U., Aoyagi, H., & Uchiyama, H. (2008). Photobioreactors for mass cultivation of algae. *Bioresource Technology*, *99*(10), 4021–4028. https://doi.org/10.1016/j.biortech.2007.01.046.

Verma, J. P., & Jaiswal, D. K. (2016). Book review: Advances in biodegradation and bioremediation of industrial waste. *Frontiers in Microbiology*. https://doi.org/10.3389/fmicb.2015.01555.

Verma, R., Kundu, L. M., & Pandey, L. M. (2021). *Decontamination of distillery spent wash through advanced oxidation methods* (pp. 103–117). Elsevier BV. https://doi.org/10.1016/b978-0-12-821011-6.00006-2.

Villar-Navarro, E., Baena-Nogueras, R. M., Paniw, M., Perales, J. A., & Lara-Martin, P. A. (2018). Removal of pharmaceuticals in urban wastewater: High rate algae pond (HRAP) based technologies as an alternative to activated sludge based processes. *Water Research*, *139*, 19–29. https://doi.org/10.1016/j.watres.2018.03.072.

Vymazal, J. (2011). Constructed wetlands for wastewater treatment: Five decades of experience. *Environmental Science and Technology*, *45*(1), 61–69. https://doi.org/10.1021/es101403q.

Wang, J. L., & Wang, S. Z. (2016). Removal of pharmaceuticals and personal care products (PPCPs) from wastewater: A review. *Journal of Environmental Management*, *182*, 620–640. https://doi.org/10.1016/j.jenvman.2016.07.049.

Wang, P., Wong, Y. S., & Tam, N. F. Y. (2017). Green microalgae in removal and biotransformation of estradiol and ethinylestradiol. *Journal of Applied Phycology*, *29*, 263–273. https://doi.org/10.1007/s10811-016-0960-y.

Wang, Y., Liu, J., Kang, D., Wu, C., & Wu, Y. (2017). Removal of pharmaceuticals and personal care products from wastewater using algae-based technologies: A review. *Reviews in Environmental Science and Biotechnology*, *16*(4), 717–735. https://doi.org/10.1007/s11157-017-9446-x.

Wen, X., Du, K., Wang, Z., Peng, X., Luo, L., Tao, H., et al. (2016). Effective cultivation of microalgae for biofuel production: A pilot-scale evaluation of a novel oleaginous microalga Graesiella sp. WBG-1. *Biotechnology for Biofuels*, *9*(1). https://doi.org/10.1186/s13068-016-0541-y.

Wu, X., Li, H., Lei, L., Ren, J., Li, W., & Liu, Y. (2020). Tolerance to short-term saline shocks by aerobic granular sludge. *Chemosphere*, *243*. https://doi.org/10.1016/j.chemosphere.2019.125370.

Xie, B., Tang, X., Ng, H. Y., Deng, S., Shi, X., Song, W., et al. (2020). Biological sulfamethoxazole degradation along with anaerobically digested centrate treatment by immobilized microalgal-bacterial consortium: Performance, mechanism and shifts in bacterial and microalgal communities. *Chemical Engineering Journal*, *388*. https://doi.org/10.1016/j.cej.2020.124217.

Xiong, J. Q., Govindwar, S., Kurade, M. B., Paeng, K. J., Roh, H. S., Khan, M. A., & Jeon, B. H. (2019). Toxicity of sulfamethazine and sulfamethoxazole and their removal by a green microalga, *Scenedesmus obliquus*. *Chemosphere*, *218*, 551–558. https://doi.org/10.1016/j.chemosphere.2018.11.146.

Xiong, J. Q., Kurade, M. B., & Jeon, B. H. (2017a). Biodegradation of levofloxacin by an acclimated freshwater microalga, Chlorella vulgaris. *Chemical Engineering Journal*, *313*, 1251–1257. https://doi.org/10.1016/j.cej.2016.11.017.

Xiong, J. Q., Kurade, M. B., & Jeon, B. H. (2017b). Ecotoxicological effects of enrofloxacin and its removal by monoculture of microalgal species and their consortium. *Environmental Pollution*, *226*, 486–493. https://doi.org/10.1016/j.envpol.2017.04.044.

Xiong, J. Q., Kurade, M. B., & Jeon, B. H. (2018). Can microalgae remove pharmaceutical contaminants from water? *Trends in Biotechnology*, *36*(1), 30–44. https://doi.org/10.1016/j.tibtech.2017.09.003.

Xiong, J. Q., Kurade, M. B., Kim, J. R., Roh, H. S., & Jeon, B. H. (2017). Ciprofloxacin toxicity and its co-metabolic removal by a freshwater microalga *Chlamydomonas mexicana*. *Journal of Hazardous Materials*, *323*, 212–219. https://doi.org/10.1016/j.jhazmat.2016.04.073.

Xiong, Q., Liu, Y. S., Hu, L. X., Shi, Z. Q., Cai, W. W., He, L. Y., et al. (2020). Co-metabolism of sulfamethoxazole by a freshwater microalga Chlorella pyrenoidosa. *Water Research*, *175*. https://doi.org/10.1016/j.watres.2020.115656.

Xu, L., Weathers, P. J., Xiong, X. R., & Liu, C. Z. (2009). Microalgal bioreactors: Challenges and opportunities. *Engineering in Life Sciences*, *9*(3), 178–189. https://doi.org/10.1002/elsc.200800111.

Yan, W., Xiao, Y., Yan, W., Ding, R., Wang, S., & Zhao, F. (2019). The effect of bioelectrochemical systems on antibiotics removal and antibiotic resistance genes: A review. *Chemical Engineering Journal*, *358*, 1421–1437. https://doi.org/10.1016/j.cej.2018.10.128.

Yang, K., Lu, J., Jiang, W., Jiang, C., Chen, J., Wang, Z., et al. (2017). An integrated view of the intimate coupling UV irradiation and algal treatment on antibiotic: Compatibility, efficiency and microbic impact assessment. *Journal of Environmental Chemical Engineering*, *5*(5), 4262–4268. https://doi.org/10.1016/j.jece.2017.08.028.

Yu, Y., Zhou, Y., Wang, Z., Torres, O. L., Guo, R., & Chen, J. (2017). Investigation of the removal mechanism of antibiotic ceftazidime by green algae and subsequent microbic impact assessment. *Scientific Reports*, *7*(1). https://doi.org/10.1038/s41598-017-04128-3.

Zangar, R. C., Davydov, D. R., & Verma, S. (2004). Mechanisms that regulate production of reactive oxygen species by cytochrome P450. *Toxicology and Applied Pharmacology*, *199* (3), 316–331. https://doi.org/10.1016/j.taap.2004.01.018.

Zhang, J., Fu, D., & Wu, J. (2012). Photodegradation of Norfloxacin in aqueous solution containing algae. *Journal of Environmental Sciences*, *24*(4), 743–749. https://doi.org/10.1016/S1001-0742(11)60814-0.

Zhang, Y., Habteselassie, M. Y., Resurreccion, E. P., Mantripragada, V., Peng, S., Bauer, S., et al. (2014). Evaluating removal of steroid estrogens by a model alga as a possible sustainability benefit of hypothetical integrated algae cultivation and wastewater treatment systems. *ACS Sustainable Chemistry & Engineering*, *2*(11), 2544–2553. https://doi.org/10.1021/sc5004538.

Zhang, D. Q., Hua, T., Gersberg, R. M., Zhu, J., Ng, W. J., & Tan, S. K. (2013). Carbamazepine and naproxen: Fate in wetland mesocosms planted with Scirpus validus. *Chemosphere*, *91*(1), 14–21. https://doi.org/10.1016/j.chemosphere.2012.11.018.

Zhang, Y., Noori, J. S., & Angelidaki, I. (2011). Simultaneous organic carbon, nutrients removal and energy production in a photomicrobial fuel cell (PFC). *Energy and Environmental Science*, *4*(10), 4340–4346. https://doi.org/10.1039/c1ee02089g.

Zhang, K., Xue, Y., Zhang, J., & Hu, X. (2020). Removal of lead from acidic wastewater by biomineralized bacteria with pH self-regulation. *Chemosphere*, *241*. https://doi.org/10.1016/j.chemosphere.2019.125041.

Zhou, W., Li, Y., Min, M., Hu, B., Chen, P., & Ruan, R. (2011). Local bioprospecting for high-lipid producing microalgal strains to be grown on concentrated municipal wastewater for biofuel production. *Bioresource Technology*, *102*(13), 6909–6919. https://doi.org/10.1016/j.biortech.2011.04.038.

An overview of recent progress in membrane-based treatment for pharmaceutical wastewaters

10

Amir Aslam[a,b] *and Saif Ullah Khan*[a]
[a]Department of Civil Engineering, Zakir Husain College of Engineering & Technology, Aligarh Muslim University, Aligarh, India, [b]Department of Civil and Environmental Engineering, TU Munich, Munich, Germany

1. Introduction

Recently, pharmaceuticals have engrossed much attention and have been termed as contaminants of emerging concern (Daughton & Ternes, 1999). Pharmaceuticals are consumed to treat illness so as to improve quality of life, and ultimately are excreted out through urine and feces. Hospital wastewaters contain a complex mixture of active pharmaceutical ingredients and microorganisms (Khan, Khan, Ahmed, Dhingra, et al., 2020). In most cases, these wastewaters are discharged to municipal wastewater treatment plants (WWTPs) without any pretreatment. As a consequence, excreted pharmaceuticals enter sewage treatment plants through sewage lines. Several classes of pharmaceuticals exist namely: antibiotics, beta blockers, analgesics, antiepileptic, antihistamine, and many more (Khan, Khan, Ahmed, Farooqi, Hussain, et al., 2020). Conventional design of WWTP does not include the treatment for pharmaceuticals and the local WWTPs are not even equipped to eradicate pharmaceuticals (Alam et al., 2021). As a result, through WWTP discharges, pharmaceuticals enter surface and groundwater systems (Li, Cabassud, & Guigui, 2015).

As per WHO, PhACs concentrations are generally <0.1 μg/L in groundwater, surface waters, and partially treated water while they are found in concentrations <0.05 μg/L in treated water (Khan, Ahmed, et al., 2020). A German-based commission performed comprehensive research in this regard in 2014 and discovered that out of 713 drugs tested, 631 drugs were present in a higher amount than the detection limits. These pharmaceutical compounds were discovered in all sources of water including ground water, surface water, and drinking water (Khan et al., 2017). When treated wastewater is utilized for first- and second-generation agriculture, the medicinal compounds are absorbed by the plants, which can obstruct their development and growth. Researchers are unable to establish whether the detrimental effect on plants is due to their direct intake or because the soil bacteria' antimicrobial function affects the symbiosis (Chander, Kumar, Goyal, & Gupta, 2005; Fatta-Kassinos, Meric, & Nikolaou, 2011). A research carried out stated that humans will be exposed to antibiotic-resistant bacteria through their food and water if crops are watered or

The Treatment of Pharmaceutical Wastewater. https://doi.org/10.1016/B978-0-323-99160-5.00015-1

irrigated with untreated water containing bacteria or genes resistant to antibiotics (Kümmerer, 2010). This happens because wastewater is a way for resistant genes and antibiotics to enter the natural bacterial ecosystem, with nonpathogenic bacteria acting as a reservoir for resistant genes.

2. Adverse effects on human health

Active pharmaceutical compounds are a major source of concern since they have inherent biological activity and can disrupt normal human physiology, resulting in severe outcomes. Approximately half of all pharmaceutical wastewaters produced by hospitals, laboratories, diagnosis centers, etc., worldwide are estimated to be dumped without treatment (Khan, Rameez, Basheer, & Farooqi, 2021). The presence of such PhAs in the environment has a variety of consequences, including the development of antibiotic-resistant microbes in the aquatic environment, a reduction in nitrite oxidation and methanogenesis, and the potential for increased toxicity of the chemical combination of metabolites (Kümmerer, 2001). In his review, Liu and Wong discussed the contamination status of various potential emerging pollutants (PEPs) by Pharmaceuticals and personal care products (PPCPs). They looked at the negative impact of the PEPs, on the health and hygiene of human beings. The authors, on the other hand, did not go into great detail concerning membrane technologies, and instead focused on other treatment technologies that are influenced by the different chemicals used, as well as assessing the potential environmental and health hazards (Liu & Wong, 2013).

The consumption of pharmaceutical-contaminated vegetables and fruits may have a negative influence on human health. Antibiotic-resistant bacteria can develop in the human body as a result of everyday intake of pharmaceutical compounds (particularly antibiotics) infected vegetables, putting humans at risk of serious health problems. Other medical concerns have also been observed, including low estrogenic activity, acute systemic hypersensitivity reactions, and suppression of the enzyme accountable for central nervous system activity (Khan, Khan, Ahmed, Farooqi, Yousefi, et al., 2020). Drinking water obtained from contaminated groundwater, in combination with badly treated fermentation and hospital effluent, may cause numerous health problems including enteropathogenic, helminthiasis, and enterovirus infections in humans (Kumaraswamy et al., 2014).

Although as far as humans are concerned, various researches have been conducted and no such direct effects of pharmaceutical compounds have been observed as yet, but with the rapid increase of these compounds in the environment, cautions must be exercised. As conventional treatment could not treat pharmaceuticals effectively, attention is diverted toward membrane or adsorption based technologies (Li, Chen, & Chen, 2017; Lv, Wu, Li, Liao, & Liu, 2014). Membrane bioreactors (MBR) have emerged as promising technology to treat pharmaceuticals as it offers dual advantages of biodegradation and membrane filtration (Shi et al., 2018). MBR possesses several advantages over conventional activated sludge treatment. Many studies have shown that MBR is superior to conventional biological treatment in removing pharmaceuticals

(Alvarino, Suárez, Garrido, Lema, & Omil, 2016). At the same time, the biggest drawback associated with membrane bioreactors is fouling (Zhang et al., 2018).

3. Membrane-based separation processes (MBSPs)

Membrane-based separation processes serve a variety of purposes. The type of membrane process used, depends on the nature of the membrane utilized, as they are synthesized by a variety of materials such as polymer, ceramic, zeolites, and so on, each having unique filtering capabilities based on the pore size of the membrane, charge on their surface, structure of the membrane used and their response to water (repelling or attracting). All of the MBS methods use different types of mechanisms for the segregation of waste from the water, ranging from dispersion of molecules to the size exclusion principles (Munk & Aminabhavi, 2002). There are different membrane-based processes such as microfiltration, ultrafiltration, nanofiltration, reverse osmosis, electrodialysis, and other hybrid methods (such as MBR).

Following are some of the MBSPs explained along with their removal efficiencies for the possible removal of pharmaceutical compounds.

3.1 Ultrafiltration

Ultrafiltration (UF) is a membrane-based process that works according to the size exclusive principle. Generally, the pore size of the UF membranes ranges between 0.01 and 0.1 μm, and are capable of eliminating contaminants that fall in the size range of 0.01–0.2 μm. Membranes utilized for ultrafiltration are usually made up of materials that are either inorganic such as Al_2O_3, TiO_2, polymers like polyethersulfone, polyacrylonitrile, polyvinylidene fluoride, etc., or are derivatives of cellulose (Wang & Bai, 2017). UF has been employed extensively in various industries for the treatment of wastewater, surfactant recovery in industries, separation of protein, and other applications.

Although at times, UF alone had proven to be insufficient in fulfilling the requisite quality of purification, therefore many researchers in the past have designed, developed, and tested different hybrid technologies. There have been several instances, where the UF membrane technology has been used for the removal of pharmaceutical compounds. A Korean research team, with the objective of eliminating PhACs namely, carbamazepine (CBZ), atenolol, caffeine, etc., integrated ultrafiltration/reverse osmosis membranes with the coagulation and disc filtration process, developed a large structure for the purpose of recycling of water. Nevertheless, this process did not produce any effective result, since RO itself can attain better retention capabilities (Chon, Cho, & Shon, 2013).

Ultrafiltration processes enhanced by the micelles could be an economically convenient way of eliminating pharmaceutical compounds. Another group of researchers (Acero, Benitez, Real, & Teva, 2017) used different micelles, along with the ultrafiltration membranes, for the efficient removal of the pharmaceutical compounds. The compounds that were eliminated include caffeine, sulfamethoxazole (SFX), atrazine, and isoproturon. Moreover, they explored using cetylpyridinium chloride (CPC), Triton X-100 (TX-100), cetyltrimethylammonium bromide (CTAB), Tween-20 (TW-20), and sodium dodecyl sulfate as the micelles. Micelles such as CTAB and CPC, which were cationic in nature successfully eradicated the water repelling pharmaceutical compounds (PhACs), possessing negative charges. CPC proved to outclass all others with best separation up to 95% as compared to all the micelles analyzed.

In another experimental work, 11 different PhACs were targeted by utilizing activated carbon in a sequential ultrafiltration treatment unit (Acero et al., 2017). They utilized activated carbon in a stepwise UF treatment plant to remove PhACs, by using an adsorption-based combination treatment strategy. The PhACs eliminated in this study were, sulfamethoxazole (SFX), antipyrine, caffeine, isoproturon, diclofenac, ketorolac, 2-hydroxy phenyl, flumequine, acetaminophen, isoproturon, atrazine. Low PhACs doses of 10–50 mg/L were found to be sufficient for removing potential emerging pollutants (PEPs) from wastewater in this study, and the relevance of a combined treatment method was emphasized upon.

Another group of researchers used a UF-NF hybrid system to separate four different pharmaceutical compounds namely phenacetin, naproxen, amoxicillin, and metoprolol, from the subsequent effluents (Javier Benitez, Acero, Real, Roldán, & Rodriguez, 2011). Structure of the membrane, pressure applied, and temperature at which the membrane operates are the factors which affect the permeate flow in both ultrafiltration and nanofiltration. Naproxen had a higher retention coefficient in UF membrane than metoprolol, with phenacetin being the lowest. Due to tight pore architectures, which resulted in direct rejection and electrostatic repulsion associated with membranes on commercial scale, the trend was maximum for amoxicillin and minimum for phenacetin. Except for phenacetin, the highest retention of PhACs in this study was achieved by using nanofiltration, with an average of 80%. Furthermore, UF had a larger flux decrease and membrane fouling, especially when treating the secondary effluents. Another group of researchers (Boleda, Galceran, & Ventura, 2011) used a UF-RO procedure to remove 29 PhACs in a comparative review of already existing treatment strategies, demonstrating that enhanced treatment methods can be up to 94% more efficient than standard treatment procedures in eradicating these medicinal compounds. This is largely due to carbon filtering, which is effective in eradicating traditional contaminants, but is unable to do the same in case of pharmaceutical compounds or other ECs. This method has proven to be particularly useful in the treatment of wastewater containing oxy chlorinated chemicals in drinking water treatment plants (Escher et al., 2011).

There have been instances where ultrasonic treatment processes were unable to eliminate pharmaceutical compounds like CBZ, amoxicillin and diclofenac from the WWTPs (Escher et al., 2011). However, combinations of hybrid ultrafiltration and ultrasonic irradiation are often used as they have proven to be capable of rejecting PhACs from wastewater, since UF alone proved to be inefficient with efficiency up to 10% only. At a frequency of 35 kHz, 99.5% of PEPs could be eliminated by using ultrasonic irradiation (Secondes, Naddeo, Belgiorno, & Ballesteros Jr, 2014).

Table 1 Few recent works carried out on the removal of pharmaceutical compounds using ultrafiltration membrane processes.

Membrane	PhACs retained	Retention (%)	Remarks	References
ZnO particle coated UF	Ibuprofen, Atenolol	99, 96	Material used is ceramic	Bhattacharya, Mukherjee, Deb, Swarnakar, and Banerjee (2020)
UF embedded with choline chloride-urea on silica nanoparticle	Paracetamol, Ibuprofen	99.9, 96	Material used is polyimide	Ali, Abi Jaoude, and Alhseinat (2021)
UF membrane	Nicotine, Atenolol, Carbamazepine, Diclofenac, Ibuprofen, Gemfibrozil	10–55		Yoon, Westerhoff, Snyder, Wert, and Yoon (2007)
Ultrafiltration	Diclofenac, CBZ, AMOXICILLIN	25, 20, and 30		Secondes et al. (2014)
Ultrafiltration	SFX, diclofenac, metoprolol, acetaminophen	Less than 50%		Acero, Benitez, Teva, and Leal (2010)

Beside these research works, many other scholars have also investigated the successful use of ultrafiltration process as a PhACs removal, few of which are mentioned in Table 1.

3.2 Microfiltration

The pore diameter of the microfiltration (MF) membrane is much higher (0.1–5 μm) as compared to UF membranes, which usually reject material in the 0.1–10 μm size range. They have been extensively used in the treatment of wastewater having dyes, colloidal particles, etc. It can also be used for the removal of soluble emerging contaminants (ECs) having high molecular weight.

Diclofenac and ibuprofen are the medicinal compounds that can be deteriorated utilizing a UV-based TiO_2 photocatalytic method if they are available in the range of 0.14–1.48 μg/L (for diclofenac) and 160–169 μg/L (for ibuprofen) (Kümmerer, 2010; Zhang, Geißen, & Gal, 2008). However, to get purified water, the photocatalyst utilized in these procedures must be removed from the polluted wastewater. In these applications, a hybrid MF could be effective for separating not only the above PhACs but also other PEPs and will also recycle the photocatalyst TiO_2 (Gerrity & Snyder, 2012). Few other researchers in their study proposed an on-site TiO_2 production

Table 2 Few recent studies carried out on other microfiltration based membrane for retaining PhACs.

Membrane	PhACs retained	Retention	References
Microfiltration	Naproxen, nicotine, caffeine	15%–30%	Guedes-Alonso, Montesdeoca-Esponda, Pacheco-Juárez, Sosa-Ferrera, and Santana-Rodríguez (2020)
Microfiltration membrane	Ibuprofen, gemfibrozil, naproxen, ketoprofen, salicylic acid	6.8–0.8 μg/L	Khan, Wintgens, Sherman, Zaricky, and Schäfer (2004)
Microfiltration	Diclofenac	65%	Nghiem, Schäfer, and Elimelech (2005)
Microfiltration	Triclosan, ibuprofen, and acetaminophen	10%–95%	Radjenović, Petrović, Ventura, and Barceló (2008)

approach of developing a composite membrane with polyvinylidene fluoride polymers utilizing titanium tetraisopropoxide (TTIP), as a continuation of previous effort (Fischer et al., 2015). These composite membranes are induced with TiO_2, which results in the removal of already decomposed compounds such as diclofenac, ibuprofen, and methylene blue.

Another group of researchers, Wang and Huang (2019) used composite membranes loaded with carbon nanotubes having single-wall and multiwall (SWCNT and MWCNT) onto polyvinylidene fluoride membrane surfaces to considerably increase the elimination of PPCPs, highlighting the viability of microfiltration process in the treatment of water. The elimination of PhACs such as triclosan, acetaminophen, and ibuprofen ranged from 10% to 95% that increased with the increase in the quantity of aromatic rings and also the fillers' specific surface area. The PPCP removal rate was altered by a change in the solution pH from 4 to 10 by up to 70%, and neutral PPCPs were removed at a higher rate than ions due to lower electrostatic repulsion. It is possible that the membranes' capacity to remove PPCPs is owing to favorable PPCP-CNT interaction. Some other works carried out in past on microfiltration are enlisted in Table 2.

3.3 Nanofiltration

Nanofiltration membrane excludes particles such as biomacromolecules, organic and some of the metallic salts, (size varies between 0.001 and 0.01 μm). Nanofiltration allows a better recovery rate of divalent salts and textile wastewater, because NF membranes offer a substantially narrower pore structure as compared to ultrafiltration membranes. Ruíz, Arévalo, Álvarez, and Cisneros (2012) took into account the polluted wastewater that he obtained from agricultural and municipal waste, consisting of five emerging contaminants namely carbamazepine (CBZ), triclosan (TCS), 4-nonylphenol, and butyl benzyl phthalate. He used a commercial NF-270 and exerted the pressure of 800 kPa to remove these ECs. A pretreatment step was used in the study to prevent scaling. Adsorption onto membranes increased ECs removal for hydrophobic chemicals, but water solubility reduced BPA retention. In another experiment (López-Muñoz, Sotto, & Arsuaga, 2012) utilized NF to eliminate numerous pharmaceutical compounds namely hydrochlorothiazide, nicotine, 4-acetamidoantipyrine, and ranitidine HCL. At low pressure, PhACs rejection varied almost linearly, and the required rejection was accomplished by dynamic interaction and steric resistance offered by the PhACs (solute) and the interface of the membrane. NF-90 had a high solute retention rate, i.e., about more than 95%; however, NF-270 had a high solute retention rate, as the minimum retention efficiency achieved by it is for nicotine, i.e., 75% and the maximum retention efficiency being 95% in case of ranitidine hydrochloride. Above findings indicate that, in addition to transmembrane pressure, the physicochemical features of pharmaceutical compounds and membranes influenced permeate flux and PhACs rejection rate.

In another study carried out to separate bisphenol A (BPA), ibuprofen, and salicylic acid, polyamide membranes grafted with methacrylic acid cross linked with ethylene diamine showed BPA rejection percentage of 95, while pristine membranes exhibited elimination of 74% (Kim, Park, Lee, & Kwon, 2008). In addition, the removal of ibuprofen and salicylic acid by ED-modified membranes was a little lesser than that of acid-modified metallic membranes. Surprisingly, succinic acid membranes regained their electronegative surface, allowing all of the PEPs in the concentrate to be retained. In another experiment, Sun, Hatton, and Chung (2011) utilized hollow fiber nanofiltration membranes having a charged surface, to remove another pharmaceutical compound, namely, cyclophosphamide efficiently. To manufacture this membrane, a hyperbranched polyethyleneimine (PEI) was used to cross-link on the backbone of polyamide-imide. Even at high pressures, the porous membrane support provided a high flux with good stability for the water to permeate. Membrane pore size shrank dramatically following cross linking with PEI, and the surface becomes positively charged and more water-attaining in nature. The increased rejection of ciprofloxacin was due to these synergistic effects. Furthermore, the high molecular weight of PEL (60 K) has a significant impact to show the highest rejection. Using nanofiltration (NF), Nghiem, Schäfer, and Elimelech (2006) investigated the removal of CBZ (nonionizable) in the presence of pharmaceutical compounds namely, ibuprofen and sulfamethoxazole (SFX) which were ionizable, in order to demonstrate, that the eradication of CBZ present, was not dependent on the feed chemistry. With a rise in pK_a value, the provided substances altered from neutral (no charge) to negatively charged, which enhanced PhAC retention. As a result of the enhanced ionic strength, the removal of compounds having negative charges, i.e., ibuprofen and SFX, increases.

The issue of developing innovative membranes for the cleansing of PhACs from genotoxic contaminants has proven to be a difficult practice. To achieve it, Martínez, Van der Bruggen, Negrin, and Alconero (2012) employed PES-NF membranes in a cross flow nanofiltration setting to extract 1-(5-bromo-fur-2-il)-2-bromo-2-nitroethane with retention of more than 80%. Membranes that were made up of commercial NF-270 membranes, as well as 25% and 75% PES produced the best results in terms of ethanol flux. Organic solvent NF (OSNF), also known as organophilic NF, is a new method for treating organic based effluents. OSNF was utilized by Székely, Bandarra, Heggie, Sellergren, and Ferreira (2011) to eliminate genotoxic contaminants from nine active pharmaceutical ingredients (API) model feeds containing amides and macrolides. The main goal of this research was to replace the extraction in typical filtration processes, which resulted in up to 80% API recovery. Furthermore, Székely, Bandarra, Heggie, Sellergren, and Ferreira (2012) investigated the viability of the organic solvent-NF approach for the eradication of 1,3-diisopropylurea (IPU) at a dilution rate of 3, to achieve 90% with a loss of only 2.5% API in the continuing study on the application of OSNF. Another new IPU selective molecularly imprinted polymer (MIP) had also been utilized for the removal of the trace levels of IPU, attaining 83% elimination in a single step for a feed having concentration of 100 ppm. OSNF and diafiltration (DF) were combined together at a MIP dilution ratio of 3:1, in order to decrease IPU from 100 mg IPU/g to 2 mg IPU/g of API.

Few other researchers, Ilyas, Abtahi, Akkilic, Roesink, and de Vos (2017) created hollow fiber NF membranes based on weak polyelectrolyte multilayer (PEM). To make PEM-based NF membranes, weak polyelectrolytes like poly allylamine hydrochloride (PAH) and polyacrylic acid (PAA) were put onto the UF support. These membranes were also utilized to test the retention of charged and hydrophilic micro-PEPs of various sizes (200–400 g/mol). Bezafibrate, atenolol, naproxen, atrazine, and sulfamethoxazole (SFX) were among the PhACs tested, and the maximum removal efficiency of up to 80% was displayed by the membranes devised, when the pH value was taken as 6. Few other studies on nanofiltration available in literature are enlisted in Table 3.

Table 3 Few recent works done on the removal of pharmaceutical compounds using nanofiltration membrane processes.

Membrane	PhACs retained	Retention	Remark	References
NF membrane	Fluconazole, phenylbutazone, prednisone, and metformin	>96.5%, >94.1%, >84.1%, and >93.3%	NF 90, NF 270, MPF34, and DK membranes used	Foureaux et al. (2019)
NF30F, NF90, NF270, Desal 51 HL	1-(5-bromo-fur-2-il)-2-nitroethane	(In %) 90.42, 91.73, 91.32, 92.2	Material used is PES (NF30F) and PA (NF90, NF270, Desal 51 HL)	Martínez et al. (2012)

Table 3 Continued

Membrane	PhACs retained	Retention	Remark	References
NF membrane	Erythromycin-H_2O, oxybenzone, diazepam, CBZ	30%–60%		Yoon et al. (2007)
NF membrane	Ibuprofen, gemfibrozil, neproxen, ketoprofen, and salicylic acid	(In μg/L) 2.83, 4.65, 2.76, 3.09, and 0.57		Khan et al. (2004)
NF270, NF90	Ibuprofen, sulfamethoxazole, carbamazepine	(In %) For NF270 (90, 25, 85) For NF90 (100, 98, 98)	Material used is polyamide	Nghiem et al. (2005)
Nanofiltration	Ketoprofen, diclofenac, sulfamethoxazole, CBZ	More than 85%		Radjenović et al. (2008)

3.4 Reverse osmosis

Reverse osmosis is a type of membrane-based separation process, which is being used extensively for the desalination of water, for a long period of time. But in recent times reverse osmosis alone, or in the form of hybrid processes (MF-RO, NF-RO, etc.), is also being considered for the elimination of emerging contaminants such as pharmaceutical compounds.

A novel family of PEPs, beta-blockers has been found to pose serious health concerns to humans. Some of the most regularly used medications, like propranolol and metoprolol, have been designated to possibly be hazardous for creatures living in water bodies. The effect that ozonation has on the diminution of beta-blocker was investigated by Benner, Salhi, Ternes, and von Gunten (2008) in his study. The removal of beta-blockers was achieved by ozonating RO brine effluents. However, experiments performed on effluents both nonchlorinated and chlorinated, obtained from wastewater treatment plants, revealed higher stability of ozone, resulting in the reduction in the exposure of OH-radical, demonstrating RO's efficacy in removing beta-blockers.

In another study, Abdelmelek, Greaves, Ishida, Cooper, and Song (2011) investigated the elimination of pharmaceutical compounds by reverse osmosis, followed by AOP treatment of the retentates. Excitation-emission matrix spectroscopy was used to monitor degradation, and the presence of the OH-radical linked to protein in the retentate of reverse osmosis revealed that PhACs were removed efficiently. The

Table 4 Comprises some of the other efforts made by researchers to eliminate PhACs using reverse osmosis and forward osmosis.

Membrane	PhACs retained	Retention	Remarks	References
RO membrane	Metformin, cotinine, trimethoprim, caffeine, venlafaxine, carbamazepine, erythromycin, and fluoxetine	92.6%, 99%, 99.6%, 97.8%, 99%, 99.6%, 99.9%, and 99.2%		Hollman, Khan, Dominic, and Achari (2020)
RO membrane	Fluconazole, phenylbutazone, prednisone, and metformin	>96.5%, >94.1%, >84.1%, and >93.3%	DOW Filmtec BW30, SG permeate	Foureaux et al. (2019)
RO membrane	Nicotine, atenolol, carbamazepine, and diclofenac	50%–95%		Urtiaga, Pérez, Ibáñez, and Ortiz (2013)
RO membrane	Ibuprofen, gemfibrozil, neproxen, ketoprofen, and salicylic acid	(In μg/L) 1.04, 1.95, 1.32, 0.82, and 1.44	After 45 min, ozone online	Khan et al. (2004)
RO membrane	Ketoprofen, diclofenac, sulfamethoxazole, CBZ	More than 85%		Radjenović et al. (2008)

findings of AOP treatment demonstrated that the membrane-based separation process can effectively eradicate the PhACs found in the given effluents, even if the inorganic and organic elements are present in them.

In another work, Alonso, El Kori, Melián-Martel, and Del Río-Gamero (2018) in his experiment employed a commercial membrane wounded with spiral on a small scale, for the efficient removal of the target pharmaceutical compound (ciprofloxacin) from effluents having high ionic strength, utilizing the process of reverse osmosis, that was able to eliminate more than 90% of ciprofloxacin present. Some of the research works carried out in passed based on reverse osmosis and forward osmosis are listed in Table 4.

3.5 *Forward osmosis*

Besides its vast applications, RO has several disadvantages such as the high-pressure requirement for its working and membrane fouling. As a result, several alternatives have been considered, one of which being forward osmosis (FO). The forward osmosis approach makes use of the osmotic gradients established deliberately throughout the membrane by a highly concentrated draw solution. In a comparative study done on FO

Table 5 Few other studies carried out based on forward osmosis method.

Membrane	PhACs retained	Retention %	Remarks	References
FO membrane	Oxytetracycline	Approx. 98%		Farid et al. (2022)
FO membrane	DCF, IBP, CBZ, and naproxen	99, 95, 93, 93	Cellulose triacetate membrane used	Jin, Shan, Wang, Wei, and Tang (2012)
FO membrane	SFX, trimethoprim, norfloxacin, roxithromycin	More than 98%	FO integrated with electrochemical oxidation (ECO)	Liu, Zhang, Feng, Shen, and Yang (2015)
FO membrane	Ampicillin, erythromycin, ROX, norfloxacin, nizatidine	100, 99.8, 99.8, 99.7, 99.6	HTI-NW (Nonwoven) and HTI-ES (embedded support) membrane used	Kong, Yang, Wu, Wang, and Xie (2015)
FO membrane	Trimethoprim, SFX, DCF, triclosan	92, 90, 93, 98	TFC membrane is used	Xie, Nghiem, Price, and Elimelech (2014)

and RO, FO displayed a better removal efficiency of about more than 99% removal for pharmaceutical compounds such as amitriptyline than reverse osmosis (RO) (Hancock, Xu, Heil, Bellona, & Cath, 2011).

In another study (Yao, Duan, Song, & Hermanowicz, 2021), utilized forward osmosis (FO) along with the membrane bioreactor (MBR), for the degradation of a nonsteroidal pharmaceutical compound Ibuprofen, and the average elimination efficiency obtained for the pollutants was >96.32%. Salamanca et al. (2021) investigated the removal of 24 different ECs, with the help of a biomimetic hollow fiber module having an active layer of polyamide (PAD) thin film composite (TFC), with embedded aquaporin protein. All of the pollutants investigated displayed the rejection rate of >93%, out of which 19 pollutants have elimination rate of up to 99%. Some other studies based on forward osmosis method are shown in Table 5.

3.6 Electrodialysis

In recent times, electrodialysis (ED) is being widely regarded as an effective electrochemical membrane technique for the nutrient concentration and recovery from wastewater. It is a type of separation technique that has the ability to demineralize a wide range of effluents obtained from the pharmaceutical, chemical, and food industries. The benefit of using ED process is that it can be carried out at room temperature without causing any effect to the constituent's thermal breakdown. Transport of ionizable species can also be done without impacting the concentration of non-ionizable species.

In another study, Ravikumar and Sridhar (2013) performed a research, in which they reported a hybrid process which uses an electrodialysis (ED) method for removing salt from the effluent containing the PhACs, and followed it by dimethyl sulfoxide (DMSO) organic solvent recovery via distillation. When DMSO is employed in the production of bulk drugs, the effluents produced contain some unwanted components such as salts (sodium azide and ammonium chloride) and a tiny amount of color-inducing substances. To process a 10L batch size, an indigenously built ED system with a membrane area of $1.05\,m^2$ was used. Vacuum distillation was used in two stages to recover pure DMSO from desalted liquor, with the first stage obtaining a DMSO concentration of 60%–70% and the second stage recovering pure DMSO. The drug made with this recycled DMSO was determined to meet all of the desired requirements. And since the industry, which was the base of this research work, recovered around 30 metric tons of solvent, the design of ED-distillation process plan to treat 7500L/day of effluent indicated that it was technically and economically feasible. Beside these, few more research works have also been carried out as tabulated (José et al., 2021) in Table 6.

Table 6 Summarizes a few more work done for the removal of pharmaceutical compounds using electrodialysis.

Membrane	PhACs retained	Retention (in %)	Remarks	References
Electrodialysis	Ibuprofen	57	pH between 6.81 and 6.34	José et al. (2021)
Electrodialysis	Ibuprofen, atenolol, naproxen, diclofenac	48–58		Guedes-Alonso et al. (2020)
Electrodialysis	Atenolol, gemfibrozil, ibuprofen, ofloxacin	More than 90%	EDR system used	Gabarrón et al. (2016)
Electrodialysis	Ibuprofen, diclofenac, propranolol	60, 92, 91		Pronk, Biebow, and Boller (2006)

3.7 Hybrid membrane-based separation processes

Nowadays, researchers have also shown interest in the hybrid approaches, where two or more membrane-based processes or a membrane-based process along with the traditional procedures followed one after the other together in order to attain better removal efficiency. Such combined methods help in addressing several issues associated with membrane such as membrane fouling, high installation costs and maintenance cost issues.

Another study, Urtiaga et al. (2013) used a hybrid system consisting of ultrafiltration and reverse osmosis (RO) to remove PhACs utilizing the electrochemical oxidation (ECO) approach. UF was beneficial for major PhACs like atenolol, bezafibrate, caffeine, and diclofenac. PEP concentrations in the permeate ranged from 4 to 44 ng/L. The overall PhACs content was reduced from 149 to 10 μg/L by ECO of RO reject using diamond electrode. According to the author, PhACs concentrations were found to be dramatically lowered at high electro-oxidation intensity.

Furthermore, utilizing a hybrid MF-RO approach, PhACs like CBZ, atenolol, azithromycin, diclofenac, pesticides and erythromycin, which were present in large amounts in the wastewater obtained in a WWTP were eradicated up to 98% (Rodriguez-Mozaz et al., 2015). This study reflects upon the effectiveness of pilot WWTPs for effluents from a municipal wastewater plant. Another group of researchers, Wang et al. (2009) utilized nanofiltration (NF), reverse osmosis (RO), and membrane bioreactor (MBR) methods to examine the rejection of cyclophosphamide (CP), which is a regularly used chemotherapeutic medication that has a negative impact on living organisms if present in water bodies. RO was found to be effective in retaining CP up to 90% of the time; however, NF rejected CP 20%–40% of the time. The removal rate of cyclophosphamide by nanofiltration increased in MBR effluent treatment, implying that the hybrid systems (MBR-RO and MBR-NF) are both a reliable technique for treating effluents having cyclophosphamide.

4. Conclusions

This review mainly focuses on the treatment performance of membrane-based techniques in separating out pharmaceutical compounds from wastewater streams. Since these pharmaceutical compounds cannot be effectively removed by conventional treatment methods, efforts are being made to develop various effective membrane methods for their elimination. Most of the MBSPs such as MF, UF and NF have been utilized for a long time, but methods like electrodialysis have recently received increasing attention. Another membrane-based method which is being under consideration is the forward osmosis (FO), Although its applications are presently limited, it could be a better alternative to MBSPs in the future since it is less susceptible to membrane fouling. Similarly, nanofiltration (NF) requires certain modifications in order to minimize the impact of the natural organic matter on the lifespan and rejection efficiency of the membrane. As individual MBSPs are not sufficient for the complete removal of pharmaceutical substances, the researchers are now emphasizing more on combining multiple MBSPs or other treatment techniques to attain considerably higher removal efficiency. However, one major drawback associated with MBR is fouling which is caused by extracellular polymeric substance and soluble microbial product secretion. Recent advances carried on improving MBR's performance may have cost related issues. Moreover, adding of PAC is not suggested unless a high removal of pharmaceuticals is strictly required. MBR-NF/RO seems to have cost related issues as well and these technologies are out of reach for the developing countries. This study is an attempt to bring together past and present day research works, in order to provide a better operational understanding of the different membrane-based separation methods in future, which can be better utilized in future for the elimination of the pharmaceutical compounds from the wastewater streams.

References

Abdelmelek, S. B., Greaves, J., Ishida, K. P., Cooper, W. J., & Song, W. (2011). Removal of pharmaceutical and personal care products from reverse osmosis retentate using advanced oxidation processes. *Environmental Science & Technology*, *45*(8), 3665–3671.

Acero, J. L., Benitez, F. J., Real, F. J., & Teva, F. (2017). Removal of emerging contaminants from secondary effluents by micellar-enhanced ultrafiltration. *Separation and Purification Technology*, *181*, 123–131.

Acero, J. L., Benitez, F. J., Teva, F., & Leal, A. I. (2010). Retention of emerging micropollutants from UP water and a municipal secondary effluent by ultrafiltration and nanofiltration. *Chemical Engineering Journal*, *163*(3), 264–272.

Alam, R., Sheob, M., Saeed, B., Khan, S. U., Shirinkar, M., Frontistis, Z., et al. (2021). Use of electrocoagulation for treatment of pharmaceutical compounds in water/wastewater: A review exploring opportunities and challenges. *Water*, *13*(15), 2105.

Ali, J. K., Abi Jaoude, M., & Alhseinat, E. (2021). Polyimide ultrafiltration membrane embedded with reline-functionalized nanosilica for the remediation of pharmaceuticals in water. *Separation and Purification Technology*, *266*, 118585.

Alonso, J. J. S., El Kori, N., Melián-Martel, N., & Del Río-Gamero, B. (2018). Removal of ciprofloxacin from seawater by reverse osmosis. *Journal of Environmental Management*, *217*, 337–345.

Alvarino, T., Suárez, S., Garrido, M., Lema, J. M., & Omil, F. (2016). A UASB reactor coupled to a hybrid aerobic MBR as innovative plant configuration to enhance the removal of organic micropollutants. *Chemosphere*, *144*, 452–458.

Benner, J., Salhi, E., Ternes, T., & von Gunten, U. (2008). Ozonation of reverse osmosis concentrate: Kinetics and efficiency of beta blocker oxidation. *Water Research*, *42*(12), 3003–3012.

Bhattacharya, P., Mukherjee, D., Deb, N., Swarnakar, S., & Banerjee, S. (2020). Application of green synthesized ZnO nanoparticle coated ceramic ultrafiltration membrane for remediation of pharmaceutical components from synthetic water: Reusability assay of treated water on seed germination. *Journal of Environmental Chemical Engineering*, *8*(3), 103803.

Boleda, M. R., Galceran, M. T., & Ventura, F. (2011). Behavior of pharmaceuticals and drugs of abuse in a drinking water treatment plant (DWTP) using combined conventional and ultrafiltration and reverse osmosis (UF/RO) treatments. *Environmental Pollution*, *159*(6), 1584–1591.

Chander, Y., Kumar, K., Goyal, S. M., & Gupta, S. C. (2005). Antibacterial activity of soil bound antibiotics. *Journal of Environmental Quality*, *34*(6), 1952–1957.

Chon, K., Cho, J., & Shon, H. K. (2013). A pilot-scale hybrid municipal wastewater reclamation system using combined coagulation and disk filtration, ultrafiltration, and reverse osmosis: Removal of nutrients and micropollutants, and characterization of membrane foulants. *Bioresource Technology*, *141*, 109–116.

Daughton, C. G., & Ternes, T. A. (1999). Pharmaceuticals and personal care products in the environment: Agents of subtle change? *Environmental Health Perspectives*, *107*(Suppl 6), 907–938.

Escher, B. I., Baumgartner, R., Koller, M., Treyer, K., Lienert, J., & McArdell, C. S. (2011). Environmental toxicology and risk assessment of pharmaceuticals from hospital wastewater. *Water Research*, *45*(1), 75–92.

Farid, M. U., Choi, P. J., Kharraz, J. A., Lao, J. Y., St-Hilaire, S., Ruan, Y., et al. (2022). Hybrid nanobubble-forward osmosis system for aquaculture wastewater treatment and reuse. *Chemical Engineering Journal*, 135164.

Fatta-Kassinos, D., Meric, S., & Nikolaou, A. (2011). Pharmaceutical residues in environmental waters and wastewater: Current state of knowledge and future research. *Analytical and Bioanalytical Chemistry*, *399*(1), 251–275.

Fischer, K., Grimm, M., Meyers, J., Dietrich, C., Gläser, R., & Schulze, A. (2015). Photoactive microfiltration membranes via directed synthesis of TiO_2 nanoparticles on the polymer surface for removal of drugs from water. *Journal of Membrane Science*, *478*, 49–57.

Foureaux, A. F. S., Reis, E. O., Lebron, Y., Moreira, V., Santos, L. V., Amaral, M. S., et al. (2019). Rejection of pharmaceutical compounds from surface water by nanofiltration and reverse osmosis. *Separation and Purification Technology*, *212*, 171–179.

Gabarrón, S., Gernjak, W., Valero, F., Barceló, A., Petrovic, M., & Rodríguez-Roda, I. (2016). Evaluation of emerging contaminants in a drinking water treatment plant using electrodialysis reversal technology. *Journal of Hazardous Materials*, *309*, 192–201.

Gerrity, D., & Snyder, S. (2012). Wastewater and drinking water treatment technologies. In *Human pharmaceuticals in the environment* (pp. 225–255). New York, NY: Springer.

Guedes-Alonso, R., Montesdeoca-Esponda, S., Pacheco-Juárez, J., Sosa-Ferrera, Z., & Santana-Rodríguez, J. J. (2020). A survey of the presence of pharmaceutical residues in wastewaters. Evaluation of their removal using conventional and natural treatment procedures. *Molecules*, *25*(7), 1639.

Hancock, N. T., Xu, P., Heil, D. M., Bellona, C., & Cath, T. Y. (2011). Comprehensive bench- and pilot-scale investigation of trace organic compounds rejection by forward osmosis. *Environmental Science & Technology*, *45*(19), 8483–8490.

Hollman, J., Khan, M. F., Dominic, J. A., & Achari, G. (2020). Pilot-scale treatment of neutral pharmaceuticals in municipal wastewater using reverse osmosis and ozonation. *Journal of Environmental Engineering*, *146*(11), 04020121.

Ilyas, S., Abtahi, S. M., Akkilic, N., Roesink, H. D. W., & de Vos, W. M. (2017). Weak polyelectrolyte multilayers as tunable separation layers for micro-pollutant removal by hollow fiber nanofiltration membranes. *Journal of Membrane Science*, *537*, 220–228.

Javier Benitez, F., Acero, J. L., Real, F. J., Roldán, G., & Rodriguez, E. (2011). Ultrafiltration and nanofiltration membranes applied to the removal of the pharmaceuticals amoxicillin, naproxen, metoprolol and phenacetin from water. *Journal of Chemical Technology & Biotechnology*, *86*(6), 858–866.

Jin, X., Shan, J., Wang, C., Wei, J., & Tang, C. Y. (2012). Rejection of pharmaceuticals by forward osmosis membranes. *Journal of Hazardous Materials*, *227*, 55–61.

José, C., Briand, L. E., Michlig, N., Repetti, M. R., Benedetich, C., Cornaglia, L. M., et al. (2021). Isolation of ibuprofen enantiomers and racemic esters through electrodialysis. *Journal of Membrane Science, 618*, 118714.

Khan, A. H., Khan, N. A., Ahmed, S., Dhingra, A., Pratap Singh, C., Khan, S. U., et al. (2020). Application of advanced oxidation processes followed by different treatment technologies for hospital wastewater treatment. *Journal of Cleaner Production, 269*, 122411. https://doi.org/10.1016/j.jclepro.2020.122411.

Khan, N. A., Ahmed, S., Farooqi, I. H., Ali, I., Vambol, V., Changani, F., et al. (2020). Occurrence, sources and conventional treatment techniques for various antibiotics present in hospital wastewaters: A critical review. *Trends in Analytical Chemistry, 129*, 115921.

Khan, N. A., Khan, S. U., Ahmed, S., Farooqi, I. H., Hussain, A., Vambol, S., et al. (2020). Smart ways of hospital wastewater management, regulatory standards and conventional treatment techniques: A short review. *Smart and Sustainable Built Environment, 9*(4), 727–736.

Khan, N. A., Khan, S. U., Ahmed, S., Farooqi, I. H., Yousefi, M., Mohammadi, A. A., et al. (2020). Recent trends in disposal and treatment technologies of emerging-pollutants—A critical review. *Trends in Analytical Chemistry, 122*, 115744.

Khan, A., Wang, J., Li, J., Wang, X., Chen, Z., Alsaedi, A., et al. (2017). The role of graphene oxide and graphene oxide-based nanomaterials in the removal of pharmaceuticals from aqueous media: A review. *Environmental Science and Pollution Research, 24*(9), 7938–7958.

Khan, S. U., Rameez, H., Basheer, F., & Farooqi, I. H. (2021). Eco-toxicity and health issues associated with the pharmaceuticals in aqueous environments: A global scenario. *Pharmaceutical wastewater treatment technologies: Concepts and implementation strategies* (pp. 145–179). IWA.

Khan, S. J., Wintgens, T., Sherman, P., Zaricky, J., & Schäfer, A. I. (2004). Removal of hormones and pharmaceuticals in the advanced water recycling demonstration plant in Queensland, Australia. *Water Science and Technology, 50*(5), 15–22.

Kim, J. H., Park, P. K., Lee, C. H., & Kwon, H. H. (2008). Surface modification of nanofiltration membranes to improve the removal of organic micro-pollutants (EDCs and PhACs) in drinking water treatment: Graft polymerization and cross-linking followed by functional group substitution. *Journal of Membrane Science, 321*(2), 190–198.

Kong, F. X., Yang, H. W., Wu, Y. Q., Wang, X. M., & Xie, Y. F. (2015). Rejection of pharmaceuticals during forward osmosis and prediction by using the solution–diffusion model. *Journal of Membrane Science, 476*, 410–420.

Kumaraswamy, R., Amha, Y. M., Anwar, M. Z., Henschel, A., Rodríguez, J., & Ahmad, F. (2014). Molecular analysis for screening human bacterial pathogens in municipal wastewater treatment and reuse. *Environmental Science & Technology, 48*(19), 11610–11619.

Kümmerer, K. (2001). Drugs in the environment: Emission of drugs, diagnostic aids and disinfectants into wastewater by hospitals in relation to other sources—A review. *Chemosphere, 45*(6–7), 957–969.

Kümmerer, K. (2010). Pharmaceuticals in the environment. *Annual Review of Environment and Resources, 35*, 57–75.

Li, C., Cabassud, C., & Guigui, C. (2015). Evaluation of membrane bioreactor on removal of pharmaceutical micropollutants: A review. *Desalination and Water Treatment, 55*(4), 845–858.

Li, C. M., Chen, C. H., & Chen, W. H. (2017). Different influences of nanopore dimension and pH between chlorpheniramine adsorptions on graphene oxide-iron oxide suspension and particle. *Chemical Engineering Journal, 307*, 447–455.

Liu, J. L., & Wong, M. H. (2013). Pharmaceuticals and personal care products (PPCPs): A review on environmental contamination in China. *Environment International*, *59*, 208–224.

Liu, P., Zhang, H., Feng, Y., Shen, C., & Yang, F. (2015). Integrating electrochemical oxidation into forward osmosis process for removal of trace antibiotics in wastewater. *Journal of Hazardous Materials*, *296*, 248–255.

López-Muñoz, M. J., Sotto, A., & Arsuaga, J. M. (2012). Nanofiltration removal of pharmaceutically active compounds. *Desalination and Water Treatment*, *42*(1–3), 138–143.

Lv, G., Wu, L., Li, Z., Liao, L., & Liu, M. (2014). Binding sites of chlorpheniramine on 1:1 layered kaolinite from aqueous solution. *Journal of Colloid and Interface Science*, *424*, 16–21.

Martínez, M. B., Van der Bruggen, B., Negrin, Z. R., & Alconero, P. L. (2012). Separation of a high-value pharmaceutical compound from waste ethanol by nanofiltration. *Journal of Industrial and Engineering Chemistry*, *18*(5), 1635–1641.

Munk, P., & Aminabhavi, T. M. (2002). *Introduction to macromolecular science*. Wiley.

Nghiem, L. D., Schäfer, A. I., & Elimelech, M. (2005). Pharmaceutical retention mechanisms by nanofiltration membranes. *Environmental Science & Technology*, *39*(19), 7698–7705.

Nghiem, L. D., Schäfer, A. I., & Elimelech, M. (2006). Role of electrostatic interactions in the retention of pharmaceutically active contaminants by a loose nanofiltration membrane. *Journal of Membrane Science*, *286*(1–2), 52–59.

Pronk, W., Biebow, M., & Boller, M. (2006). Electrodialysis for recovering salts from a urine solution containing micropollutants. *Environmental Science & Technology*, *40*(7), 2414–2420.

Radjenović, J., Petrović, M., Ventura, F., & Barceló, D. (2008). Rejection of pharmaceuticals in nanofiltration and reverse osmosis membrane drinking water treatment. *Water Research*, *42*(14), 3601–3610.

Ravikumar, Y. V. L., & Sridhar, S. (2013). Development of an electrodialysis–distillation integrated process for separation of hazardous sodium azide to recover valuable DMSO solvent from pharmaceutical effluent. *Separation and Purification Technology*. https://doi.org/10.1016/j.seppur.2013.02.031.

Rodriguez-Mozaz, S., Ricart, M., Köck-Schulmeyer, M., Guasch, H., Bonnineau, C., Proia, L., et al. (2015). Pharmaceuticals and pesticides in reclaimed water: Efficiency assessment of a microfiltration–reverse osmosis (MF–RO) pilot plant. *Journal of Hazardous Materials*, *282*, 165–173.

Ruíz, F. N., Arévalo, A. A., Álvarez, J. C. D., & Cisneros, B. J. (2012). Operating conditions and membrane selection for the removal of conventional and emerging pollutants from spring water using nanofiltration technology: The Tula Valley case. *Desalination and Water Treatment*, *42*(1–3), 117–124.

Salamanca, M., López-Serna, R., Palacio, L., Hernández, A., Prádanos, P., & Peña, M. (2021). Study of the rejection of contaminants of emerging concern by a biomimetic aquaporin hollow fiber forward osmosis membrane. *Journal of Water Process Engineering*, *40*, 101914. https://doi.org/10.1016/j.jwpe.2021.101914.

Secondes, M. F. N., Naddeo, V., Belgiorno, V., & Ballesteros, F., Jr. (2014). Removal of emerging contaminants by simultaneous application of membrane ultrafiltration, activated carbon adsorption, and ultrasound irradiation. *Journal of Hazardous Materials*, *264*, 342–349.

Shi, Y., Huang, J., Zeng, G., Gu, Y., Hu, Y., Tang, B., et al. (2018). Evaluation of soluble microbial products (SMP) on membrane fouling in membrane bioreactors (MBRs) at the fractional and overall level: A review. *Reviews in Environmental Science and Bio/Technology*, *17*(1), 71–85.

Sun, S. P., Hatton, T. A., & Chung, T. S. (2011). Hyperbranched polyethyleneimine induced cross-linking of polyamide – imide nanofiltration hollow fiber membranes for effective removal of ciprofloxacin. *Environmental Science & Technology*, *45*(9), 4003–4009.

Székely, G., Bandarra, J., Heggie, W., Sellergren, B., & Ferreira, F. C. (2011). Organic solvent nanofiltration: A platform for removal of genotoxins from active pharmaceutical ingredients. *Journal of Membrane Science*, *381*(1–2), 21–33.

Székely, G., Bandarra, J., Heggie, W., Sellergren, B., & Ferreira, F. C. (2012). A hybrid approach to reach stringent low genotoxic impurity contents in active pharmaceutical ingredients: Combining molecularly imprinted polymers and organic solvent nanofiltration for removal of 1,3-diisopropylurea. *Separation and Purification Technology*, *86*, 79–87.

Urtiaga, A. M., Pérez, G., Ibáñez, R., & Ortiz, I. (2013). Removal of pharmaceuticals from a WWTP secondary effluent by ultrafiltration/reverse osmosis followed by electrochemical oxidation of the RO concentrate. *Desalination*, *331*, 26–34.

Wang, L., Albasi, C., Faucet-Marquis, V., Pfohl-Leszkowicz, A., Dorandeu, C., Marion, B., et al. (2009). Cyclophosphamide removal from water by nanofiltration and reverse osmosis membrane. *Water Research*, *43*(17), 4115–4122.

Wang, J., & Bai, Z. (2017). Fe-based catalysts for heterogeneous catalytic ozonation of emerging contaminants in water and wastewater. *Chemical Engineering Journal*, *312*, 79–98.

Wang, Y., & Huang, H. (2019). Carbon nanotube composite membranes for microfiltration of pharmaceuticals and personal care products. In *Advanced nanomaterials for membrane synthesis and its applications* (pp. 183–202). Elsevier.

Xie, M., Nghiem, L. D., Price, W. E., & Elimelech, M. (2014). Relating rejection of trace organic contaminants to membrane properties in forward osmosis: Measurements, modelling and implications. *Water Research*, *49*, 265–274.

Yao, M., Duan, L., Song, Y., & Hermanowicz, S. W. (2021). Degradation mechanism of Ibuprofen via a forward osmosis membrane bioreactor. *Bioresource Technology*, *321*, 124448. https://doi.org/10.1016/j.biortech.2020.124448.

Yoon, Y., Westerhoff, P., Snyder, S. A., Wert, E. C., & Yoon, J. (2007). Removal of endocrine disrupting compounds and pharmaceuticals by nanofiltration and ultrafiltration membranes. *Desalination*, *202*(1–3), 16–23.

Zhang, Y., Geißen, S. U., & Gal, C. (2008). Carbamazepine and diclofenac: Removal in wastewater treatment plants and occurrence in water bodies. *Chemosphere*, *73*(8), 1151–1161.

Zhang, M., Hong, H., Lin, H., Shen, L., Yu, H., Ma, G., et al. (2018). Mechanistic insights into alginate fouling caused by calcium ions based on terahertz time-domain spectra analyses and DFT calculations. *Water Research*, *129*, 337–346.

Sustainable and eco-friendly treatment of pharmaceuticals wastewater

11

Md. Eman Talukder[a], *Md. Nahid Pervez*[a,b], *Mst. Monira Rahman Mishu*[c], *K.M. Faridul Hasan*[a], *Faizan Shafiq*[a], *Yaping Zhao*[d], *Yingjie Cai*[a], *Hongchen Song*[e], *Md. Ibrahim H. Mondal*[f], *George K Stylios*[g], *Vincenzo Naddeo*[b], *Afzal Husain Khan*[h], and *Alberto Figoli*[i]
[a]Hubei Provincial Engineering Laboratory for Clean Production and High-Value Utilisation of Bio-based Textile Materials, Wuhan Textile University, Wuhan, Hubei, China, [b]Sanitary Environmental Engineering Division (SEED), Department of Civil Engineering, University of Salerno, Fisciano, Italy, [c]Faculty of Nutrition and Food Science, Patuakhali Science and Technology University, Patuakhali, Bangladesh, [d]School of Ecological and Environmental Science, Shanghai Key Laboratory for Urban Ecological Process and Eco-Restoration, East China Normal University, Shanghai, China, [e]Water Science Center, Guangzhou Institute of Advanced Technology, Chinese Academy of Sciences, Guangzhou, China, [f]Polymer and Textile Research Lab, Department of Applied Chemistry and Chemical Engineering, Rajshahi University, Rajshahi, Bangladesh, [g]Research Institute for Flexible Materials, School of Textiles and Design, Heriot-Watt University, Galashiels, United Kingdom, [h]Civil Engineering Department, College of Engineering, Jazan University, Jazan, Saudi Arabia, [i]Institute on Membrane Technology (CNR-ITM), Rende, Italy

1. Introduction

Pharmaceutical technology can be categorized by product and process or technique. Biopharmaceutical engineering deals with fermentation, cell engineering, enzyme engineering, and genetic engineering. Fermentation engineering is the most popular, which is dependent on the actions of microbes to ferment, filter, extract antibiotics, vitamins, and amino acids from organic source materials (Massima Mouele et al., 2021). The chemical pharmacy sector is involved with the production of organic or inorganic materials utilizing multilayer chemical processes. There is also a natural medicines sector, which is concerned with a way of directly extracting medicinal ingredients from plants and animals (Shigei et al., 2021).

The pharmaceutical business has grown in recent years, and as a result, environmental degradation is becoming increasingly significant. Pharmaceutical wastewater comes in a number of forms depending on the wide range of products and their methods of processing. Highly concentrated antibiotic wastewater is one of the major sources of biopharmaceutical wastewater, which has high chroma, low C/N, high suspended solids (SS) and sulfate concentration, complex composition, biological

The Treatment of Pharmaceutical Wastewater. https://doi.org/10.1016/B978-0-323-99160-5.00012-6

toxicity, and high concentration of salt, which are difficult to biodegrade, and are hazardous to microbes (Frascaroli et al., 2021; Guo et al., 2017; Husain Khan et al., 2020).

Traditional pharmaceutical wastewater treatment technologies, and with tight regulatory requirements, are impossible to meet consumer demand. Hence advanced wastewater treatment for pharmaceuticals is critical. This book chapter explains and critically analyses the fundamentals of advanced pharmaceutical wastewater treatment technologies.

2. Pharmaceutical wastewater characteristics

Water is necessary for our eco system and paramount for the continual existence of humans, animals, and plants. However, since industrialization its continued usage is increasing exponentially for many industrial processes resulting in a considerable volume of wastewater. Pharmaceutical wastewater (PWW) is defined as liquid and solid waste released from pharmaceutical industries, and related agricultural operations, and containing a wide range of poisons. It is most commonly referred to as municipal wastewater, which contains a wide range of toxins as a consequence of the mixed wastewaters from various sources (Fall et al., 2011; Khan et al., 2021a). Municipal wastewaters cannot be classified by a single set of effluent, because of various factors and their variety of characteristics. They can be socioeconomic, water consumption need, degree of infiltration, type, and volume of industrial component contaminants, and so on (Fall et al., 2011). Nevertheless water, makes up the majority of wastewater (about 99.9%), with the remainder consisting of dissolved and suspended organic and inorganic solids (carbohydrates, fats, soaps and lignin, and metals), industrial chemical substances, etc. (Santos, Aparicio, & Alonso, 2007). These contaminants may have deleterious consequences on humans, aquatic life and plants, resulting in poor health, and, low quality of life (Santos et al., 2009). As a result, it is important to understand the specifics of the various physicochemical properties used to assess PWW quality such as total suspended solids (TSS), biochemical oxygen demand (BOD), chemical oxygen demand (COD), total dissolved solids (TDS), temperature, pH, and electrical conductivity of PWW.

It is because of the vast volume of PWW, as well as its complexity and hazards, that very serious concerns have been raised concerning its successful treatment. Pharmaceutically active compounds (PhACs) include hormones, poisonous chemicals, antibiotics, surfactants, and volatile organic compounds (VOCs), all of which are harmful to the environment. The pharmaceutical enterprises worldwide create massive amounts of pharmaceutical waste (PW) indiscriminately (Carballa, Omil, & Lema, 2008). As a result, the scientific community has a significant challenge in effectively managing and disposing of these PWW, especially considering PWW found in wastewater effluent treatment plants. However, the growing occurrence of improper drug disposal waste into water systems from hospitals and households have also turned them into another potential source of PWW (Strade, Kalnina, & Kulczycka, 2020). As a result, pharmaceutical wastewater contains waste from both the pharmaceutical business, households, and hospitals that collectively pollute our environment. The different characteristics of pharmaceutical wastewater are listed in Table 1.

Table 1 Pharmaceutical wastewater effluent characteristics.

Serial no.	Parameters	Standard effluent values
1	Chemical oxygen demand (COD)	2000–6000 ppm
2	Biological oxygen demand (BOD)	900–400 ppm
3	Total dissolved solids (TDS)	1350–7250 ppm
4	Total suspended solids (TSS)	500–2000 ppm
5	pH	1–9
6	Temperature (°C)	25–30

Due to the fact that the composition of pharmaceutical raw materials varies substantially from one product class to another, the dynamically fluctuating nature of PWW is a prevalent difficulty. Many pharmaceutical corporations hide their composition under the guise of secrecy to avoid strict pollution control requirements. As a result, intentionally pharmaceutical industries continue to leak PW into the environment without fully reporting the size and composition of the harmful compounds. All industrial wastewater, residential sewage, and hospital PW are routinely mixed together in publicly owned wastewater treatment facilities, compounding the issue.

Due to insufficient experimental data investigating the relationships between pharmaceutical compounds, effluent characteristics (such as COD, BOD, TSS, total phosphorus (TP), and oil content), and different types of operating parameters (such as wastewater flow and hydraulic retention time (HRT)), generalized wastewater treatment processes are not effective (Ganiyu, van Hullebusch, et al., 2015; Massima Mouele et al., 2021). This complicates the remediation procedures even further, and is getting even more perplexed because studies have shown that medications like ibuprofen or paracetamol, ketoprofen, indomethacin, etoricoxib, mefenamic acid, and acetaminophen may be removed completely (80%–100%) after a solid retention duration of 10–20 days, rendering complacency (Jones, Voulvoulis, & Lester, 2005). However, their recurrence and persistency are dangerous and need to be removed from water (Hedgespeth et al., 2012).

3. Conventional pharmaceutical wastewater treatments

3.1 Membrane bioreactor (MBR)

The presence of pharmaceutical ingredients in aquatic systems and their associated effects on living organisms have been a problem for a long time (Heberer, 2002). MBR is a specialized stage of suspended and organic solid removal combining both membrane filtration and traditional biological treatment methods like activated sludge. A polymeric membrane configured on an ultra- or microfiltration unit instead of a secondary clarifier, like a cyclic activated sludge unit (CASP), is used in MBR systems to separate biosolids. This system could also be designed to facilitate waste nutrient removal, like BOD. The pores of the membrane (0.035–0.4 μm range) are typically submerged in an aerobic biological reactor, such as in MBR systems of wastewater treatments (Gupta, Sati, & Gupta, 2019). A higher quality of effluent can be generated by eliminating sedimentation and filtration methods, which are generally used for discharged wastewater from the pharmaceutical industry, and because biological processes operate at higher liquor concentrations (mixed), sedimentation steps could be omitted. However, the general range of liquor ratio to solids is ranging at 1.0%–1.2%, which is still nearly four times higher than traditional wastewater plants. Therefore, optimum scouring and aeration could be obtained in the case of MBR effluent treatments. Fig. 1 shows a schematic illustration of an anaerobic filtration system. In a recent study, it was seen that an anaerobic MBR system used for chemical and pharmaceutical production discharged effluents of 78%–97% COD removal efficiency depending on influent concentrations and methanol addition (Svojitka et al., 2017). Moreover, in another pilot study, it was reported that the COD removal was 95% for discharged effluents from Taiwanese pharmaceutical industries monitored over a 140-day period using a MBR system.

3.2 Activated carbon

Commercially, activated carbon is produced from different biobased materials like woods, naturally derived coir materials, coals, lignites, so on (Demirbas, 2009; Mohan, Singh, & Singh, 2005). The main characteristic features of activated carbons are larger surface area and porosity and surface chemistry enabling convenient reactions. Interestingly, the utilization of biowaste is considered as a low-cost technology

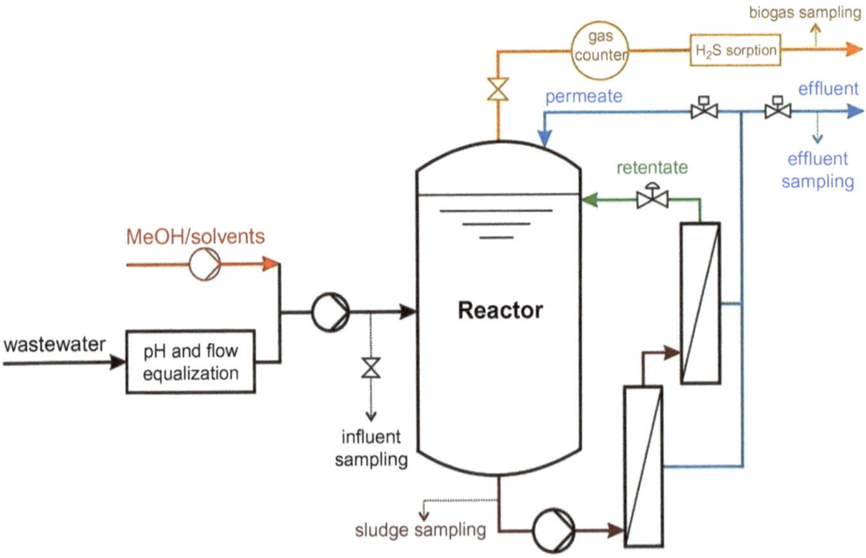

Fig. 1 A schematic flow diagram for an anaerobic membrane bioreactor system. Reprinted with permission from Elsevier (Svojitka, J., et al. (2017). Performance of an anaerobic membrane bioreactor for pharmaceutical wastewater treatment. *Bioresource Technology, 229*, 180–189). Copyright Elsevier 2017.

implemented for effluent treatments to remove contaminants efficiently (Mohan et al., 2005). However, the flow process of the activated carbon-based technology entails sample pretreatment, crushing, drying, and sieving to attain smaller particle dimensions (Wong et al., 2018). The volatile compounds are eliminated through carbonization at elevated temperatures and an inert atmosphere (generally around 300–500 °C) to form biochar. Sometimes, gaseous pollutants like SO_2, CO_2, and NO_2 must also be eliminated so that they do not react with water in acidic conditions. A schematic representation of activated carbon synthesis routes from biowaste material is shown in Fig. 2.

According to prevailing opinion, activated carbon is recommended for the removal of pollutants because it ensures that contaminants are adsorbed on the filter surfaces. Some specific pollutants include organics (unwanted odor, micropollutants, and tastes), radon, fluorine, chlorine, and so on. Nevertheless, this method is inefficient for the removal of inorganic contaminants, nitrates, and metallic and microbial pollutants (Wong et al., 2018). However, the efficiency of adsorption depends on water composition, operating parameters, and the nature of the activated carbons. Interestingly, activated carbon is easy to install industrially (Wong et al., 2018). However, it possesses certain limitations such as higher energy consumption, and skilled manpower requirements that may incur higher expenditure due to requiring regular filter material changes. The activated carbon-based technology of wastewater treatment can be implemented in industry, households, and the community. In a recent study (Dwivedi et al., 2018), it was shown that carbamazepine manufactured by the pharmaceutical industry contains significant amounts of recalcitrant compounds found in discharged effluents. However, the detoxification of such effluents (Fenton-treated) granulated activated carbon shows $99.51\% \pm 0.02\%$ carbamazepine removal (Dwivedi et al., 2018). Similarly, in another study by Cunha et al. (2020), it was reported that activated carbon employed for pharmaceutical wastewater treatment provided 99% efficiency for pollutant removal.

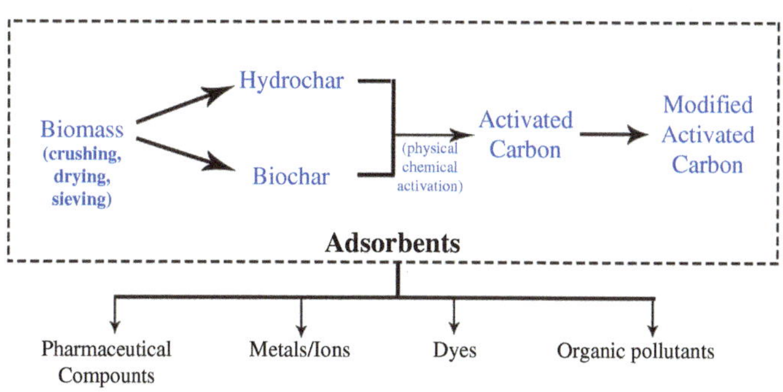

Fig. 2 Activated biocarbon synthesis routes from biobased wastes.
Reprinted with permission from Elsevier (Wong, S., et al. (2018). Recent advances in applications of activated carbon from biowaste for wastewater treatment: A short review. *Journal of Cleaner Production, 175*, 361–375). Copyright Elsevier 2018.

3.3 Coagulation and sedimentation

The process of introducing chemical agents to wastewater, rapidly mixing them to distribute them, and then transforming stable pollutants into unstable precipitates is known as coagulation, but it is a complicated PWW treatment. The ability to remove and compress PWW around hydrophilic colloids is crucial for sedimentation to improve the PWW treatment. Charge neutralization and polymer bridges are the key methods by which natural coagulants remove pharmaceuticals from water and wastewater. The nature of flocculent (polymer or inorganic salt) is a crucial factor affecting the coagulation process.

There are many types of coagulants such as chemical and natural plant-based coagulants, often used to remove PW (SS, chromaticity, and hazardous organic materials) from PWW. Natural coagulants help to biodegrade PW more quickly (Alazaiza et al., 2022). Coagulant dose, pH, beginning turbidity, mixing speed, and duration are all factors that may influence the coagulation process and the speed with which PW is removed. Sedimentation is the most typical process that occurs after coagulation. Gravity separation may be used to remove pollutants that have a greater density than typical wastewater. As shown in Fig. 3, coagulation and sedimentation have various advantages, including the convenience of easy use and being of proven technology, although eliminating dissolved organic detritus is recognized as difficult (Teh et al., 2016).

The benefits of the coagulation sedimentation technique include improved PWW biodegradation efficiency and lower pollutant concentrations, but the downside is that secondary contamination is easily formed. As a result, electrolyzed floating gas and gas-filled air flotation are frequently utilized to remove sediments. Treatment of PWW containing oxytetracycline, gentamicin, and tetracycline is more effective in the pharmaceutical business.

Ferric chloride and sulfate ($FeCl_3$, and $Fe_2(SO_4)_3$), aluminum chloride and sulfate ($AlCl_3$, $Al_2(SO_4)_3$), Alum, and magnesium chloride are used as inorganic chemical coagulants. On the other hand, *Cicer arietinum*, *Moringa oleifera*, and *Dolichos lablab* are used as excellent natural coagulant alternatives of chemical coagulants. PolyAMINEs and PolyDADMACs are most commonly used as cationic natural

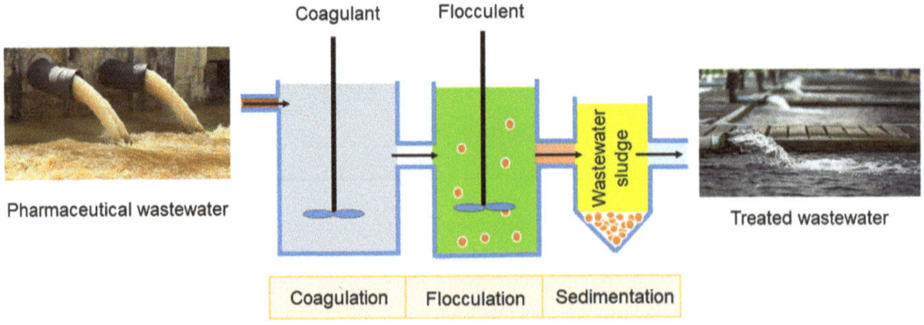

Fig. 3 Coagulation-flocculation process for wastewater treatment (Teh et al., 2016). Adapted and Copyright American Chemical Society 2016.

organic coagulants and work for charge neutralization alone. Recent research by Nonfodji et al. (2020) has shown the effectiveness of a natural coagulant derived from *M. oleifera* seeds in the treatment of PWW in a hospital setting. In another study, Maharani, Setiawan, and Ningsih (2021) investigated that the moringa seed coagulant can remove 90% and 71% of BOD and COD. In contrast, a tapioca starch coagulant can remove 95% and 94% of BOD and COD from PWW, respectively. Shirafkan et al. (2016) reported that polyaluminum chloride can remove 97%–98% turbidity and 70%–78% COD and TDS from PWW.

4. Advanced pharmaceutical wastewater treatment

4.1 Membrane and membrane-integrated hybrid systems

Membrane technology is now commercially available as a substitute to conventional methods such as coagulation, sedimentation, extraction, distillation, and adsorption processes (using sand filters and activated carbon air-filters and ion exchangers). The membrane process is considered a versatile process due to its separation mechanism, enabling it to be used in a wide range of industries. Furthermore, this technique is simple to implement and consumes very little energy. Membrane technology for PWW treatment, on the other hand, is prone to membrane fouling, which results in reduced overall performance. Usually, membrane fouling results in a loss of solvent permeability, a change in solute transfer, and a drop in wastewater membrane filtration performance, all of which are highly expensive to overcome (Rosman et al., 2018).

Pharmaceutical wastewater is also treated using membrane distillation (MD) methods using hydrophobic membranes to remove hazardous organic components. The removal effectiveness of organics (such as diclofenac, atenolol, and ibuprofen) was better than 90% in a pilot study using Xzero AGMD modules (Hussain et al., 2021). The prototype system's thermal energy was provided via district heating. Another study used MD in conjunction with an enzymatic bioreactor to extract traces of organic compounds from pharmaceutical residues. 4-*tert*-Octylphenol, octocrylene, 4-tertbutylphenol, benzophenone, and oxybenzone were all eliminated in 99% of the cases. The distillate was nontoxic and had no effect on membrane characteristics (Hussain et al., 2021).

In the purification of mixed waste effluents, the pressure-driven membrane filtration process is a standard separation technique that has high efficiency. As seen in Fig. 4, pressure-driven membrane filtration is generally categorized into four categories: microfiltration (MF), ultrafiltration (UF), nanofiltration (NF), and reverse osmosis (RO). The procedure necessitates a variety of specie sizes and pressures. By reducing the membrane's pore size from MF to UF to NF/RO and eliminating the molecular weight cut-off (MWCO), the membrane's mass transfer resistance rose from 1 bar for UF to 60—80 bar for RO when the same flux was produced. Nanofiltration (NF)/reverse osmosis (RO), which uses membranes with tiny holes, is an effective technology for removing pharmaceutical residues and other dissolved

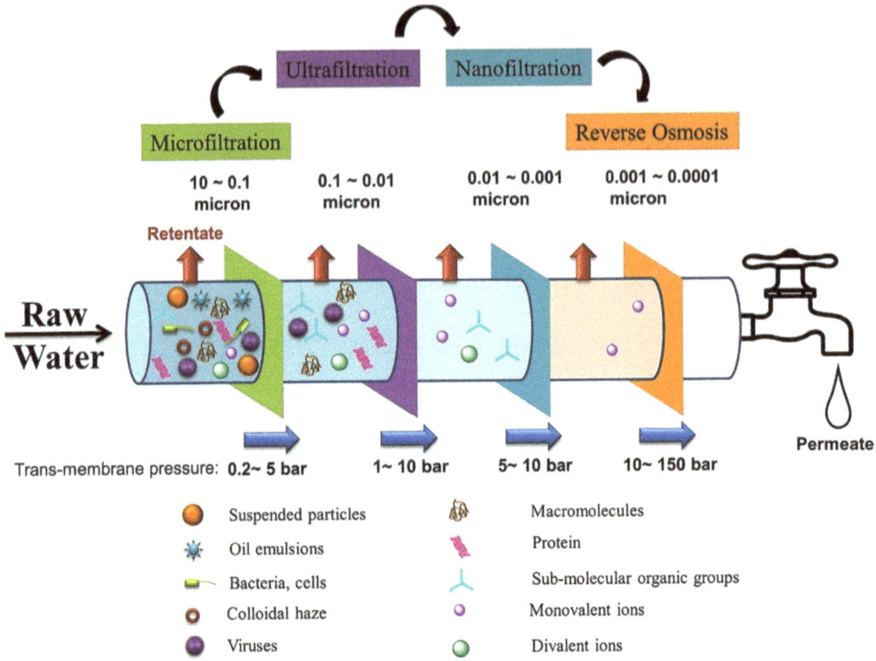

Fig. 4 Schematic representation of pressure-driven membrane filtration principles (Liao et al., 2018).
Copyright Elsevier 2018.

pollutants (Radjenović et al., 2008). The MWCO of NF and RO membranes is between 0.001 and 0.008 μm, with a lower limit of 0.001 μm. As a result, these membranes can easily reject medicinal substances of a particle size in the micron range (Hussain et al., 2021; Rosman et al., 2018). Nanomaterial membranes (NMs) promise to be a long-term solution for improving membrane properties and improving the performance of MBRs to treat pharmaceutical wastewater (Pervez et al., 2020).

Low-pressure membrane systems like MF and UF make water treatment simpler. However, since pharmaceutical molecular sizes are often smaller than the MWCO, MF and UF membranes are seldomly used for water filtration of pharmaceuticals. Pore sizes in MF and UF are approximately 0.1 and 0.01–0.10 μm, respectively. Because these procedures are often employed as pretreatment processes to remove colloids and natural colloid matter (NOM), xenobiotics with tiny particles as small as 1 μm may easily pass through the MF and UF membranes, indicating that they are effective at only removing colloids and NOM. Despite the fact that MF/UF treatment is inefficient at eliminating pharmaceuticals and PWW, the concentration of various pharmaceutical target contaminants is equivalent to that of NF treatment (Molinari, Lavorato, & Argurio, 2020; Park & Cho, 2005). For example, Frascaroli et al. (2021) used a dead-end stirred-cell filtration method to study 52 EDC/pharmaceutical and PCPs and obtained comparable results (Yoon et al., 2006). EDC/PPCPs were computed and separated into two groups in these experiments: (i) lower log water partition coefficient (K_{OWc}) and (ii) higher log K_{OWc}. The findings revealed that the UF membrane successfully removed hydrophobic compounds with aromatic ring configurations by aliphatic substitutes (CH, CH_2, CH_3) (neutral or positively charged) and showed high pK_a and $\log K_{OW}$ values in group II (Yoon et al., 2006). These findings confirmed that the pharmaceutical's hydrophobicity and polarity significantly effect UF membrane wastewater treatment (Table 2).

Hydrophobic compounds that are adsorbed onto particles or colloids and are unable to flow through the pores of the membrane are rapidly rejected. The polarity of the medicinal solute has an impact on membrane electrostatic interaction rejection. Nghiem, Schäfer, and Elimelech (2005) investigated electrostatic interactions in NF membrane rejection of pharmaceuticals from PWW that were negatively charged above 5 pH values. Researchers discovered that ionizable medications, such as ibuprofen and sulfamethoxazole, were shown to have higher retention/rejection values when pH values rose from neutral to negative charge above their acid dissociation constant (pK_a) value. At neutral pH, the polyamide membrane utilized in the previous work had a negative charge, which improved the permeability and selectivity of NF/RO membranes (Bellona & Drewes, 2005). The acid functional group of the membrane was deprotonated, resulting in a negatively charged membrane. Electrostatic repulsions between positively charged and negatively charged solutes impeded absorption and diffusion through negatively charged membranes (Radjenović et al., 2008). Snyder shows examples of various pharmaceutical contamination elimination percentages utilizing membrane and MBR procedures (Snyder, 2007).

Despite the fact that membrane procedures have been acknowledged to have excellent properties for water cleanup, using membrane systems as a standalone separation technique has significant drawbacks. This includes the following: (a) the presence of a hydrophobic or absorptive mechanism on the membrane surface; (b) electrostatic interaction on the membrane surface that contributes to electrostatic concentration polarization; (c) the passage of possibly smaller-sized molecules via the sponge-like pore membrane; and (d) the fact that a single cycle of membrane filtration operation is

Table 2 Performance of membrane filtration for PWW treatment.

Type of membrane filtration	Types	Rejection	References
NF	Organic matter, organic acids	Above 95%	Košutić et al. (2007) and Snyder et al. (2007)
NF/RO	COD	>94%	Cséfalvay, Imre, and Mizsey (2008)
RO	Water, salt	Above 95%	Košutić et al. (2007)
Forward osmosis (FO)	DCF, CBZ, and IBP	≥93%	Jin et al. (2012)
Membrane bioreactor (MBR)	COD	99.9%	Shahbeig et al. (2017)

DCF, diclofenac; *CBZ*, carbamazepine; *IBP*, ibuprofen.

not capable of eliminating all contaminated pharmaceutical waste. Another area of concern is fouling, which happens when a cake develops on the membrane surface, decreasing permeability and complete separation performance, which is detrimental to membrane performance (Koops et al., 1994). To counteract the fouling phenomenon, membrane back-flushing and chemical cleaning are extensively utilized (Schlichter, Mavrov, & Chmiel, 2004). The disposal of concentrates following membrane separation operations, on the other hand, is a challenge that has yet to be resolved.

4.2 Adsorption

Pharmaceutical research has exploded in the last three decades, making it possible to live a safer and healthier life. However, increasing demand and widespread use of these hazardous medications is having a negative influence on natural ecosystems. As a result, successfully cleaning pharmaceutically-polluted water streams is a worldwide priority to sustain a safer environment and better living. In the field of wastewater and PWW treatment, nanotechnology is promising for the creation of innovative high-tech materials such as membranes, adsorbents, nanocatalysts, functional surfaces, coatings, and reagents that are more efficient than conventional materials. Nano-sorbent materials, are widely regarded as the most effective wastewater treatment approaches, because of their simplicity of application and wide variety of adsorbents that can use. Several standard approaches have been used to treat residential wastewater, but they are ineffective for removing drugs. Nanoadsorbents have recently seen significant breakthroughs in PWW treatment because of their low manufacturing cost, sustainability, improved physicochemical features, and excellent pharmaceutical removal efficacy (Khan et al., 2021b; Talukder et al., 2021a). There has been an investigation on the effectiveness of a sustainable nanoadsorbent for the detoxification of pharmaceutical contaminants in water in the literature recently (Singh et al., 2021). The sorption parameters that have most influence and the process interactions have also been explored.

Nanoscale materials, often known as nanoparticles, have immense potential for mineralizing pharmaceutical compounds in aqueous solutions with low cost, ease of handling, and with more effectiveness than currently available methods. Nanoparticles may also absorb PW by-products and pollutants at a minimal cost because they can be easily recycled. With more research being done in this area, the value of PWs and products will continue to rise as we move away from traditional methodologies and nanotechnology. The use of nanomaterials in PWW treatment was investigated, with an emphasis on value creation, energy production, and waste reduction, as well as its role in the circular economy, which supports long-term development and sustainability (Singh et al., 2021). The adsorption of a PWW (could adsorb 70% of fluoxetine) in an aqueous solution was investigated using alkali lignin and poly(vinyl alcohol) blended nanofibers.

Because of their large surface area, exceptionally high porous structure, adaptable pore size/pore size distribution, range of material options, and flexibility to blend with other functional materials, nanofiber-based membrane filtration is a promising technology in PWW cleanup, as shown in Fig. 5. Fouling, poor mechanical qualities, shrinkage, and deformation are essential limitations of nanofiber membranes for PWW treatment, despite their unique features (Pervez & Stylios, 2018; Talukder et al., 2021b).

A novel kind of functional hybrid material known as metal-organic framework (MOFs) has been discovered. MOFs are formed by creating strong connections between metal-containing and organic components (reticular synthesis). Thousands of distinct MOFs have been created due to the flexibility of their elements; in terms of size, functionality, and geometry. Recently, MOFs are the most potent adsorbent to adsorb ineffective molecules and metal particles from PWW (de Andrade et al., 2018). Cr-based MOF (MIL-101) has been used to adsorb naproxen (maximum adsorption capacity (Q_m 132 mg/g) and clofibric acid (Q_m 244 mg/g) from PWW (Hasan, Jeon, & Jhung, 2012). Chloroform activated Zr-based MOF (UiO-66) has been used to adsorb sulfachlorpyridazine (Q_m 417 mg/g) from PWW (Azhar et al., 2017).

4.3 Advanced oxidation processes

In wastewater treatment, advanced oxidation processes (AOPs), which use hydroxyl radicals (OH) to oxidize organic (and in some instances inorganic) pollutants, are recognized as chemical treatment procedures (Glaze, Kang, & Chapin, 1987; Khan et al., 2020). AOPs are divided into two categories: homogeneous and heterogeneous. Processes that consume energy and those that do not are both homogeneous. Fig. 6 depicts this categorization.

Nonenergy homogenous processes include hydrogen peroxide (H_2O_2) and catalyst methods, ozonation with H_2O_2, and ozonation in an alkaline medium, to name a few examples. OH• radical generation is smaller in the H_2O_2 process than in catalysis with H_2O_2/UV, indicating a significant difference between the two methods (Khan et al., 2021c; Poyatos et al., 2010). Catalysts are often utilized in heterogeneous AOPs to generate contaminated compound degradation. Heterogeneous process catalysts have the advantage of separating the product more efficiently than homogeneous process catalysts (Bermúdez et al., 2021; Poyatos et al., 2010). AOPs, such as TiO_2/UV systems, H_2O_2/UV systems, ozonation, radiation and photo-Fenton electrochemical oxidation have been employed to combat these pollutants, as shown in Fig. 7. Because of their high reactivity, hydroxyl radicals may be employed to convert organic compounds into innocuous chemicals slowly. Pharmaceuticals, which are regularly found in natural and PWW bodies and in drinking water, are receiving a lot of attention since they do not biodegrade quickly, can linger longer and are harmful. As a result, pharmaceutical residues continue to represent a risk to the environment and human health (Kanakaraju, Glass, & Oelgemöller, 2018).

H_2O_2 added to PWW treatment resulted in nonselective, more powerful hydroxyl radicals, which were more effective than ozone in removing PWW from the environment. Anticancer, antipsychotic, and pain reliever drugs make up the majority of the pharmaceutical industry's effluent. This approach decreased COD by 75%–88.5% in

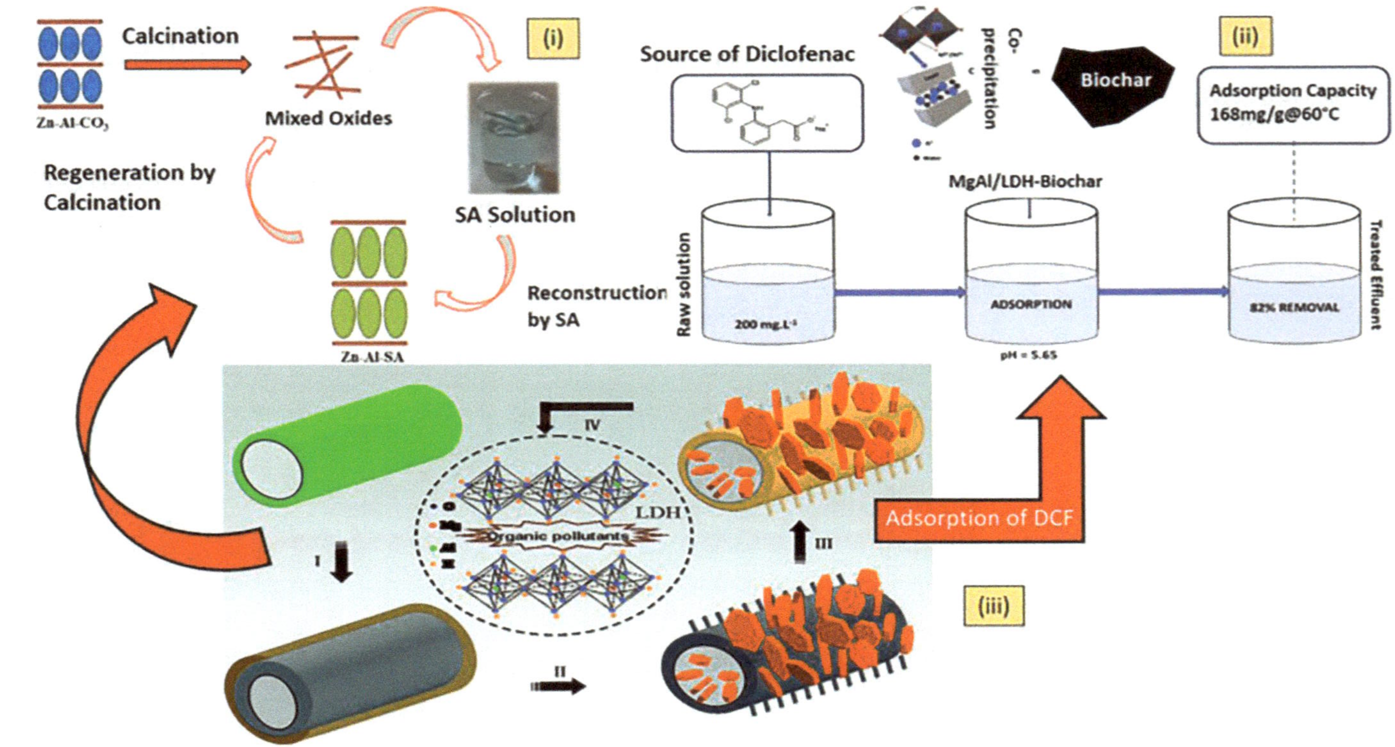

Fig. 5 Nanoadsorbent for the removal of pharmaceutical waste (Khan et al., 2021b). Copyright Elsevier 2021.

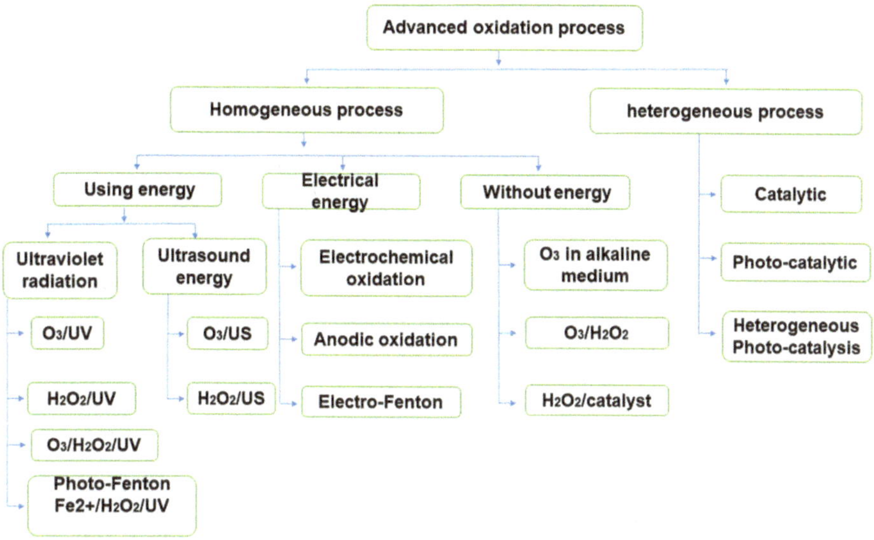

Fig. 6 Schematic representation of advanced oxidation processes.

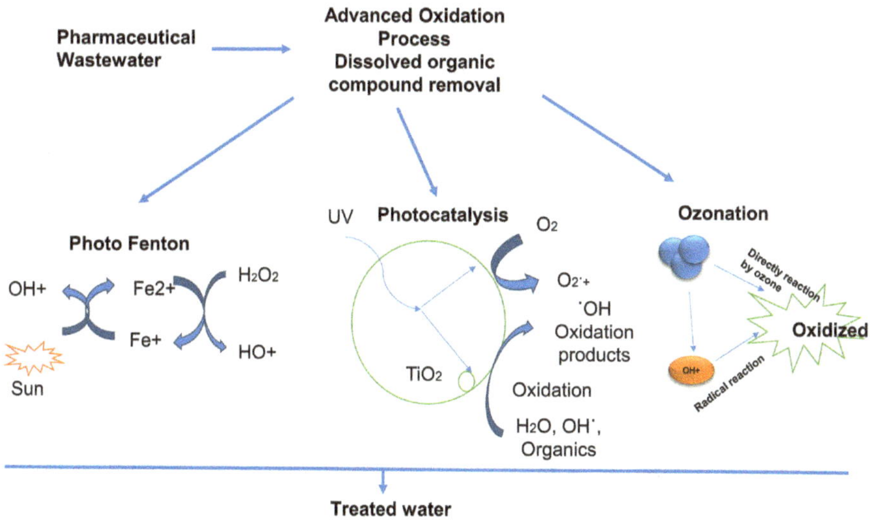

Fig. 7 Schematic representation of advanced oxidation processes (Jiménez et al., 2019). Under a Creative Commons license, Open access, Elsevier 2019.

under 3h at a pH of 5–11 and COD was decreased by 85.4%–92.7% in 2.5h using activated char adsorption at pH 5–11. All other measures of water quality were significantly lowered (>73% removal) after ozonation. Additionally, the effects of modifying important operational factors (such as pH and the concentration of H_2O_2) were studied (Patel et al., 2020).

4.4 Photocatalytic oxidation

After secondary contamination, photocatalytic oxidation can be used to clean wastewater. It has the ability to rapidly remove hazardous organic compounds from wastewater. Additionally, it may aid in biodegradation and reduce microbial toxicity after modifying its composition. An example of this approach is the biochemical method. Anaerobic and aerobic treatments are often used in conjunction with one another to reduce the content of organic matter in wastewater. Highly concentrated organic pharmaceutical effluent may be treated immediately using aerobic biological treatment techniques, for example. After anaerobic treatment, the organic pharmaceutical wastewater may be recycled, with the residual sludge being small even if the water extract and COD are high after the treatment (Feng, 2019).

5. Conclusions and future perspectives

It is known that the presence of pharmaceutical compounds in effluents causes severe problems for humans, animals and plants, and for our planet. Although conventional technologies claim that they eliminate pharmaceutical compounds, they do not meet the strict regulation requirements for effectively treating effluent. The main points that the community should focus in improving wastewater filtration performance, as outlined below:

- In membrane technology, by far, the most significant issue that is restricting the effective use of membrane separation is membrane fouling. This was appropriately explained by Borea et al. (2019), who noted that membrane fouling in electromembrane bioreactor operation could be controlled by the applied current density during pharmaceutical removal.
- Adsorption has long been regarded as an effective way to treat wastewater; however, when commercially available adsorbents are applied, costs associated with this method may be substantial. Poor adsorption effectiveness has been seen in emerging contaminant solutions because of the presence of competing ions.
- Specific challenges to the advanced oxidation process involve the fragility of the catalyst structure, low mass transfer rate, and potential flow of the effluent stream in heterogeneous catalysts. Hence, the development of binding-integrated catalysts could be a possible route to deal with this scheme.
- Although photocatalytic procedures have the potential to treat large volumes of wastewater, one of their major drawbacks is the prohibitive cost of photocatalyst production in comparison to homogeneous catalysts. The careful and appropriate set up of system-level engineered photocatalysts is essential.
- Ultimately, technical advancements will be necessary to provide more dependable and affordable cleaning so that we can have access to clean water. The scale-up of recent promising techniques may provide the solution we are aiming for.

References

Alazaiza, M. Y., et al. (2022). Application of natural coagulants for pharmaceutical removal from water and wastewater: A review. *Water*, *14*(2), 140.

Azhar, M. R., et al. (2017). Adsorptive removal of antibiotic sulfonamide by UiO-66 and ZIF-67 for wastewater treatment. *Journal of Colloid and Interface Science*, *500*, 88–95.

Bellona, C., & Drewes, J. E. (2005). The role of membrane surface charge and solute physico-chemical properties in the rejection of organic acids by NF membranes. *Journal of Membrane Science*, *249*(1–2), 227–234.

Bermúdez, L. A., et al. (2021). Effectiveness of advanced oxidation processes in wastewater treatment: State of the art. *Water*, *13*(15), 2094.

Borea, L., et al. (2019). Are pharmaceuticals removal and membrane fouling in electro-membrane bioreactor affected by current density? *Science of the Total Environment*, *692*, 732–740.

Carballa, M., Omil, F., & Lema, J. M. (2008). Comparison of predicted and measured concentrations of selected pharmaceuticals, fragrances and hormones in Spanish sewage. *Chemosphere*, *72*(8), 1118–1123.

Cséfalvay, E., Imre, P. M., & Mizsey, P. (2008). Applicability of nanofiltration and reverse osmosis for the treatment of wastewater of different origin. *Central European Journal of Chemistry*, *6*(2), 277–283.

Cunha, M. R., et al. (2020). Removal of captopril pharmaceutical from synthetic pharmaceutical-industry wastewaters: Use of activated carbon derived from Butia catarinensis. *Journal of Environmental Chemical Engineering*, *8*(6), 104506.

de Andrade, J. R., et al. (2018). Adsorption of pharmaceuticals from water and wastewater using nonconventional low-cost materials: A review. *Industrial & Engineering Chemistry Research*, *57*(9), 3103–3127.

Demirbas, A. (2009). Agricultural based activated carbons for the removal of dyes from aqueous solutions: A review. *Journal of Hazardous Materials*, *167*(1–3), 1–9.

Dwivedi, K., et al. (2018). Evaluation and optimization of Fenton pretreatment integrated with granulated activated carbon (GAC) filtration for carbamazepine removal from complex wastewater of pharmaceutical industry. *Journal of Environmental Chemical Engineering*, *6*(3), 3681–3689.

Fall, C., et al. (2011). Divergence between respirometry and physicochemical methods in the fractionation of the chemical oxygen demand in municipal wastewater. *Water Environment Research*, *83*(2), 162–172.

Feng, H. (2019). Research Progress in pharmaceutical wastewater treatment technology. In *118. E3S Web of Conferences. EDP Sciences* (p. 04019).

Frascaroli, G., et al. (2021). Pharmaceuticals in wastewater treatment plants: A systematic review on the substances of greatest concern responsible for the development of antimicrobial resistance. *Applied Sciences*, *11*(15), 6670.

Ganiyu, S. O., van Hullebusch, E. D., et al. (2015). Coupling of membrane filtration and advanced oxidation processes for removal of pharmaceutical residues: A critical review. *Separation and Purification Technology*, *156*, 891–914.

Glaze, W. H., Kang, J. W., & Chapin, D. H. (1987). The chemistry of water treatment processes involving ozone, hydrogen peroxide and ultraviolet radiation. *Ozone: Science & Engineering*, *9*(4), 335–352.

Guo, Y., Qi, P. S., & Liu, Y. Z. (2017). A review on advanced treatment of pharmaceutical wastewater. *IOP Conference Series: Earth and Environmental Science*, *63*(1), 012025. IOP Publishing.

Gupta, R., Sati, B., & Gupta, A. (2019). Treatment and recycling of wastewater from pharmaceutical industry. In *Advances in biological treatment of industrial waste water and their recycling for a sustainable future* (pp. 267–302). Singapore: Springer.

Hasan, Z., Jeon, J., & Jhung, S. H. (2012). Adsorptive removal of naproxen and clofibric acid from water using metal-organic frameworks. *Journal of Hazardous Materials*, *209*, 151–157.

Heberer, T. (2002). Occurrence, fate, and assessment of polycyclic musk residues in the aquatic environment of urban areas—A review. *Acta Hydrochimica et Hydrobiologica, 30*(506), 227–243.

Hedgespeth, M. L., et al. (2012). Pharmaceuticals and personal care products (PPCPs) in treated wastewater discharges into Charleston Harbor, South Carolina. *Science of the Total Environment, 437*, 1–9.

Husain Khan, A., et al. (2020). Pharmaceuticals of emerging concern in hospital wastewater: Removal of ibuprofen and Ofloxacin drugs using MBBR method. *International Journal of Environmental Analytical Chemistry*, 1–15.

Hussain, A., et al. (2021). Membrane distillation: Recent technological developments and advancements in membrane materials. *Emergent Materials, 5*, 1–21.

Jiménez, S., et al. (2019). Produced water treatment by advanced oxidation processes. *Science of the Total Environment, 666*, 12–21.

Jin, X., et al. (2012). Rejection of pharmaceuticals by forward osmosis membranes. *Journal of Hazardous Materials, 227*, 5–61.

Jones, O. A. H., Voulvoulis, N., & Lester, J. N. (2005). Human pharmaceuticals in wastewater treatment processes. *Critical Reviews in Environmental Science and Technology, 35*(4), 401–427.

Kanakaraju, D., Glass, B. D., & Oelgemöller, M. (2018). Advanced oxidation process-mediated removal of pharmaceuticals from water: A review. *Journal of Environmental Management*, 189–207.

Khan, A. H., et al. (2020). Application of advanced oxidation processes followed by different treatment technologies for hospital wastewater treatment. *Journal of Cleaner Production, 269*, 122411.

Khan, A. H., et al. (2021a). Effect of seasonal variation on the occurrences of high-risk pharmaceutical in drain-laden surface water: A risk analysis of Yamuna River. *Science of the Total Environment, 794*, 148484.

Khan, A. H., et al. (2021b). Sustainable green nanoadsorbents for remediation of pharmaceuticals from water and wastewater: A critical review. *Environmental Research, 204*, 112243.

Khan, N. A., et al. (2021c). New insights into the integrated application of Fenton-based oxidation processes for the treatment of pharmaceutical wastewater. *Journal of Water Process Engineering, 44*, 102440.

Koops, G. H., et al. (1994). Integrally skinned polysulfone hollow fiber membranes for pervaporation. *Journal of Applied Polymer Science, 54*(3), 385–404.

Košutić, K., et al. (2007). Removal of antibiotics from a model wastewater by RO/NF membranes. *Separation and Purification Technology, 53*(3), 244–249.

Liao, Y., et al. (2018). Progress in electrospun polymeric nanofibrous membranes for water treatment: Fabrication, modification and applications. *Progress in Polymer Science, 77*, 69–94.

Maharani, A., Setiawan, D., & Ningsih, E. (2021). Comparison of the effectiveness of natural coagulant performance on% BOD removal and% COD removal in pharmaceutical industry waste. *Tibuana, 4*(1), 55–60.

Massima Mouele, E. S., et al. (2021). Removal of pharmaceutical residues from water and wastewater using dielectric barrier discharge methods—A review. *International Journal of Environmental Research and Public Health, 18*(4), 1683.

Mohan, D., Singh, K. P., & Singh, V. K. (2005). Removal of hexavalent chromium from aqueous solution using low-cost activated carbons derived from agricultural waste materials and activated carbon fabric cloth. *Industrial & Engineering Chemistry Research*, *44*(4), 1027–1042.

Molinari, R., Lavorato, C., & Argurio, P. (2020). Application of hybrid membrane processes coupling separation and biological or chemical reaction in advanced wastewater treatment. *Membranes*, *10*(10), 281.

Nghiem, L. D., Schäfer, A. I., & Elimelech, M. (2005). Pharmaceutical retention mechanisms by nanofiltration membranes. *Environmental Science & Technology*, *39*(19), 7698–7705.

Nonfodji, O. M., et al. (2020). Performance of *Moringa oleifera* seeds protein and *Moringa oleifera* seeds protein-polyaluminum chloride composite coagulant in removing organic matter and antibiotic resistant bacteria from hospital wastewater. *Journal of Water Process Engineering*, *33*, 101103.

Park, G. Y., & Cho, J. (2005). Transport of pharmaceutical and NOM in NF and tight UF membranes. In *Proceedings of AWWA Membrane Technology Conference, Phoenix*.

Patel, S., et al. (2020). Treatment of a pharmaceutical industrial effluent by a hybrid process of advanced oxidation and adsorption. *ACS Omega*, *5*(50), 32305–32317.

Pervez, M., & Stylios, G. K. (2018). Investigating the synthesis and characterization of a novel "green" H2O2-assisted, water-soluble chitosan/polyvinyl alcohol nanofiber for environmental end uses. *Nanomaterials*, *8*(6), 395.

Pervez, M. N., et al. (2020). A critical review on nanomaterials membrane bioreactor (NMs-MBR) for wastewater treatment. *NPJ Clean Water*, *3*(1), 1–21.

Poyatos, J. M., et al. (2010). Advanced oxidation processes for wastewater treatment: State of the art. *Water, Air, and Soil Pollution*, *205*(1), 187–204.

Radjenović, J., et al. (2008). Rejection of pharmaceuticals in nanofiltration and reverse osmosis membrane drinking water treatment. *Water Research*, *42*(14), 3601–3610.

Rosman, N., et al. (2018). Hybrid membrane filtration-advanced oxidation processes for removal of pharmaceutical residue. *Journal of Colloid and Interface Science*, *532*, 236–260.

Santos, J. L., Aparicio, I., & Alonso, E. (2007). Occurrence and risk assessment of pharmaceutically active compounds in wastewater treatment plants a case study: Seville city (Spain). *Environment International*, *33*(4), 596–601.

Santos, J. L., et al. (2009). Occurrence of pharmaceutically active compounds during 1-year period in wastewaters from four wastewater treatment plants in Seville (Spain). *Journal of Hazardous Materials*, *164*(2–3), 1509–1516.

Schlichter, B., Mavrov, V., & Chmiel, H. (2004). Study of a hybrid process combining ozonation and microfiltration/ultrafiltration for drinking water production from surface water. *Desalination*, *168*, 307–317.

Shahbeig, H., et al. (2017). Pharmaceutical wastewater treatment using membrane bioreactor-ozonation system. *Water Environment Journal*, *31*(1), 57–63.

Shigei, M., et al. (2021). Pharmaceutical and antibiotic pollutant levels in wastewater and the waters of the Zarqa River, Jordan. *Applied Sciences*, *11*(18), 8638.

Shirafkan, A., et al. (2016). Hybrid coagulation/ozonation treatment of pharmaceutical wastewater using ferric chloride, polyaluminum chloride and ozone. *International Journal of Environmental Science and Technology*, *13*(6), 143–1452.

Singh, S., et al. (2021). Adsorption and detoxification of pharmaceutical compounds from wastewater using nanomaterials: A review on mechanism, kinetics, valorization and circular economy. *Journal of Environmental Management*, *300*, 113569.

Snyder, S. A. (2007). *Removal of EDCs and pharmaceuticals in drinking and reuse treatment processes*. AWWA Research Foundation.

Snyder, S. A., et al. (2007). Role of membranes and activated carbon in the removal of endocrine disruptors and pharmaceuticals. *Desalination*, *202*(1–3), 156–181.

Strade, E., Kalnina, D., & Kulczycka, J. (2020). Water efficiency and safe re-use of different grades of water-topical issues for the pharmaceutical industry. *Water Resources and Industry*, *24*, 100132.

Svojitka, J., et al. (2017). Performance of an anaerobic membrane bioreactor for pharmaceutical wastewater treatment. *Bioresource Technology*, *229*, 180–189.

Talukder, M. E., et al. (2021a). Chitosan-functionalized sodium alginate-based electrospun nanofiber membrane for As (III) removal from aqueous solution. *Journal of Environmental Chemical Engineering*, *9*, 106693.

Talukder, M. E., et al. (2021b). Novel fibrin functionalized multilayered electrospun nanofiber membrane for burn wound treatment. *Journal of Materials Science*, *56*(22), 12814–12834.

Teh, C. Y., et al. (2016). Recent advancement of coagulation–flocculation and its application in wastewater treatment. *Industrial & Engineering Chemistry Research*, *55*(16), 4363–4389.

Wong, S., et al. (2018). Recent advances in applications of activated carbon from biowaste for wastewater treatment: A short review. *Journal of Cleaner Production*, *175*, 361–375.

Yoon, Y., et al. (2006). Nanofiltration and ultrafiltration of endocrine disrupting compounds, pharmaceuticals and personal care products. *Journal of Membrane Science*, *270*(1–2), 88–100.

Natural polymer-based sustainable adsorbents for pharmaceutical wastewater treatment

12

Javed Alam[a], *Arun Kumar Shukla*[a], *Fekri Abdulraqeb Ahmed Ali*[c], *Omar Daoud*[b], and *Mansour Alhoshan*[a,b,c]
[a]King Abdullah Institute for Nanotechnology, King Saud University, Riyadh, Saudi Arabia, [b]Chemical Engineering Department, College of Engineering, King Saud University, Riyadh, Saudi Arabia, [c]K.A. CARE Energy Research and Innovation Center, Riyadh, Saudi Arabia

1. Introduction

Emerging contaminants (ECs) are organic pollutants whose widespread occurrence in the aquatic environment has been investigated widely in the last few years at relatively low concentration levels (typically in nanogram, ng/L and microgram, μg/L levels) (Al-Riyami, Ahmed, Al-Busaidi, & Choudri, 2018; Bottoni, Caroli, & Barra Caracciolo, 2010; Jiang, Lee, Der Fang, Boyd, & Gibb, 2015; Petrie, Barden, & Kasprzyk-Hordern, 2015; Sousa et al., 2018). ECs can refer to many different classes of chemicals found in pharmaceuticals (PhCs), personal care products (PCPs), pesticides, household and industrial products, micro plastics, metals, surfactants, and industrial additives. These pollutants are extremely toxic, and carcinogenic, and their accumulation poses severe threats to aquatic environments including, inland surface waters, and ground water sources (Afzal et al., 2020; Ahuja, 2020; Cabeza, Candela, Ronen, & Teijon, 2012; Deblonde, Cossu-Leguille, & Hartemann, 2011; Khan et al., 2021; Naidu, Espana, Liu, & Jit, 2016; Patel et al., 2019; Rasheed, Bilal, Nabeel, Adeel, & Iqbal, 2019) (Fig. 1).

In ECs, pharmaceuticals have received increasing attention in recent years for their persistent input and possible threat to ecological, aquatic environment and public health. Recent studies indicated that PhCs are extensively detected in surface water, tap water and groundwater in many countries. It is a very shocking condition that >100 PhCs are identified above their detection limit in the developed countries. Hence, there is urgent need for their removal from natural and wastewater systems (Bottoni et al., 2010; Jiang et al., 2015; Montes-Grajales, Fennix-Agudelo, & Miranda-Castro, 2017; Petrie et al., 2015; Talib & Randhir, 2016) (Fig. 2).

Thus, controlling PhCs concentration in water resources is a critical issue that requires attention especially in rapidly developing countries. Toward this, both conventional and advanced methods have succeeded, but they have demonstrated mere

The Treatment of Pharmaceutical Wastewater. https://doi.org/10.1016/B978-0-323-99160-5.00007-2

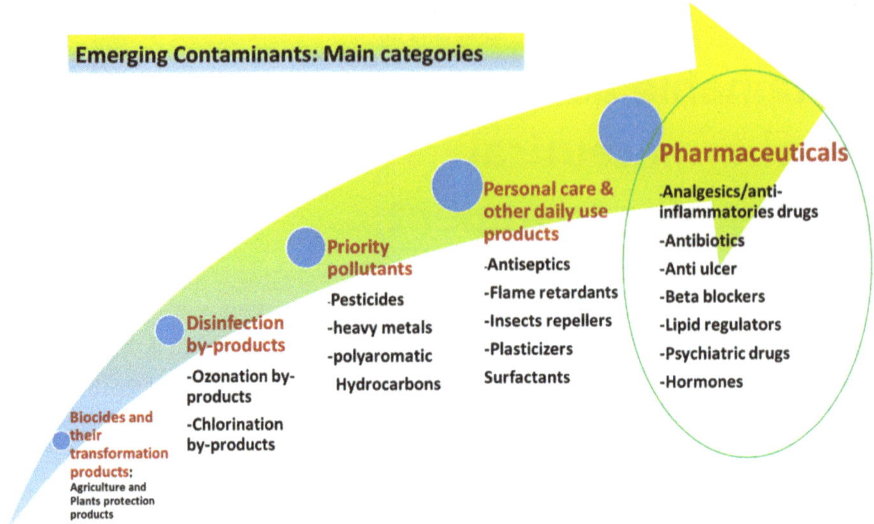

Fig. 1 At a glance: Emerging contaminants-main categories.

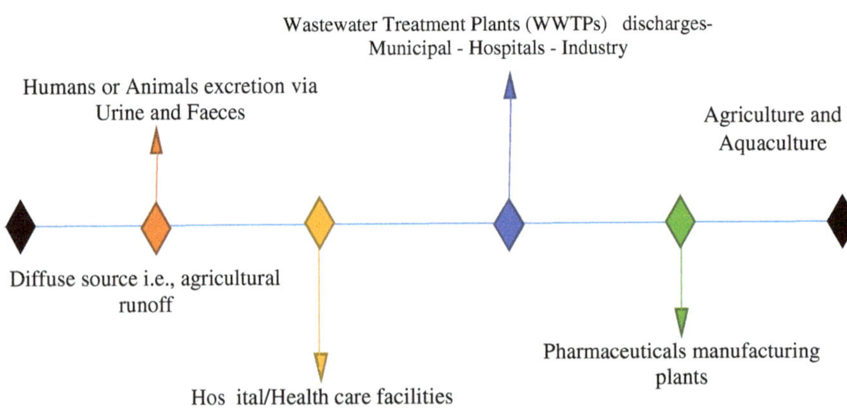

Fig. 2 Source of Pharmaceuticals in aquatic environments/water bodies.

removal of PhCs from wastewater at sink level. This is probably, due to the many reasons: first, the existing municipal wastewater treatment plants are not designed for the removal of (PhCs); second, the complex nature of PhCs results from their different functionalities, physicochemical, and biological properties (Maycock & Watts, 2011; OECD, 2019a, 2019b) (Fig. 3).

Hence to investigate other technologies for reliable treatment of PhCs wastewater steams at the watershed scale is crucial. Toward to this, many nonconventional approaches such as constructed wetlands systems have been designed to remove PhCs from WW, however, they can only remove certain PhCs, to certain extents, and also they produce harmful by-products with as-yet unknown public health or

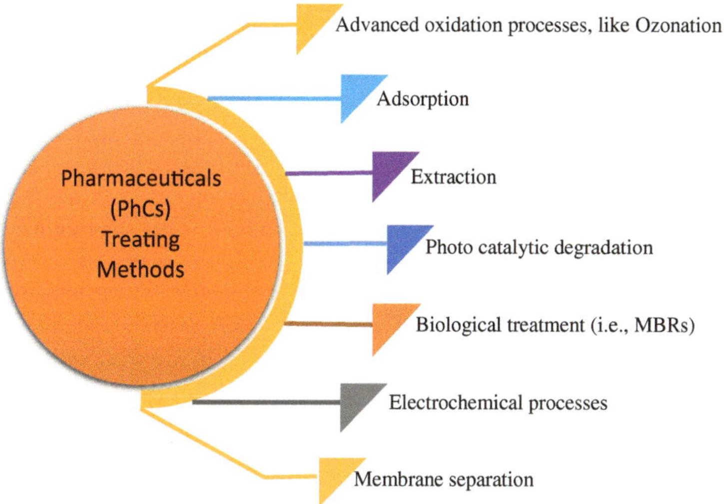

Fig. 3 Illustrates a summary of the techniques applied with focus on both conventional and advanced water treatment methods (Fast, Gude, Truax, Martin, & Magbanua, 2017; Ibáñez et al., 2013; Nguyen, Hai, Kang, Price, & Nghiem, 2013; Pal, 2018; Rout, Zhang, Bhunia, & Surampalli, 2021; Sadegh, Mazloumbilandi, & Chahardouri, 2019; Wei et al., 2021).

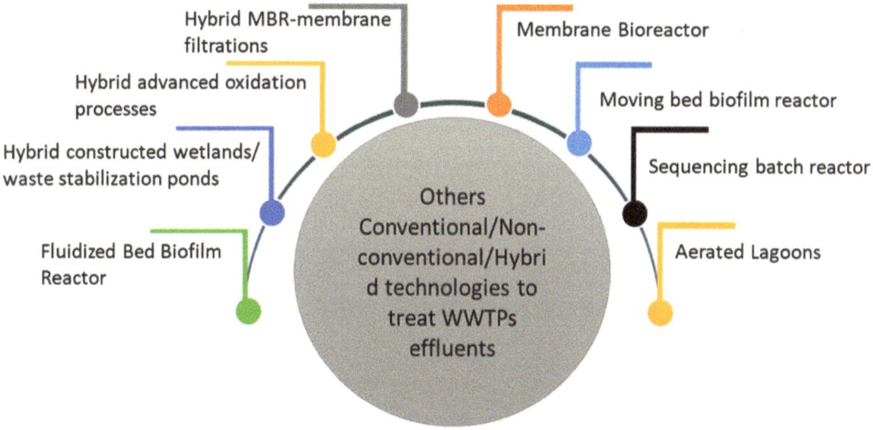

Fig. 4 State of art technologies used in PhCs treatment of WWTPs effluents (Kyzas, Jie, Lazaridis, Bikiaris, & Matis, 2015; Lin, Tsai, Yu, Wang, & Lin, 2011; Mohapatra & Kirpalani, 2019; Patel et al., 2019; Petrović, Gonzalez, & Barceló, 2003; Rizzo et al., 2019; Taheran et al., 2016; Zaied, Rashid, Mohd Nasrullah, Zularisam, & Singh, 2020).

environmental effects. In addition, there are some challenges associated with the wastewater treatment plant size/footprint and other design criteria (Dhangar & Kumar, 2020; Singh, Nagpal, Agrawal, & Rachna., 2018) (Fig. 4).

Although biological treatment is adopted as best available technology to remove some extent of a wide range of PhCs, WWTP has high power consumption in biological processes. Hence, developing innovative and improved methods for the removal of PhCs pollutants from WWTP effluents and drinking water sources is, therefore, a persistent societal concern. Adsorption process is demonstrated to be an attractive and effective technique for the removal of PhCs, and micropollutants, even at trace levels from the wastewater sources, thanks to its affordable cost, simplicity of design, and viable many renewal of adsorbents. In recent years, the application of new adsorbent materials obtained from natural resources, and environmental-friendly products, instead of the conventional synthetic ones, has known a rapid increase. In addition, adsorption process is a faster, cleaner and more efficient at removing PhCs from drinking and wastewater. Over other water-treatment technologies, adsorption technique has developed rapidly because it offers several advantages such as, no chemicals are needed, high efficiency, no by-products are formed, no lengthy exposure times are required and workers are not needed to handle any reactive liquids/or gases. Apart from these, ease of operation make it a promising and practical choice for the removal of PhCs pollutants from WWTP effluents (Ali & Gupta, 2007; Crini, Lichtfouse, Wilson, & Morin-Crini, 2019; Khan et al., 2022; Khan, Khan, Tiwari, Zubair, & Naushad, 2021; Mu'azu et al., 2022; Singh et al., 2018; Siyal, Shamsuddin, Low, & Rabat, 2020) (Fig. 5).

Since adsorption technology introduced, activated carbon (AC) in the form of GAC, granular activated carbon, and PAC, powdered activated carbon has been the primary choice for the treatment or recycling of wastewater steams because of its excellent adsorption capacity as a result of active free valencies and small particle size. In spite of this, AC cannot be applied as an adsorbent on a large scale for water treatment systems because of its high cost and longtime consumption, difficulties in their regeneration and reuse. These limitations attracted researchers to use the low-cost adsorbent materials that could approximately be regenerated. In this respect, many alternative adsorbents, e.g., natural adsorbents, industrial wastes, agricultural

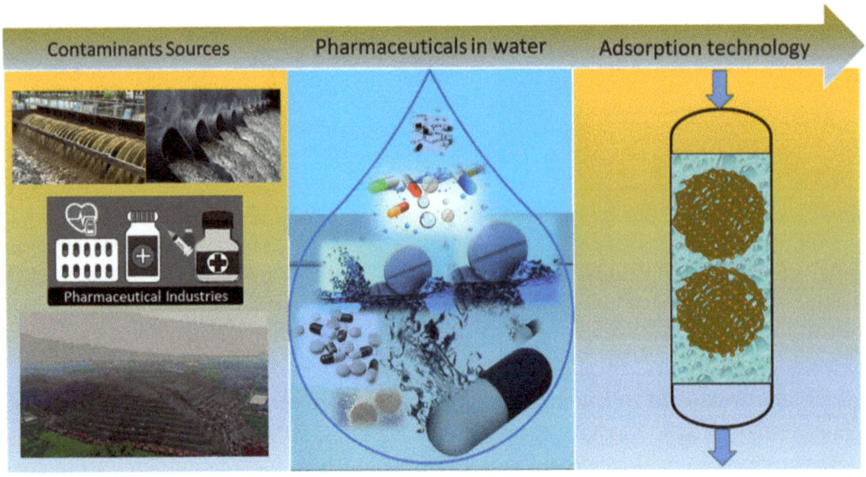

Fig. 5 Graphical representation of PhCs adsorption process.

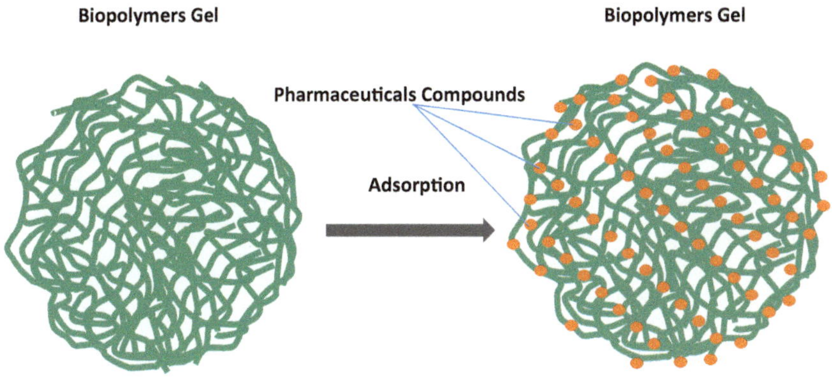

Fig. 6 Graphical illustrations of pharmaceuticals compounds adsorption in hydrogel.

wastes, biomass, nanomaterials of noble metal, metal oxide, dendritic polymers, zeolites, and natural polymers have been investigated for the adsorption of a great variety of PhCs pollutants. Recent years have witnessed a spurt in research activities on natural or bio polymers for removal of PhCs-pollutants. From a literature survey of recently published articles, it is very evident that natural polymers have demonstrated outstanding pharmaceutical compounds—removal capabilities and their great ability of regenerating and reusing in multiple adsorption-desorption cycles. Also, little modifications in their chemical structure can alter the original properties of natural polymers, making them better suited for the adsorption of various PhCs types (Adewuyi, 2020; Biswas & Pal, 2021; Crini, 2005; Fouda-Mbanga, Prabakaran, & Pillay, 2021; Khademian, Salehi, Sanaeepur, Galiano, & Figoli, 2020; Oladoja, Unuabonah, Amuda, & Kolawole, 2017; Qureshi, Nishat, Jadoun, & Ansari, 2020; Tambosi et al., 2010; Zhang et al., 2016) (Fig. 6).

From the wide variety of existing biopolymers, polysaccharides are mostly more abundant, low cost, and ease to extract; therefore, polysaccharides are ideal biopolymers with which to develop PhCs remediation target systems.

2. Biopolymers at a glance: Classification and mechanisms of PhCs adsorption

Some of the features that make biopolymers to be unique as adsorbent materials are:

(i) Abundance
(ii) Biopolymers are cheap materials
(iii) They are renewable resource
(iv) Stable and hydrophilic material
(v) They are modifiable materials that can be easily tailored to selective or specific applications
(vi) Biopolymers possess favorable biochemical features such as biocompatibility, non-toxicity, biodegradability, high chemical reactivity, polyfunctionality, chirality and chelation, and great adsorption capacities (Fig. 7, Table 1).

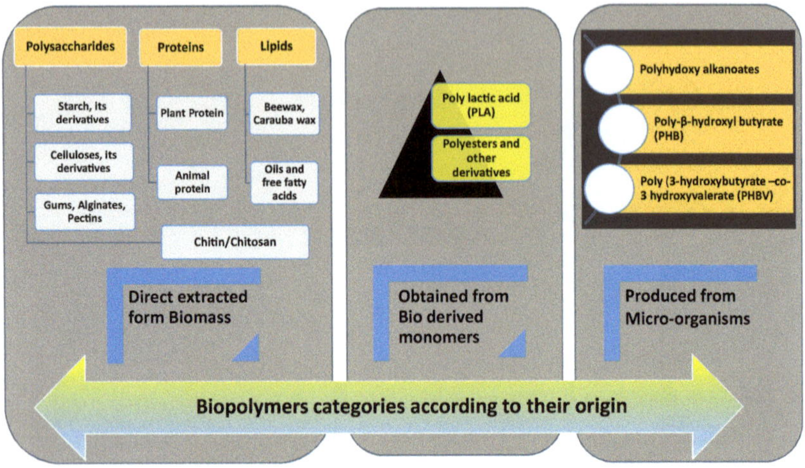

Fig. 7 Biopolymers categories used for PhCs wastewater treatment.

Table 1 Biopolymers frequently used and modification thereof for PhCs wastewater treatment.

Biopolymers	Modification of biopolymers	Crosslinking agents
Chitin Chitosan Cellulose and its derivatives Collagen Gelatin Lignin Alginate Agar Carrageenan Pectin Xanthan Dextran Tannin, Starch Keratin and Silk Arabic gum Guar gum	Grafting Crosslinking Ionic crosslinking Covalently crosslinking Blending Coating Composite/nanocomposite Biopolymers	Glutaraldehyde Formaldehyde Epichlorohydrin Ethylene glycol Diglycidyl ether Iminodiacetic acid Nitriloacetic acid Organic diisocyanates Organic phosphate Metal and metal oxides Sodium borate/boric acid Glyoxal *N*, *N′*-methylenebisacrylamide Dextran

The various biopolymers, their classification, mechanisms of adsorption action, factors and their application in PhCs wastewater treatment in the scientific literature are compiled. Also, this book chapter condenses the research carried out on the biopolymers usage to treat PhCs wastewater. Cellulose, alginate chitin, and carrageenans are the most abundant among biopolymers. Of biopolymers, chitin is a large, structural

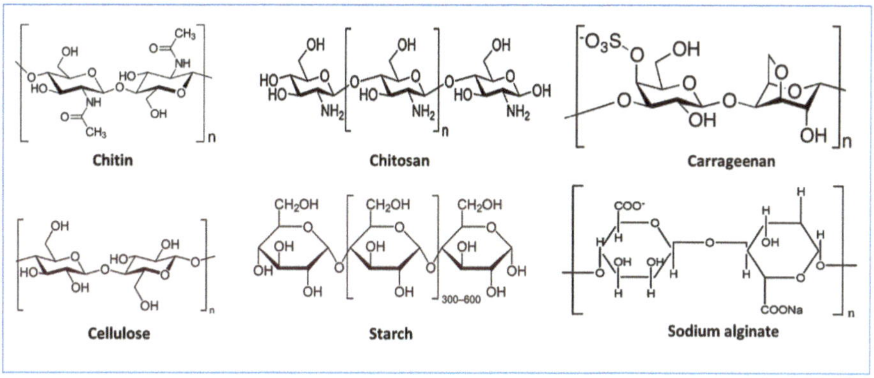

Fig. 8 Presents different bio-biopolymers extensively reported in the literature.

polysaccharide consisting an amide derivative of glucose. It can be deacetylated further to become chitosan (CS), which retains most of the characteristics that is favorable for water treatment applications; at the same time, it increases the chemical reactivity of the CS molecule, as the amino functional groups are present now in the CS structure and have been found to have adsorbent properties, which is convenient for PhCs wastewater treatment applications (Fig. 8).

Even though CS has been shown to be a decent adsorbent because of the surface charge resulted by the amine and acetyl amine functional groups in backbone, its interactions to pollutants are not precise; for this, further modification has been done to incorporate moieties/or structures that add as the selective attractors to the PhCs pollutants. For instance, carboxymethyl functional groups were grafted to the surface of CS to make it more selective to PhCs and other micropollutants. Also, another important consideration of using CS as an adsorbent material is its solubility in acidic solutions. Usually, CS loses its ability to adhere as a solid adsorbent with adsorbates at the lower pH because of its dissolution which can restrict its application in water treatment. Toward this challenge, the functionalization and cross-linking are two commonly used chemical modification approaches to stabilize CS against its dissolution in acidic media. A cross-linking ensures CS remains chemically and mechanically robust over a wide range of pH values from pH 1 to 14 (Biswas & Pal, 2021; Distantina, Rochmadi, Fahrurrozi, & Wiratni, 2012; Guibal, 2004) (Fig. 9).

3. Cross-linked biopolymers: Especially cross-linked chitosan

Crosslinking is generally carried out to enable biopolymers to be used as stable and polyfunctional adsorbent materials. Cross-links are bonds that can either be ionic/or covalent in nature, and these bonds link one polymer chain to another. Several literature reports have revealed that the cross-linking density, polymer molecular weight, an electrical charge of the polymers and other factors have a significant effect on the

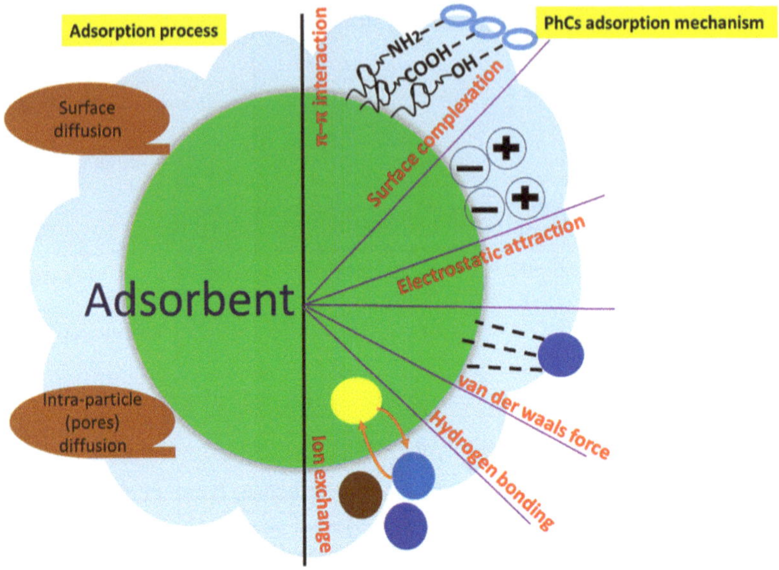

Fig. 9 Proposed key adsorption mechanisms for the sorption of PhCs on biopolymer adsorbents.

adsorption efficiency of modified/or cross-linked biopolymers. There are two types of crosslinking, physical and chemical (Hong et al., 2007; Klein & Poverenov, 2020; Patil & Jadge, 2008).

3.1 *Physical cross-linking*

Physical cross-linking in biopolymer hydrogel chemistry is being evolved as an important due to the absence of external cross-linking agents. Due to low toxicity and biocompatibility property, physical crosslinking is preferable and there are many reports in the literature that describe in the detail how to achieve physically cross-linked biopolymer applicable for PhCs adsorption applications (Berger et al., 2004; Reddy, Reddy, & Jiang, 2015). Generally, it involves many physical interactions among the polymer chains like:

(i) Ionic-cross-linking or ionic-interaction
(ii) Hydrophobic interaction
(iii) Stereo complex formation
(iv) Hydrogen bonding

Low toxicity and biocompatibility property are two features of using physical crosslinking, however, there is a major disadvantage: the gelation rate is difficult to control, with nonuniformity/or imperfect in the structure due to the free polymer chain ends. Also, it provides reversible hydrogel and also there are structural inhomogeneities. Therefore, to counter such problems, chemical cross-linking is found to be a better substituent for physical cross-linking to prepare irreversible and mechanically stable biopolymer adsorbent hydrogels particularly in water treatment adsorption applications.

3.2 Chemical crosslinking

Chemically cross-linked biopolymer hydrogels are mostly mechanically strong, prevent diffusion of hydrogel, and resist dilution/or dissolution of hydrogel matrix. It involves chemical modification of polymer chains by using additional cross-linking agents to bond two polymer chains. Different chemical crosslinking are reported, and out of the reported in the literature two main chemical techniques are grafting and covalent cross-linking. The grafting is further categorized into radiation grafting and chemical grafting, depending on the activation source. Adding chemical crosslinks affect the physical properties of the biopolymer, depending upon the cross linking degree and absence and presence of crystallinity domain. Chemical cross linking results in:

(i) Elasticity.
(ii) Decrease in the viscosity.
(iii) Insolubility of the polymer.
(iv) Increased glass transition temperature and increase strength and toughness.
(v) Lower melting point.
(vi) Transformation of thermoplasts into thermosets (Fig. 10, Tables 2–4).

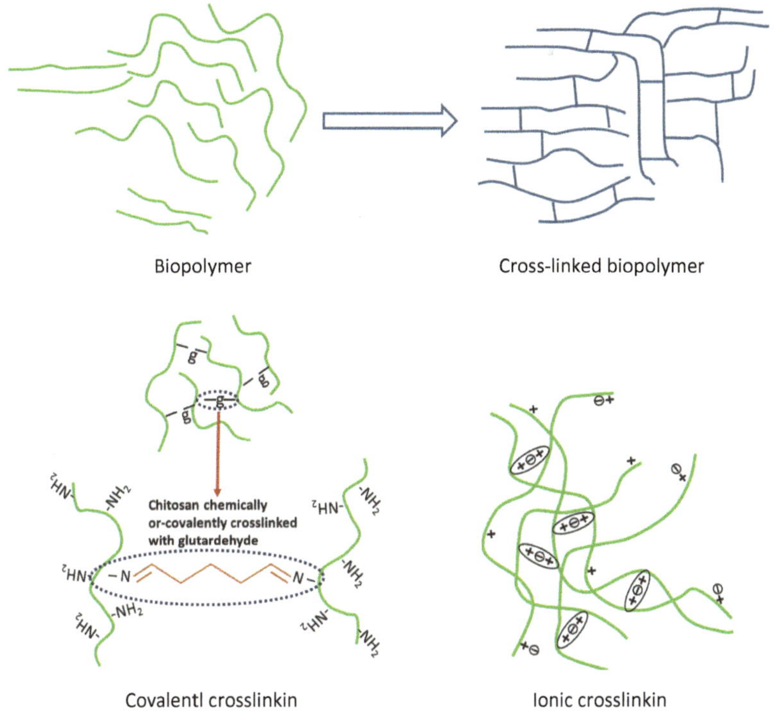

Fig. 10 Graphical illustrations of different types of cross-linking in biopolymer hydrogel adsorbents.

Table 2 Summery antibiotics pharmaceutical adsorption onto chitosan-based adsorbents.

Pharmaceuticals name	Adsorbent	References
Amoxicillin	Chitosan beads	Adriano, Veredas, Santana, and Gonçalves (2005)
Amoxicillin	Chitosan stabilized bimetallic Fe/Ni nanoparticles	Weng, Lin, Zhong, and Chen (2013)
Enrofloxacin	Grafted chitosan hydrogel with acrylic acid	Wang et al. (2019)
Norfloxacin	MgO/chitosan/graphene oxide	Nazraz, Yamini, and Asiabi (2019)
Norfloxacin	Chitosan	Meler, Grimling, and Pluta (2012)
Norfloxacin	Magnetic molecular Imprinted chitosan/g-Fe_2O_3	Wu, Huang, Zhou, and Mao (2016)
Norfloxacin	polyanions (poly(sodium *p*-styrenesulfonate)) branches modified chitosan/Fe_3O_4	Zhou et al. (2018)
Cephalosporin	Chitosan-*g*-poly(glycidyl methacrylate)	Cigu et al. (2016)
Erythromycin	Magnetic imprinted chitosan-stabilized pickering emulsion	Ou et al. (2015)
Chloramphenicol	Magnetic imprinted polymers based on chitosan microspheres	Ma, Dai, Dai, Da, and Yan (2015)
Chlortetracycline	Carboxymethyl-chitosan-modified namontmorillonite	Ma et al. (2019)
Cefotaxime	Magnetic $NiFe_2O_4$eCOF-chitosanterephthalaldehyde film	Li et al. (2020)
Cefotaxime	Chitosan particles	Li, Wang, Zhang, Yang, and Duan (2021)
Cefotaxime	Chitosan films	Li et al. (2021)
Ofloxacin	Horseradish peroxidase immobilized chitosan crosslink with graphene-oxide nanocomposite	Sun et al. (2017)
Levofloxacin	Nanotitanium oxide/chitosan/nano-bentonite	Mahmoud, El-Ghanam, Mohamed, and Saad (2020)
Ceftriaxone	Nanotitanium oxide/chitosan/nano-bentonite	Mahmoud et al. (2020)

Table 3 Pharmaceutical adsorption onto chitosan and chitosan composites-based adsorbents.

Pharm. compound	Adsorbent	References
Diclofenac sodium	Copolymerized chitosan/ Fe_3O_4 composite	Zhang et al. (2016)
Metamizol Acetylsalicylic acid Acetaminophen Caffeine	Chitosan/waste coffee-ground composite	Lessa, Nunes, and Fajardo (2018)
Diclofenac sodium	Chitosan microspheres	Carvalho, Matias, Braga, Evangelista, and Prado (2011)
Diclofenac sodium	Polycations (poly(*p*-vinylbenzyl trimethylammonium chloride)) modified chitosan/Fe_3O_4	Zhou et al. (2018)
Diclofenac	Glutaraldehyde cross-linked magnetic chitosan	Zhang, Shen, Dai, and Zhou (2014)
Diclofenac	$ZnFe_2O_4$/chitosan	dos Santos, Pereira, Foletto, and Dotto (2019)
Diclofenac	Poly(vinyl alcohol)/chitosan/amino-grafted silica@ polye-thylenimine (PVA/CS/SAP@PEI) gel beads	Lu, Fan, Yang, Huang, and Ouyang (2020)
Diclofenac	Chitosan-modified waste tire crumb rubber	Phasuphan, Praphairaksit, and Imyim (2019)
Diclofenac	Amine-functionalized chitosan coated with Fe_3O_4 nanoparticles	Liang et al. (2019)
Diclofenac	Carboxyl-grafted chitosan	Tzereme et al. (2019)
Diclofenac	Magnetic chitosan/activated biochar	Mojiri, Kazeroon, and Gholami (2019)
Ibuprofen	Magnetic chitosan/activated biochar	Mojiri et al. (2019)
Ibuprofen	Chitosan-modified waste tire crumb rubber	Phasuphan et al. (2019)
Ibuprofen	b-Cyclodextrin grafted chitosan	Bany-Aiesh, Banat, and Al-Sou'od (2015)
Ibuprofen	*O*-carboxymethyl-*N*-laurylchitosan/ g-Fe_2O_3	Chahm and Rodrigues (2017)
Ibuprofen	Silver nanoparticles loaded chitosan microspheres	Pereira et al. (2020)
Naproxen	Chitosan-modified waste tire crumb rubber	Phasuphan et al. (2019)
Naproxen	Magnetic chitosan/activated biochar	Mojiri et al. (2019)
Dipyrone	Chitosan microspheres	Carvalho et al. (2011)

Table 4 Pharmaceuticals adsorption performance of chitosan and derivatives thereof.

Pharm. compound	**Adsorbent**	**References**
Permethrin	Chitosan-zinc oxide nanoparticles	Shahram, Rahmanifar, Moradi, and Azar (2014)
Pramipexole	Sulfonic acid grafted chitosan cross-linked with glutaraldehyde	Kyzas, Bikiaris, and Lambropoulou (2017)
Pramipexole	Chitosan	Kyzas, Kostoglou, Lazaridis, Lambropoulou, and Bikiaris (2013)
Pramipexole	Glutaraldehyde cross-linked chitosan which grafted with sulfonate (CsSLF)	Kyzas et al. (2013)
Pramipexole	Glutaraldehyde cross-linked chitosan which grafted with Ne(2ecarboxybenzyl) groups (CsNCB)	Kyzas et al. (2013)
Clofibric acid	Glutaraldehyde cross-linked magnetic chitosan	Zhang et al. (2014)
Clofibric acid	PEI-modified chitosan beads	Nie, Deng, Wang, Huang, and Gang (2014)
Penicillin acylase	*N*-benzamide and *N*-phenylacetamide derivatives of chitosan beads	Patel and Gaikar (2004)
Carbamazepine	Magnetic molecularly imprinted chitosan-Fe_3O_4	Zhang, Zhang, Dai, Zhou, and Liu (2013)
Tylosin	Polyanions (poly(sodium *p*-styrenesulfonate)) branches modified chitosan/Fe_3O_4	Zhou et al. (2018)
Triclosan	Chitosan film	Noirod, Lamangthong, and Ninjiaranai (2016)
Allopurinol	Chitosan	Alkhamis, Obeidat, and Najib (2001)
Ketotifen	Chitosan	Alkhamis et al. (2001)
Dorzolamide	Chitosan	Kyzas, Bikiaris, Seredych, Bandosz, and Deliyanni (2014)
Dorzolamide	Graphite oxide/poly(acrylic acid) grafted chitosan nanocomposite	Kyzas et al. (2014)
Hydroquinone	Graphene oxide functionalized with magnetic cyclodextrin-chitosan	Li et al. (2013)

4. Conclusions

Today, the entire global communities confirmed to be contaminated with the highly toxic and low biodegradable compounds known as emerging contaminants continues to grow rapidly at an alarming rate. In emerging contaminants, pharmaceuticals have received increasing consideration recently for their persistent input and possible threat to ecological, aquatic environment and public health. Hence, there is urgent need for their removal from natural and wastewater resources. Treatment technologies to remove pharmaceuticals from water and wastewater bodies exist on both municipal and household. Activated carbon filters media and reverse osmosis membrane process are two main options, but they are costly and high-maintenance. As an alternative, the adsorption mode using biopolymers has been found an attractive, simple and dominant tool for the elimination of these toxic compounds. Biopolymers, as a sustainable adsorbent materials class, show significant promise for effective adsorption of these pharmaceutical compounds. Hence, this book chapter aims to help researchers for choosing the biopolymers through assessing the published research reports dealing with the success of them for adsorptive removal of the pharmaceutical compounds.

Acknowledgment

The authors extend their appreciation to the Deanship of Scientific Research at King Saud University for funding this work through research group no (RG-1439-085).

References

Adewuyi, A. (2020). Chemically modified biosorbents and their role in the removal of emerging pharmaceutical waste in the water system. *Water*, *12*(6), 1–31. https://doi.org/10.3390/W12061551.

Adriano, W. S., Veredas, V., Santana, C. C., & Gonçalves, L. R. B. (2005). Adsorption of amoxicillin on chitosan beads: Kinetics, equilibrium and validation of finite bath models. *Biochemical Engineering Journal*, *27*(2), 132–137. https://doi.org/10.1016/j.bej.2005.08.010.

Afzal, H. K., Aziz, H. A., Khan, N. A., Ahmed, S., Mehtab, M. S., Vambol, S., et al. (2020). Pharmaceuticals of emerging concern in hospital wastewater: Removal of ibuprofen and ofloxacin drugs using MBBR method. *International Journal of Environmental Analytical Chemistry*, 1–15. https://doi.org/10.1080/03067319.2020.1855333.

Ahuja, S. (2020). Overview: Current status of environmental research on water contaminants. *ACS Symposium Series*, *1352*, 1–11. https://doi.org/10.1021/bk-2020-1352.ch001.

Ali, I., & Gupta, V. K. (2007). Advances in water treatment by adsorption technology. *Nature Protocols*, *1*(6), 2661–2667. https://doi.org/10.1038/nprot.2006.370.

Alkhamis, K. A., Obeidat, W. M., & Najib, N. M. (2001). Adsorption of allopurinol and ketotifen by chitosan. *AAPS PharmSciTech*, *2*(1), 27–33. https://doi.org/10.1208/pt020103.

Al-Riyami, I. M., Ahmed, M., Al-Busaidi, A., & Choudri, B. S. (2018). Antibiotics in wastewaters: A review with focus on Oman. *Applied Water Science*, *8*(7), 1–10. https://doi.org/10.1007/s13201-018-0846-z.

Bany-Aiesh, H., Banat, R., & Al-Sou'od, K. (2015). Kinetics and adsorption isotherm of ibuprofen onto grafted β-CD/chitosan polymer. *American Journal of Applied Sciences*, *12*(12), 917–930. https://doi.org/10.3844/ajassp.2015.917.930.

Berger, J., Reist, M., Mayer, J. M., Felt, O., Peppas, N. A., & Gurny, R. (2004). Structure and interactions in covalently and ionically crosslinked chitosan hydrogels for biomedical applications. *European Journal of Pharmaceutics and Biopharmaceutics*, *57*(1), 19–34. https://doi.org/10.1016/S0939-6411(03)00161-9.

Biswas, S., & Pal, A. (2021). Application of biopolymers as a new age sustainable material for surfactant adsorption: A brief review. *Carbohydrate Polymer Technologies and Applications*, *2*, 100145. https://doi.org/10.1016/j.carpta.2021.100145.

Bottoni, P., Caroli, S., & Barra Caracciolo, A. (2010). Pharmaceuticals as priority water contaminants. *Toxicological and Environmental Chemistry*, *92*(3), 549–565. https://doi.org/10.1080/02772241003614320.

Cabeza, Y., Candela, L., Ronen, D., & Teijon, G. (2012). Monitoring the occurrence of emerging contaminants in treated wastewater and groundwater between 2008 and 2010. The Baix Llobregat (Barcelona, Spain). *Journal of Hazardous Materials*, *239–240*, 32–39. https://doi.org/10.1016/j.jhazmat.2012.07.032.

Carvalho, T. O., Matias, A. E. B., Braga, L. R., Evangelista, S. M., & Prado, A. G. S. (2011). Calorimetric studies of removal of nonsteroidal anti-inflammatory drugs diclofenac and dipyrone from water. *Journal of Thermal Analysis and Calorimetry*, *106*(2), 475–481. https://doi.org/10.1007/s10973-010-1243-5.

Chahm, T., & Rodrigues, C. A. (2017). Removal of ibuprofen from aqueous solutions using O-carboxymethyl-N-laurylchitosan/γ-Fe2O3. *Environmental Nanotechnology, Monitoring & Management*, *7*, 139–148. https://doi.org/10.1016/j.enmm.2017.03.001.

Cigu, T. A., Vasiliu, S., Racovita, S., Lionte, C., Sunel, V., Popa, M., et al. (2016). Adsorption and release studies of new cephalosporin from chitosan-g-poly(glycidyl methacrylate) microparticles. *European Polymer Journal*, *82*, 132–152. https://doi.org/10.1016/j.eurpolymj.2016.07.011.

Crini, G. (2005). Recent developments in polysaccharide-based materials used as adsorbents in wastewater treatment. *Progress in Polymer Science*, *30*(1), 38–70. https://doi.org/10.1016/j.progpolymsci.2004.11.002.

Crini, G., Lichtfouse, E., Wilson, L. D., & Morin-Crini, N. (2019). Conventional and non-conventional adsorbents for wastewater treatment. *Environmental Chemistry Letters*, *17*(1), 195–213. https://doi.org/10.1007/s10311-018-0786-8.

Deblonde, T., Cossu-Leguille, C., & Hartemann, P. (2011). Emerging pollutants in wastewater: A review of the literature. *International Journal of Hygiene and Environmental Health*, *214*(6), 442–448. https://doi.org/10.1016/j.ijheh.2011.08.002.

Dhangar, K., & Kumar, M. (2020). Tricks and tracks in removal of emerging contaminants from the wastewater through hybrid treatment systems: A review. *Science of the Total Environment*, *738*(336), 140320. https://doi.org/10.1016/j.scitotenv.2020.140320.

Distantina, S., Rochmadi, Fahrurrozi, M., & Wiratni. (2012). Preparation of hydrogel based on glutaraldehyde-crosslinked carrageenan. In *Vol. 38. 3rd international conference on chemistry and chemical engineering* (pp. 150–154).

dos Santos, J. M. N., Pereira, C. R., Foletto, E. L., & Dotto, G. L. (2019). Alternative synthesis for ZnFe2O4/chitosan magnetic particles to remove diclofenac from water by adsorption. *International Journal of Biological Macromolecules*, *131*, 301–308. https://doi.org/10.1016/j.ijbiomac.2019.03.079.

Fast, S. A., Gude, V. G., Truax, D. D., Martin, J., & Magbanua, B. S. (2017). A critical evaluation of advanced oxidation processes for emerging contaminants removal. *Environmental Processes*, *4*(1), 283–302. https://doi.org/10.1007/s40710-017-0207-1.

Fouda-Mbanga, B. G., Prabakaran, E., & Pillay, K. (2021). Carbohydrate biopolymers, lignin based adsorbents for removal of heavy metals (Cd2+, Pb2+, Zn2+) from wastewater, regeneration and reuse for spent adsorbents including latent fingerprint detection: A review. *Biotechnology Reports*, *30*, e00609. https://doi.org/10.1016/j.btre.2021.e00609.

Guibal, E. (2004). Interactions of metal ions with chitosan-based sorbents: A review. *Separation and Purification Technology*, *38*(1), 43–74. https://doi.org/10.1016/j.seppur.2003.10.004.

Hong, Y., Song, H., Gong, Y., Mao, Z., Gao, C., & Shen, J. (2007). Covalently crosslinked chitosan hydrogel: Properties of in vitro degradation and chondrocyte encapsulation. *Acta Biomaterialia*, *3*(1), 23–31. https://doi.org/10.1016/j.actbio.2006.06.007.

Ibáñez, M., Gracia-Lor, E., Bijlsma, L., Morales, E., Pastor, L., & Hernández, F. (2013). Removal of emerging contaminants in sewage water subjected to advanced oxidation with ozone. *Journal of Hazardous Materials*, *260*, 389–398. https://doi.org/10.1016/j.jhazmat.2013.05.023.

Jiang, J. J., Lee, C. L., Der Fang, M., Boyd, K. G., & Gibb, S. W. (2015). Source apportionment and risk assessment of emerging contaminants: An approach of pharmaco-signature in water systems. *PLoS One*, *10*(4), 1–21. https://doi.org/10.1371/journal.pone.0122813.

Khademian, E., Salehi, E., Sanaeepur, H., Galiano, F., & Figoli, A. (2020). A systematic review on carbohydrate biopolymers for adsorptive remediation of copper ions from aqueous environments-part A: Classification and modification strategies. *Science of the Total Environment*, *738*, 139829. https://doi.org/10.1016/j.scitotenv.2020.139829.

Khan, A. H., Aziz, H. A., Khan, N. A., Dhingra, A., Ahmed, S., & Naushad, M. (2021). Effect of seasonal variation on the occurrences of high-risk pharmaceutical in drain-laden surface water: A risk analysis of Yamuna River. *Science of the Total Environment*, *794*, 148484. https://doi.org/10.1016/j.scitotenv.2021.148484.

Khan, N. A., Khan, A. H., Tiwari, P., Zubair, M., & Naushad, M. (2021). New insights into the integrated application of Fenton-based oxidation processes for the treatment of pharmaceutical wastewater. *Journal of Water Process Engineering*, *44*, 102440. https://doi.org/10.1016/j.jwpe.2021.102440.

Khan, A. H., Khan, N. A., Zubair, M., Shaida, M. A., Manzar, M. S., Abutaleb, A., et al. (2022). Sustainable green nanoadsorbents for remediation of pharmaceuticals from water and wastewater: A critical review. *Environmental Research*, *204*, 112243. https://doi.org/10.1016/j.envres.2021.112243.

Klein, M., & Poverenov, E. (2020). Natural biopolymer-based hydrogels for use in food and agriculture. *Journal of the Science of Food and Agriculture*, *100*(6), 2337–2347. https://doi.org/10.1002/jsfa.10274.

Kyzas, G. Z., Bikiaris, D. N., & Lambropoulou, D. A. (2017). Effect of humic acid on pharmaceuticals adsorption using sulfonic acid grafted chitosan. *Journal of Molecular Liquids*, *230*, 1–5. https://doi.org/10.1016/j.molliq.2017.01.015.

Kyzas, G. Z., Bikiaris, D. N., Seredych, M., Bandosz, T. J., & Deliyanni, E. A. (2014). Removal of dorzolamide from biomedical wastewaters with adsorption onto graphite oxide/poly (acrylic acid) grafted chitosan nanocomposite. *Bioresource Technology*, *152*, 399–406. https://doi.org/10.1016/j.biortech.2013.11.046.

Kyzas, G. Z., Jie, F., Lazaridis, N. K., Bikiaris, D. N., & Matis, K. A. (2015). New approaches on the removal of pharmaceuticals from wastewaters with adsorbent materials. *Journal of Molecular Liquids*, *209*, 87–93. https://doi.org/10.1016/j.molliq.2015.05.025.

Kyzas, G. Z., Kostoglou, M., Lazaridis, N. K., Lambropoulou, D. A., & Bikiaris, D. N. (2013). Environmental friendly technology for the removal of pharmaceutical contaminants from wastewaters using modified chitosan adsorbents. *Chemical Engineering Journal*, *222*, 248–258. https://doi.org/10.1016/j.cej.2013.02.048.

Lessa, E. F., Nunes, M. L., & Fajardo, A. R. (2018). Chitosan/waste coffee-grounds composite: An efficient and eco-friendly adsorbent for removal of pharmaceutical contaminants from water. *Carbohydrate Polymers*, *189*, 257–266. https://doi.org/10.1016/j.carbpol.2018.02.018.

Li, L., Fan, L., Sun, M., Qiu, H., Li, X., Duan, H., et al. (2013). Adsorbent for hydroquinone removal based on graphene oxide functionalized with magnetic cyclodextrin–chitosan.

International Journal of Biological Macromolecules, *58*, 169–175. https://doi.org/10.1016/j.ijbiomac.2013.03.058.

Li, Z., Liu, Y., Zou, S., Chunbo, L., Bai, H., Haibo, M., et al. (2020). Removal and adsorption mechanism of tetracycline and cefotaxime contaminants in water by NiFe2O4-COF-chitosan-terephthalaldehyde nanocomposites film. *Chemical Engineering Journal*, *382*, 123008. https://doi.org/10.1016/j.cej.2019.123008.

Li, Z., Wang, X., Zhang, X., Yang, Y., & Duan, J. (2021). A high-efficiency and plane-enhanced chitosan film for cefotaxime adsorption compared with chitosan particles in water. *Chemical Engineering Journal*, *413*, 127494. https://doi.org/10.1016/j.cej.2020.127494.

Liang, X. X., Omer, A. M., Zhao-hong, H., Wang, Y.-g., Di, Y., & Ouyang, X.-k. (2019). Efficient adsorption of diclofenac sodium from aqueous solutions using magnetic amine-functionalized chitosan. *Chemosphere*, *217*, 270–278. https://doi.org/10.1016/j.chemosphere.2018.11.023.

Lin, A. Y. C., Tsai, Y. T., Yu, T. H., Wang, X. H., & Lin, C. F. (2011). Occurrence and fate of pharmaceuticals and personal care products in Taiwan's aquatic environment. *Desalination and Water Treatment*, *32*(1–3), 57–64. https://doi.org/10.5004/dwt.2011.2678.

Lu, Y., Fan, L., Yang, L.-Y., Huang, F., & Ouyang, X.-k. (2020). PEI-modified core-shell/bead-like amino silica enhanced poly (vinyl alcohol)/chitosan for diclofenac sodium efficient adsorption. *Carbohydrate Polymers*, *229*, 115459. https://doi.org/10.1016/j.carbpol.2019.115459.

Ma, W., Dai, J., Dai, X., Da, Z., & Yan, Y. (2015). Core–shell molecularly imprinted polymers based on magnetic chitosan microspheres for chloramphenicol selective adsorption. *Monatshefte Für Chemie—Chemical Monthly*, *146*(3), 465–474. https://doi.org/10.1007/s00706-014-1351-1.

Ma, J., Lei, Y., Khan, M. A., Wang, F., Chu, Y., Lei, W., et al. (2019). Adsorption properties, kinetics & thermodynamics of tetracycline on carboxymethyl-chitosan reformed montmorillonite. *International Journal of Biological Macromolecules*, *124*, 557–567. https://doi.org/10.1016/j.ijbiomac.2018.11.235.

Mahmoud, M. E., El-Ghanam, A. M., Mohamed, R. H. A., & Saad, S. R. (2020). Enhanced adsorption of levofloxacin and ceftriaxone antibiotics from water by assembled composite of nanotitanium oxide/chitosan/nano-bentonite. *Materials Science and Engineering: C*, *108*, 110199. https://doi.org/10.1016/j.msec.2019.110199.

Maycock, D. S., & Watts, C. D. (2011). Pharmaceuticals in drinking water. In *Encyclopedia of environmental health* (pp. 472–484). https://doi.org/10.1016/B978-0-444-52272-6.00457-8.

Meler, J., Grimling, B., & Pluta, J. (2012). Studies of norfloxacin adsorption on chitosan. *Progress on Chemistry and Application of Chitin and Its Derivatives*, *2012*(January 2012), 99–106.

Mohapatra, D. P., & Kirpalani, D. M. (2019). Advancement in treatment of wastewater: fate of emerging contaminants. *Canadian Journal of Chemical Engineering*, *97*(10), 2621–2631. https://doi.org/10.1002/cjce.23533.

Mojiri, A., Kazeroon, R. A., & Gholami, A. (2019). Cross-linked magnetic chitosan/activated biochar for removal of emerging micropollutants from water: Optimization by the artificial neural network. *Water*, *11*(3), 551. https://doi.org/10.3390/w11030551.

Montes-Grajales, D., Fennix-Agudelo, M., & Miranda-Castro, W. (2017). Occurrence of personal care products as emerging chemicals of concern in water resources: A review. *Science of the Total Environment*, *595*, 601–614. https://doi.org/10.1016/j.scitotenv.2017.03.286.

Mu'azu, N. D., Manzar, M. S., Zubair, M., Alhajri, E. G., Essa, M. H., Meili, L., et al. (2022). Volcanic Ashe and its NaOH modified adsorbent for superb cationic dye uptake from water: Statistical evaluation, optimization, and mechanistic studies. *Colloids and Surfaces A: Physicochemical and Engineering Aspects*, *634*, 127879. https://doi.org/10.1016/j.colsurfa.2021.127879.

Naidu, R., Espana, V. A. A., Liu, Y., & Jit, J. (2016). Emerging contaminants in the environment: Risk-based analysis for better management. *Chemosphere*, *154*, 350–357. https://doi.org/10.1016/j.chemosphere.2016.03.068.

Nazraz, M., Yamini, Y., & Asiabi, H. (2019). Chitosan-based sorbent for efficient removal and extraction of ciprofloxacin and norfloxacin from aqueous solutions. *Microchimica Acta*, *186*(7), 459. https://doi.org/10.1007/s00604-019-3563-x.

Nguyen, L. N., Hai, F. I., Kang, J., Price, W. E., & Nghiem, L. D. (2013). Removal of emerging trace organic contaminants by MBR-based hybrid treatment processes. *International Biodeterioration and Biodegradation*, *85*, 474–482. https://doi.org/10.1016/j.ibiod.2013.03.014.

Nie, Y., Deng, S., Wang, B., Huang, J., & Gang, Y. (2014). Removal of clofibric acid from aqueous solution by polyethylenimine-modified chitosan beads. *Frontiers of Environmental Science & Engineering*, *8*(5), 675–682. https://doi.org/10.1007/s11783-013-0622-0.

Noirod, P., Lamangthong, J., & Ninjiaranai, P. (2016). Application of chitosan film for the removal of triclosan from aqueous solutions by adsorption. *Key Engineering Materials*, *675–676*, 455–458. https://doi.org/10.4028/www.scientific.net/KEM.675-676.455.

OECD. (2019a). Pharmaceutical residues in freshwater. In *OECD studies on water* (p. 136).

OECD. (2019b). Pharmaceutical residues in freshwater hazards and policy responses. In *OECD studies on water* (pp. 1–17).

Oladoja, N. A., Unuabonah, E. I., Amuda, O. S., & Kolawole, O. M. (2017). Mechanistic insight into the coagulation efficiency of polysaccharide-based coagulants. *SpringerBriefs in Molecular Science* (pp. 13–35). https://doi.org/10.1007/978-3-319-56599-6_2.

Ou, H., Chen, Q., Pan, J., Zhang, Y., Huang, Y., & Qi, X. (2015). Selective removal of erythromycin by magnetic imprinted polymers synthesized from chitosan-stabilized pickering emulsion. *Journal of Hazardous Materials*, *289*, 28–37. https://doi.org/10.1016/j.jhazmat.2015.02.030.

Pal, P. (2018). Treatment and disposal of pharmaceutical wastewater: Toward the sustainable strategy. *Separation and Purification Reviews*, *47*(3), 179–198. https://doi.org/10.1080/15422119.2017.1354888.

Patel, C. B., & Gaikar, V. G. (2004). Adsorption and immobilization of penicillin acylase on chitosan beads. *Separation Science and Technology*, *39*(11), 2655–2675. https://doi.org/10.1081/SS-200026758.

Patel, M., Kumar, R., Kishor, K., Mlsna, T., Pittman, C. U., & Mohan, D. (2019). Pharmaceuticals of emerging concern in aquatic systems: Chemistry, occurrence, effects, and removal methods. *Chemical Reviews*, *119*(6), 3510–3673. https://doi.org/10.1021/acs.chemrev.8b00299.

Patil, S., & Jadge, D. (2008). Crosslinking of polysaccharides: Methods and applications. *Pharmaceutical Reviews*, *6*, 51–76 (September).

Pereira, A. K.d. S., Reis, D. T., Barbosa, K. M., Scheidt, G. N., da Costa, L. S., & Santos, L. S. S. (2020). Antibacterial effects and ibuprofen release potential using chitosan microspheres loaded with silver nanoparticles. *Carbohydrate Research*, *488*, 107891. https://doi.org/10.1016/j.carres.2019.107891.

Petrie, B., Barden, R., & Kasprzyk-Hordern, B. (2015). A review on emerging contaminants in wastewaters and the environment: Current knowledge, understudied areas and recommendations for future monitoring. *Water Research, 72*, 3–27. https://doi.org/10.1016/j.watres.2014.08.053.

Petrović, M., Gonzalez, S., & Barceló, D. (2003). Analysis and removal of emerging contaminants in wastewater and drinking water. *Trends in Analytical Chemistry, 22*(10), 685–696. https://doi.org/10.1016/S0165-9936(03)01105-1.

Phasuphan, W., Praphairaksit, N., & Imyim, A. (2019). Removal of ibuprofen, diclofenac, and naproxen from water using chitosan-modified waste tire crumb rubber. *Journal of Molecular Liquids, 294*, 111554. https://doi.org/10.1016/j.molliq.2019.111554.

Qureshi, M. A., Nishat, N., Jadoun, S., & Ansari, M. Z. (2020). Polysaccharide based superabsorbent hydrogels and their methods of synthesis: A review. *Carbohydrate Polymer Technologies and Applications, 1*(October), 100014. https://doi.org/10.1016/j.carpta.2020.100014.

Rasheed, T., Bilal, M., Nabeel, F., Adeel, M., & Iqbal, H. M. N. (2019). Environmentally-related contaminants of high concern: Potential sources and analytical modalities for detection, quantification, and treatment. *Environment International, 122*(November 2018), 52–66. https://doi.org/10.1016/j.envint.2018.11.038.

Reddy, N., Reddy, R., & Jiang, Q. (2015). Crosslinking biopolymers for biomedical applications. *Trends in Biotechnology, 33*(6), 362–369. https://doi.org/10.1016/j.tibtech.2015.03.008.

Rizzo, L., Malato, S., Antakyali, D., Beretsou, V. G., Đolić, M. B., Gernjak, W., et al. (2019). Consolidated vs new advanced treatment methods for the removal of contaminants of emerging concern from urban wastewater. *Science of the Total Environment, 655*(August 2018), 986–1008. https://doi.org/10.1016/j.scitotenv.2018.11.265.

Rout, P. R., Zhang, T. C., Bhunia, P., & Surampalli, R. Y. (2021). Treatment technologies for emerging contaminants in wastewater treatment plants: A review. *Science of the Total Environment, 753*, 141990. https://doi.org/10.1016/j.scitotenv.2020.141990.

Sadegh, H., Mazloumbilandi, M., & Chahardouri, M. (2019). Low-cost materials with adsorption performance. *Vol. 5. Handbook of ecomaterials* (pp. 1–33). Cham: Springer.

Shahram, M. D., Rahmanifar, B., Moradi, A. M., & Azar, P. A. (2014). Removal of permethrin pesticide from water by chitosan–zinc oxide nanoparticles composite as an adsorbent. *Journal of Saudi Chemical Society, 18*(4), 348–355. https://doi.org/10.1016/j.jscs.2014.01.004.

Singh, N. B., Nagpal, G., Agrawal, S., & Rachna. (2018). Water purification by using adsorbents: A review. *Environmental Technology and Innovation, 11*, 187–240. https://doi.org/10.1016/j.eti.2018.05.006.

Siyal, A. A., Shamsuddin, M. R., Low, A., & Rabat, N. E. (2020). A review on recent developments in the adsorption of surfactants from wastewater. *Journal of Environmental Management, 254*(October 2019), 109797. https://doi.org/10.1016/j.jenvman.2019.109797.

Sousa, J. C. G., Ribeiro, A. R., Barbosa, M. O., Fernando, M., Pereira, R., & Silva, A. M. T. (2018). A review on environmental monitoring of water organic pollutants identified by EU guidelines. *Journal of Hazardous Materials, 344*, 146–162. https://doi.org/10.1016/j.jhazmat.2017.09.058.

Sun, K., Shi, Y., Wei, X., Potter, N., Li, Z., & Zhu, J. (2017). Modification of clays and zeolites by ionic liquids for the uptake of chloramphenicol from water. *Chemical Engineering Journal, 313*, 336–344. https://doi.org/10.1016/j.cej.2016.12.083.

Taheran, M., Brar, S. K., Verma, M., Surampalli, R. Y., Zhang, T. C., & Valero, J. R. (2016). Membrane processes for removal of pharmaceutically active compounds (PhACs) from water and wastewaters. *Science of the Total Environment, 547*, 60–77. https://doi.org/10.1016/j.scitotenv.2015.12.139.

Talib, A., & Randhir, T. O. (2016). Managing emerging contaminants: Status, impacts, and watershed-wide strategies. *Exposure and Health*, *8*(1), 143–158. https://doi.org/10.1007/s12403-015-0192-4.

Tambosi, J. L., Yamanaka, L. Y., José, H. J., De Fátima, R., Moreira, P. M., & Schröder, H. F. (2010). Recent research data on the removal of pharmaceuticals from sewage treatment plants (STP). *Quimica Nova*, *33*(2), 411–420. https://doi.org/10.1590/S0100-40422010000200032.

Tzereme, A., Christodoulou, E., Kyzas, G., Kostoglou, M., Bikiaris, D., & Lambropoulou, D. (2019). Chitosan grafted adsorbents for diclofenac pharmaceutical compound removal from single-component aqueous solutions and mixtures. *Polymers*, *11*(3), 497. https://doi.org/10.3390/polym11030497.

Wang, N., Xiao, W., Niu, B., Duan, W., Zhou, L., & Zheng, Y. (2019). Highly efficient adsorption of fluoroquinolone antibiotics using chitosan derived granular hydrogel with 3D structure. *Journal of Molecular Liquids*, *281*, 307–314. https://doi.org/10.1016/j.molliq.2019.02.061.

Wei, X., Zhang, Q., Cao, S., Xufeng, X., Yi Chen, L., Liu, R. Y., et al. (2021). Removal of pharmaceuticals and personal care products (PPCPs) and environmental estrogens (EEs) from water using positively charged hollow fiber nanofiltration membrane. *Environmental Science and Pollution Research*, *28*(7), 8486–8497. https://doi.org/10.1007/s11356-020-11103-5.

Weng, X., Lin, S., Zhong, Y., & Chen, Z. (2013). Chitosan stabilized bimetallic Fe/Ni nanoparticles used to remove mixed contaminants-amoxicillin and Cd (II) from aqueous solutions. *Chemical Engineering Journal*, *229*, 27–34. https://doi.org/10.1016/j.cej.2013.05.096.

Wu, X., Huang, M., Zhou, T., & Mao, J. (2016). Recognizing removal of norfloxacin by novel magnetic molecular imprinted chitosan/γ-Fe2O3 composites: Selective adsorption mechanisms, practical application and regeneration. *Separation and Purification Technology*, *165*, 92–100. https://doi.org/10.1016/j.seppur.2016.03.041.

Zaied, B. K., Rashid, M., Mohd Nasrullah, A. W., Zularisam, D. P., & Singh, L. (2020). A comprehensive review on contaminants removal from pharmaceutical wastewater by electrocoagulation process. *Science of the Total Environment*, *726*, 138095. https://doi.org/10.1016/j.scitotenv.2020.138095.

Zhang, S., Dong, Y., Yang, Z., Yang, W., Jiaqing, W., & Dong, C. (2016). Adsorption of pharmaceuticals on chitosan-based magnetic composite particles with core-brush topology. *Chemical Engineering Journal*, *304*, 325–334. https://doi.org/10.1016/j.cej.2016.06.087.

Zhang, Y., Shen, Z., Dai, C., & Zhou, X. (2014). Removal of selected pharmaceuticals from aqueous solution using magnetic chitosan: Sorption behavior and mechanism. *Environmental Science and Pollution Research*, *21*(22), 12780–12789. https://doi.org/10.1007/s11356-014-3212-1.

Zhang, Y.-L., Zhang, J., Dai, C.-M., Zhou, X.-F., & Liu, S.-G. (2013). Sorption of carbamazepine from water by magnetic molecularly imprinted polymers based on chitosan-Fe3O4. *Carbohydrate Polymers*, *97*(2), 809–816. https://doi.org/10.1016/j.carbpol.2013.05.072.

Zhou, X., Dong, C., Yang, Z., Tian, Z., Lishi, L., Yang, W., et al. (2018). Enhanced adsorption of pharmaceuticals onto core-brush shaped aromatic rings-functionalized chitosan magnetic composite particles: Effects of structural characteristics of both pharmaceuticals and brushes. *Journal of Cleaner Production*, *172*, 1025–1034. https://doi.org/10.1016/j.jclepro.2017.10.207.

Innovative and eco-friendly technologies for the upgradation of pharmaceutical wastewater treatment processes

13

Abdullahi Haruna Birniwa[a], *Umar Ali*[a], *Shamsul Rahman Mohamed Kutty*[b], *Ahmad Hussaini Jagaba*[b,c], *and Azmatullah Noor*[b]

[a]Department of Chemistry, Sule Lamido University, Kafin-Hausa, Nigeria, [b]Department of Civil and Environmental Engineering, Universiti Teknologi PETRONAS, Bandar Seri Iskandar, Perak Darul Ridzuan, Malaysia, [c]Department of Civil Engineering, Abubakar Tafawa Balewa University, Bauchi, Nigeria

1. Introduction

As an essential resource, the demand for water is growing dramatically because of the huge expansion of the world's population (Al-Mahbashi et al., 2022). Water is a vital ingredient of life (Lawal et al., 2021; Nasara et al., 2021). The significant degradation of wastewater supplies, insufficient treatment before disposal in water bodies, and the appearance of a new set of pollutants considered to be emerging contaminants (ECs), have also contributed to poor water quality (Ji, Liu, Xu, Zheng, & Zhu, 2010). The degradation in water quality as a result of such circumstances has become a global concern that calls for urgent and sustainable remedies to be given immediate attention (Abdullahi, Musa, Habibu, Birniwa, & Mohammad, 2022; Geissen et al., 2015; Jagaba, Kutty, Naushad, et al., 2022).

Globally, water quality issues have been targeted at the level of organic matter, nutrients, fecal pollution, and heavy metals, polymers for which permissible emission limits have been designated by virtually every legislation throughout the world (Abdullahi, Musa, Habibu, Birniwa, & Mohammad, 2021; Birniwa, 2019; Simazaki et al., 2008). The interest in their sources, incidence, and fate in the environment has risen with respect to the newly growing category of "EC" pollutants reported by Daughton and Ternes (1999). Antibiotics are among the most regularly discovered pharmaceutical-based contaminants (tetracycline), along with antiinflammatory products (ibuprofen), antidepressants (carbamezepines), hormones (estradiol), anticonvulsant substances (phenytoin), beta-blockers (propranolol), lipid regulators (fenophric acid), and X-ray contrast agents (Rasheed et al., 2020). Fig. 1 shows the major pathways for release of pharmaceutical wastewater and their constituents into water sources, and its sequence in the aqueous system.

The Treatment of Pharmaceutical Wastewater. https://doi.org/10.1016/B978-0-323-99160-5.00006-0

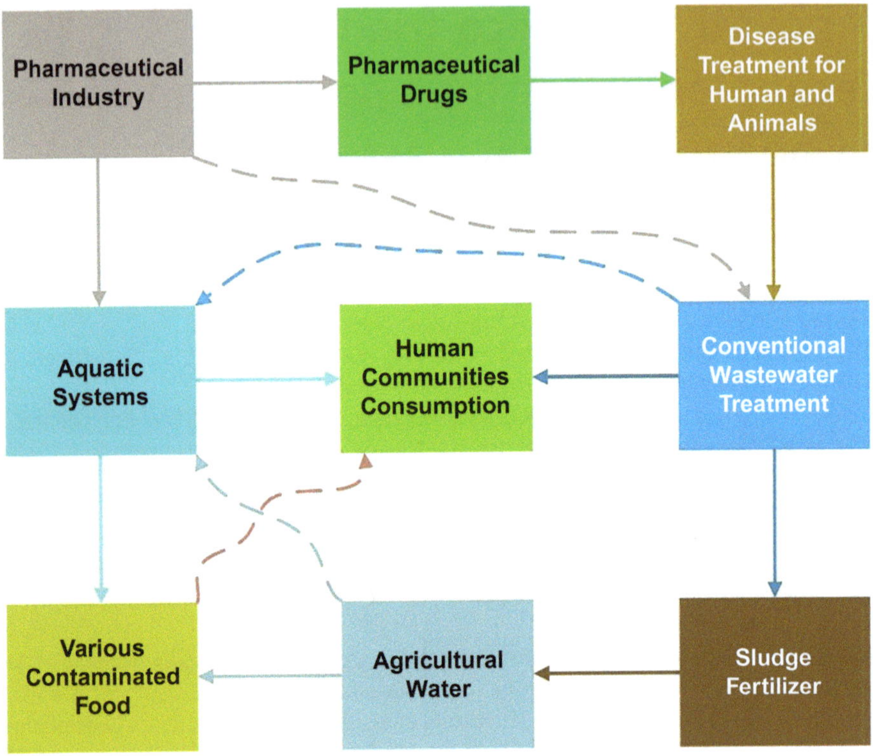

Fig. 1 The major mechanisms for the release of pharmaceutical effluent into water sources, as well as their succession in an aquatic system.

Publications on the subject in the literature show that all personal and pharmaceutical care products (PPCPs) cannot be eliminated successfully in existing traditional wastewater treatment processes (Verlicchi, Al Aukidy, & Zambello, 2012). The activated sludge method, which was initially conceived to eliminate persistent organic and solids wastes to comply with minimal release standards (Tsang, 2015), is commonly used in wastewater treatment plants (WWTPs) for secondary treatment. However, these treatments are inefficient for disposing of PPCPs, which makes WWTPs the main source of release to surface water (Grover, Zhou, Frickers, & Readman, 2011). The presence of veterinary medicines residues in animal feces and subsequent release can leach into groundwater (Archana, Dhodapkar, & Kumar, 2017) and contaminate potable water sources (Ebele, Abou-Elwafa Abdallah, & Harrad, 2017). The discharge of antibiotics in water systems can spread, and hence, create antibiotic-resistant bacteria and genes (Kim, Cho, Kim, Vanderford, & Snyder, 2007). Consequently, efficient, scalable, environmentally friendly, and cost-effective techniques of treatment for removing PPCPs from wastewater need to immediately be developed (Jagaba, Kutty, Isa, Affam, et al., 2022; Jagaba, Kutty, Lawal, Aminu, et al., 2022; Jagaba, Kutty, Noor, Affam, et al., 2022; Jagaba, Kutty, Noor, Isah, et al., 2022).

Pharmaceuticals and personal care products (PPCPs) containing a wide range of bioactive compound products, including antibiotics, antiinflammatories, antiseptics, analgesics, natural and synthetic hormones, are used as, disinfectants, fragrances, lotions, and preservatives, for personal health, hygiene, cosmetics, and veterinary purposes, among other things (Snyder et al., 2007). Pharmaceutical wastes (PWs) in water are also linked to endocrine disorders, chronic illness, allergies, malignancies, antibiotic resistance, imbalances in microbiota, metabolic disorders, etc. (Yu, Peldszus, & Huck, 2008). Meanwhile, a number of physiological processes are impacted in aquatic life, harming fish spawning, decreasing fertilization, and causing bioaccumulation of body tissue pollutants and eutrophication (Abdullahi, Musa, Habibu, Birniwa, & Mohammad, 2022; Rasheed et al., 2020). In order to avoid or limit harmful effects on human health or environmental damage, identification, measurement, control, and disposal of pollutants are crucial issues (Saharan, Chaudhary, Mehta, & Umar, 2014).

The Quintiles IMS Institute forecasts that by 2021, the pharmaceutical industry is expected to reach a value of over US$1485 billion, an increase of US$350–380 billion above the 2016 US$1105 billion (Mathur, Sanyal, & Das, 2021). The main environmental sources for PPCPs include pharmaceutical wastewater, healthcare institutions, and homes. PPCP is resistant to natural degradation in contrast to other regulated pollutants, and thus, accumulates in receiving waters and finally reaches wastewater-processing facilities (WWTPs), where most PPCPs are very little processed or wholly unaffected (Tarpani & Azapagic, 2018). This results in variable concentrations of these pollutants in the range of micrograms per gallon or nanograms per liter in treated effluent (Blair, Nikolaus, Hedman, Klaper, & Grundl, 2015; Kumar, Sarmah, & Padhye, 2019; Vergili, Kaya, Gönder, Boergers, & Tuerk, 2019; Wu, Witter, Spongberg, & Czajkowski, 2009).

Biological procedures that have been utilized for many years and have demonstrated substantial efficiency in organic matter and nutrient removal are among the treatment techniques accessible (Noor, Kutty, Baloo, et al., 2021). The biodegradation of microorganisms involves biological processes such as sequential batch reactor systems, as shown in Fig. 2.

Agricultural activities can contribute to the presence and incidence of PPCPs in the environment. This contribution might be explained by the contribution of animal waste, which can include veterinary waste, used as a fertilizer, present in wastewater or water reused for irrigation, and runoff from fields into surface water (Blair et al., 2015; Wu et al., 2009).

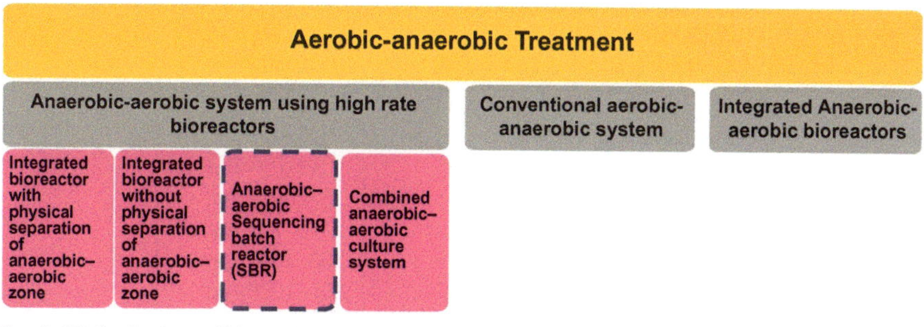

Fig. 2 Biological conditions for PPCPs treatments.

PPCP manufacturers emit quantities of PPCP into the aquatic environment by discharging treated wastewater or by discharging directly into waterbodies. Industries such as pharmaceuticals, cosmetics, brewing, and food processing are the primary contributors of PPCP in wastewater (Subedi, Lee, Moon, & Kannan, 2014; Tran, Li, Hu, & Ong, 2014). However, there is a need for better wastewater treatment methods since it may be highly challenging to detect, handle, and/or eliminate dangerous biological or chemical pollutants in pharmaceutical processes (Cincinelli, Martellini, Coppini, Fibbi, & Katsoyiannis, 2015). These chemicals tend to escape treatment and to accumulate in the environment because to their low volatility properties (Rasheed, Adeel, Nabeel, Bilal, & Iqbal, 2019). These substances generally occur in low quantities (μg/L or ng/L). Moreover, certain traditional water treatment techniques, such as chlorination or ozonation, result in hazardous secondary products (Lee, Wong, Tan, & Lai, 2018). Table 1 shows the various different characteristics of pharmaceutical effluent.

2. Strategies for removal of pharmaceutical products from wastewater

Enhanced treatment technologies established by various researchers include biological processes (algae, bacteria, enzymes) (Jagaba, Kutty, Noor, Birniwa, et al., 2021; Noor, Kutty, Almahbashi, et al., 2021), physical processes (filtration, membrane, sorption), or advanced oxidative processes (AOPs) (oxidizers, photocatalysis, strong sonolysis), and multitreatment. Research in this field includes the development of pharmaceutical products and treatment systems (Hokkanen, Bhatnagar, & Sillanpää, 2016; Silva, Jaria, Otero, Esteves, & Calisto, 2018; Silva, Otero, & Esteves, 2012). The efficiency and cost of these technologies varies, and each has its own benefits and limits (Altmann, Ruhl, Zietzschmann, & Jekel, 2014).

2.1 Removal by adsorption

Adsorption utilizing standard and nonconventional adsorbents, as depicted in Fig. 3, is an advantageous technique utilized extensively in wastewater treatment because it is a physiochemically efficient, quick, and inexpensive procedure (Birniwa, Abdullahi,

Table 1 Typical characteristics of pharmaceutical effluents.

Parameter standard	Effluent values
Biological oxygen demand (BOD)	900–400 ppm
Chemical oxygen demand (COD)	2000–6000 ppm
Acidity (pH)	1.5–6.0 (variable)
Oil and grease	35–2000 ppm
Total dissolved solids (TDS)	1350–7350 ppm
Total suspended solids (TSS)	500–2000 ppm
Total Kjeldahl nitrogen (TKN)	800–1000 ppm

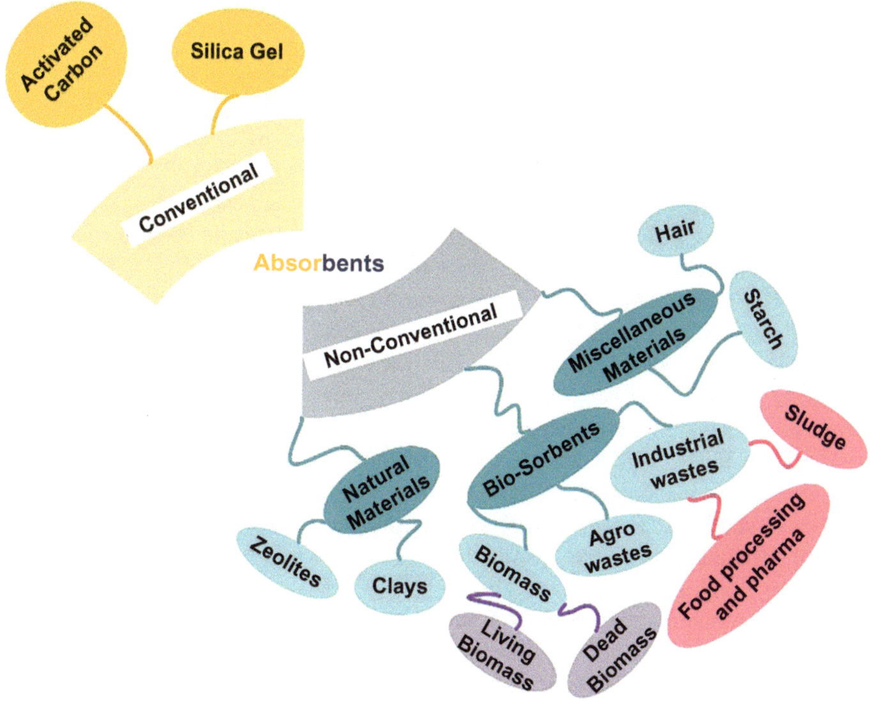

Fig. 3 Types of conventional and nonconventional adsorbents.

Yakasai, & Ismaila, 2021; Jagaba, Kutty, Isa, Affam, et al., 2022; Jagaba, Kutty, Lawal, Aminu, et al., 2022; Jagaba, Kutty, Noor, Affam, et al., 2022; Jagaba, Kutty, Noor, Isah, et al., 2022). The main benefit is the absence of a danger of pollution by hazardous by-products (Baccar, Sarrà, Bouzid, Feki, & Blánquez, 2012). Carbon-based materials with large surface area, porosity, and strong reaction activity (Abdullahi, Birniwa, & Audu, 2020; Abdullahi, Birniwa, Mohammad, Mamman, & Chadi, 2020) are used for this purpose, and are regarded as an efficient and adaptable approach due to their minimal initial investment, operational simplicity, nonselective nature, and lack of creation of hazardous chemicals (Akhtar, Amin, & Shahzad, 2016; Jagaba, Kutty, Baloo, et al., 2022; Rafatullah et al., 2013).

2.2 *Activated carbon*

Activated carbon (AC) is a black, porous, microporous, and mesoporeal or macroporous activated solid. It is frequently utilized as an adsorbent for air and water filtration at industrial scale, thanks to its large surface area. The characteristics of AC vary in relation to the substance employed as a source of carbon, as raw materials resources are typically utilized, such as timber, nut shells, charcoal, and lignite (Jagaba, Kutty, Hayder, Baloo, et al., 2021; Jagaba, Kutty, Hayder, Elsadig, et al., 2021; Jagaba, Kutty, Salih, Noor, et al., 2021). AC is a powerful candidate for adsorbent elimination of PPCPs because of its characteristics. A comparison research to assess the effectiveness of the removal of contaminants by means of AC treatment and ozonation in municipal wastewater reveals that removal efficiency is superior for AC (Margot et al., 2013). Initially, for this purpose, activated carbon (AC) was employed; however, the utilization of carbon nanotubes (CNTs) and graphene-based substances as well as manufactured biochar is the latest methodology. In this area, you can find several materials for the adsorbent elimination of PPCPs (Rivera-Utrilla, Prados-Joya, Sánchez-Polo, Ferro-García, & Bautista-Toledo, 2009).

In several studies, PPCP elimination was observed using powdered AC (PAC) and granular AC (GAC). Activated carbon (AC), which has already demonstrated a capacity to eliminate selected pharmaceuticals (e.g., beza-fibrate, carbamacepines, caffeine, clofibric acid, and diclofenac). is commonly employed as an adsorbent to remove both organic and inorganic contaminants (Coimbra, Calisto, Ferreira, Esteves, & Otero, 2019; Sotelo et al., 2014; Ternes et al., 2002). The preferred characteristics for AC include: (i) high surface area (BET; generally more than $400\,m^2\,g^{-1}$) (Álvarez-Torrellas, Rodríguez, Ovejero, & García, 2016); (ii) ease of market availability; (iii) a unique combination of highly developed pore system and heteroatoma responsiveness to a variety of structural and surface functionalities (Cabrita et al., 2021; Gupta, Dinesh Mohan, Suhas, & Singh, 2006); and (iv) high removal efficiency with less generating of harmful by-products (Nielsen, Biggs, Skinner, & Bandosz, 2021).

AC can be manufactured out of a number of raw materials (coal being the most often utilized) that have qualities such as high carbon content and low inorganic content, as well as negligible aging degradation (Dias, Alvim-Ferraz, Almeida, Rivera-Utrilla, & Sánchez-Polo, 2007). Two primary activation techniques, physical and chemical activation, are often used for the manufacture of ACs. One of the characteristics of physical activation is that additives/contaminants from the activating agents are not included, whereas chemical activation requires lower temperatures and less activation time. Both microporous carbon and ultramicroporous carbon cause physical activation, while chemical activation carbon often has larger and greater pore volumes than microporous carbon (Almahbashi et al., 2021). The major disadvantages of ACs derived from traditional precursors are their prolific employment in the adsorption method, despite their expensive production and regeneration costs, which restricts their large-scale or industrial implementation (Alsbaiee et al., 2016; Baloo et al., 2021; Mestrea, Piresa, Nogueiraa, Parra, & Carvalhoa, 2021). These factors have driven the quest for environmentally acceptable and inexpensive materials comprising waste treatments and/or biosorbents in recent years. Table 2 shows removal of different contaminants obtained from PW by activated carbon.

2.3 Agricultural waste

Solid farm waste is an easy-to-use, replaceable, and cheap resource, which can be utilized for removal of PPCPs (Abdullahi, Birniwa, & Audu, 2020; Abdullahi, Birniwa, Mohammad, et al., 2020). Lignin, cellulose, lipids, hemicellulose, starch, proteins, and simple sugars are major constituents of agricultural waste (Jagaba, Kutty, Hayder, Baloo, et al., 2021; Jagaba, Kutty, Hayder, Elsadig, et al., 2021; Jagaba, Kutty, Salih, Noor, et al., 2021). Because of the existence of several functional groups, they can remove a wide variety of pollutants through the production of chemical

Table 2 Removal of pharmaceuticals by activated carbon.

Contaminant	Adsorbent used	Initial concentration of the contaminant	Source of water sample	References
Antibiotics				
Nitroimidazole	AC a & b	50 mg/L	Aqueous solution	Rivera-Utrilla et al. (2009)
Sulfamethoxazole	PAC	600 ng/L	WWTPs effluents	Altmann et al. (2014)
	PAC	100 ng/L	Synthetic water	
Tylosin	PAC	0.13 mmol/L	Synthetic water	Ji et al. (2010)
Tetracycline	PAC	0.19 mmol/L	Synthetic water	Ji et al. (2010)
Anticancer				
Tamoxifen	GAC	1 mg/L	Sewage effluents	Grover et al. (2011)
Anticonvulsants				
Dilantin	GAC	1 ng/L	Surface water	Kim et al. (2007)
Carbamazepine	PAC	100 ng/L	Surface water	Snyder et al. (2007)
	GAC	500 ng/L	Synthetic water	Yu et al. (2008)
	PAC	2.5 μg/L	WWTPs effluents	Altmann et al. (2014)
Primidone	PAC	900 ng/L	WWTPs effluents	Altmann et al. (2014)
Antidiarrheal				
Lopromide	GAC	1 ng/L	Surface water	Kim et al. (2007)
Antidepressant				
Diazepam	PAC	100 ng/L	Surface water	Snyder et al. (2007)
Thioridazine	GAC	1 mg/L	Sewage effluents	Grover et al. (2011)
Antihypertensive				
Propranolol	GAC	1 mg/L	Sewage effluents	Grover et al. (2011)

Continued

Table 2 Continued

Contaminant	Adsorbent used	Initial concentration of the contaminant	Source of water sample	References
Antiinflammatory nonsteroidal drugs				
Diclofenac	PAC	100 ng/L	Surface water	Snyder et al. (2007)
	PAC	5.8 μg/L	WWTPs effluents	Snyder et al. (2007)
	PAC	100 ng/L	Synthetic water	Yu et al. (2008)
Paracetamol	PAC	100 ng/L	Surface water	Snyder et al. (2007)
	PAC	100 ng/L	Synthetic water	Yu et al. (2008)
Ketoprofen	ACc	19.28 mg/L	Surface water	Baccar et al. (2012)
Fenoprofen	PAC	0.1 mg/L	Synthetic water	Simazaki et al. (2008)
Naproxen	PAC	100 ng/L	Surface water	Snyder et al. (2007)
	GAC	500 ng/L	Synthetic water	Yu et al. (2008)
	PAC	100 ng/L	Synthetic water	Rivera-Utrilla et al. (2009)
Antiinfective				
Thimerosal (organic mercury compound)	GAC	0.2 μg/L	Pharmaceutical wastewater	Simazaki et al. (2008)
Hormone				
Estradiol	PAC	100 ng/L	Surface water	Snyder et al. (2007)
Estriol	GAC	5 ng/L	Surface water	Kim et al. (2007)
Lipid regulator				
Bezafibrate	PAC	1.3 μg/L	WWTPs effluents	Altmann et al. (2014)
Gemfibrozil	PAC	100 ng/L	Surface water	Snyder et al. (2007)

complexes. Rice and wheat husks provide a high level of commercially viable adsorbent production (Jagaba, Kutty, Isa, Affam, et al., 2022; Jagaba, Kutty, Lawal, Aminu, et al., 2022; Jagaba, Kutty, Noor, Affam, et al., 2022; Jagaba, Kutty, Noor, Isah, et al., 2022). They are commonly available as agricultural byproducts and are thus a cost-effective option. Tea bark, a byproduct of the wood business, has a large quantity of tannins and can be an adsorbent, as tannin-related hydroxyl groups are the active ingredient for molecular adsorption (Birniwa & Abdullahi, 2019). In one investigation, 92.4% sugar cane bagasse was utilized for the production of AC for diclofenac alternate sodium removal. The material has been shown to be quickly adsorbed and binding to diclofenac, and to be recyclable with acetone washing (Birniwa et al., 2022; Ghaleb et al., 2020).

2.4 Water-contaminated waste-based adsorbents with pharmaceutical

In general, waste utilized in the context of adsorption by pharmaceutical products as adsorbent precursors is of one of two kinds: farm or industrial. This categorization is thus utilized in this section. For agricultural waste, as this is readily accessible and widely available, the manufacturing carbons cost made from these wastes is estimated to be cheap, making it possible for adsorbents generated from nonrenewable sources to be commercially available (Ioannidou & Zabaniotou, 2007). Moreover, due to high hardness and low ash content, this sort of residue is regarded as a favorable precursor to adsorbent manufacturing (Dias et al., 2007). Better adsorbents made from agricultural waste (BET up to 1260 $m^2\ g^{-1}$) are also discussed in terms of BETs (BET up to 1260 $m^2\ g^{-1}$) when compared to other abundant waste items like used tires.

With regard to urban and industrial waste, their build-up is the outcome of contemporary lifestyles and of industrial operations, which have gradually developed a pattern of huge production, consumption, and disposal (Birniwa, Abubakar, Huq, & Mahmud, 2021). For example, the paper and pulp industry in Europe is a well-developed, profitable business that contributes greatly to economic progress. However, it is also a hugely demanding energy and water sector, as well as one of the world's greatest sources of industrial effluents (Coimbra et al., 2019). As regards the huge quantity of water it needs, significant quantities of wastewater are created, and sludge formation is always the outcome of treatment. However, this sludge can be reused to form biomass-based hydrogels for adsorption, as shown in Fig. 4.

By 2050, it is expected that there will be 550 million tons of paper produced worldwide from the existing 150 million tons, that is increase by 400 million tons. This rise may mean increases of 48%–86% in the output of pulp and paper mill sludge (Faubert, Barnabé, Bouchard, Côté, & Villeneuve, 2016). This sector is clearly facing major problems in the field of sludge management, which is a perfect candidate for a residue valorization policy. Antibiotic adsorption patterns and that of nonsteroidal antiinflammatory drugs (NSAIDs) are being studied by researchers. Other medication kinds were absorbed more often (e.g., analgesics, anticonvulsants, antidepressants).

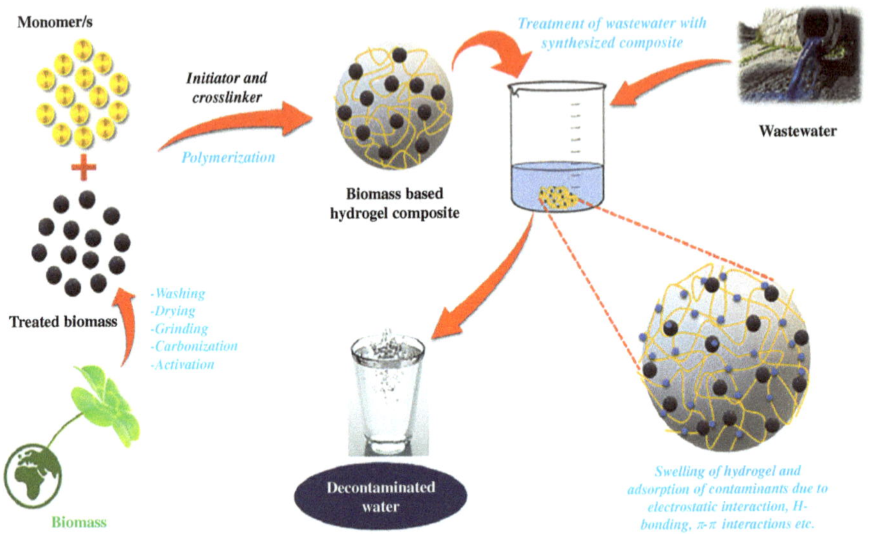

Fig. 4 Synthesis and application of biomass-based hydrogels for water treatment (Yadav, Bagotia, Sharma, & Kumar, 2021).

2.5 Adsorption of antibiotics

The adsorption of antibiotic products on different adsorbents is one of the most widely researched pharmaceutical fields (Mamman, Mohammad, Abdullahi, Chadi, & Birniwa, 2020). This is because they are regarded as highly dangerous chemicals that may have impacts on nontarget species, including bacterial antibiotic resistance, due to their environmental presence (Zheng, Wang, Zhao, Herbert, & Xing, 2013). Studies are also readily found in the antibiotic class, assessing the effect of metal ions and pH on the adsorption pattern of adsorbents derived from agricultural/vegetable waste. This is due to the fact that adsorbent molecular donor aptitude (and hydrophobicity), depending on pH, can be anionic, cationic, or neutral, i.e., (i) pH changes sampling and, therefore, the functional interaction between the adsorbate and the metal ions present in the aqueous solution and (ii) sulfonamides contain various functional groups) (Liu et al., 2011; Tan, Liu, et al., 2015; Xie, Chen, Xu, Zheng, & Zhu, 2014).

Zhou et al. studied the capacity of a wasted mushroom substrate to extract sulfamethyl antibiotics from the aqueous solution without any heat or chemical treatment (Zhou, Zhang, Li, Su, & Zhang, 2016). Sulfathiazole > sulfamethoxazol > sulfamethyldiazine > sulfamethhazine, which was linked to the variations in chemicals structure and functionality of the antibiotics, have been shown to reduce the adsorption capability. The effect of the initial adsorbent concentration and the influence of pH was assessed by Rajapaksha et al. (2014), the adsorption capacity rose by 3–11 with the initial antibiotic concentration and was decreased by pH increases. The dependency on the pH for sulfamethazine adsorption in pyrolyzed and activated tea waste was likewise found, but the greatest adsorption under low pH conditions was checked, contrary to that in the prior example.

Sulfamethoxazole adsorption on agricultural/vegetable residues (bamboo, brazilwood for peppers, bagasse sugarcane, and hickory wood) was also investigated for the purpose of creating a new adsorbent that adsorbs drug contaminants in water (both ground- and soil-saving) when used in soils irrigated with reclaimed water (Yao et al., 2012).

Adsorbents that perform better were used in column trials, bagasse, and bamboo sugarcane: around 60% of sulfamethoxazole was transferred from recalled water to soils unchanged with adsorbents and just 2%–14% was detected in adsorbent-amended soils. In studies on tetracycline, it was revealed that sodium hydroxide-activated macadamia shells-adsorbents exhibited better adsorptive ability compared to poly-walled carbon nanotubes in tetracycline (Kong et al., 2016; Martins et al., 2015). For macadamia shell adsorbents, a BET of $1524\,m^2\,g^{-1}$ and the adsorption capacity of 455 mg/g were observed (in comparison with $1839\,m^2\,g^{-1}$ and 309 mg/g for BET and adsorption capacity, respectively, for carbon nanotubes).

Although tetracycline was still considered, Jing, Wang, Liu, Wang, and Jiang (2014) conducted an investigation on the modification of a rice husk-based methanol adsorbent. Fixed-bed tests have been carried out in order to evaluate this adsorbent potential for practical application. A concentration of less than 0.1 mg/L was measured in the effluent of the fixed-bed experiment with an influential solution of 5.0 mg/L tetracycline. In the treatment of the precursor, however, the use of methanol may be environmentally problematic. = Used precursors for tetracycline adsorption, rice husk, and fishing stones, which are connected to interactions and the generation of H, –COOH, and C=O bonds, to achieve high removal rates (Álvarez-Torrellas et al., 2016). The adsorbed fish stone adsorbent, in terms of BET ($1521\,m^2\,g^{-1}$ compared to $1102\,m^2\,g^{-1}$ for GAC), demonstrated its usefulness as an alternative adsorbent compared with the commercial activated granular activated carbon (GAC) adsorbent. Torres-Pérez, Gérente, and Andrès (2012) has transformed beet pulp and peanut hulls in agricultural wastes into adsorbents and its tetracycline adsorption capability has been examined. Adsorption results from genuine spring waters, in contrast to the vast majority of research, showed that the maximum adsorption capacity was somewhat lower than that of the synthetic matrix. However, when a commercial GAC was used, a further reduction (one-third) occurred.

2.6 Nonsteroidal antiinflammatory drugs (NSAIDs)

NSAIDs are among the most prescribed drugs in the world and are known to cause harmful consequences on invertebrates and algae in the environment once released into it (Baccar et al., 2012). Analyzed the interaction energies between NSAID (ibuprofen, naproxen, and diclofenac) and loblolly pine chip adsorbent while taking into account the adsorbent and adsorbent confirmative and chemical features. Molecular modeling research aims not only to evaluate adsorption capacity but also the adsorbent's contribution to adsorption capacity via surface function groups. Two kinds of adsorbents (N-bio-adsorbent) and (O-bio-adsorbent) have been developed.

Certain hydrophobic interactions are shown to be prominent in the adsorption of NSAIDs. The adsorbents exhibited variations in adsorption capabilities for the pyrolytic-based carbonaceous structure (290–372 mg/g, for N-adsorbents and 214–286 mg/g, for O-adsorbents). The authors also examined many adsorption systems (diclofenac/naproxen, naproxen/ibuprofen, and ibuprofen/diclofenac), a topic that has been seldom evaluated for literature, as an added benefit of this work. Ibuprofen adsorption reduced with dichlofenac (58%–62%); naproxen adsorbing decreased with ibuprofen and dichlofenac correspondingly 27%–35% and 27%–38%, respectively. The adsorption of both diclofenac and naproxen was similar in the multicomponent system but somewhat lower than the binary system, whereas ibuprofen experienced a notably reduced adsorption, linked to its decreased molecular size at the adsorption sites. Some findings of diclofenac, naproxen, ibuprofen, and ketoproven adsorption patterns on an olive waste preparation adsorbent were associated with pK_a, kow, and pharmaceutical compounds structural characteristics (Silva et al., 2018). For example, diclofenac is more adsorbed than naproxen and, because the pK_a is similar, this has to do with its hydrophobicity; ibuprofen has poor pH 4.1 (contrary to what is expected, because of the lowest repulsive electrostatic interaction between the deprotonated form in this pH value and that of the adsorbing agent which has negative charges) due to its chemical structure. Ketoprofen has a greater attraction to adsorbent compared to other chemicals probably because of the ketone and carboxylate groups. This can cluster through interaction hydrophobicity, making access to the adsorbent harder.

In comparison with the adsorption of diclofenac or ibuprofen in commercially produced powdered AC, Jung et al. (2021) utilized a loblolly pinewood precursor to generate N-bioand O-bio-adsorbents (PAC). The adsorbent generated was both greater than the PAC used for comparison and provided with a significantly reduced ash content. It was also good in the presence of organic materials, and the advertising saturation was slightly lower. This contradicts the common belief that the presence of organic materials might diminish adsurge by blocking pores or contesting accessible pores. When compared to a single commercial sample, the adsorption of ibuprofen by sisal-based activated carbons was proven to be efficient and effective; however, when diclofenac was used, Isabel grape bagasse was not efficiently handled (Antunes et al., 2012), the authors found that pH 2, considering the use of adsorber in an actual wastewater system, was the optimal value for adsorption, which was quite unrealizing.

2.7 Microwave-assisted adsorption of pharmaceutical waste

Adsorption of PW from wastewater also involves microwave-assisted NSAID adsorption using agricultural waste and vegetables (Reza, Ahmaruzzaman, Sil, & Gupta, 2014). Reza and coworkers have examined ibuprofen adsorption by microwave-assisted activation using adsorbents made from bamboo waste. Maximum adsorption was achieved at pH 2–5, which might be linked with the ionic ibuprofen condition in this pH range, as well as adsorbent surface functional groups. In comparison with those adsorbents, bamboo-derived material was identified as an effective means to remove ibuprofen from the water phase. For adsorption of diclofenac and nimesulide, an adsorbent generated from coca shell using microwave assistance was used.

Simulative effluents from hospital, containing diclofenac (pH 8), were used as a means for assessing the performance of the adsorbent, which was demonstrated to be able to remove nimesulide and diclofenac efficiently regardless of the complexity of the matrix (Saucier et al., 2015). Radiation techniques for microwave heating enable volumetric heating and may help minimize the cost of preparing AC. Microwave-assisted activation with K_2CO_3 and KOH has already attracted attention for the adsorption of metronidazole on adsorbents made from siris seed shells. Ciprofloxacin adsorbent behavior, using the precursors *Cyperus alternifolius*, and that of norfloxacin with lotus stalk-based adsorbents was further elucidated by Ahmed and Theydan (2014). These authors revealed the possibility for microwave activation to produce an adsorbent with good surface characteristics.

2.8 Other sources of adsorbent for pharmaceutical and personal care products

Other adsorbing sources include MWCNT carbon nanotubes adsorbing diclofenac and triclosan, which showed an initial concentration of 50 mg/L, a maximum adsorption rate of 19.9 mg/g, and a maximal adsorption capacity of 19.7 mg/g, respectively, in a volume of 350 $m^2\ g^{-1}$ of the adsorbent (Hu, Cheng, Sun, & Zhu, 2017). Graphene and graphene oxide have also been utilized, and research has also been undertaken by Kyzas, Koltsakidou, Nanaki, Bikiaris, and Lambropoulou (2015), showing GOs were adsorbed to the initial 1 mg/g concentration and to adsorptions to 67 and 116 mg/g correspondingly [Removal of beta-blockers from aqueous]. In this context, graphene and graphene oxide are employed in a research, the effectiveness of Fe_3O_4-activated coconut shell biochar for the removal of carbamazepine and tetracycline was examined. Biochar is a stable substance that results from plant biomass pyrolysis or carbonization. An initial concentration of 200 mg/g and a total surface area of 365 $m^2\ g^{-1}$ provided a maximum adsorption capacity of 94.2 mg/g. The biochar was produced by ball milling and was claimed to be magnetically separable.

3. Bioremediation

The bioremediation of PPCPs is one of the most significant environmental treatment techniques. It has an advantage compared to other methods, since the running costs are minimal and the operating conditions are moderate. It is a technique that uses microorganic toxins such as bacteria and fungus to remove pollutants from the environment in terms of their metabolic activities. Mineralized contamination can be completely removed, degraded to a reduced molecular weight, or the contaminated molecule can be chemically modified. Microorganisms are not well-versed in the transformation of pollutants (Petrie, Barden, & Kasprzyk-Hordern, 2015). Extracellular enzymes that sporadically break down water-soluble pollutants to high pH levels, their features that make them a candidate for their potential candidate include the specific nature of their enzymes to help pollutant degrade, filament hyphal growth that allows the potential for pollutants to access and colonize at higher rates, and so on (Asif et al., 2021). This can be metabolic or co-metabolistic. The main substrate activates catabolic enzymes in the cometabolism process. The compound is not the major energy supply in this situation.

Targeted therapy can be highly promising for approaches employing isolated enzymes. The metabolic process is catalyzed by enzymes in microbial transformation. There has been considerable interest in applications employing oxidoreductase enzymes (lignase, laccase, peroxidase, tyrosinase) (Naghdi et al., 2018). Because of high specific reaction levels, free enzymes are used for a more controlled technique. The structure of the substrate surface, enzyme characteristics, and other system requirements such as pH and temperature are the determining factors behind these enzyme reactions. It is necessary, however, to create diverse and easy screening methods and to identify particular enzyme requirements for simple process control (Jagaba, Kutty, Abubakar, et al., 2022).

While enzymes offer an enormous potentiality for many biochemical processes, their use at various temperatures and pH levels are limited because of their unstable nature (Iyer & Ananthanarayan, 2008). The immobilization of enzymes is a frequently utilized technique for overcoming this issue. This enables the stability and endurance of enzymes to be improved, making possible many applications in industry. It also makes the repetitive use of enzymes cost-effective. Various studies have shown that immobilization can enhance stability and enzyme activity. A straightforward technique is physical adsorption on the carrier. The immobilization of nano-charcoal laccase for carbamazepine removal was demonstrated by Naghdi et al. (2017). The results showed that the immobilized laccase was better stored on functional nano-charcoal than on free-laccases, with pH and temperature stability. The laccases maintained 70% of the original activity after 3 cycles as a result of immobilization. Carbamazepine showed an 83% and 86% elimination of immobilized degradation laccases in spiking water and secondary effluent, respectively. The covalent effectiveness of immobilized enzymes in the elimination of PPCPs has also been demonstrated. Diclofenac removal from polyvinyl alcohol/chitosan nanofibers by laccases immobilized by covalent bonding was reported by Xu, Tang, Zhou, Li, and Zhang (2015). They stated that diclofenac had been completely removed within 6h. In addition, interconnection is an immobilization approach that may be utilized to improve stability.

4. Membranes separation

Processes in the field of removal of wastewater are gradually applied using membranes, such as microfiltration (MF), nanofiltration (NF), ultrafiltration (UF), and reverse osmosis (RO) methods. Since the membranes cannot remove all PPCPs, MF and UF can only be utilized during pretreatment. NF/RO has been the most efficient for removing PPCPs from wastewater, since for most PPCPs the size and weight match the separating range of NF. The NF membrane's porosity is generally between 1 and 2nm and the molecular permeability is about 200–1000Da, the membrane surface is negative charged and the retention rate of single ions is less than 90%;

electrospun nanofibrous membrane of the retention rate is greater than 90%. Two NF membranes were produced in research by Rana et al. (2012) for ibuprofen reduction, one with charged macromolecular additions altering the surface and the other with a hydrophobic additive. Due to the load repellence, the NF's performance with a loaded additive was superior to the ibuprofen-acidic hydrophobic additive. In addition, we have researched the effectiveness of the NF membrane in flow filtration systems to remove the PPCPs diclofenac, carbamazepine, and ibuprofen from a drinking water plant, which have been poured into water. As a consequence, raw water and raw water spiked with various amounts of PhACs both showed a quick drop in flux. Raw water showed flux losses between 25.5% and 27%, but raw water spiked with DIC showed considerable flux decreases between 37.2% and 40.8%. In actuality, a change in water flow owing to an increase in PhACs in raw water was not anticipated due to the extremely low PhAC concentrations (0.0250.1 mg L^{-1}) (Vergili, 2013).

4.1 Membrane bioreactor technology (MBR)

In the wastewater processing sector, bioreactor membrane technology is commonly employed for water production that may be recycled. MBR is a process that blends aerobiological biological activities with separation by membranes. MBR treatment generates a readily disinfected effluent. The procedure is dependent on the pressure and the membrane's pores for the filtering of wastewater components. In wastewater treatment, submerged MBR is the most common membrane design employed. In this instance, through pretreatment is needed to prevent membrane obstruction. The determining parameters include hydrodynamic conditions, membrane type, flow rate, and module arrangement. MBRs incorporate membrane separation into the aerobic biological processes. They are now an option for wastewater treatment by conventional activated sludge (CAS) methods (Jagaba, Kutty, Isa, Ghaleb, et al., 2022).

A total of 99 PPCPs were investigated in research by Kim et al. (2014) in influent and MBR effluent applications, removals of PPCP ranged from 34% to >99%, the 23 PPCPs were removed by 90%. Osmotic membrane bioreactor (OMBR) is an emerging technology with great potentials for wastewater treatment and reclamation (OMBR) offers better advantages than standard MBR in terms of water quality and energy use. In order to eliminate persistent hydrophilic pollutants, which can no longer be removed by standard MBR, osmotic MBR is very beneficial. In this situation, the biodegradation contact time rises (Tan, Qiu, & Ting, 2015). A recent study has been carried out in municipal wastewater for the application of RO and NF membrane treatment for the removal of PPCPs, developed and tested on 27 PPCPs. Trimethoprim had the lowest concentration of caffeine (18.4 ng/L). The maximum concentration was found out. The findings showed that PPCPs may be successfully removed by MBR systems at rates ranging from 41.08% to 95.41%, while greater removal rates of about 95% were attained when PPCPs were added to NF membrane systems, this is due to the different elimination techniques, due to the variations in techniques for elimination.

4.2 Enzymatic membrane reactors (EMRs)

Recent research has been conducted on EMRs to remove PPCPs from wastewater, EMRs may be utilized in industry since they use immobilized enzymes, which, in light of the biocatalyst's reuse, are vital, cost-effective and allow continuous processing. The membrane functions in this situation as a selective obstacle and also as an aid to enzyme immobilization. An EMR with covalently grafted laccases on ceramic membranes has been utilized in a recent investigation on tetracycline degradation. Grafted laccases were more effective than free enzymes. The enzyme process in the reactor has been shown for over 200h. During this period, no catalytic activity was lost (de Cazes, Abejón, Belleville, & Sanchez-Marcano, 2014).

5. Removal of pharmaceuticals in advanced oxidation processes (AOPs)

Conventional water treatment systems do not easily breakdown pharmaceuticals in wastewater. The development of AOPs has offered a viable alternative for the complete breakdown of these chemicals (Jagaba, Kutty, Isa, Affam, et al., 2022; Jagaba, Kutty, Lawal, Aminu, et al., 2022; Jagaba, Kutty, Noor, Affam, et al., 2022; Jagaba, Kutty, Noor, Isah, et al., 2022). The strength of AOP's often relies on hydrogen peroxide (H_2O_2), intermediate high-reactivity oxygen or free-radical species (O_3), and auxiliary power supplies (i.e., ultraviolet-visible (UV-vis) radiation, g-radiation electronic current, and ultrasound) in processes that lead to degraded by-product and ultimately nontoxic finished products that target contaminants. AOPs include many methods like ozonation, Fenton processes, electrochemical oxidation, and combinations with some form of photocatalysis (e.g., photo-Fenton processes and electro-Fenton). The merits and demerits of AOPs are highlighted in Table 3. AOPs may comprise three categories of process: photochemical, hybrid or mixed, and non-photochemical. The approach used in the production of hydroxyl radicals, these processes can also be classified as homogenous processes, this homogenous process takes place in a single phase, generating ozone, hydroxyl, and also other oxidizing radicals. In contrast, the use of solids as catalysts is characteristic of the second method, for example, carbon-supported metals or semiconductors such as ZnO or TiO_2, AOPs provide the most promise in this regard have been identified for pharmaceutical degradation.

The most often used forms of AOP are the homogeneous types:

- Ozone-based, containing (O_3) processes, such as ozone alone, ozone with ultraviolet (O_3+UV), ozone with hydrogen peroxide ($O_3+H_2O_2$), and ozone with ultraviolet with hydrogen peroxide (O_3+UV+H_2O_2).
- Oxidation of wet peroxides and application of H_2O_2 as oxidant.
- Fenton-based processes, which include Fenton hydrogen peroxide and iron (ii), Fenton-like hydrogen peroxide and iron (iii), sono-Fenton (ultrasound with hydrogen peroxide and iron (ii), photo-Fenton (ultraviolet with hydrogen peroxide and iron (ii), sono-electro-Fenton, sono-photo-Fenton, electro-Fenton, and photo-electro-Fenton.

Table 3 AOPs—benefits and drawbacks.

AOP	Merits	Demerits
Ozonation		
O_3	❖ Ozone has a high oxidation rate and a rapid response time ❖ Ozone has the ability to eliminate a wide range of organic and inorganic contaminants ❖ There is no added chemical in the water	❖ Higher installation and maintenance costs ❖ High energy use ❖ Requires pH adjustment, which is impractical ❖ Potential fire dangers and health concerns related with ozone production
O_3-H_2O_2	❖ Extremely efficient ❖ More powerful than O_3 or UV alone ❖ More efficient than H_2O_2-UV at producing OH at the same oxidant concentration ❖ Disinfection ❖ Developed technology for remediation	❖ Bromate production is possible, but it requires a lot of energy and money ❖ Because of the possibility for microbial development, excess H_2O_2 may need to be treated ❖ Controlling the right quantity of O_3-H_2O_2 concentration is required ❖ Low pH is harmful to the processes
O_3-UV	❖ The residual oxidant will decompose quickly ❖ Ozone absorbs more ultraviolet light than H_2O_2 (more efficient than H_2O_2-UV) ❖ UV radiation acts as a disinfectant	❖ Expensive in terms of energy and money ❖ UV irradiation requires a custom reactor ❖ The technique removes volatile chemicals ❖ Turbidity has an impact on UV light penetration ❖ Mass transfer might be limited due to ozone diffusion ❖ When combined with chlorine disinfection, there is a potential rise in the generation of trihalomethanes (THM) and haloaceticacids (HAA)
Photolysis H_2O_2-UV	❖ H_2O_2 is relatively stable and may be kept for extended periods of time ❖ The presence of a full-scale drinking water treatment facility	❖ Turbidity has an impact on UV light passage ❖ UV lighting necessitates the use of specialized reactors ❖ H_2O_2 residue must be treated
Photocatalysis semiconductor-UV	❖ It is activated by near UV light (300–380 nm), as opposed to other UV oxidation processes that need a shorter wavelength	❖ Catalyst nonfunctionality is possible ❖ When employed as a slurry, TiO_2 must be recovered

Continued

Table 3 Continued

AOP	Merits	Demerits
Fenton	❖ Some ground waters contain enough iron to cause Fenton's reaction ❖ Does not require as much energy as O_3, UV ❖ There are commercial procedures available	❖ Necessitates maintaining a low pH (2.5) as needed to prevent iron precipitation ❖ Iron extraction system is required ❖ The operational expenses of pH adjustment
Electrochemical	❖ Extremely efficient ❖ More efficient than the alternative oxidation method ❖ No light radiation is necessary	❖ Expensive in terms of energy and money ❖ Requires sludge management in conjunction with electro-coagulant and indirect oxidation

Some examples of heterogeneous processes are:

- Heterogeneous photocatalysis in a photocatalytic semiconductor when UV and/or visible light is present.
- The catalytic oxidation of wet peroxide (CWPO) using heterogeneous catalysts at temperatures of between 323 and 353 K.
- The heterogeneous operating of catalytic ozonation (COz), which promotes highly reactive generation of species and hence leads to greater rates of mineralization.

In short, hybrid AOPs in various combinations have been explored.

5.1 Electrochemical treatment

Electrochemical methods of oxidation are novel techniques to reduce pollution from medicines by producing electrically reactive species that do not require chemical products. Loos et al. (2018) studied the effects on synthetic wastewater and hospital wastewater of electrical degradation with boron-doped diamond (BDD) electrodes in four pharmaceutical molecules: iopromide, sulfamethoxazole, 17-alpha ethinylestradiol, and diclofenac. They studied the impacts on electrochemical cells of the flow rate, current, and starting compound levels and wastewater matrix. This study indicated that BDD electrochemical oxidation might be a viable method for eliminating drug components from the effluent of a hospital wastewater treatment plant. Garcia and coworkers tested carbamazepine degradation in three different water media: demineralized water, tap water, and treated wastewater. In demineralized water and municipal wastewater, 14 mM NaCl Nb/BDD-electro-oxidation performed well (García-Espinoza, Gortáres-Moroyoqui, & Orta-Ledesma, 2016).

5.2 Ozonation

Ozonation and the combinations of ozone with ultraviolet (O_3/UV) and ozone with hydrogen peroxide (O_3/H_2O_2) are highly effective at drug removal and have thus been adopted and utilized in preference to other water treatments, such as the oxidation technique, preoxidation, and/or disinfection. Research conducted by Zhao et al. (2017) indicated that four separate doses of ozone were used for 7 min for high levels of the antiinflammatory drug indomethacin with 2–35 g/L, in 30 min at the maximum ozone concentration (35 mg/L), 50% of total organic carbon (TOC) had nonetheless been removed. Alharbi, Price, Kang, Fujioka, and Nghiem (2016) showed the successful treatment of ozone degradation at different levels of ozone in four distinct APIs, including dilofenac, carbamazepine, sulfamethoxazole, and trimethoprim. A study also examined the possibility for ozonation and ozone and H_2O_2 for the elimination at various temperatures and pH values of two persistent medicinal products, gemfibrozil and ibuprofen. The results reveal that gemfibrozil was fully destroyed when an ozone-dosage of 1.5 mg/L with an ozone:H_2O_2 ratio of 1:0.25 was administered, with a maximum ibuprofen elimination of 80%. The temperature effect for gemfibrozil and ibuprofen removal from 25 to 40 μC was also shown to be insignificant in this investigation. Whereas pH has been demonstrated to have a minimal influence on the elimination efficiency of gemfibrozil from 6 to 9, by comparison, the elimination by the ozone at pH 6 of ibuprofen was significantly lower. Similarly, Hu et al. (2021) examined the degradation by ozonation of salicylic acid in samples of water of acidic and basic environments. The scientists discovered that the target chemical was removed significantly and quickly at pH 4 compared to pH 8 and 10. At pH 4 and 1 mg/L of ozone, salicylic acid clearance rate reached 95% while the removal rate was around 10%–15% of salicylic acid at pH 7–10.

5.3 Fenton and photo-Fenton

Fenton and photo-Fenton ultrasonic methods are AOPs that have become more prominent, and both approaches are efficient ways to degrade a wide range of pharmaceuticals. Fenton is process based on iron salts (Fe^{2+}/H_2O_2/dark), which leads to the oxidation of iron salts (Fe^{2+}) and H_2O_2, and to the production of hydroxyl radicals (HO). Fenton is likewise based upon the high capability of oxidation of OH radicals following ozone oxidation. Fenton-like processes, in this respect, have been created and developed in the last several years; for example photo-Fenton oxidation and electro-Fenton. According to Eq. (1), free hydroxyl radicals may arise concurrently without light, and the Fe^{2+} catalyst can be regenerated:

$$Fe^{2+} + H_2O_2 \rightarrow Fe^{3+} + HO^{\cdot} + HO \quad (1)$$

The iron (iii) reacts with H_2O_2 Fe^{2+} at a pH of less than 3 and with H_2O_2, as shown in Eq. (2):

$$Fe^{3+} + H_2O_2 \rightarrow Fe^{2+} + HO^{\cdot}{}_2 + H^+ \quad (2)$$

The elimination of drug products with Fenton oxidation and Fenton processes have been investigated in this area. For example, Romero, Acevedo, Marco, Giménez, and Esplugas (2016) showed that the entire Fenton process degraded MET with concentrations of 5 mg/L for Fe^{2+} and 125 mg/L for H_2O_2. The Fenton reaction takes place in slightly acidic circumstances in the presence of iron ions with hydrogen peroxide, which generates hydroxyl radicals. Another research was conducted by Giraldo-Aguirre et al. (2018), which also observed solar-photo-Fenton deterioration up to 96% in light output of 10 W and 90 μmol/L in Fe^{2+} and 10 mmol/L of H_2O_2, resulting in solar-photo degradation.

5.4 Sonolysis

Sonolysis is an advanced process of oxidation (AOP). Frequency and strength of applied ultrasound are two crucial factors impacting the effectiveness of the AOP to deconstruct new molecules, while the presence of pharmaceuticals in wastewater has not been fully studied. The sonochemical methods efficiently remove ibuprofen, as shown by several studies (Ince, 2018; Ziylan-Yavas & Ince, 2018). It has been shown that single frequencies of 200 kHz have generated 93% ibuprofen removal, whereas complete ibuprofen breakdown and 33% TOC elimination has been obtained at high frequency—577 kHz. During the same investigation, ibuprofen suppression was shown to be increased even at low single frequencies by the addition of zerovalent iron. While sonolysis largely requires no extra chemicals, this process is more energy-intensive, leads to poor mineralization, and is only applicable at laboratory-scale. To address these disadvantages, the operational costs and the synergistic effects of these AOPs can be reduced in conjunction with other AOPs. For example, a group of researchers (Ghafoori, Mowla, Jahani, Mehrvar, & Chan, 2015) have done sonolysis and photolysis in pharmaceutical wastewaters (UV/H_2O_2)—which is known as sonophotolytic degradation—of substances consisting of diclofenac, chloramphenicol, paracetamol, and salicylic acid. In the use of sonophotolysis, only 3% and 8% of the TOC removal was observed by photolysis and UV, respectively.

6. Constructed wetlands (CWs)

Because of their cheap cost, eco-friendliness, and ease of installation and maintenance, constructed wetlands have received increased attention in recent decades for removing pharmaceutical compounds (Zhang et al., 2017). They are regarded as an excellent method of removing wastewater contaminants (Matamoros, Arias, Brix, & Bayona, 2007). Before being absorbed by gravity, polluted water travels across a layer of swaying or emergently rooted marsh plants. CW is a remediation strategy used to remove or modify various chemicals, through mechanisms such as biological degradation, photodegradation, sorption, or phytoremediation (Sgroi et al., 2017). Many studies have shown the significance of wetlands in the elimination of developing contaminants. For instance, ibuprofen, diclofenac, and bisphenol A were used to test how well HFCWs removed naproxen, tonalide, and bisphenol A. The clearance effectiveness ranged from 85% to 99%. For paracetamol, removal efficiencies of more than 90% were achieved, with minimal difference between horizontal subsurface flow wetlands (HSSF)-CWs with different plant species (Ranieri, Verlicchi, & Young, 2011). However, despite all the advantages of the CW and its efficiency in removing organic molecules, including pharmaceutical products, this technique also has certain inconveniences, such as a large land demand and substratum fouling. Therefore, to create a more integrated technology, chemical processes should be replaced with alternative methods.

7. Treatment of pharmaceutical residues in wastewater

Drugs are the most important category of the numerous developing water contaminants. These include antibiotics, analgesics, steroids, antidepressant medications, stimulants, antibacterial medications, pain medication, hormones, antiinflammatory medication, β-blockers, cholesterol control medication, contrast agents, and impotence medicines (Balakrishna, Rath, Praveenkumarreddy, Guruge, & Subedi, 2017; Birniwa et al., 2022; Wu et al., 2018). Drug residues are currently detected in nearly all environmental media, including surface water (lakes, rivers, streams, estuaries, and seawater) (Feitosa-Felizzola & Chiron, 2009), groundwater, and wastewater effluent and sludge (Fent, Weston, & Caminada, 2006; Jagaba, Kutty, Lawal, Birniwa, et al., 2022), and are also present on every continent (Fick et al., 2009). The permanence, possible toxicity, and ongoing release from wastewater to water bodies causes potential long-term hazards to aquatic and terrestrial creatures (Patel et al., 2019). For example, since the 1990s, antibiotics have attracted special attention as one of the most widely used medicines present in the environment. Antibiotics are substances having antibacterial activity that can be either natural or manufactured chemicals. Recent research has shown worldwide consumption of 100,000–200,000 tons of antibiotics per annum (Kumar & Pal, 2018). A high number of different antibiotic residues have been detected in water bodies. These antibiotics might render the microbial genome resistant, even at low doses (González, Sans, Esplugas, & Malato, 2009), and induce long-term irreversible alterations. Residual pharmaceutical wastewater must thus be treated (Watkinson, Murby, Kolpin, & Costanzo, 2009).

8. Conclusions

Wastewater effluents containing pharmaceuticals have a significant impact on our environment due to their carcinogenic effects. Pharmaceuticals in wastewater contain complex chemical compositions that are poisonous to humans and pose a threat to aquatic organisms. This chapter includes a significant number of studies that have resulted in extensive and intensive evaluation of the subject matter. Many traditional approaches have failed to treat pharmaceuticals-contaminated water due to shortcomings including lower efficiencies, high initial, operational, and maintenance costs, crude formulation, lower selectivity and mass transfer rate, high toxicity, operating complexity, large amounts of organic solvents, harmful effects on health, high energy requirements, generation of voluminous sludge, low trace metals removal rate, and negative environmental effects. To address issues in PW treatment systems, this chapter discussed innovative technologies required to upgrade the existing treatment units. The chapter highlighted the common contaminants found in PW along with mechanisms for their release into water sources. It also explained how agricultural activities contribute to the release of pharmaceutical and personal care products (PPCPs). The chapter also discussed existing physical, chemical, and biological techniques for removal of contaminants from PW. Biological techniques for the remediation of PW have high efficiency, adaptability, and capability. However, the poor biodegradability of several PW contaminants is the reason why they do not provide a high removal yield in these biological treatment techniques. Most of the technologies are not too effective in removing pharmaceuticals, as they require more advance treatment methods to efficiently eradicate or reduce the concentration of pharmaceuticals without having adverse effects on the microorganisms. These techniques include adsorption, bioremediation, advanced oxidation processes such as Fenton and photo-Fenton, ozonation, electrochemical treatment, and sonolysis. Other strategies are membrane separation though either membrane bioreactor technology and or enzymatic membrane reactors, and constructed wetlands. The effectiveness of agro-biomass/byproducts from plants as well as their function as cosubstrates to promote the biodegradation of pharmaceuticals is another subject of this chapter. The chapter extensively discussed the contributions of agricultural-based biomass for the production of adsorbent materials used for PW treatment. In recent times, much effort has been invested in combined AOPs/biological systems, and research findings have resulted in significant success in encouraging the use of these appealing innovative treatment combination methods. These combined procedures produce much better results than AOPs or biological treatment methods alone.

References

Abdullahi, S., Musa, H., Habibu, S., Birniwa, A., & Mohammad, R. (2021). Facile synthesis and dyeing performance of some disperse monomeric and polymeric dyes on nylon and polyester fabrics. *Bulletin of the Chemical Society of Ethiopia*, *35*(3), 485–497.

Abdullahi, S. S., Musa, H., Habibu, S., Birniwa, A. H., & Mohammad, R. E. A. (2022). Comparative study and dyeing performance of as-synthesized azo heterocyclic monomeric, polymeric, and commercial disperse dyes. *Turkish Journal of Chemistry*, *46*, 1–12. https://doi.org/10.55730/kim-2202-51.

Abdullahi, S., Birniwa, A., & Audu, A. (2020). Modification of Ashaka and tango bentonites with tetraphenyl phosphonium and hexadecyl pyridinium bromides for possible use in polymer-clay nanocomposite preparation. *Nanotechnology, Science and Applications*, *22*, 15–22.

Abdullahi, S. S., Birniwa, A., Mohammad, R., Mamman, S., & Chadi, A. (2020). Impact of fibre reinforced polyester composites on tensile strength of baobab (Adansonia digitata) stem. *Caliphate Journal of Science & Technology (CaJoST)*, *2*, 94–100.

Ahmed, M. J., & Theydan, S. K. (2014). Fluoroquinolones antibiotics adsorption onto microporous activated carbon from lignocellulosic biomass by microwave pyrolysis. *Journal of the Taiwan Institute of Chemical Engineers*, *45*(1), 219–226. https://doi.org/10.1016/j.jtice.2013.05.014.

Akhtar, J., Amin, N. A. S., & Shahzad, K. (2016). A review on removal of pharmaceuticals from water by adsorption. *Desalination and Water Treatment*, *57*(27), 12842–12860. https://doi.org/10.1080/19443994.2015.1051121.

Al-Mahbashi, N., Kutty, S. R. M., Jagaba, A. H., Al-Nini, A., Ali, M., Saeed, A. A. H., et al. (2022). Column study for adsorption of copper and cadmium using activated carbon derived from sewage sludge. *Advances in Civil Engineering*. https://doi.org/10.1155/2022/3590462.

Alharbi, S. K., Price, W. E., Kang, J., Fujioka, T., & Nghiem, L. D. (2016). Ozonation of carbamazepine, diclofenac, sulfamethoxazole and trimethoprim and formation of major oxidation products. *Desalination and Water Treatment*, *57*(60), 29340–29351. https://doi.org/10.1080/19443994.2016.1172986.

Almahbashi, N., Kutty, S., Ayoub, M., Noor, A., Salihi, I., Al-Nini, A., et al. (2021). Optimization of preparation conditions of sewage sludge based activated carbon. *Ain Shams Engineering Journal*, *12*(2), 1175–1182.

Alsbaiee, A., Smith, B. J., Xiao, L., Ling, Y., Helbling, D. E., & Dichtel, W. R. (2016). Rapid removal of organic micropollutants from water by a porous β-cyclodextrin polymer. *Nature*, *529*(7585), 190–194. https://doi.org/10.1038/nature16185.

Altmann, J., Ruhl, A. S., Zietzschmann, F., & Jekel, M. (2014). Direct comparison of ozonation and adsorption onto powdered activated carbon for micropollutant removal in advanced wastewater treatment. *Water Research*, *55*, 185–193. https://doi.org/10.1016/j.watres.2014.02.025.

Álvarez-Torrellas, S., Rodríguez, A., Ovejero, G., & García, J. (2016). Comparative adsorption performance of ibuprofen and tetracycline from aqueous solution by carbonaceous materials. *Chemical Engineering Journal*, *283*, 936–947. https://doi.org/10.1016/j.cej.2015.08.023.

Antunes, M., Esteves, V. I., Guégan, R., Crespo, J. S., Fernandes, A. N., & Giovanela, M. (2012). Removal of diclofenac sodium from aqueous solution by Isabel grape bagasse. *Chemical Engineering Journal*, *192*, 114–121. https://doi.org/10.1016/j.cej.2012.03.062.

Archana, G., Dhodapkar, R., & Kumar, A. (2017). Ecotoxicological risk assessment and seasonal variation of some pharmaceuticals and personal care products in the sewage treatment plant and surface water bodies (lakes). *Environmental Monitoring and Assessment*, *189*(9). https://doi.org/10.1007/s10661-017-6148-3.

Asif, M., Hai, F., Kang, J., Van De Merwe, J., Leusch, Asif, et al. (2021). Removal of pharmaceuticals and personal care products using native fungal enzymes extracted during the ligninolytic process. *Environmental Research*, *195*, 110878.

Baccar, R., Sarrà, M., Bouzid, J., Feki, M., & Blánquez, P. (2012). Removal of pharmaceutical compounds by activated carbon prepared from agricultural by-product. *Chemical Engineering Journal*, *211–212*, 310–317. https://doi.org/10.1016/j.cej.2012.09.099.

Balakrishna, K., Rath, A., Praveenkumarreddy, Y., Guruge, K. S., & Subedi, B. (2017). A review of the occurrence of pharmaceuticals and personal care products in Indian water bodies. *Ecotoxicology and Environmental Safety*, *137*, 113–120. https://doi.org/10.1016/j.ecoenv.2016.11.014.

Baloo, L., Isa, M. H., Sapari, N. B., Jagaba, A. H., Wei, L. J., Yavari, S., et al. (2021). Adsorptive removal of methylene blue and acid orange 10 dyes from aqueous solutions using oil palm wastes-derived activated carbons. *Alexandria Engineering Journal*, *60*(6), 5611–5629.

Birniwa, A. H. (2019). Polymers and the environment. *Vol. 1. The fundamental of polymer science and technology* (1st ed., pp. 153–161). Zaria: ABU Press.

Birniwa, A., & Abdullahi, S. (2019). Study on physico-mechanical behaviour of Acacia nilotica (gum tree) and glass fiber blend reinforced epoxy resin composite. *ChemSearch Journal*, *10*(2), 46–53.

Birniwa, A., Abubakar, A., Ekramul, H. N. M., Kutty, S., Jagaba, A., Abdullahi, S., & Zango, Z. (2022). Application of agricultural wastes for cationic dyes removal from wastewater. *Textile wastewater treatment* (pp. 239–274). Berlin/Heidelberg, Germany: Springer.

Birniwa, A. H., Mahmud, H. N. M. E., Abdullahi, S. S., HAbibu, S., Jagaba, A. H., Ibrahim, M. N. M., et al. (2022). Adsorption behavior of methylene blue cationic dye in aqueous solution using polypyrrole-polyethylenimine nano-adsorbent. *Polymers*, *14*(16), 3362. https://doi.org/10.3390/polym14163362.

Birniwa, A. H., Abdullahi, S. S., Yakasai, M. Y., & Ismaila, A. (2021). Studies on physico-mechanical behaviour of kenaf/glass fiber reinforced epoxy hybrid composites. *Bulletin of the Chemical Society of Ethiopia*, *35*(1), 171–184.

Birniwa, A. H., Abubakar, A. S., Huq, A. O., & Mahmud, H. N. M. E. (2021). Polypyrrole-polyethyleneimine (PPy-PEI) nanocomposite: An effective adsorbent for nickel ion adsorption from aqueous solution. *Journal of Macromolecular Science, Part A*, *58*(3), 206–217.

Blair, B., Nikolaus, A., Hedman, C., Klaper, R., & Grundl, T. (2015). Evaluating the degradation, sorption, and negative mass balances of pharmaceuticals and personal care products during wastewater treatment. *Chemosphere*, *134*, 395–401. https://doi.org/10.1016/j.chemosphere.2015.04.078.

Cabrita, Ruiz, Mestrea, Fonseca, I., Carvalhoa, & Ania. (2021). Removal of an analgesic using activated carbons prepared from urban and industrial residues. *Chemical Engineering Journal*, *163*, 249–255.

Cincinelli, A., Martellini, T., Coppini, E., Fibbi, D., & Katsoyiannis, A. (2015). Nanotechnologies for removal of pharmaceuticals and personal care products from water and wastewater. A review. *Journal of Nanoscience and Nanotechnology*, *15*(5), 3333–3347. https://doi.org/10.1166/jnn.2015.10036.

Coimbra, R. N., Calisto, V., Ferreira, C. I. A., Esteves, V. I., & Otero, M. (2019). Removal of pharmaceuticals from municipal wastewater by adsorption onto pyrolyzed pulp mill sludge. *Arabian Journal of Chemistry*, *12*(8), 3611–3620. https://doi.org/10.1016/j.arabjc.2015.12.001 Daughton.

Daughton, C. G., & Ternes, T. A. (1999). Pharmaceuticals and personal care products in the environment: Agents of subtle change? *Environmental Health Perspectives*, *107*(6), 907–938. https://doi.org/10.1289/ehp.99107s6907.

de Cazes, M., Abejón, R., Belleville, M. P., & Sanchez-Marcano, J. (2014). Membrane bioprocesses for pharmaceutical micropollutant removal from waters. *Membranes*, *4*(4), 692–729. https://doi.org/10.3390/membranes4040692.

Dias, J. M., Alvim-Ferraz, M. C. M., Almeida, M. F., Rivera-Utrilla, J., & Sánchez-Polo, M. (2007). Waste materials for activated carbon preparation and its use in aqueous-phase treatment: A review. *Journal of Environmental Management*, *85*(4), 833–846. https://doi.org/10.1016/j.jenvman.2007.07.031.

Ebele, A. J., Abou-Elwafa Abdallah, M., & Harrad, S. (2017). Pharmaceuticals and personal care products (PPCPs) in the freshwater aquatic environment. *Emerging Contaminants*, *3*(1), 1–16. https://doi.org/10.1016/j.emcon.2016.12.004.

Faubert, P., Barnabé, S., Bouchard, S., Côté, R., & Villeneuve, C. (2016). Pulp and paper mill sludge management practices: What are the challenges to assess the impacts on greenhouse gas emissions? *Resources, Conservation and Recycling*, *108*, 107–133. https://doi.org/10.1016/j.resconrec.2016.01.007.

Feitosa-Felizzola, J., & Chiron, S. (2009). Occurrence and distribution of selected antibiotics in a small Mediterranean stream (Arc River, Southern France). *Journal of Hydrology*, *364*(1–2), 50–57. https://doi.org/10.1016/j.jhydrol.2008.10.006.

Fent, K., Weston, A. A., & Caminada, D. (2006). Ecotoxicology of human pharmaceuticals. *Aquatic Toxicology*, *76*(2), 122–159. https://doi.org/10.1016/j.aquatox.2005.09.009.

Fick, J., Söderström, H., Lindberg, R. H., Phan, C., Tysklind, M., & Larsson, D. G. J. (2009). Contamination of surface, ground, and drinking water from pharmaceutical production. *Environmental Toxicology and Chemistry*, *28*(12), 2522–2527. https://doi.org/10.1897/09-073.1.

García-Espinoza, J. D., Gortáres-Moroyoqui, P., & Orta-Ledesma, M. T. (2016). Electrochemical removal of carbamazepine in water with Ti/PbO_2 cylindrical mesh anode. *Water Science & Technology*, *73*, 1155–1165. https://doi.org/10.2166/wst.2015.591.

Geissen, V., Mol, H., Klumpp, E., Umlauf, G., Nadal, M., van der Ploeg, M., et al. (2015). Emerging pollutants in the environment: A challenge for water resource management. *International Soil and Water Conservation Research*, *3*(1), 57–65. https://doi.org/10.1016/j.iswcr.2015.03.002.

Ghafoori, S., Mowla, A., Jahani, R., Mehrvar, M., & Chan, P. K. (2015). Sonophotolytic degradation of synthetic pharmaceutical wastewater: Statistical experimental design and modeling. *Journal of Environmental Management, 150*, 128–137. https://doi.org/10.1016/j.jenvman.2014.11.011.

Ghaleb, A., Kutty, S., Ho, Y., Jagaba, A., Noor, A., Al-Sabaeei, A., et al. (2020). Anaerobic co-digestion for oily-biological sludge with sugarcane bagasse for biogas production under mesophilic condition. *IOP Conference Series: Materials Science and Engineering, 991* (1), 012084. IOP Publishing.

Giraldo-Aguirre, A. L., Serna-Galvis, E. A., Erazo-Erazo, E. D., Silva-Agredo, J., Giraldo-Ospina, H., Flórez-Acosta, O. A., et al. (2018). Removal of β-lactam antibiotics from pharmaceutical wastewaters using photo-Fenton process at near-neutral pH. *Environmental Science and Pollution Research, 25*(21), 20293–20303. https://doi.org/10.1007/s11356-017-8420-z.

González, O., Sans, C., Esplugas, S., & Malato, S. (2009). Application of solar advanced oxidation processes to the degradation of the antibiotic sulfamethoxazole. *Photochemical and Photobiological Sciences, 8*(7), 1032–1039. https://doi.org/10.1039/b822658j.

Grover, D. P., Zhou, J. L., Frickers, P. E., & Readman, J. W. (2011). Improved removal of estrogenic and pharmaceutical compounds in sewage effluent by full scale granular activated carbon: Impact on receiving river water. *Journal of Hazardous Materials, 185*(2–3), 1005–1011. https://doi.org/10.1016/j.jhazmat.2010.10.005.

Gupta, V. K., Dinesh Mohan, S., Suhas, & Singh, K. P. (2006). Removal of 2-aminophenol using novel adsorbents. *Industrial and Engineering Chemistry Research, 45*(3), 1113–1122. https://doi.org/10.1021/ie051075k.

Hokkanen, S., Bhatnagar, A., & Sillanpää, M. (2016). A review on modification methods to cellulose-based adsorbents to improve adsorption capacity. *Water Research, 91*, 156–173. https://doi.org/10.1016/j.watres.2016.01.008.

Hu, Li, X., Chen, Z., Cui, Y., Gu, Jia, F., et al. (2021). Performance and extracellular polymers substance analysis of a pilot scale anaerobic membrane bioreactor for treating tetrahydrofuran pharmaceutical wastewater at different HRTs. *Journal of Hazardous Materials, 342*, 383–391.

Hu, X., Cheng, Z., Sun, Z., & Zhu, H. (2017). Adsorption of diclofenac and triclosan in aqueous solution by purified multi-walled carbon nanotubes. *Polish Journal of Environmental Studies, 26*(1), 87–95. https://doi.org/10.15244/pjoes/63885.

Ince, N. H. (2018). Ultrasound-assisted advanced oxidation processes for water decontamination. *Ultrasonics Sonochemistry, 40*, 97–103.

Ioannidou, & Zabaniotou, A. (2007). Agricultural residues as precursors for activated carbon production—A review. *Renewable and Sustainable Energy Reviews, 11*.

Iyer, P. V., & Ananthanarayan, L. (2008). Enzyme stability and stabilization-aqueous and non-aqueous environment. *Process Biochemistry, 43*(10), 1019–1032. https://doi.org/10.1016/j.procbio.2008.06.004.

Jagaba, A., Kutty, S., Hayder, G., Elsadig, E. O., Lawal, I., Sayed, K., et al. (2021). Evaluation of the physical, chemical, bacteriological and trace metals concentrations in different brands of packaged drinking water. *Vol. 29. Engineering letters* (4th). International Association of Engineers.

Jagaba, A. H., Kutty, S. R. M., Noor, A., Birniwa, A. H., Affam, A. C., Lawal, I. M., et al. (2021). A systematic literature review of biocarriers: Central elements for biofilm formation, organic and nutrients removal in sequencing batch biofilm reactor. *Journal of Water Process Engineering, 42*, 102178.

Jagaba, A. H., Kutty, S. R. M., Salih, G. H. A., Noor, A., bin Md Hafiz, M. F. U., Yaro, N. S. A., et al. (2021). Palm oil clinker as a waste by-product: utilization and circular economy potential. *Elaeis guineensis*. Intech Open.

Jagaba, A. H., Kutty, S. R. M., Hayder, G., Baloo, L., Noor, A., Yaro, N. S. A., et al. (2021). A systematic literature review on waste-to-resource potential of palm oil clinker for sustainable engineering and environmental applications. *Materials*, *14*(16), 4456.

Jagaba, A. H., Kutty, S. R. M., Baloo, L., Birniwa, A. H., Lawal, I. M., Aliyu, M. K., et al. (2022). Combined treatment of domestic and pulp and paper industry wastewater in a rice straw embedded activated sludge bioreactor to achieve sustainable development goals. *Case Studies in Chemical and Environmental Engineering*, *6*, 100261.

Jagaba, A. H., Kutty, S. R. M., Noor, A., Isah, A. S., Lawal, I. M., Birniwa, A. H., et al. (2022). Kinetics of pulp and paper wastewater treatment by high sludge retention time activated sludge process. *Journal of Hunan University Natural Sciences*, *49*(2).

Jagaba, A. H., Kutty, S. R. M., Isa, M. H., Affam, A. C., Aminu, N., Abubakar, S., et al. (2022). Effect of environmental and operational parameters on sequential batch reactor systems in dye degradation. In *Dye biodegradation, mechanisms and techniques* (pp. 193–225). Springer.

Jagaba, AH., Kutty, S. R. M., Abubakar, S., Birniwa, A. H., Lawal, I. M., Umaru, I., et al. (2022). Synthesis, characterization, and performance evaluation of hybrid waste sludge biochar for cod and color removal from agro-industrial effluent. *Separation*, *9*(9), 258. https://doi.org/10.3390/separations9090258.

Jagaba, A. H., Kutty, S. R. M., Naushad, M, Lawal, I. M., Noor, A., Affam, A. C., et al. (2022). Removal of nutrients from pulp and paper biorefinery effluent: Operation, kinetic modelling and optimization by response surface methodology. *Environmental Research*, *214*, 114091.

Jagaba, A. H., Kutty, S. R. M., Isa, M. H., Ghaleb, A. A. S., Lawal, I. M., Usman, A. K., et al. (2022). Toxic effects of xenobiotic compounds on the microbial community of activated sludge. *ChemBio Review*, 497–535. https://doi.org/10.1002/cben.202100055.

Jagaba, A. H., Kutty, S. R. M., Noor, A., Affam, A. C., Ghfar, A. A., Usman, A. K., et al. (2022). Parametric optimization and kinetic modelling for organic matter removal from agro-waste derived paper packaging biorefinery wastewater. *Biomass Conversion and Biorefinery*, 1–18.

Jagaba, A. H., Kutty, S. R. M., Lawal, I. M., Aminu, N., Noor, A., Al-dhawi, B. N. S., et al. (2022). Diverse sustainable materials for the treatment of petroleum sludge and remediation of contaminated sites: A review. *Cleaner Waste Systems*. https://doi.org/10.1016/j.clwas.2022.100010.

Jagaba, A. H., Kutty, S. R. M., Lawal, I. M., Birniwa, A. H., Affam, A. C., Yaro, N. S. A., et al. (2022). Circular economy potential and contributions of petroleum industry sludge utilization to environmental sustainability through engineered processes-A review. *Cleaner and Circular Bioeconomy*, *3*. https://doi.org/10.1016/j.clcb.2022.100029.

Ji, L., Liu, F., Xu, Z., Zheng, S., & Zhu, D. (2010). Adsorption of pharmaceutical antibiotics on template-synthesized ordered micro- and mesoporous carbons. *Environmental Science and Technology*, *44*(8), 3116–3122. https://doi.org/10.1021/es903716s.

Jing, X. R., Wang, Y. Y., Liu, W. J., Wang, Y. K., & Jiang, H. (2014). Enhanced adsorption performance of tetracycline in aqueous solutions by methanol-modified biochar. *Chemical Engineering Journal*, *248*, 168–174. https://doi.org/10.1016/j.cej.2014.03.006.

Jung, Park, J., Lim, K., Park, Heo, J., et al. (2021). Adsorption of selected endocrine disrupting compounds and pharmaceuticals on activated biochars. *Journal of Hazardous Materials*, *263*, 702–710.

Kim, M., Guerra, P., Shah, A., Parsa, M., Alaee, M., & Smyth, S. A. (2014). Removal of pharmaceuticals and personal care products in a membrane bioreactor wastewater treatment plant. *Water Science and Technology*, *69*(11), 2221–2229. https://doi.org/10.2166/wst.2014.145.

Kim, S. D., Cho, J., Kim, I. S., Vanderford, B. J., & Snyder, S. A. (2007). Occurrence and removal of pharmaceuticals and endocrine disruptors in South Korean surface, drinking, and waste waters. *Water Research*, *41*(5), 1013–1021. https://doi.org/10.1016/j.watres.2006.06.034.

Kong, W., Zhao, F., Guan, H., Zhao, Y., Zhang, H., & Zhang, B. (2016). Highly adsorptive mesoporous carbon from biomass using molten-salt route. *Journal of Materials Science*, *51*(14), 6793–6800. https://doi.org/10.1007/s10853-016-9966-8.

Kumar, A., & Pal, D. (2018). Antibiotic resistance and wastewater: Correlation, impact and critical human health challenges. *Journal of Environmental Chemical Engineering*, *6*(1), 52–58. https://doi.org/10.1016/j.jece.2017.11.059.

Kumar, R., Sarmah, A. K., & Padhye, L. P. (2019). Fate of pharmaceuticals and personal care products in a wastewater treatment plant with parallel secondary wastewater treatment train. *Journal of Environmental Management*, *233*, 649–659. https://doi.org/10.1016/j.jenvman.2018.12.062.

Kyzas, G. Z., Koltsakidou, A., Nanaki, S. G., Bikiaris, D. N., & Lambropoulou, D. A. (2015). Removal of beta-blockers from aqueous media by adsorption onto graphene oxide. *Science of the Total Environment*, *537*, 411–420. https://doi.org/10.1016/j.scitotenv.2015.07.144.

Lawal, I. M., Bertram, D., White, C. J., Jagaba, A. H., Hassan, I., & Shuaibu, A. (2021). Multi-criteria performance evaluation of gridded precipitation and temperature products in data-sparse regions. *Atmosphere*, *12*(12), 1597.

Lee, K. M., Wong, C. P. P., Tan, T. L., & Lai, C. W. (2018). Functionalized carbon nanotubes for adsorptive removal of water pollutants. *Materials Science & Engineering, B: Solid-State Materials for Advanced Technology*, *236–237*, 61–69. https://doi.org/10.1016/j.mseb.2018.12.004.

Liu, Y., Liu, Z., Gao, J., Dai, J., Han, J., Wang, Y., et al. (2011). Selective adsorption behavior of Pb (II) by mesoporous silica SBA-15-supported Pb (II)-imprinted polymer based on surface molecularly imprinting technique. *Journal of Hazardous Materials*, *186*(1), 197–205.

Loos, G., Scheers, T., Van Eyck, K., Van Schepdael, A., Adams, E., Van der Bruggen, B., et al. (2018). Electrochemical oxidation of key pharmaceuticals using a boron doped diamond electrode. *Separation and Purification Technology*, *195*, 184–191. https://doi.org/10.1016/j.seppur.2017.12.009.

Mamman, S., Mohammad, R., Abdullahi, S., Chadi, A., & Birniwa, A. (2020). Synthesis, characterization and optimization of magnetite molecularly imprinted polymer for application in the removal of non-steroidal anti inflammatory drugs (NSAIDS). *Chem-Search Journal*, *11*(1), 1–8.

Margot, J., Kienle, C., Magnet, A., Weil, M., Rossi, L., de Alencastro, L. F., et al. (2013). Treatment of micropollutants in municipal wastewater: Ozone or powdered activated carbon? *Science of the Total Environment*, *461–462*, 480–498. https://doi.org/10.1016/j.scitotenv.2013.05.034.

Martins, A. C., Pezoti, O., Cazetta, A. L., Bedin, K. C., Yamazaki, D. A. S., Bandoch, G. F. G., et al. (2015). Removal of tetracycline by NaOH-activated carbon produced from macadamia nut shells: Kinetic and equilibrium studies. *Chemical Engineering Journal*, *260*, 291–299. https://doi.org/10.1016/j.cej.2014.09.017.

Matamoros, V., Arias, C., Brix, H., & Bayona, J. M. (2007). Removal of pharmaceuticals and personal care products (PPCPs) from urban wastewater in a pilot vertical flow constructed wetland and a sand filter. *Environmental Science and Technology*, *41*(23), 8171–8177. https://doi.org/10.1021/es071594+.

Mathur, P., Sanyal, D., & Das, R. K. (2021). *Treatment of pharmaceutical and personal care products in wastewater* (pp. 451–474). Elsevier BV. https://doi.org/10.1016/b978-0-12-822956-9.00024-6.

Mestrea, A., Piresa, J., Nogueiraa, J., Parra, J., Carvalhoa, A., et al. (2021). Waste-derived activated carbons for removal of ibuprofen from solution: role of surface chemistry and pore structure. *Bioresource Technology*, *100*, 1720–1726.

Naghdi, M., Taheran, M., Brar, S. K., Kermanshahi-pour, A., Verma, M., & Surampalli, R. Y. (2017). Immobilized laccase on oxygen functionalized nanobiochars through mineral acids treatment for removal of carbamazepine. *Science of the Total Environment*, *584–585*, 393–401. https://doi.org/10.1016/j.scitotenv.2017.01.021.

Naghdi, M., Taheran, M., Brar, S. K., Kermanshahi-pour, A., Verma, M., & Surampalli, R. Y. (2018). Removal of pharmaceutical compounds in water and wastewater using fungal oxidoreductase enzymes. *Environmental Pollution*, *234*, 190–213. https://doi.org/10.1016/j.envpol.2017.11.060.

Nasara, M. A., Zubairu, I., Jagaba, A. H., Azare, A. A., Yerima, Y. M., & Yerima, B. (2021). Assessment of non-revenue water management practices in Nigeria (A case study of Bauchi state water and sewerage cooperation). *American Journal of Engineering Research*, *10*(5), 390–401.

Nielsen, Biggs, M., Skinner, & Bandosz, T. J. (2021). The effects of activated carbon surface features on the reactive adsorption of carbamazepine and sulfamethoxazole. *Carbon*, *80*, 419–432.

Noor, A., Kutty, S., Baloo, L., Almahbashi, N., Kumar, V., & Ghaleb, A. (2021). Bio-kinetics of organic removal in EAAS reactor for co-treatment of refinery wastewater with municipal wastewater. *IOP Conference Series: Materials Science and Engineering*, *1092*(1), 012068. IOP Publishing.

Noor, A., Kutty, S. R., Almahbashi, N. M., Kumar, V., Ghaleb, A. A., & Al-Dhawi, B. N. (2021). Integrated submerged media extended aeration activated sludge (ISmEAAS) reactor start-up and biomass acclimatization. *Journal of Hunan University Natural Sciences*, *48*(9).

Patel, M., Kumar, R., Kishor, K., Mlsna, T., Pittman, C. U., & Mohan, D. (2019). Pharmaceuticals of emerging concern in aquatic systems: Chemistry, occurrence, effects, and removal methods. *Chemical Reviews*, *119*(6), 3510–3673. https://doi.org/10.1021/acs.chemrev.8b00299.

Petrie, B., Barden, R., & Kasprzyk-Hordern, B. (2015). A review on emerging contaminants in wastewaters and the environment: Current knowledge, understudied areas and recommendations for future monitoring. *Water Research*, *72*, 3–27. https://doi.org/10.1016/j.watres.2014.08.053.

Rafatullah, M., Ahmad, T., Ghazali, A., Sulaiman, O., Danish, M., & Hashim, R. (2013). Oil palm biomass as a precursor of activated carbons: A review. *Critical Reviews in Environmental Science and Technology*, *43*(11), 1117–1161. https://doi.org/10.1080/10934529.2011.627039.

Rajapaksha, A. U., Vithanage, M., Zhang, M., Ahmad, M., Mohan, D., Chang, S. X., et al. (2014). Pyrolysis condition affected sulfamethazine sorption by tea waste biochars. *Bioresource Technology*, *166*, 303–308. https://doi.org/10.1016/j.biortech.2014.05.029.

Rana, D., Scheier, B., Narbaitz, R. M., Matsuura, T., Tabe, S., Jasim, S. Y., et al. (2012). Comparison of cellulose acetate (CA) membrane and novel CA membranes containing surface modifying macromolecules to remove pharmaceutical and personal care product micropollutants from drinking water. *Journal of Membrane Science*, *409*, 346–354.

Ranieri, E., Verlicchi, P., & Young, T. M. (2011). Paracetamol removal in subsurface flow constructed wetlands. *Journal of Hydrology*, *404*(3–4), 130–135. https://doi.org/10.1016/j.jhydrol.2011.03.015.

Rasheed, T., Bilal, M., Hassan, A. A., Nabeel, F., Bharagava, R. N., Romanholo Ferreira, L. F., et al. (2020). Environmental threatening concern and efficient removal of pharmaceutically active compounds using metal-organic frameworks as adsorbents. *Environmental Research*, *185*, 109436. https://doi.org/10.1016/j.envres.2020.109436.

Rasheed, T., Adeel, M., Nabeel, F., Bilal, M., & Iqbal, H. M. N. (2019). TiO2/SiO2 decorated carbon nanostructured materials as a multifunctional platform for emerging pollutants removal. *Science of the Total Environment*, *688*, 299–311. https://doi.org/10.1016/j.scitotenv.2019.06.200.

Reza, R. A., Ahmaruzzaman, M., Sil, A. K., & Gupta, V. K. (2014). Comparative adsorption behavior of ibuprofen and clofibric acid onto microwave assisted activated bamboo waste. *Industrial and Engineering Chemistry Research*, *53*(22), 9331–9339. https://doi.org/10.1021/ie404162p.

Rivera-Utrilla, J., Prados-Joya, G., Sánchez-Polo, M., Ferro-García, M. A., & Bautista-Toledo, I. (2009). Removal of nitroimidazole antibiotics from aqueous solution by adsorption/bioadsorption on activated carbon. *Journal of Hazardous Materials*, *170*(1), 298–305. https://doi.org/10.1016/j.jhazmat.2009.04.096.

Romero, V., Acevedo, S., Marco, P., Giménez, J., & Esplugas, S. (2016). Enhancement of Fenton and photo-Fenton processes at initial circumneutral pH for the degradation of the β-blocker metoprolol. *Water Research*, *88*, 449–457. https://doi.org/10.1016/j.watres.2015.10.035.

Saharan, P., Chaudhary, G. R., Mehta, S. K., & Umar, A. (2014). Removal of water contaminants by iron oxide nanomaterials. *Journal of Nanoscience and Nanotechnology*, *14*(1), 627–643. https://doi.org/10.1166/jnn.2014.9053.

Saucier, C., Adebayo, M. A., Lima, E. C., Cataluña, R., Thue, P. S., Prola, L. D. T., et al. (2015). Microwave-assisted activated carbon from cocoa shell as adsorbent for removal of sodium diclofenac and nimesulide from aqueous effluents. *Journal of Hazardous Materials*, *289*, 18–27. https://doi.org/10.1016/j.jhazmat.2015.02.026.

Sgroi, M., Pelissari, C., Ávila, C., Sezerino, P. H., Vagliasindi, F. G. A., García, J., et al. (2017). Removal of conventional water quality parameters, emerging contaminants and fluorescing organic matter in a hybrid constructed wetland system. In *Vol. 4. Lecture Notes in Civil Engineering* (pp. 313–317). Springer. https://doi.org/10.1007/978-3-319-58421-8_50.

Silva, C. P., Jaria, G., Otero, M., Esteves, V. I., & Calisto, V. (2018). Waste-based alternative adsorbents for the remediation of pharmaceutical contaminated waters: Has a step forward already been taken? *Bioresource Technology*, *250*, 888–901. https://doi.org/10.1016/j.biortech.2017.11.102.

Silva, C. P., Otero, M., & Esteves, V. (2012). Processes for the elimination of estrogenic steroid hormones from water: A review. *Environmental Pollution*, *165*, 38–58. https://doi.org/10.1016/j.envpol.2012.02.002.

Simazaki, D., Fujiwara, J., Manabe, S., Matsuda, M., Asami, M., & Kunikane, S. (2008). Removal of selected pharmaceuticals by chlorination, coagulation-sedimentation and powdered activated carbon treatment. *Water Science and Technology*, *58*(5), 1129–1135. https://doi.org/10.2166/wst.2008.472.

Snyder, S. A., Adham, S., Redding, A. M., Cannon, F. S., DeCarolis, J., Oppenheimer, J., et al. (2007). Role of membranes and activated carbon in the removal of endocrine disruptors and pharmaceuticals. *Desalination*, *202*(1–3), 156–181. https://doi.org/10.1016/j.desal.2005.12.052.

Sotelo, J. L., Ovejero, G., Rodríguez, A., Álvarez, S., Galán, J., & García, J. (2014). Competitive adsorption studies of caffeine and diclofenac aqueous solutions by activated carbon. *Chemical Engineering Journal*, *240*, 443–453. https://doi.org/10.1016/j.cej.2013.11.094.

Subedi, B., Lee, S., Moon, H. B., & Kannan, K. (2014). Emission of artificial sweeteners, select pharmaceuticals, and personal care products through sewage sludge from wastewater treatment plants in Korea. *Environment International*, *68*, 33–40. https://doi.org/10.1016/j.envint.2014.03.006.

Tan, X., Liu, Y., Zeng, G., Wang, X., Hu, X., Gu, Y., et al. (2015). Application of biochar for the removal of pollutants from aqueous solutions. *Chemosphere*, *125*, 70–85. https://doi.org/10.1016/j.chemosphere.2014.12.058.

Tan, J.-M., Qiu, G., & Ting, Y.-P. (2015). Osmotic membrane bioreactor for municipal wastewater treatment and the effects of silver nanoparticles on system performance. *Journal of Cleaner Production*, *88*, 146–151.

Tarpani, R. R. Z., & Azapagic, A. (2018). A methodology for estimating concentrations of pharmaceuticals and personal care products (PPCPs) in wastewater treatment plants and in freshwaters. *Science of the Total Environment*, *622–623*, 1417–1430. https://doi.org/10.1016/j.scitotenv.2017.12.059.

Ternes, T. A., Meisenheimer, M., McDowell, D., Sacher, F., Brauch, H. J., Haist-Gulde, B., et al. (2002). Removal of pharmaceuticals during drinking water treatment. *Environmental Science and Technology*, *36*(17), 3855–3863. https://doi.org/10.1021/es015757k.

Torres-Pérez, J., Gérente, C., & Andrès, Y. (2012). Sustainable activated carbons from agricultural residues dedicated to antibiotic removal by adsorption. *Chinese Journal of Chemical Engineering*, *20*(3), 524–529. https://doi.org/10.1016/S1004-9541(11)60214-0.

Tran, N. H., Li, J., Hu, J., & Ong, S. L. (2014). Occurrence and suitability of pharmaceuticals and personal care products as molecular markers for raw wastewater contamination in surface water and groundwater. *Environmental Science and Pollution Research*, *21*(6), 4727–4740. https://doi.org/10.1007/s11356-013-2428-9.

Tsang, Y. F. (2015). *Environmental protection and pollution management in China* (pp. 309–334). Springer Science and Business Media LLC. https://doi.org/10.1057/9781137373809_14.

Vergili, I. (2013). Application of nanofiltration for the removal of carbamazepine, diclofenac and ibuprofen from drinking water sources. *Journal of Environmental Management*, *127*, 177–187.

Vergili, I., Kaya, Y., Gönder, Z. B., Boergers, A., & Tuerk, J. (2019). Occurrence and prioritization of pharmaceutical active compounds in domestic/municipal wastewater treatment plants. *Bulletin of Environmental Contamination and Toxicology*, *102*(2), 252–258. https://doi.org/10.1007/s00128-019-02550-z.

Verlicchi, P., Al Aukidy, M., & Zambello, E. (2012). Occurrence of pharmaceutical compounds in urban wastewater: Removal, mass load and environmental risk after a secondary treatment—A review. *Science of the Total Environment*, *429*, 123–155.

Watkinson, A. J., Murby, E. J., Kolpin, D. W., & Costanzo, S. D. (2009). The occurrence of antibiotics in an urban watershed: From wastewater to drinking water. *Science of the Total Environment*, *407*(8), 2711–2723. https://doi.org/10.1016/j.scitotenv.2008.11.059.

Wu, S., Lyu, T., Zhao, Y., Vymazal, J., Arias, C. A., & Brix, H. (2018). Rethinking intensification of constructed wetlands as a green eco-technology for wastewater treatment. *Environmental Science and Technology*, *52*(4), 1693–1694. https://doi.org/10.1021/acs.est.8b00010.

Wu, C., Witter, J. D., Spongberg, A. L., & Czajkowski, K. P. (2009). Occurrence of selected pharmaceuticals in an agricultural landscape, western Lake Erie basin. *Water Research*, *43*(14), 3407–3416. https://doi.org/10.1016/j.watres.2009.05.014.

Xie, M., Chen, W., Xu, Z., Zheng, S., & Zhu, D. (2014). Adsorption of sulfonamides to demineralized pine wood biochars prepared under different thermochemical conditions. *Environmental Pollution*, *186*, 187–194. https://doi.org/10.1016/j.envpol.2013.11.022.

Xu, R., Tang, R., Zhou, Q., Li, F., & Zhang, B. (2015). Enhancement of catalytic activity of immobilized laccase for diclofenac biodegradation by carbon nanotubes. *Chemical Engineering Journal*, *262*, 88–95. https://doi.org/10.1016/j.cej.2014.09.072.

Yadav, A., Bagotia, N., Sharma, A., & Kumar, S. (2021). Advances in decontamination of wastewater using biomass-based composites: A critical review. *Science of the Total Environment*, *784*, 147108.

Yao, Y., Gao, B., Chen, H., Jiang, L., Inyang, M., Zimmerman, A. R., et al. (2012). Adsorption of sulfamethoxazole on biochar and its impact on reclaimed water irrigation. *Journal of Hazardous Materials*, *209–210*, 408–413. https://doi.org/10.1016/j.jhazmat.2012.01.046.

Yu, Z., Peldszus, S., & Huck, P. M. (2008). Adsorption characteristics of selected pharmaceuticals and an endocrine disrupting compound-naproxen, carbamazepine and nonylphenol-on activated carbon. *Water Research*, *42*(12), 2873–2882. https://doi.org/10.1016/j.watres.2008.02.020.

Zhang, L., Lv, T., Zhang, Y., Stein, O. R., Arias, C. A., Brix, H., et al. (2017). Effects of constructed wetland design on ibuprofen removal—A mesocosm scale study. *Science of the Total Environment*, *609*, 38–45. https://doi.org/10.1016/j.scitotenv.2017.07.130.

Zhao, Y., Kuang, J., Zhang, S., Li, X., Wang, B., Huang, J., et al. (2017). Ozonation of indomethacin: Kinetics, mechanisms, and toxicity. *Journal of Hazardous Materials*, *323*, 460–470.

Zheng, H., Wang, Z., Zhao, J., Herbert, S., & Xing, B. (2013). Sorption of antibiotic sulfamethoxazole varies with biochars produced at different temperatures. *Environmental Pollution*, *181*, 60–67. https://doi.org/10.1016/j.envpol.2013.05.056.

Zhou, A., Zhang, Y., Li, R., Su, X., & Zhang, L. (2016). Adsorptive removal of sulfa antibiotics from water using spent mushroom substrate, an agricultural waste. *Desalination and Water Treatment*, *57*(1), 388–397. https://doi.org/10.1080/19443994.2014.979239.

Ziylan-Yavas, A., & Ince, N. H. (2018). Single, simultaneous, and sequential applications of ultrasonic frequencies for the elimination of ibuprofen in water. *Ultrasonics Sonochemistry*, *40*, 17–23. https://doi.org/10.1016/j.ultsonch.2017.01.032.

Biohazardous effect associated with various pharma-effluent discharge in a biotic system

14

Vivek Sharma[a], *Monalisha Ghosh Dastidar*[b], *and Sharmili Roy*[c]
[a]Department of Chemical Engineering, Indian Institute of Technology Bombay, Mumbai, Maharashtra, India, [b]Research School of Engineering, College of Engineering and Computer Science, Australian National University, Canberra, ACT, Australia, [c]Division of Oncology, School of Medicine, Stanford University, Stanford, CA, United States

1. Introduction

Water is essential for sustaining each and every life form on earth. It is a natural resource which covers 70% of the earth's surface. Out of the 70%, which consists of rives, springs, ponds, and ocean waters, only 3% is portable water suitable for human consumption. In general, lakes and rivers are the sources of drinking water since the water from these sources is less salty as compared to the ocean water. The lakes and rivers consist of much lesser percentage of water as compared to the oceans (Jackson et al., 2001). Therefore, these water bodies are extremely crucial for the survival of living organisms. Due to the rapid growth of population, the usage of freshwater has increased by 10% (Rijsberman, SuffixJr, & Boland, 2004; Vörösmarty, Ellen, Green, & Revenga, 2005). A huge part of the fresh water goes for agriculture fields which is around ~40% (Mushtaq et al., 2020). Moreover, the water bodies are getting polluted day-by-day due to urbanization and industrialization. The daily usage and disposal of plastics bags, chemicals, heavy metals, expired medicines, industrials waste, etc., are directly/indirectly contaminating the rivers and freshwater streams (Bhat et al., 2014). Most of these contaminants are toxic to the aquatic and terrestrial ecosystems. The amount of pollutant is measured by calculating the number of pollutants discharged vs the amount of dissolved oxygen. The conversion between organic substances to inorganic elements present in the water bodies are transformed which is known as the biological oxygen demand (BOD) (Beeby, 1993). Additionally, not only human lives, but other terrestrial animals, fishes and other aquatic organisms are also getting immensely affected, especially by the discharge of heavy metal ions from the industries directly to the water bodies. Furthermore, many pathogens are emerging in the untreated sewage and water bodies causing health problems and disorders in humans and animals (American Public Health Association, American Water Works Association, Water Pollution Control Federation, Water Environment Federation, 1912; Helmer & Hespanhol, 1997). For instance, it has been

The Treatment of Pharmaceutical Wastewater. https://doi.org/10.1016/B978-0-323-99160-5.00002-3

reported that due to the presence of inorganic pollutants in community water, ~16 million cases/year of gastroenteritis are reported in the USA alone (Allaire, Wu, & Lall, 2018).

During the recent outbreak of SARS-CoV-2, the pollution initially reduced for some time due to the phase of complete lockdown. But the huge surge in the usage of masks, gloves and personal protective equipment (PPE) contaminated the water bodies more than expected (Ormaza-Gonzaìlez, Castro-Rodas, & Statham, 2021). Although many health managements sectors have taken huge steps toward proper handling and disposal procedures, contamination of water bodies still needs to be taken care of for all kinds of lives and right balance of the ecosystem (Baranwal, Mahapatra, Purohit, Roy, & Chandra, 2020; Chandra & Roy, 2020; Dastidar & Roy, 2021; Khan et al., 2021; Roy & Baranwal, 2020; Roy & Ramadoss, 2021; Sampath Kumar, Chintagunta, Jeevan Kumar, Roy, & Kumar, 2020). Various health issues and disorders in human life such as different types of cancer, lung diseases, heart issues, low birth rates, abdominal problems, respiratory issues and many more have resulted due to the consumption of water and fishes from polluted water bodies (Nel & Markotter, 2009; Ullah, Javed, Shafique, & Khan, 2014). Neutralizing the chemical substances in freshwater is challenging. Therefore, in order to reduce the effluents, industries, health management sectors, and governments need to take serious measures to stop the water from being more hazardous. In this chapter, we have discussed elaborately about various types of effluents, their sources, and their impact on human, animal, and aquatic lives. Additionally, this chapter summarizes the metal ions that are usually present in water as effluents and their effects on the ecosystem in a tabular format. Besides, we have also briefly discussed about the challenges that are being faced by the researchers for removing these pollutants from water.

2. Types of effluent exposures, their sources, and impact

Since the 1980s, the wastes discharged from various industries have become an increased environmental concern for the public (Kringstad & Lindström, 1984). Few studies from previous years have shown the evidence of effluents from pulp mills, paper industries, pharmaceutical, and agricultural wastes are capable of inducing many harmful chemicals in the biotic system which has chronic responses on fish's reproductive system even at very minimal concentration. This scenario is disturbing the ecosystem and has environmental consequences (Mark Hewitt, Parrott, & McMaster, 2006). Looking at the intensity of this issue, many researchers have worked and are still working toward determining the detrimental effects of these effluents on the environment and the ecosystem. Due to this problem, a new regulatory package is introduced for the pulp and paper industry to establish new levels of biological oxygen demand (BOD) and total suspended solid waste (TSS) (Munkittrick, McMaster, McCarthy, Servos, & Van Der Kraak, 1998). Not only that, this regulatory system limits huge number of process and treatment changes required in all types of industries and forcibly implemented to all the related business hubs (Servos, Kelly, John, & Glen, 1996). However, there are many more hazardous effluents that consist of many complex compounds which are unidentified and their effects on the biotic systems are unknown. Researchers are trying to identify the origin and environmental fate of these effluents and the treatment procedures to reduce their hazardous effects in the ecosystem. The objective of this chapter is to provide an overview of chemical wastes produced by various kinds of industries, understand their adverse effects on the environment and ecosystem, and look into the challenges of their removal.

2.1 Heavy metal discharge

Since 1966, copper, lead, zinc, etc., are considered as heavy metals which are distressing the environment as well as the biotic system. Later in 1980 and 1981 the levels of heavy metal concentrations increased in the lakes and oceans which was disturbing both the species diversity as well as the population of planktons in the water bodies. This demanded the biological recovery processes to improve the water quality (Deniseger, Erickson, Austin, Roch, & Clark, 1990). Notably, heavy metals give rise to apoptosis in soil sediments along with wastewater and in various animals that live near the disposal sites. With the continuous discharge of significant amount of metal pollutants in water bodies, a huge number of metal particles accumulate in the river and ocean beds, which is increasing the environmental contamination and poisoning the ecosystems (Ali, Khan, & Ilahi, 2019; Raj et al., 2021). Most of the metallic effluents in water bodies comes from the lead in gasoline, acid rain leaching from the soil, precipitation, copper smelting, mining, landfill, nuclear fuel processing, industrial and atmosphere pollution (Varjani, Rakholiya, Shindhal, Shah, & Ngo, 2021). Due to this, heavy metals have been found inside the liver, muscle tissues of many fish species in the polluted water where industrial wastes are dumped (Sobhanardakani, Tayebi, & Farmany, 2011). Not only that, most of the humans who used to work in those factories or people living near those areas are directly getting exposed daily with these hazardous heavy metals. Heavy metals above the permissible levels may have huge impact in human body and other animal species (Mishra et al., 2020) (Table 1). However, there are few technologies available to remove heavy metal and one of them is phytoremediation. This technology is a bioremediation process which helps to purify water from contaminated ions of heavy metals using natural and genetically modified plants (Govere, 2021). Specially, there are many genetically modified microbes which have the efficiency and effectiveness to improve the quality of biotic system using various processes such as phytoaccumulation, rhizofiltration, phytofiltration, phytovolatilization, phytodegradation, etc. (Wang, Chen, Jia, & Chen, 2021). The use of bioremediation for the removal of heavy metals is a cost effective, efficient, solar-driven, and eco-friendly process compared to other engineering processes (Ali, Khan, & Sajad, 2013). Regardless of these advantages, bioremediation has some limitations such as (a) long time to clean up, (b) usage of plants are sometimes restricted by their low biomass, (c) slow growth rate of plants, (d) contamination from multiple pollutants cannot be detected by one specific species, (e) it is very difficult to mobilize the tightly bound fraction of metal ions in water, and (f) exogenous applications of microbes can create problem for the stability of biological treatments (Ashraf, Ali, Zahir, Ashraf, & Asghar, 2019; Chojnacka, 2010; Gerhardt, Gerwing, & Greenberg, 2017). On the other

Table 1 Types of heavy metal residuals released in the water and their effects on plants, human health and in biotic systems.

Heavy metals	Toxicity form	Toxic effects on humans	Toxic effects on plants	Toxic effects on aquatic animals	References
Arsenic (As)	Monomethylarsonic acid (MMA), dimethylarsinic acid (DMA)	Causes cancer, leads to miscarriage in pregnant females and skin problems	Stunts the growth, affects the health of leaves	Damage the vegetative stage of planktons, and damage fish's reproduction	Nizam et al. (2013), Heck, Park, Qiu, Cockburn, and Ritz (2014), Brombach, Lancaster, Dexter, Thompson, and Corns (2021), and Jaishankar, Tseten, Anbalagan, Mathew, and Beeregowda (2014)
Cadmium (Cd)	Cd^{2+}	Damage kidneys and bones	Stunts the root's growth, stop the seed germination, reduce stems	Contaminate the drinking water which is huge threat to humans and animals	Aziz, Adlan, and Ariffin (2008), Cimrin and Yilmaz (2005), Gaur et al. (2021), and Nordberg (2004)
Copper (Cu)	Cu salts	Damage liver, kidneys, create dizziness, nausea, stomach pain, and vomiting	Reduce polypeptide, root deformation, and change the lipid content	Decrease the oxidation potential in water which disturbs the fished and water plants	Babilas, Muszyński, Milewski, Leśniak-Ziółkowska, and Dydo (2021), Kodama, Fujisawa, and Bhadhprasit (2012), and Shkolnik (2012)

Chromium (Cr)	Cr salts	Acute renal failure, cancers, hyperemia, and lymphocytic infiltration	Spoil the nutrients of the plants	Damage fish's growth, poison the water with contamination	Malik, Bashir, Qureashi, and Pandith (2019) and Singh, Gautam, Mishra, and Gupta (2011)
Lead (Pb)	Organic lead	Damage brain, causes Alzheimer's disease, dementia, neurodegenerative disease, kidney damage	Spoil the plants and make them poisonous	Poison water bodies affecting the fishes, insects, and water plants, disturb the aquatic ecosystem	Malik et al. (2019) and Ramos-Guivar, Flores-Cano, and Caetano Passamani (2021)
Zinc (Zn)	Zn salts	Respiratory disorders, neuronal damages, prostate cancer risks	Stunts the growth of plants	Damage the growth of fish and contaminate the drinking water	Forcella et al. (2022) and Liu, Zhang, Cao, Peng, and Barvor (2021)

hand, spectroscopic techniques are also frequently used to identify the elemental composition of heavy metals from aqueous samples due to their electromagnetic characteristics (Helaluddin, Khalid, Alaama, & Abbas, 2016; Malik et al., 2019). However, various natural, organic, and inorganic nanomaterials help in removal of heavy metals, salts, and metal oxides (Kumar et al., 2019, 2020; Roy & Ahmed, 2021; Roy et al., 2021; Roy, Malode, Shetti, & Chandra, 2019; Roy, Rahman, & Ahmed, 2019; Saleh, 2020; Shetti et al., 2020). Many electrochemical sensing methods are also implemented for the detection of contaminated water samples due to their user-friendly, reliable, and economic approach (Munirah, Roy, Ying, Rahman, & Ahmed, 2016; Pujol et al., 2014; Roy, Abd Rahman, & Ahmed, 2016; Roy, Hossain, et al., 2017; Roy, Mohd-Naim, Safavieh, & Ahmed, 2017; Roy, Rahman, & Ahmed, 2016; Roy, Rahman, Santos, & Ahmed, 2016; Roy, Wei, Ying, Safavieh, & Ahmed, 2016). A major advantage of using electrochemical sensors is shorter assay time as compared to spectroscopy and bioremediation technologies. To remove the heavy metals and their ions, typical biosensing methods are commonly used such as potentiometry, amperometry, voltammetry, pulsed voltammetry, etc. (Devi, Sasidharan, & Sundramoorthy, 2018; Szłyk & Szydłowska-Czerniak, 2004). Nevertheless, these technologies also have notable drawbacks like sensitivity and specificity issues, need to improve certain designs for heavy metal detection such as modifications of sensing electrodes, materials of electrode, design of a compatible sensor unit, current or voltage issues for sensor unit, etc. (Bansod, Kumar, Thakur, Rana, & Singh, 2017; Cui, Wu, & Ju, 2015). Hence, detection and removal of heavy metal ions from biotic systems requires many more developments for human health safety purposes. Many advanced and adopted methods are showing great performances but are having limitations as well. Such processes are being modified from time-to-time in order to improve their performance to prevent these above-mentioned drawbacks in future.

2.1.1 Challenges in heavy metal pollutant removal

Conventional heavy metal removal techniques such as chemical precipitation, ion exchange, electrocoagulation is found to be very expensive and therefore can't be employed as sustainable solution. A much cheaper and effective alternative method, biosorption has been extensively explored currently (Sharma, Shahnaz, Subbiah, & Narayanasamy, 2020). However, this method is also sensitive to a lot of environmental factors such as pH and temperature of the water body, contact time, initial solute concentration, and amount of adsorbent used. Another well studied method involves the use of microalgae to digest the pollutants. This method also faces some bottlenecks such as high nutrient demands, foreign microorganism contamination, harvesting difficulties, etc. (Yan, Qu, Wang, Cao, & Han, 2022).

2.2 Agricultural wastes

According to the definition of United Nations, agricultural wastes are the waste products produced because of various agricultural operation including wastes from farms, poultry houses, and slaughterhouses. It also includes wastes from the harvest, salt and silt drained from the fields, and run-off fertilizers and pesticides from the fields that end up in water bodies, soil, and air. The composition of agricultural wastes produced from different crops in shown in Table 1. This would give us an idea about what kind of wastes one must deal with when these crops are grown.

2.2.1 Sources

The agricultural wastes can be broadly classified into four categories: crop waste, animal waste, processing waste, and hazardous waste (Agamuthu, 2009). The crop wastes include the wastes generated from the agricultural produce such as straw, husk, sugarcane bagasse, and molasses. The processing wastes consists of consists of packaging materials and the hazardous wastes consists of chemical toxins such as pesticides and insecticides. These are prey-predator specific chemical formulations which along with their benefits, has negative effects as well (Table 2).

2.2.2 Effects of agricultural wastes on the environment

Agriculture can lead to the pollution of land, air, and water. The land pollution is contributed by the agricultural solid wastes such as the ones mentioned in Table 1. When these solid agricultural wastes are burned in open areas, it leads to air pollution due to emission of methane, nitrogen oxide, and ammonia (Kumar & Joshi, 2013). This can cause ozone depletion as well as human health problems such as respiratory distress (e.g., cough, asthma, and bronchitis), eye and skin irritation. The precipitation of chemicals present in pesticides, insecticides, and fertilizers also contributes to the contamination of the land masses used for agriculture (Abler & Shortle, 1995; Galbraith, Amerasinghe, & Huber-Lee, 2005). Since the water bodies are majorly affected by the agricultural wastes, and require multidisciplinary approaches for understanding and purifying the polluted water, a hierarchical complexity of agriculturally related water contaminants was developed by Rickert (1993) (Fig. 1).

The influence of Agricultural wastes on water quality is summarized in Table 3 (Ongley, 1996). This causes an adverse impact on human, animal, and marine health.

Fertilizers are the most important contributors for the enhancement in world agricultural productivity. Uncontrolled application of nitrogen and phosphorous

Table 2 Agricultural waste composition based on their sources.

S. no.	Source	Waste composition
1	Rice[a]	Husk, bran
2	Wheat[b]	Bran, straw
3	Oat[b]	Straw
4	Maize[a]	Stover, husk, skin, cobs
5	Millet[a]	Stover
6	Fruits and vegetables[a]	Outer peel, crown, bud ends, fruit trimming
7	Sugarcane[a]	Sugarcane tops, bagasse, molasses

[a] Phonbumrung and Khemsawas (1998).
[b] Arvanitoyannis and Tserkezou (2008).

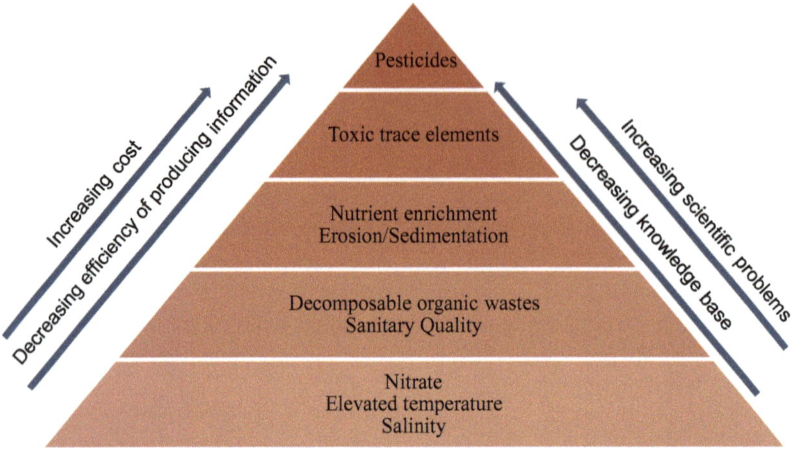

Fig. 1 Hierarchical complexity of agriculturally related water contaminants.
This figure is reproduced from Rickert, D. (1993). *Water quality assessment to determine the nature and extent of water pollution by agriculture and related activities*. Informes sobre Temas Hidricos (FAO) with permission.

Table 3 Influence of agriculture on the quality of water.

Agricultural activity	Influence on surface water	Influence on ground water
Tillage/plowing	Sediment/turbidity: the sediment from these activities consists of remnants of pesticides and biocides, leading to silting of the bottom of water bodies and reduction in aquatic life	Kumar et al. (2021)
Fertilizing	Erosion or surface runoff leads to eutrophication and deoxygenation of water bodies due to excessive phosphorous. This results in reduction of aquatic animals and healthy microbes. It also causes a bad taste and smell in public water supply	Percolation of nitrate to groundwater. This poses a risk to human health (Liu, Zheng, Wei, Kai, & Xu, 2021)
Manure spreading	Surface runoff leads to contamination of water sources due to high levels of pathogens, metals, phosphorus, ammonia, and nitrogen. This results in eutrophication and killing of fishes	Nitrogen contaminates the groundwater (El-Sheekh et al., 2021)

Agricultural activity	Influence on surface water	Influence on ground water
Pesticides	Chemicals containing phosphorous and nitrogen are either carried as dust by the wind or washes off from the soil surface into water bodies. This contamination affects the growth and reproduction of the fishes and other living organisms in the water body resulting disturbance of the aquatic ecosystem. Consumption of the fishes from these water sources can take a toll on human health as well	Percolation of pesticides into groundwater can take a toll on human health when the water is consumes from wells and tube wells (Riyaz, Shah, & Sivasankaran, 2021; Singh, Sharma, Sarma, Misra, & Brar, 2021)
Feedlots/ animal corrals	The nearby water bodies get contaminated by various pathogens such as bacteria, viruses, etc., which can lead to an outbreak of chronic human health concern. The discharge of metals from feces can also lead to water pollution	Percolation of nitrogen, metals, etc., lead to contamination of groundwater (Mehmood, Hakeem, Bhat, & Dar, 2021)
Irrigation	Contamination of water bodies due to surface runoff of salts, fertilizers, and pesticides. This causes disturbance in the aquatic ecosystem and bioaccumulation of trace elements such as selenium in the fishes. Consumption of these fishes can pose threat to public health	Percolation of salts and nitrate into groundwater (Liu, Wang, et al., 2021)
Clear cutting	Land abrasion can cause high turbidity of water bodies and silting of riverbeds. This can result in alteration of hydrologic regime, frequently with disappearance of perennial streams. As a result, there can be shortage of portable water causing a lot of inconvenience in human lifestyle	Increase in the level of groundwater (Löfgren, Ring, von Brömssen, Sørensen, & Högbom, 2009)

containing fertilizers to the crops lead to eutrophication of nearby water bodies and contaminate the ground water to some extent. Another serious issue worth mentioning is Hypoxia, the worst of which was reported in the Gulf of Mexico called the "dead zone." Hypoxia is basically a condition in which the water body is unable to support any life form due to extreme depletion of oxygen (typically, oxygen concentration $<2\,mg/L$). In the Gulf of Mexico, seasonally from late spring the dissolved oxygen starts to deplete and becomes worst during winter. This results in huge premature deaths of fishes. It has been reported to the immense agricultural discharge of nitrogen and phosphorus from the Atchafalaya and Mississippi rivers. This results in sudden increase in algae, which further get eaten by bacteria which reduce the dissolved oxygen to almost zero (Rabalais, Turner, & Wiseman, 2002). To avoid the chemical/inorganic fertilizers, farmers are employing the use of organic fertilizers which are mostly derived from human and animal excreta. They might also contain heavy metals in case the excreta are derived from diseased animals or humans adding onto more problems.

Contrasting to fertilizers, the negative effects of pesticides are compounded by their individual ingredients or ingredients that are reacting with one another such as emulsifiers, wetting agents, and solvents (Pohanish, 2014). Some of the chemical components used in the biocides degrade over time to form harmful by-products. These toxic chemicals ultimately lead to the contamination of soil and water bodies causing a number of unwanted ecological and human health problems (Ongley, 1996). Therefore, it is extremely important to control and scientifically use the fertilizers and biocides in order to minimize their adverse effects on the ecological balance of nature as well as human and animal health.

2.2.3 Challenges in the removal of agricultural wastes

As discussed above, the usage of fertilizers and pesticides create lots of adverse effects in the biotic system as well as in livestock practices and quality of agricultural crops. The use of water from contaminated sources for the irrigation of agricultural lands ends up in the food chain of humans and animals affecting their health (Mali, Sanyal, Bhatt, & Pathak, 2015). Therefore, the main challenge for farmers is to find fertilizers and pesticides which is affordable and has minimal chemical content (Dapena-Mora et al., 2007). The pesticides usually contain high concentration of nitrate which poses severe threat to the biotic systems. Not only nitrate, but the presence of ammonia, sulfur, and similar harmful ions found in contaminated water also comes from agricultural fields (Ahmed, Scholz, Al-Faraj, & Niaz, 2016; Bhat et al., 2017). Researchers have been able to design few device prototypes which can be used for removing the contaminants which are also useful for detecting pathogens in biotic system in large scale (Bhardwaj et al., 2019; Li et al., 2019; Roy et al., 2022; Wei et al., 2014).

2.3 Pharmaceutical wastes

Like heavy metals, pharmaceutical wastes such as expired medicine/products, dispensed drugs, and contaminated medications are also contributing to hazardous effluent discharge. These wastes play a major role in the degradation of the environment.

There are two kinds of drugs; one is prescribed prophylactically for acute and/or chronic illness, known as metabolic drugs and the other one is nonmetabolic drugs. Specially, the nonmetabolic drugs and its compounds spoil the natural aquatic environment when disposed into water bodies. Hence, contaminating the drinking water, treatment plants, lakes, and oceans (Kadam, Patil, Patil, & Tumkur, 2016; Ternes, 1998).

Many endocrine related drugs (EDCs) are getting released by humans into the environment. These kinds of drugs are synthetic organic chemicals which are made up of surfactants, pesticides, poly aromatic hydrocarbons, biophenyls, etc., which if not disposed appropriately can end up polluting the environment with their chemical compounds (Chander et al., 2016). However, in the past few decades, numerous analytical processes have been developed to detect the pharmaceutical elements in drinking water samples (Glassmeyer et al., 2005). One of the main techniques used to detect these compounds in water is high performance liquid chromatography (HPLC). Later, some research also used mass spectrometry along with HPLC to detect the chemical substances in water ranging from μg/L to ng/L concentration (Ebele, Abou-Elwafa Abdallah, & Harrad, 2017; Santos, Aparicio, Callejón, & Alonso, 2009).

2.3.1 Source

The majority (above 90%) of pharmaceutical pollutants come from the medicines used by patients (both household and hospitals combined). The rest comes from improper disposal of the drugs and their initiators from the industries. Reports suggest that the concentration of drugs like enoxacin, norfloxacin, cetirizine, ofloxacin, ciprofloxacin, etc., found in sewage treatment plants near the pharmaceutical companies is around 90–31,000ng/L (Chander et al., 2016). Many new national guidelines have been made by the governments of different countries which includes proper protocols for minimizing the disposal of pharmaceutical wastes in the environment (Suruchi & Rattan, 2014).

2.3.2 Health impacts due to the discharge of drugs in biotic system

Some pharmaceutical compounds are hydrophobic, and some are hydrophilic in nature. The main problem occurs due to hydrophilic drugs which gets dissolved in water easily. Due to this, many health problems such as respiratory disorders, nerve issues, cancers, and reproductive problems occur in humans. Not only humans, but animals, plants, and fishes are also losing their lives and existence due to disposal of heavy drugs and in the water and land resources (India, 2004; Wang, Wang, & Liu, 2015). The long-term exposure of these chemicals inside human body is considered as highly hazardous. The release of EDCs into the biotic system may lead endocrine related diseases, which are affecting reproduction system in human body. This is leading to breast cancer, fertility issues, birth defects, and testicular cancers (Nikolaou, Meric, & Fatta, 2007; Schulman et al., 2002). The pharmaceutical substances like tamoxifen and cyclophosphamide are used for prevention of breast cancer, and these elements are found on surface water which may lead metabolic problems and other serious problems in normal people (Buerge, Buser, Poiger, &

Müller, 2006; Johnson et al., 2008). The significant amount of drug disposal is affecting both human and environmental health. It is disturbing the eco system of a healthy environment. However, the recent advancements in analytical systems can help in detecting the complex pharmaceutical compounds and prevent these issues in future by helping the researchers to form proper guidelines for the disposal of hazardous drugs in the biotic system.

2.3.3 Challenges in pharmaceutical wastes removal

The major problem with the detection of pharmaceutical wastes lies in their concentration in wastewaters, since the amount of the molecules present are much lower than the detection limits of the assays (Dogan, Płotka-Wasylka, Kempińska-Kupczyk, Namieśnik, & Kot-Wasik, 2020). Also, the available methods for detection are not real time and require much sophisticated instrumentation facilities like tandem mass spectrometry and GC-MS (Kolpin et al., 2002). A group of researchers have used and compared different reactors with different configurations in terms of removal of pharmaceutical micropollutants from wastewater samples. According to their study, a significant amount of transformation of wastewater to clean water was possible by removing the pharmaceutical wastes. These reactors could efficiently remove complex pharmaceutical compounds such as trimethoprim, clarithromycin, and nonylphenolethoxylates. The authors were able to detect clear transformation of the contaminated water samples within 15 days (Joss et al., 2005). However, large scale removal of these compounds is still difficult to achieve. Therefore, the management sector of pharmaceutical industries needs to take care of their waste disposal methods and invest in the removal of hazardous compounds into the nearby water bodies to reduce their contamination.

2.4 Microplastics

Microplastics are basically plastic particles within the size range of 5 mm to 1 μm (Hartmann et al., 2019). The term microplastics was first coined in 2004 by Richard Thomson and was referred to plastic particles smaller than 5 mm across. In the past 2 decades, extensive amount of research has been done to quantify the potential adverse effects of microplastics in the marine ecosystem, wildlife and human health (Cole, Lindeque, Halsband, & Galloway, 2011; Zhang et al., 2020). The initial studies were focused on investigating the fate of the microbeads in the personal-care products, which are washed into the water bodies (Fendall & Sewell, 2009). Surprisingly, researchers have since then found microplastics in almost all the places and commodities possible, viz. arctic and Antarctic snow, deep seas and commodities like, drinking water, table salt, etc. The worst part is these pollutants can take decades for complete degradation. Considering just the marine ecosystem, it was reported that up to 51 trillion microplastics could be present on the surface water bodies (Lim, 2021). Therefore, it is very important to discuss the sources of microplastic generation, their fate in that ecosystem and ways to curb their presence in the environment. In this part, we discuss these topics with a detailed focus on the effect of microplastics in a biotic system.

2.4.1 Sources

Microplastics are found in almost everywhere, be it air, water, or soil. The major source of these tiny remnants of plastics is the plastics materials used in everyday life. For instance, it has been reported that the plastic food containers used in day-to-day life leach out a significant amount of microplastics at elevated temperature (Li et al., 2020). Microplastics can be categorized majorly into two categories based on their sources, viz. primary and secondary microplastics.

Primary sources are comprised of micron sized plastics produced for tailored applications like abrasion of tires, laundering synthetic clothes, facial scrubs, drug vectors, etc. (De Falco, Di Pace, Cocca, & Avella, 2019; Gregory, 1996; Kole, Löhr, Van Belleghem, & Ragas, 2017; Santos, Rodríguez-Mozaz, & Barceló, 2021). Synthetic cloth laundering and type wear are the highest contributor of microplastics in this category. De Falco et al. (2019) conducted multiple wash trials on commercial clothes to simulate household wash conditions. They reported that for a kg of cloths washed the amount of microfibers released went up to 1,500,000 (De Falco et al., 2019). It has been reported that the per-capita emission of microplastics from tire wear can go up to a global average of 0.81 kg per year. These numbers come significantly from automobile industries, followed by airplane tires (Kole et al., 2017). Since the advent of the use of microplastics in facial cleansers they have been heavily used since then by the commercial names of microexfoliates or microbeads (Fendall & Sewell, 2009). The rinsed off microplastics go into the water bodies through the wastewater channels. Due to their micron sizes they usually escape the wastewater screening (which is typically up to the minimum value of 1.5 mm) (Vesilind, 2003).

Secondary microplastics are mostly comprised of the fragments of plastics formed from the degradation of larger plastic debris (Cole et al., 2011). Due to the prolonged exposure to the ultraviolet rays of the sun, the polymer chains degrade due to oxidation. Since these plastic materials lose their structural integrity, they turn brittle and develop cracks. These plastics therefore leach out a huge number of plastics in the ocean under the action of waves and turbulences (Cole et al., 2011). Since plastics do not degrade completely and the process can take decades, a huge number of research were conducted to develop plastics called biodegradable plastics. However, such materials couldn't find a practical application because of the cost and their environmental implications (O'Brine & Thompson, 2010). The problem lies in its composition, i.e., the constituting materials for biodegradable polymers consist of synthetic polymers along with starch and other biomaterials. Upon disposal the decomposition happens at a much faster rate because of the presence of biodegradable components but this also leaves behind a huge number of specks of synthetic polymers (O'Brine & Thompson, 2010).

2.4.2 Effects of microplastics in environment

There has been a lot of development recently, on determining the effect of presence of microplastics in an organism. It has been pointed out that if the specks are smaller than the size of cells or tissues, it can enter the cells causing irritations in the cells. For such small specks, a direct analogy can be drawn for their effects with sooty specks as air pollutants. It is speculated that particles of such sizes can enter respiratory tracts and cause respiratory problems (Prata, 2018). These can be the most dangerous category because in a study conducted on pregnant mice showed that specks were present in almost every organ of the fetuses (Prince et al., 2020). However, the mode of harm in case of larger particles are essentially chemical toxicity. Commercial plastic products have a lot of additional components, viz. stabilizers, plasticizers. These additives are hazardous and can cause serious damages like disturbing the endocrine system. Another postulated mode of harm (specifically in case of marine animals) is in terms of nutritional values. When animals ingest the plastic specks, which have no nutritional value they do not eat further to fulfill their nutritional requirements (Di Renzo et al., 2021). Further, it has been found that zooplanktons reproduce less successfully and experience stunted growth due to microplastics (Cole et al., 2013).

2.4.3 Challenges in impact assessment of microplastics

When microplastics are concerned, the first challenge that comes is to detect them in a given sample. Due to the variation in size, shape, chemical composition, it is difficult to detect these pollutants using a defined method (Dey, Uddin, & Jamal, 2021). The major problem with microplastics is to setup a realistic scenario where the effects of microplastics in an organism can be studied. To simulate such scenario, researchers initially started with smaller polystyrene spheres. However, to be more realistic, fibers and specks of unusual shapes coated with biofilm mimicking chemicals are currently being used. Another difficulty is that for laboratory studies, one type, shape and polymer is used to ingest in organisms. However, in real scenario organisms ingest a mixture of specks (Fig. 2). Given the available literature on in vitro studies as of now there is no way these data can be extrapolated to organs in a whole organism (Lim, 2021).

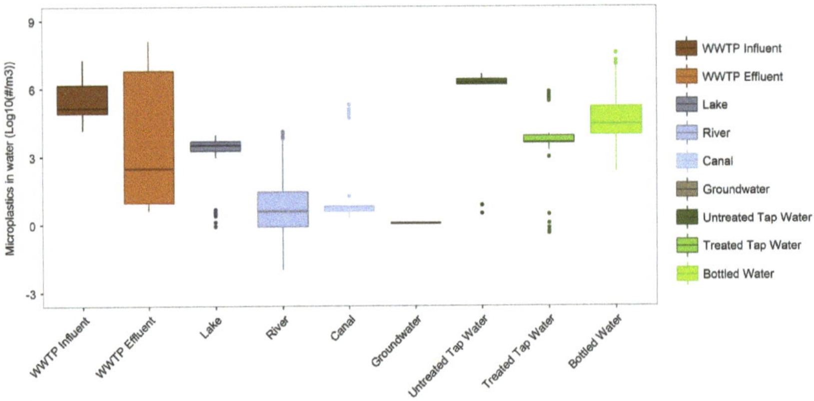

Fig. 2 Median and variation in microplastics concentration for water samples collected from different types of sources.
Reproduced with the permission from Koelmans, A. A., Nor, N. H. M., Hermsen, E., Kooi, M., Mintenig, S. M., & De France, J. (2019). Microplastics in freshwaters and drinking water: Critical review and assessment of data quality. *Water Research*, *155*, 410–422.

3. Conclusions

While protecting the ecological balance of freshwater, we need to think about the water demands of human and other life forms. It is very challenging to control the pollutants in the 21st century. The scarcity of portable water has increased rapidly, and with the increase in world population, the difference between demand and availability of water is also increasing. Hence, proper management of freshwater resources to keep them contamination free is very necessary and will help in reducing the problem of water scarcity. Although this situation has alarmed and stimulated the governments to work toward implementing new approaches for managing water systems with the help of researchers, greater efforts are needed to reserve the water and bring sustainability by reusing the water. Keeping these things in mind, in this chapter we have discussed about the hazardous pollutants from sources like agriculture, industrial wastewater, plastic wastes, and pharmaceutical wastes. We have tried to include the latest developments in this area with a major focus on their sources and effects on the ecosystem. We have also focused on the gravity of the situation including the complexities in curbing the pollutants. As a concluding remark, we would like to assert that drastic changes like reduced lifespan, are already observed in the organisms due to the presence of the pollutants. It is really difficult to degrade these pollutants naturally and therefore they remain around us for up to a time span of 500 years. Therefore, consolidated efforts from researchers, enterprises and the governments are required. On an individual scale, we need to use such products sustainably because for sure we are not using these products for the same time span.

Acknowledgments

Dr. Sharmili Roy thanks Stanford University for encouragement and providing necessary facilities for completion of this work. Vivek Sharma and Monalisha Ghosh Dastidar thanks Indian Institute of Technology Bombay and Australian National University, respectively, for providing the resources used in this work.

References

Abler, D. G., & Shortle, J. S. (1995). Technology as an agricultural pollution control policy. *American Journal of Agricultural Economics*, *77*(1), 20–32. https://doi.org/10.2307/1243885.

Agamuthu, P. (2009). Challenges and opportunities in agro-waste management: An Asian perspective. *Vol. 11, No. 12. Inaugural meeting of first regional 3R forum in Asia*.

Ahmed, T., Scholz, M., Al-Faraj, F., & Niaz, W. (2016). Water-related impacts of climate change on agriculture and subsequently on public Health: A review for generalists with particular reference to Pakistan. *International Journal of Environmental Research and Public Health*, *13*(11). https://doi.org/10.3390/ijerph13111051.

Ali, H., Khan, E., & Ilahi, I. (2019). Environmental chemistry and ecotoxicology of hazardous heavy metals: Environmental persistence, toxicity, and bioaccumulation. *Journal of Chemistry*, *2019*, 6730305. https://doi.org/10.1155/2019/6730305.

Ali, H., Khan, E., & Sajad, M. A. (2013). Phytoremediation of heavy metals—Concepts and applications. *Chemosphere*, *91*(7), 869–881. https://doi.org/10.1016/j.chemosphere.2013.01.075.

Allaire, M., Wu, H., & Lall, U. (2018). National trends in drinking water quality violations. *Proceedings of the National Academy of Sciences*, *115*(9), 2078. https://doi.org/10.1073/pnas.1719805115.

American Public Health Association, American Water Works Association, Water Pollution Control Federation, Water Environment Federation. (1912). *Standard methods for the examination of water and wastewater. Vol. 2*. American Public Health Association.

Arvanitoyannis, I. S., & Tserkezou, P. (2008). Wheat, barley and oat waste: A comparative and critical presentation of methods and potential uses of treated waste. *International Journal of Food Science & Technology*, *43*(4), 694–725. https://doi.org/10.1111/j.1365-2621.2006.01510.x.

Ashraf, S., Ali, Q., Zahir, Z. A., Ashraf, S., & Asghar, H. N. (2019). Phytoremediation: Environmentally sustainable way for reclamation of heavy metal polluted soils. *Ecotoxicology and Environmental Safety*, *174*, 714–727. https://doi.org/10.1016/j.ecoenv.2019.02.068.

Aziz, H. A., Adlan, M. N., & Ariffin, K. S. (2008). Heavy metals (Cd, Pb, Zn, Ni, Cu and Cr(III)) removal from water in Malaysia: Post treatment by high quality limestone. *Bioresource Technology*, *99*(6), 1578–1583. https://doi.org/10.1016/j.biortech.2007.04.007.

Babilas, D., Muszyński, J., Milewski, A., Leśniak-Ziółkowska, K., & Dydo, P. (2021). Electrodialysis enhanced with disodium EDTA as an innovative method for separating Cu(II) ions from zinc salts in wastewater. *Chemical Engineering Journal*, *408*, 127908. https://doi.org/10.1016/j.cej.2020.127908.

Bansod, B., Kumar, T., Thakur, R., Rana, S., & Singh, I. (2017). A review on various electrochemical techniques for heavy metal ions detection with different sensing platforms. *Biosensors and Bioelectronics*, *94*, 443–455. https://doi.org/10.1016/j.bios.2017.03.031.

Baranwal, A., Mahapatra, S., Purohit, B., Roy, S., & Chandra, P. (2020). Insights into novel coronavirus and COVID-19 outbreak. In P. Chandra, & S. Roy (Eds.), *Diagnostic strategies for COVID-19 and other coronaviruses* (pp. 1–17). Springer Singapore. https://doi.org/10.1007/978-981-15-6006-4_1.

Beeby, A. (1993). Measuring the effect of pollution. In *Applying ecology*. London, New York: Chapman and Hall.

Bhardwaj, N., Bhardwaj, S. K., Bhatt, D., Lim, D. K., Kim, K.-H., & Deep, A. (2019). Optical detection of waterborne pathogens using nanomaterials. *Trends in Analytical Chemistry*, *113*, 280–300. https://doi.org/10.1016/j.trac.2019.02.019.

Bhat, R. A., Nazir, R., Ashraf, S., Ali, M., Bandh, S. A., & Kamili, A. N. (2014). Municipal solid waste generation rates and its management at Yusmarg forest ecosystem, a tourist resort in Kashmir. *Waste Management & Research*, *32*(2), 165–169. https://doi.org/10.1177/0734242X13518089.

Bhat, R. A., Shafiq-ur-Rehman, M. M. A., Dervash, M. A., Mushtaq, N., Bhat, J. I. A., & Dar, G. H. (2017). Current status of nutrient load in dal Lake of Kashmir Himalaya. *Journal of Pharmacognosy and Phytochemistry*, *6*(6), 165–169.

Brombach, C.-C., Lancaster, S. T., Dexter, M. A., Thompson, K. C., & Corns, W. T. (2021). Stability, preservation and storage of As(iii), DMA, MMA and As(v) in water samples. *Analytical Methods*, *13*(28), 3154–3159. https://doi.org/10.1039/D1AY00743B.

Buerge, I. J., Buser, H.-R., Poiger, T., & Müller, M. D. (2006). Occurrence and fate of the cytostatic drugs cyclophosphamide and ifosfamide in wastewater and surface waters. *Environmental Science & Technology*, *40*(23), 7242–7250. https://doi.org/10.1021/es0609405.

Chander, V., Sharma, B., Negi, V., Aswal, R., Singh, P., Singh, R., et al. (2016). Pharmaceutical compounds in drinking water. *Journal of Xenobiotics*, *6*(1). https://doi.org/10.4081/xeno.2016.5774.

Chandra, P., & Roy, S. (2020). *Diagnostic strategies for COVID-19 and other coronaviruses*. Springer.

Chojnacka, K. (2010). Biosorption and bioaccumulation—The prospects for practical applications. *Environment International*, *36*(3), 299–307. https://doi.org/10.1016/j.envint.2009.12.001.

Cimrin, K. M., & Yilmaz, I. (2005). Humic acid applications to lettuce do not improve yield but do improve phosphorus availability. *Acta Agriculturae Scandinavica, Section B Soil & Plant Science*, *55*(1), 58–63.

Cole, M., Lindeque, P., Fileman, E., Halsband, C., Goodhead, R., Moger, J., et al. (2013). Microplastic ingestion by zooplankton. *Environmental Science & Technology*, *47*(12), 6646–6655. https://doi.org/10.1021/es400663f.

Cole, M., Lindeque, P., Halsband, C., & Galloway, T. S. (2011). Microplastics as contaminants in the marine environment: A review. *Marine Pollution Bulletin*, *62*(12), 2588–2597. https://doi.org/10.1016/j.marpolbul.2011.09.025.

Cui, L., Wu, J., & Ju, H. (2015). Electrochemical sensing of heavy metal ions with inorganic, organic and bio-materials. *Biosensors and Bioelectronics*, *63*, 276–286. https://doi.org/10.1016/j.bios.2014.07.052.

Dapena-Mora, A., Fernández, I., Campos, J. L., Mosquera-Corral, A., Méndez, R., & Jetten, M. S. M. (2007). Evaluation of activity and inhibition effects on Anammox process by batch tests based on the nitrogen gas production. *Enzyme and Microbial Technology*, *40*(4), 859–865. https://doi.org/10.1016/j.enzmictec.2006.06.018.

Dastidar, M. G., & Roy, S. (2021). Chapter 13—Public health management during COVID-19 and applications of point-of-care based biomolecular detection approaches. In M. Hadi Dehghani, R. R. Karri, & S. Roy (Eds.), *Environmental and health management of novel coronavirus disease (COVID-19)* (pp. 345–378). Academic Press. https://doi.org/10.1016/B978-0-323-85780-2.00009-3.

De Falco, F., Di Pace, E., Cocca, M., & Avella, M. (2019). The contribution of washing processes of synthetic clothes to microplastic pollution. *Scientific Reports*, *9*(1), 6633. https://doi.org/10.1038/s41598-019-43023-x.

Deniseger, J., Erickson, L. J., Austin, A., Roch, M., & Clark, M. J. R. (1990). The effects of decreasing heavy metal concentrations on the biota of Buttle Lake, Vancouver Island, British Columbia. *Water Research*, *24*(4), 403–416. https://doi.org/10.1016/0043-1354(90)90222-R.

Devi, N. R., Sasidharan, M., & Sundramoorthy, A. K. (2018). Gold nanoparticles-thiol-functionalized reduced graphene oxide coated electrochemical sensor system for selective detection of mercury ion. *Journal of the Electrochemical Society*, *165*(8), B3046–B3053. https://doi.org/10.1149/2.0081808jes.

Dey, T. K., Uddin, M. E., & Jamal, M. (2021). Detection and removal of microplastics in wastewater: Evolution and impact. *Environmental Science and Pollution Research*, *28*(14), 16925–16947. https://doi.org/10.1007/s11356-021-12943-5.

Di Renzo, L., Mascilongo, G., Berti, M., Bogdanović, T., Listeš, E., Brkljača, M., et al. (2021). Potential impact of microplastics and additives on the health status of loggerhead turtles (Caretta caretta) stranded along the central Adriatic coast. *Water, Air, & Soil Pollution*, *232*(3), 98. https://doi.org/10.1007/s11270-021-04994-8.

Dogan, A., Płotka-Wasylka, J., Kempińska-Kupczyk, D., Namieśnik, J., & Kot-Wasik, A. (2020). Detection, identification and determination of chiral pharmaceutical residues in wastewater: Problems and challenges. *Trends in Analytical Chemistry*, *122*, 115710. https://doi.org/10.1016/j.trac.2019.115710.

Ebele, A. J., Abou-Elwafa Abdallah, M., & Harrad, S. (2017). Pharmaceuticals and personal care products (PPCPs) in the freshwater aquatic environment. *Emerging Contaminants*, *3*(1), 1–16. https://doi.org/10.1016/j.emcon.2016.12.004.

El-Sheekh, M., Abdel-Daim, M. M., Okba, M., Gharib, S., Soliman, A., & El-Kassas, H. (2021). Green technology for bioremediation of the eutrophication phenomenon in aquatic ecosystems: A review. *African Journal of Aquatic Science*, *46*(3), 274–292. https://doi.org/10.2989/16085914.2020.1860892.

Fendall, L. S., & Sewell, M. A. (2009). Contributing to marine pollution by washing your face: Microplastics in facial cleansers. *Marine Pollution Bulletin*, *58*(8), 1225–1228. https://doi.org/10.1016/j.marpolbul.2009.04.025.

Forcella, M., Lau, P., Fabbri, M., Fusi, P., Oldani, M., Melchioretto, P., et al. (2022). Is cadmium toxicity tissue-specific? Toxicogenomics studies reveal common and specific pathways in pulmonary, hepatic, and neuronal cell models. *International Journal of Molecular Sciences*, *23*(3). https://doi.org/10.3390/ijms23031768.

Galbraith, H., Amerasinghe, P., & Huber-Lee, A. (2005). *The effects of agricultural irrigation on wetland ecosystems in developing countries: A literature review*. https://hdl.handle.net/10535/5054.

Gaur, V. K., Sharma, P., Gaur, P., Varjani, S., Ngo, H. H., Guo, W., et al. (2021). Sustainable mitigation of heavy metals from effluents: Toxicity and fate with recent technological advancements. *Bioengineered*, *12*(1), 7297–7313. https://doi.org/10.1080/21655979.2021.1978616.

Gerhardt, K. E., Gerwing, P. D., & Greenberg, B. M. (2017). Opinion: Taking phytoremediation from proven technology to accepted practice. *Plant Science*, *256*, 170–185. https://doi.org/10.1016/j.plantsci.2016.11.016.

Glassmeyer, S. T., Furlong, E. T., Kolpin, D. W., Cahill, J. D., Zaugg, S. D., Werner, S. L., et al. (2005). Transport of chemical and microbial compounds from known wastewater discharges: Potential for use as indicators of human fecal contamination. *Environmental Science & Technology*, *39*(14), 5157–5169. https://doi.org/10.1021/es048120k.

Govere, E. M. (2021). Environmental phytoremediation and analytical technologies for heavy metal removal and assessment. In A. Ricroch, S. Chopra, & M. Kuntz (Eds.), *Plant biotechnology: Experience and future prospects* (pp. 203–213). Springer International Publishing. https://doi.org/10.1007/978-3-030-68345-0_14.

Gregory, M. R. (1996). Plastic 'scrubbers' in hand cleansers: A further (and minor) source for marine pollution identified. *Marine Pollution Bulletin*, *32*(12), 867–871. https://doi.org/10.1016/S0025-326X(96)00047-1.

Hartmann, N. B., Hüffer, T., Thompson, R. C., Hassellöv, M., Verschoor, A., Daugaard, A. E., et al. (2019). Are we speaking the same language? Recommendations for a definition and categorization framework for plastic debris. *Environmental Science & Technology*, *53*(3), 1039–1047. https://doi.org/10.1021/acs.est.8b05297.

Heck, J. E., Park, A. S., Qiu, J., Cockburn, M., & Ritz, B. (2014). Risk of leukemia in relation to exposure to ambient air toxics in pregnancy and early childhood. *International Journal of Hygiene and Environmental Health*, *217*(6), 662–668. https://doi.org/10.1016/j.ijheh.2013.12.003.

Helaluddin, A. B. M., Khalid, R. S., Alaama, M., & Abbas, S. A. (2016). Main analytical techniques used for elemental analysis in various matrices. *Tropical Journal of Pharmaceutical Research*, *15*(2), 427–434.

Helmer, R., & Hespanhol, I. (1997). *Water pollution control: A guide to the use of water quality management principles*. CRC Press.

India, G. (2004). *State of community health at Medak district*. Retrieved November, 24, 2009.

Jackson, R. B., Carpenter, S. R., Dahm, C. N., McKnight, D. M., Naiman, R. J., Postel, S. L., et al. (2001). Water in a changing. *Ecological Applications*, *11*(4), 1027–1045. https://doi.org/10.1890/1051-0761(2001)011[1027:WIACW]2.0.CO;2.

Jaishankar, M., Tseten, T., Anbalagan, N., Mathew, B. B., & Beeregowda, K. N. (2014). Toxicity, mechanism and health effects of some heavy metals. *Interdisciplinary Toxicology*, *7* (2), 60–72. https://doi.org/10.2478/intox-2014-0009.

Johnson, A. C., Jürgens, M. D., Williams, R. J., Kümmerer, K., Kortenkamp, A., & Sumpter, J. P. (2008). Do cytotoxic chemotherapy drugs discharged into rivers pose a risk to the environment and human health? An overview and UK case study. *Journal of Hydrology*, *348*(1), 167–175. https://doi.org/10.1016/j.jhydrol.2007.09.054.

Joss, A., Keller, E., Alder, A. C., Göbel, A., McArdell, C. S., Ternes, T., et al. (2005). Removal of pharmaceuticals and fragrances in biological wastewater treatment. *Water Research*, *39* (14), 3139–3152. https://doi.org/10.1016/j.watres.2005.05.031.

Kadam, A., Patil, S. S., Patil, S. V., & Tumkur, A. (2016). Pharmaceutical waste management: An overview. *Indian Journal of Pharmacy Practice*, *9*, 2–8.

Khan, A. H., Tirth, V., Fawzy, M., Mahmoud, A. E. D., Khan, N. A., Ahmed, S., et al. (2021). COVID-19 transmission, vulnerability, persistence and nanotherapy: A review. *Environmental Chemistry Letters*, *19*(4), 2773–2787. https://doi.org/10.1007/s10311-021-01229-4.

Kodama, H., Fujisawa, C., & Bhadhprasit, W. (2012). Inherited copper transport disorders: Biochemical mechanisms, diagnosis, and treatment. *Current Drug Metabolism*, *13*(3), 237–250.

Kole, P. J., Löhr, A. J., Van Belleghem, F. G. A. J., & Ragas, A. M. J. (2017). Wear and tear of tyres: A stealthy source of microplastics in the environment. *International Journal of Environmental Research and Public Health*, *14*(10). https://doi.org/10.3390/ijerph14101265.

Kolpin, D. W., Furlong, E. T., Meyer, M. T., Thurman, E. M., Zaugg, S. D., Barber, L. B., et al. (2002). Pharmaceuticals, hormones, and other organic wastewater contaminants in U.S. streams, 1999 – 2000: A national reconnaissance. *Environmental Science & Technology*, *36*(6), 1202–1211. https://doi.org/10.1021/es011055j.

Kringstad, K. P., & Lindström, K. (1984). Spent liquors from pulp bleaching. *Environmental Science & Technology*, *18*(8), 236A–248A. https://doi.org/10.1021/es00126a714.

Kumar, P., & Joshi, L. (2013). Pollution caused by agricultural waste burning and possible alternate uses of crop stubble: A case study of Punjab. In S. Nautiyal, K. S. Rao, H. Kaechele, K. V. Raju, & R. Schaldach (Eds.), *Knowledge systems of societies for adaptation and mitigation of impacts of climate change* (pp. 367–385). Springer Berlin Heidelberg. https://doi.org/10.1007/978-3-642-36143-2_22.

Kumar, N., Kumar, A., Marwein, B. M., Verma, D. K., Jayabalan, I., Kumar, A., et al. (2021). Agricultural activities causing water pollution and its mitigation—A review. *International Journal of Modern Agriculture*, *10*(1), 590–609.

Kumar, A., Purohit, B., Mahato, K., Roy, S., Srivastava, A., & Chandra, P. (2020). Design and development of ultrafast sinapic acid sensor based on electrochemically nanotuned gold nanoparticles and solvothermally reduced graphene oxide. *Electroanalysis*, *32*(1), 59–69.

Kumar, A., Roy, S., Srivastava, A., Naikwade, M. M., Purohit, B., Mahato, K., et al. (2019). Chapter 10—Nanotherapeutics: A novel and powerful approach in modern healthcare system. In P. K. Maurya, & S. Singh (Eds.), *Nanotechnology in modern animal biotechnology* (pp. 149–161). Elsevier. https://doi.org/10.1016/B978-0-12-818823-1.00010-7.

Li, F., Hu, Y., Li, Z., Liu, J., Guo, L., & He, J. (2019). Three-dimensional microfluidic paper-based device for multiplexed colorimetric detection of six metal ions combined with use of a smartphone. *Analytical and Bioanalytical Chemistry*, *411*(24), 6497–6508. https://doi.org/10.1007/s00216-019-02032-5.

Li, D., Shi, Y., Yang, L., Xiao, L., Kehoe, D. K., Gun'ko, Y. K., et al. (2020). Microplastic release from the degradation of polypropylene feeding bottles during infant formula preparation. *Nature Food*, *1*(11), 746–754. https://doi.org/10.1038/s43016-020-00171-y.

Lim, X. (2021). *Microplastics are everywhere—But are they harmful?*. Nature Publishing Group.

Liu, Y., Wang, P., Gojenko, B., Yu, J., Wei, L., Luo, D., et al. (2021). A review of water pollution arising from agriculture and mining activities in Central Asia: Facts, causes and effects. *Environmental Pollution*, *291*, 118209. https://doi.org/10.1016/j.envpol.2021.118209.

Liu, Q., Zhang, M., Cao, Y., Peng, B., & Barvor, J. B. (2021). Zinc cyanide removal by precipitation with quaternary ammonium salts. *Separation and Purification Technology*, *274*, 119057. https://doi.org/10.1016/j.seppur.2021.119057.

Liu, L., Zheng, X., Wei, X., Kai, Z., & Xu, Y. (2021). Excessive application of chemical fertilizer and organophosphorus pesticides induced total phosphorus loss from planting causing surface water eutrophication. *Scientific Reports*, *11*(1), 23015. https://doi.org/10.1038/s41598-021-02521-7.

Löfgren, S., Ring, E., von Brömssen, C., Sørensen, R., & Högbom, L. (2009). Short-term effects of clear-cutting on the water chemistry of two boreal streams in northern Sweden: A paired catchment study. *Ambio*, 347–356.

Mali, S. S., Sanyal, S. K., Bhatt, B. P., & Pathak, H. (2015). Water pollution and agriculture. *State of Indian Agriculture-Water*, 39–47.

Malik, L. A., Bashir, A., Qureashi, A., & Pandith, A. H. (2019). Detection and removal of heavy metal ions: A review. *Environmental Chemistry Letters*, *17*(4), 1495–1521. https://doi.org/10.1007/s10311-019-00891-z.

Mark Hewitt, L., Parrott, J. L., & McMaster, M. E. (2006). A decade of research on the environmental impacts of pulp and paper mill effluents in Canada: Sources and characteristics of bioactive substances. *Journal of Toxicology and Environmental Health, Part B*, *9*(4), 341–356. https://doi.org/10.1080/15287390500195976.

Mehmood, M. A., Hakeem, K. R., Bhat, R. A., & Dar, G. H. (2021). *Pesticide contamination in freshwater and soil environs: Impacts, threats, and sustainable remediation*. CRC Press.

Mishra, B., Varjani, S., Agrawal, D. C., Mandal, S. K., Ngo, H. H., Taherzadeh, M. J., et al. (2020). Engineering biocatalytic material for the remediation of pollutants: A comprehensive review. *Environmental Technology & Innovation*, *20*, 101063. https://doi.org/10.1016/j.eti.2020.101063.

Munirah, H., Roy, S., Ying, J. L. Z., Rahman, I. A., & Ahmed, M. U. (2016). Rapid detection of pork DNA in food samples using reusable electrochemical sensor. *Scientia Bruneiana*, *15*.

Munkittrick, K. R., McMaster, M. E., McCarthy, L. H., Servos, M. R., & Van Der Kraak, G. J. (1998). An overview of recent studies on the potential of pulp-mill effluents to alter reproductive parameters in fish. *Journal of Toxicology and Environmental Health, Part B*, *1*(4), 347–371. https://doi.org/10.1080/10937409809524558.

Mushtaq, N., Singh, D. V., Bhat, R. A., Dervash, M. A., Hameed, O., & b. (2020). Freshwater contamination: Sources and hazards to aquatic biota. In H. Qadri, R. A. Bhat, M. A. Mehmood, & G. H. Dar (Eds.), *Fresh water pollution dynamics and remediation* (pp. 27–50). Springer Singapore. https://doi.org/10.1007/978-981-13-8277-2_3.

Nel, L. H., & Markotter, W. (2009). New and emerging waterborne infectious diseases. *Water and Health, 1*, 147.

Nikolaou, A., Meric, S., & Fatta, D. (2007). Occurrence patterns of pharmaceuticals in water and wastewater environments. *Analytical and Bioanalytical Chemistry, 387*(4), 1225–1234. https://doi.org/10.1007/s00216-006-1035-8.

Nizam, S., Kato, M., Yatsuya, H., Khalequzzaman, M., Ohnuma, S., Naito, H., et al. (2013). Differences in urinary arsenic metabolites between diabetic and non-diabetic subjects in Bangladesh. *International Journal of Environmental Research and Public Health, 10* (3). https://doi.org/10.3390/ijerph10031006.

Nordberg, G. F. (2004). Cadmium and health in the 21st century—Historical remarks and trends for the future. *Biometals, 17*(5), 485–489. https://doi.org/10.1023/B:BIOM.0000045726.75367.85.

O'Brine, T., & Thompson, R. C. (2010). Degradation of plastic carrier bags in the marine environment. *Marine Pollution Bulletin, 60*(12), 2279–2283. https://doi.org/10.1016/j.marpolbul.2010.08.005.

Ongley, E. D. (1996). *Control of water pollution from agriculture. Vol. 55*. Food & Agriculture Organization.

Ormaza-Gonzaìlez, F. I., Castro-Rodas, D., & Statham, P. J. (2021). COVID-19 impacts on beaches and coastal water pollution at selected sites in Ecuador, and management proposals post-pandemic. *Frontiers in Marine Science, 8*, 669374.

Phonbumrung, T., & Khemsawas, C. (1998, October). Agricultural crop residue. *Proceedings of the sixth meeting of regional working group on grazing and feed resources for Southeast Asia, Legaspi, Philippines* (pp. 5–9).

Pohanish, R. P. (2014). *Sittig's handbook of pesticides and agricultural chemicals*. William Andrew.

Prata, J. C. (2018). Airborne microplastics: Consequences to human health? *Environmental Pollution, 234*, 115–126. https://doi.org/10.1016/j.envpol.2017.11.043.

Prince, S. A., Cardilli, L., Reed, J. L., Saunders, T. J., Kite, C., Douillette, K., et al. (2020). A comparison of self-reported and device measured sedentary behaviour in adults: A systematic review and meta-analysis. *International Journal of Behavioral Nutrition and Physical Activity, 17*(1), 31. https://doi.org/10.1186/s12966-020-00938-3.

Pujol, L., Evrard, D., Groenen-Serrano, K., Freyssinier, M., Ruffien-Cizsak, A., & Gros, P. (2014). Electrochemical sensors and devices for heavy metals assay in water: The French groups' contribution. *Frontiers in Chemistry, 2*, 19. https://www.frontiersin.org/article/10.3389/fchem.2014.00019.

Rabalais, N. N., Turner, R. E., & Wiseman, W. J. (2002). Gulf of Mexico hypoxia, a.k.a. "the dead zone". *Annual Review of Ecology and Systematics, 33*(1), 235–263. https://doi.org/10.1146/annurev.ecolsys.33.010802.150513.

Raj, A., Yadav, A., Rawat, A. P., Singh, A. K., Kumar, S., Pandey, A. K., et al. (2021). Kinetic and thermodynamic investigations of sewage sludge biochar in removal of remazol brilliant blue R dye from aqueous solution and evaluation of residual dyes cytotoxicity. *Environmental Technology & Innovation, 23*, 101556. https://doi.org/10.1016/j.eti.2021.101556.

Ramos-Guivar, J. A., Flores-Cano, D. A., & Caetano Passamani, E. (2021). Differentiating nanomaghemite and nanomagnetite and discussing their importance in arsenic and lead removal from contaminated effluents: A critical review. *Nanomaterials, 11*(9). https://doi.org/10.3390/nano11092310.

Rickert, D. (1993). *Water quality assessment to determine the nature and extent of water pollution by agriculture and related activities*. Informes sobre Temas Hidricos (FAO).

Rijsberman, F., SuffixJr, H. V., & Boland, J. J. (2004). *Challenge of sanitation and water.* Copenhagen Consensus Center.

Riyaz, M., Shah, R. A., & Sivasankaran, K. (2021). Pesticide residues: Impacts on fauna and the environment. In *Biodegradation* IntechOpen.

Roy, S. X. W. S., Abd Rahman, I., & Ahmed, M. U. (2016). Based visual detection of Salmonella bacteria using isothermal DNA amplification and magnetic bead aggregation. *Malaysian Journal of Microbiology, 12*(5), 332–338.

Roy, S., & Ahmed, M. U. (2021). *System and method for immobilization free electrochemi luminescence DNA detection using a luminophore dye for multi-species detection.* Google Patents.

Roy, S., Arshad, F., Eissa, S., Safavieh, M., Alattas, S. G., Ahmed, M. U., et al. (2022). Recent developments towards portable point-of-care diagnostic devices for pathogen detection. *Sensors & Diagnostics*. https://doi.org/10.1039/D1SD00017A.

Roy, S., & Baranwal, A. (2020). Diverse molecular techniques for early diagnosis of COVID-19 and other coronaviruses. In P. Chandra, & S. Roy (Eds.), *Diagnostic strategies for COVID-19 and other coronaviruses* (pp. 135–159). Springer Singapore. https://doi.org/10.1007/978-981-15-6006-4_7.

Roy, S., Dikshit, P. K., Sherpa, K. C., Singh, A., Jacob, S., & Chandra Rajak, R. (2021). Recent nanobiotechnological advancements in lignocellulosic biomass valorization: A review. *Journal of Environmental Management, 297*, 113422. https://doi.org/10.1016/j.jenvman.2021.113422.

Roy, S., Hossain, M. M., Safavieh, M., Lubis, H. N., Zourob, M., & Ahmed, M. U. (2017). Chapter 16. Isothermal DNA amplification strategies for food biosensors. In *Food biosensors* (pp. 367–392). The Royal Society of Chemistry. https://doi.org/10.1039/9781782623908-00367.

Roy, S., Malode, S. J., Shetti, N. P., & Chandra, P. (2019). Modernization of biosensing strategies for the development of lab-on-chip integrated systems. *Bioelectrochemical Interface Engineering*, 325–342. https://doi.org/10.1002/9781119611103.ch17.

Roy, S., Mohd-Naim, N. F., Safavieh, M., & Ahmed, M. U. (2017). Colorimetric nucleic acid detection on paper microchip using loop mediated isothermal amplification and crystal violet dye. *ACS Sensors, 2*(11), 1713–1720. https://doi.org/10.1021/acssensors.7b00671.

Roy, S., Rahman, I. A., & Ahmed, M. U. (2016). Paper-based rapid detection of pork and chicken using LAMP–magnetic bead aggregates. *Analytical Methods, 8*(11), 2391–2399. https://doi.org/10.1039/C6AY00274A.

Roy, S., Rahman, I. A., & Ahmed, M. U. (2019). *Electrochemical DNA biosensor using graphene biochip for species identification.* Google Patents.

Roy, S., Rahman, I. A., Santos, J. H., & Ahmed, M. U. (2016). Meat species identification using DNA-redox electrostatic interactions and non-specific adsorption on graphene biochips. *Food Control, 61*, 70–78. https://doi.org/10.1016/j.foodcont.2015.09.029.

Roy, S., & Ramadoss, A. (2021). Chapter 1—Updated insight into COVID-19 disease and health management to combat the pandemic. In M. Hadi Dehghani, R. R. Karri, & S. Roy (Eds.), *Environmental and health management of novel coronavirus disease (COVID-19)* (pp. 3–39). Academic Press. https://doi.org/10.1016/B978-0-323-85780-2.00017-2.

Roy, S., Wei, S. X., Ying, J. L. Z., Safavieh, M., & Ahmed, M. U. (2016). A novel, sensitive and label-free loop-mediated isothermal amplification detection method for nucleic acids using luminophore dyes. *Biosensors and Bioelectronics, 86*, 346–352. https://doi.org/10.1016/j.bios.2016.06.065.

Saleh, T. A. (2020). Trends in the sample preparation and analysis of nanomaterials as environmental contaminants. *Trends in Environmental Analytical Chemistry*, *28*, e00101. https://doi.org/10.1016/j.teac.2020.e00101.

Sampath Kumar, N. S., Chintagunta, A. D., Jeevan Kumar, S. P., Roy, S., & Kumar, M. (2020). Immunotherapeutics for Covid-19 and post vaccination surveillance. *3Biotech*, *10*(12), 527. https://doi.org/10.1007/s13205-020-02522-9.

Santos, J. L., Aparicio, I., Callejón, M., & Alonso, E. (2009). Occurrence of pharmaceutically active compounds during 1-year period in wastewaters from four wastewater treatment plants in Seville (Spain). *Journal of Hazardous Materials*, *164*(2–3), 1509–1516. https://doi.org/10.1016/j.jhazmat.2008.09.073.

Santos, L. H. M. L. M., Rodríguez-Mozaz, S., & Barceló, D. (2021). Microplastics as vectors of pharmaceuticals in aquatic organisms—An overview of their environmental implications. *Case Studies in Chemical and Environmental Engineering*, *3*, 100079. https://doi.org/10.1016/j.cscee.2021.100079.

Schulman, L. J., Sargent, E. V., Naumann, B. D., Faria, E. C., Dolan, D. G., & Wargo, J. P. (2002). A human Health risk assessment of pharmaceuticals in the aquatic environment. *Human and Ecological Risk Assessment: An International Journal*, *8*(4), 657–680. https://doi.org/10.1080/20028091057141.

Servos, M. R., Kelly, R. M., John, H. C., & Glen, J. (1996). In Van Der Kraak (Ed.), *Environmental fate and effects of pulp and paper mill effluents*. Delray Beach: St. Lucie Press.

Sharma, V., Shahnaz, T., Subbiah, S., & Narayanasamy, S. (2020). New insights into the remediation of water pollutants using nanobentonite incorporated nanocellulose chitosan based aerogel. *Journal of Polymers and the Environment*, *28*(7), 2008–2019. https://doi.org/10.1007/s10924-020-01740-9.

Shetti, N. P., Malode, S. J., Roy, S., Chandra, P., Reddy, K. R., & Chatterjee, S. (2020). 11—Electroanalytical techniques for investigating biofilms: Applications in biosensing and biomolecular interfacing. In S. Kanchi, & D. Sharma (Eds.), *Nanomaterials in diagnostic tools and devices* (pp. 293–329). Elsevier. https://doi.org/10.1016/B978-0-12-817923-9.00011-0.

Shkolnik, M. Y. (2012). *Trace elements in plants*. Elsevier.

Singh, R., Gautam, N., Mishra, A., & Gupta, R. (2011). Heavy metals and living systems: An overview. *Indian Journal of Pharmacology*, *43*(3), 246–253. https://doi.org/10.4103/0253-7613.81505.

Singh, S., Sharma, S., Sarma, S. J., Misra, K., & Brar, S. K. (2021). Chapter 10—Pesticides in water. In S. Ahuja (Ed.), *Handbook of water purity and quality* (2nd ed., pp. 231–253). Academic Press. https://doi.org/10.1016/B978-0-12-821057-4.00004-5.

Sobhanardakani, S., Tayebi, L., & Farmany, A. (2011). Toxic metal (Pb, Hg and As) contamination of muscle, gill and liver tissues of Otolithes rubber, Pampus argenteus, Parastromateus niger, Scomberomorus commerson and Onchorynchus mykiss. *World Applied Sciences Journal*, *14*(10), 1453–1456.

Suruchi, A., & Rattan, A. (2014). Minimizing pharmaceutical waste: The role of the pharmacist. *Journal of Young Pharmacists*, *6*(3), 14–19. https://doi.org/10.5530/jyp.2014.3.3.

Szłyk, E., & Szydłowska-Czerniak, A. (2004). Determination of cadmium, lead, and copper in margarines and butters by galvanostatic stripping chronopotentiometry. *Journal of Agricultural and Food Chemistry*, *52*(13), 4064–4071. https://doi.org/10.1021/jf040032w.

Ternes, T. A. (1998). Occurrence of drugs in German sewage treatment plants and rivers. *Water Research*, *32*(11), 3245–3260. https://doi.org/10.1016/S0043-1354(98)00099-2.

Ullah, S., Javed, M. W., Shafique, M., & Khan, S. F. (2014). An integrated approach for quality assessment of drinking water using GIS: A case study of lower Dir. *Journal of Himalayan Earth Science*, *47*(2).

Varjani, S., Rakholiya, P., Shindhal, T., Shah, A. V., & Ngo, H. H. (2021). Trends in dye industry effluent treatment and recovery of value added products. *Journal of Water Process Engineering*, *39*, 101734. https://doi.org/10.1016/j.jwpe.2020.101734.

Vesilind, P. (2003). *Wastewater treatment plant design. Vol. 2*. IWA Publishing.

Vörösmarty, C. J., Ellen, M. D., Green, P. A., & Revenga, C. (2005). Geospatial indicators of emerging water stress: An application to Africa. *Ambio*, *34*(3), 230–236. http://www.jstor.org/stable/4315590.

Wang, M., Chen, S., Jia, X., & Chen, L. (2021). Chapter 1—Concept and types of bioremediation. In M. Hasanuzzaman, & M. N. V. Prasad (Eds.), *Handbook of bioremediation* (pp. 3–8). Academic Press. https://doi.org/10.1016/B978-0-12-819382-2.00001-6.

Wang, C., Wang, H., & Liu, Y. (2015). Separation of aluminum and plastic by metallurgy method for recycling waste pharmaceutical blisters. *Journal of Cleaner Production*, *102*, 378–383. https://doi.org/10.1016/j.jclepro.2015.04.067.

Wei, Q., Nagi, R., Sadeghi, K., Feng, S., Yan, E., Ki, S. J., et al. (2014). Detection and spatial mapping of mercury contamination in water samples using a smart-phone. *ACS Nano*, *8*(2), 1121–1129. https://doi.org/10.1021/nn406571t.

Yan, C., Qu, Z., Wang, J., Cao, L., & Han, Q. (2022). Microalgal bioremediation of heavy metal pollution in water: Recent advances, challenges, and prospects. *Chemosphere*, *286*, 131870. https://doi.org/10.1016/j.chemosphere.2021.131870.

Zhang, Y., Kang, S., Allen, S., Allen, D., Gao, T., & Sillanpää, M. (2020). Atmospheric microplastics: A review on current status and perspectives. *Earth-Science Reviews*, *203*, 103118. https://doi.org/10.1016/j.earscirev.2020.103118.

Index

Note: Page numbers followed by *f* indicate figures and *t* indicate tables.

A

Activated carbon (AC)
  adsorption technology, 350–351
  characteristics of, 371–372
  conventional treatments, PhACs, 332–333, 333*f*
  removal of pharmaceuticals by, 373–374*t*
Activated sludge treatment, 11
Adsorbents
  chitosan-based, 356*t*
  conventional and nonconventional, 371*f*
Adsorption
  activated carbon, 30
  advantages, 55
  of antibiotics, 376–377
  on binary compounds, 62
  graphene-based, 30–31
  graphene-based materials, 58
  hospital wastewater, 29–30
  pharmaceutically-polluted water streams, 338
  processes, 259–260
    fenton oxidation and, 263–265, 264*f*
    hybrid oxidation and, 259–260
    ozonation and, 265–266, 265*t*
    photocatalysis and, 260–263, 261*t*
  removal by, 370–371
  technology
    activated carbon, 350–351
    in hydrogel, 351*f*
    PhCs, 350*f*
  via intercalation, 56–58
Advanced oxidation processes (AOP), 181–197, 196–197*t*, 341*f*
  advantages and drawbacks, 198*f*
  benefits, 383–384*t*
  biotechnologies and microalgae, 295–296
  carbamazepine, 255–257
  drawbacks, 383–384*t*
  electrochemical oxidation, 258–259
  electrochemical treatment, 385
  fenton and photo-fenton, 386
  fenton oxidation, 253–257
  heterogeneous processes, 385
  homogeneous types, 382–385
  hydroxyl radicals, 339–341
  ozonation, 257–258, 258*t*, 385–386
  photocatalytic oxidation, 252–253, 252*f*, 254*f*
  purification of pharmaceutical wastewaters, 249–259
  removal of pharmaceuticals in, 382–387
  sonolysis, 386–387
  types of, 251*f*
  UV and UV/H_2O_2, 255*t*, 256*f*
Agricultural waste
  effluent, 404–408
  solid farm waste, 372–375
  water-contaminated waste-based adsorbents, 375
Alum, 334–335
Aluminum chloride and sulfate, 334–335
Anaerobic membrane bioreactors (AnMBRs), 232
  conventional treatments, PhACs, 332*f*
Antibiotic resistance bacteria (ARB), 149
Antibiotic resistance genes (ARGs), 149
Antibiotics, 220–221
  adsorption of, 376–377
  pharmaceutical adsorption, 356*t*
  physiochemical properties, PhACs, 72–78, 73–77*t*
  wastewater, 329–330
Aquatic environments
  fate of, in environment, 130–132
  occurrence of, PhACs, 120–123, 121–123*t*
  pharmaceutical residues, 219*f*
  source of pharmaceuticals in, 348*f*
Aqueous matrices, pharmaceuticals frequency in, 150*f*

B

Beta-blockers, physiochemical properties, 86–96, 92–96*t*
Bioaccumulation, PC removal mechanisms, 279–280
Bioadsorption, microalgae, 278–279
Bioaugmentation, 11
Biochemical oxygen demand (BOD), 330
Biodegradation (biotransformation), microalgae, 280–282
Bioentrapped membrane reactor (BEMR), 232–233
BiOI microspheres, 262–263, 262*f*
Biological methods, 10–12, 223–224
Biopharmaceutical engineering, 329
Biopolymers
 adsorbent materials, 351–352
 applications, 353*f*
 classification, 351–353
 cross-linked, 353–357
 modification, 352*t*
 PhCs wastewater treatment, 352*f*
 wastewater treatment, 352*t*
Bioremediation
 heavy metal discharge, 401–404
 technologies, 277
Biosolids, 6
Biotechnologies and microalgae
 advanced oxidation processes, 295–296
 constructed wetlands, 296–297
 conventional activated sludge, 297–298
 membrane photobioreactor, 298
 microbial fuel cells, 297
Biotransformation, 11–12

C

Carbamazepine, 263, 380
Carbamazepine (CBZ), 255–257
Carbon nanotube (CNT), 55–56
 electrochemical sensor, 170–171
 graphene, 60
Carbon-xerogel/GO (GO/CX) hybrids, 61
 formation of, 62*f*
Cavitation, 194–195
Chemical cross-linking, 355–357
Chemical oxygen demand (COD), 330
Chitosan-based adsorbents, 356–358*t*
Chlorination, 187–189
 by-products, 187
 degradation by-products, 188
Chromatographic techniques, 161–167
 column, 161
 detectors, 167–170
 gas, 164–167, 164*f*
 liquid, 162–163*t*, 163
 specification of, 162*t*
Coagulation, 9
 conventional treatments, PhACs, 334–335
 and flocculation process, 334*f*
Column chromatographic techniques, 161
Cometabolism, of microalgae, 295
Constructed wastelands, 10
Constructed wetlands (CWs), 387
 biotechnologies and microalgae, 296–297
Contaminants of emerging concern (CEC), 69
Conventional activated sludge (CAS) methods, 381
 biotechnologies and microalgae, 297–298
Conventional approaches, microalgae
 close culture techniques, 293–294
 immobilized culture techniques, 294
 open culture techniques, 293
 suspended culture techniques, 294
Conventional wastewater treatment systems
 fate of, in environment, 129–130
COVID-19 pandemic, vaccines and drugs, 54
Cross-linking
 biopolymers, 353–357
  hydrogel adsorbents, 355*f*
 chemical, 355–357
 chitosan, 353–357
 physical, 354

D

Degradation
 biofiltration devices, 43
 hospital wastewater, 29–30
 profile, 39–40
Derivation, 165–166
Detection scheme and restrictions, 170, 171*t*
Detectors, chromatographic techniques, 167–170
Diclofenac, 315–316
Diethylenetriaminepentaacetic acid (DDMGO) nanocomposite, 62
Drugs accumulation, in wastewater of Lucknow and Ghaziabad, 7*t*

E

Effluent
agricultural wastes, 404–408
exposures, 400–412
heavy metal discharge, 401–404
impact, 400–412
microplastics, 410–412
pharmaceutical wastes, 408–410
sources, 400–412
Electrochemical oxidation, 258–259
Electrodialysis, 321–322, 322*t*
Emerging contaminants (ECs), 53, 149
categories, 348*f*
Emerging pollutants (EPs), 149
Endocrine disrupting compounds, 97
Environment
fate, 190–191
pharmaceutical contaminant in, 150*f*
Enzymatic membrane reactors (EMRs), 382

F

Fenton oxidation
and adsorption processes, 263–265, 264*f*
advanced oxidation processes, 253–257
Fenton system, 189–191
Fermentation engineering, 329
Ferric chloride and sulfate, 334–335
Flocculation, 9
Forward osmosis, 320–321, 321*t*
Free-water surface constructed wastelands, 10
Fullerenes, 55–56

G

Gas chromatography
components, 164*f*
conventional, 164–167
large-volume injection in, 165*f*
single column, 164
Graphene
chemical vapor deposition, 57*f*
CNT composites, 60
epitaxial, 57*f*
functionalized forms of, 58–60, 59*f*
liquid exfoliation of, 57*f*
nanoparticle composites, 60
polymer composites, 59–60
scotch-tape technique, 57*f*
silicon carbide, 57*f*
structure of, 56*f*
synthesis routes of, 57*f*
Graphene-based materials, 53
graphene, 55–58
graphene oxide, 55–58
reduced graphene oxide, 55–58
removal of pharmaceuticals using, 61–62
Graphene-based quantum dot, structure of, 56*f*
Graphene oxide (GO), 55–58
structure of, 56*f*
Graphene oxide-sodium alginate hydrogel, 63*f*
Groundwater-surface water interface, 6

H

Heavy metal discharge, 401–404
challenges in pollutant removal, 404
effluent, 401–404
Heterogeneous photocatalysis, 191–194, 192*f*
Heterotrophic microalgae, 277
High-performance liquid chromatography (HPLC), 163
Hospital wastewater (HWW), 6
in Africa, 39–41
in Australia, 37
in Brazil, 41–42
challenges, 19
characteristics of pharmaceutical wastewater, 20–25
in China, 38–39
in Denmark, 35–36
effluent, 19
discharge limitations and legislation/law, 21–24*t*
in environment, route of, 34*f*
fate of pharmaceuticals, in environment, 26
in France, 36–37
health hazards, discharged pharmaceuticals, 27–28
in India, 38
in Italy, 35
in Malaysia, 41
management
in developed countries, 33–37
in developing countries, 38–43
in Mexico, 43

Hospital wastewater (HWW) *(Continued)*
occurrence
in environment, 26
in water, 25–26
pharmaceutical wastewater treatment options, 28–33
advanced treatment methods, 30–33
conventional methods, 28–30
regulations, 19
source
in environment, 26
in water, 25–26
in United Kingdom, 37
in United States, 33–34
wastewater composition and physicochemical properties, 25*t*
Hybrid membrane, 323
Hybrid oxidation and adsorption processes, 259–260
Hybrid technologies, 227–237, 228–230*t*
Hydrogen peroxide (H_2O_2), 339
Hydrolysis, in environment, 131–132
Hydrophilicity of pharmaceutical compounds, 165, 166*f*
Hydrophobic compounds, 337
Hydrophobicity of pharmaceutical compounds, 165, 166*f*

I

Ibuprofen, 315–316
Injection port derivation (IPD), 166
Ion-exchange chromatography, 163

L

Laser induced fluorescent (LIF) detection, 167–168
Lipid regulators, physiochemical properties, 78–86, 87–91*t*
Liquid chromatographic techniques, 162–163*t*, 163
Liquid-liquid extraction (LLE)
advantages, 153*t*
dispersive, 152–153
drawbacks, 153*t*

M

Magnesium chloride, 334–335
Mass analyzer, 168, 169*t*
quadrupole, 169–170
time of flight, 170
Mass spectrometry (MS), 168, 168*f*
Membrane and membrane-integrated hybrid systems, 335–338
Membrane based separation processes
electrodialysis, 321–322, 322*t*
forward osmosis, 320–321, 321*t*
on human health, adverse effect, 312–313
hybrid membrane, 323
microfiltration, 315–316, 316*t*
nanofiltration, 316–318, 318–319*t*
reverse osmosis, 319–320, 320*t*
ultrafiltration, 313–315, 315*t*
Membrane bioreactor (MBR), 381
conventional treatments, PhACs, 331–332
Membrane distillation (MD) methods, 335
Membrane photobioreactor, biotechnologies and microalgae, 298
Membranes separation, 380–382
Membrane technology, 227–234
Metabolism and transformation processes, 110–114, 112–113*t*, 114*f*
Metabolites, 218–219
Metal-organic framework (MOFs), 339
Metronidazole (MNZ), 61, 263
structure of, 61*f*
Microalgae, 277
acclimation of, 294–295
autotrophic, 277
bioremediation, 278
biotechnologies and
advanced oxidation processes, 295–296
constructed wetlands, 296–297
conventional activated sludge, 297–298
membrane photobioreactor, 298
microbial fuel cells, 297
cometabolism of, 295
conventional approaches
close culture techniques, 293–294
immobilized culture techniques, 294
open culture techniques, 293
suspended culture techniques, 294
growth condition, 291–292
heterotrophic, 277
hybrid approaches, 299–300*t*
pharmaceutical contaminants, removal mechanisms of

bioaccumulation, 279–280
bioadsorption, 278–279
biodegradation (biotransformation), 280–282
photobiodegradation, 282, 283–290*t*
transformation products of, 281*t*
volatilization, 282–291
photosynthetic and CO_2 biofixation ability, 277–278
phototrophic, 277
species, 292–293
Microbial fuel cells, 297
Microfiltration (MF), 315–316, 316*t*, 335–336
Microplastics, effluent, 410–412
Micropollutants, 180
Microwave-assisted adsorption of pharmaceutical waste, 378–379
Moxifloxacin, 252–253
Municipal wastewaters, classification, 330

N

Nanoadsorbent, 340*f*
Nanocomposites, 55
Nanofiltration (NF), 316–318, 318–319*t*, 335–336
Nanomaterial membranes (NMs), 335–336
Nanoparticle composites, graphene, 60
Natural colloid matter (NOM), 336–337
Natural polymers, properties of, 350–351
N-bioand O-bio-adsorbents, 378
Nonsteroidal antiinflammatory drugs (NSAIDS), 377–378
physiochemical properties, PhACs, 78, 79–85*t*
in wastewater, 3
Normal phase chromatography, 163

O

On-line derivation, 166
Osmosis
forward, 320–321, 321*t*
reverse, 319–320, 320*t*
Oxidation processes, 181–197
classification of, 183*f*
Oxytetracycline (OTC) wastewater, 223–224, 235–236
Ozonation, 181–182
and adsorption processes, 265–266, 265*t*
advanced oxidation processes, 257–258, 258*t*
Ozone treatment, 185–186

P

Partitioning, in environment, 130–131
Personal care products (PCPs), 69, 379
Pharmaceutical compounds (PhACs)
endocrine disrupting compounds, 97
fate of, in environment, 125–134
in aquatic environment, 130–132
conventional wastewater treatment systems, 129–130
hydrolysis and photolysis, 131–132
partitioning, 130–131
in soils, 132–134
in wastewater treatment plants, 126–130, 126*f*, 127–128*t*
metabolism and transformation processes, 110–114, 112–113*t*, 114*f*
occurrence of, 114–125, 115–116*t*
in aquatic environments, 120–123, 121–123*t*
in soil, 123–125, 124–125*t*
in wastewater treatment plants, 117–120, 117–119*t*
physiochemical properties, 71–110
antibiotics, 72–78, 73–77*t*
beta-blockers, 86–96, 92–96*t*
lipid regulators, 78–86, 87–91*t*
nonsteroidal antiinflammatory drugs, 78, 79–85*t*
properties of, 98–108*t*
release into environment, 70*f*
Pharmaceutical drugs, structure of, 54*f*
Pharmaceutically active compounds (PhACs), 69
conventional treatments
activated carbon, 332–333, 333*f*
anaerobic membrane bioreactor system, 332*f*
coagulation, 334–335
membrane bioreactor (MBR), 331–332
sedimentation, 334–335
source of, 330
wastewater effluent characteristics, 331*t*
Pharmaceutical wastes, effluent, 408–410

Pharmaceutical wastewater (PWW)
analytical techniques, 7
characteristics, 330–331
definition, 330
detection of, 8*t*
effect of, 1
metabolites, 2
negative impacts on waterbodies, 1
occurrences, 1–3
in India, 6–7
physicochemical methods, 9–10
activated sludge treatment, 11
bioaugmentation, 11
biological methods, 10–12
biotransformation, 11–12
coagulation, 9
constructed wastelands, 10
flocculation, 9
sedimentation, 9–10
vermicomposting, 11
sources, 1–7, 2*f*
structures of, 3, 7*t*
toxicological effect of, 7–8
treatment, 9–12
Pharmaceutical wastewater treatment plant (PWWTP), 225
Photobiodegradation, PC removal mechanisms, 282, 283–290*t*
Photocatalysis
and adsorption processes, 260–263, 261*t*
heterogeneous, 191–194, 192*f*
oxidation, 252–253, 252*f*, 254*f*, 342
Photodegradation, 132, 182*f*
Photo-fenton process, 189–191, 192*f*
Photolysis, in environment, 131–132
Photosynthetic and CO_2 biofixation ability, microalgae, 277–278
Phototrophic microalgae, 277
Physical cross-linking, 354
Physiochemical properties, PhACs, 71–110
antibiotics, 72–78, 73–77*t*
beta-blockers, 86–96, 92–96*t*
lipid regulators, 78–86, 87–91*t*
nonsteroidal antiinflammatory drugs, 78, 79–85*t*
PolyAMINEs, 334–335
PolyDADMACs, 334–335
Polyvinylidene fluoride (PVDF), 233–234
Preanalysis protocols
advantages, 153*t*
analytical techniques
chromatographic techniques, 161–167
separation process, 161*f*
dispersive liquid-liquid microextraction, 152–153, 153*f*
electrochemical sensor, 170–172
liquid-liquid extraction, 151–152
preservation of samples, 151
pretreatment (clean-up), 151–159
pretreatment and extraction of solid samples, 160, 161*f*
sampling method, 151
separation method, 152, 152*t*
solid-phase extraction, 151–157
Precipitation, 9
Pressure-driven membrane filtration process, 335–336, 336*f*
Purification of pharmaceutical wastewaters, 249–259

R

Reduced graphene oxide (rGO), 55–58
removal of pharmaceuticals using, 61–62
structure of, 56*f*
Reduced graphene oxide-sodium alginate (rGO-SA) hydrogel microspheres, 56–58
Reverse osmosis (RO), 319–320, 320*t*, 335–336

S

Salt-marsh sediment MBR (SMSMBR), 232–233
SARS-CoV-2 and pollution, 400. *See also* COVID-19 pandemic
Scotch-tape technique, 57*f*
Sedimentation, 9–10
conventional treatments, PhACs, 334–335
Septic tanks, 6
Sewage sludge, 6
Sewage treatment plants, 6
Silicon carbide (SiC), 57*f*
Size exclusion chromatography (SEC), 163
SODIS process, 180–181
Soil
fate of, in environment, 132–134

occurrence of, PhACs in, 123–125, 124–125*t*
Solar radiation, 180–181
Solid-phase extraction (SPE), 153–157
advantages, 156*t*
cartridge-loaded, 156
clean-up, 157–159, 158–159*t*
drawbacks, 156*t*
formats and devices in, 157*t*
methodology of, 157–159
type of phase, 155*t*
with UHPLC-MS/MS and, 156–157
Solvent extraction. *See* Liquid-liquid extraction (LLE)
Sonication, 194–195
Sustainable nanoadsorbent, 338

T

Tetraalkylammonium (TAA) salts, 167
Tetracycline, 263, 377
Thermal desorption (TD), 164
Total dissolved solids (TDS), 330
Total suspended solids (TSS), 330
Toxicity, 221–222
Transformation processes, 110–114, 112–113*t*, 114*f*
Transformation products (TP), PC removal mechanisms, 281*t*
Trimethoprim, 381

U

Ultrafiltration (UF), 313–315, 315*t*, 335–336
Ultra-high-performance liquid chromatography (UHPLC), 163
Upflow anaerobic sludge blanket (UASB), 232–233
UV and UV/H_2O_2, 183–185
advanced oxidation processes, 255*t*, 256*f*
UV/chlorine, 187–189
UV/O_3 system, 183–185

V

Vermicomposting, 11
Volatilization, PC removal mechanisms, 282–291

W

Wastewater sources, 6
Wastewater treatment plants
fate of, in environment, 126–130, 126*f*, 127–128*t*
occurrence of, PhACs in, 117–120, 117–119*t*
Water reservoirs, pharmaceuticals concentrations, 250*t*

Z

Zinc oxide, 253